Matters of Exchange

Matters of Exchange

Commerce, Medicine, and Science in the Dutch Golden Age

Harold J. Cook

Yale University Press
New Haven & London

Published with assistance from the foundation established in memory of
Philip Hamilton McMillan of the Class of 1894, Yale College.

Set in Electra Roman type by Tseng Information Systems, Inc.
Printed in the United States of America.

Library of Congress Cataloging-in-Publication Data

Cook, Harold John.
Matters of exchange : commerce, medicine, and science in the
Dutch Golden Age / Harold J. Cook.
p. cm.
Includes bibliographical references and index.
ISBN 978-0-300-11796-7 (clothbound : alk. paper)

1. Science—Netherlands—History—17th century. 2. Netherlands—
Commerce—History—17th century. 3. Medicine—Netherlands—
History—17th century. I. Title.
Q127.N2C66 2007
509.492′09032—dc22 2006026973

A catalogue record for this book is available from the British Library.

The paper in this book meets the guidelines for permanence and durability
of the Committee on Production Guidelines for Book Longevity of the
Council on Library Resources.

10 9 8 7 6 5 4 3 2 1

To the memory of

Gemma Beukers

and to all our Dutch friends
who made us so welcome in their homeland

My subject is nature, that is, life
—Pliny, *Natural History*

CONTENTS

Preface

Those who turn to the past for comfort because they suppose that we know how it all turned out travel at their ease; going by other means into this region of unexpected dangers and delights can sometimes transform the voyager. More than fifty years ago, L. P. Hartly famously wrote: "The past is a foreign country; they do things differently there." The differences are often surprising. No guidebook can fully prepare travelers for the people and events encountered in the past. On first impression, many things seem familiar, but the unfamiliar appears often enough, rending the fabric of expected order, giving off only glimmers of reasons for its appearance. It all suggests a world composed of forces hardly noticed in our comfortable lives of custom and habit.

The voyage of imagination that led to this book began when I was a student when, for some reason, I developed an ambition to integrate accounts of intellectual history with sociopolitical and economic history. I still seek to understand what the changes of the early modern period would look like were they depicted, not as a kind of mind-body dualism, with the history of "ideas" affecting the history of "society" or vice versa, but as an integrated whole that represented united lives. My first attempt at such an account used printed and manuscript documents from the English seventeenth century to explore how learned medicine was bound up with transformations of economic, political, and intellectual life in the period of the "scientific revolution," placing much emphasis on the medical marketplace. The point was not to praise the market but to show that the values inherent in it confronted those of classical wisdom and consequently greatly helped to change medical ideas and practices. In the conclusion, I ventured some comments of a comparative kind in order to point to the changes that were the same throughout Europe as well as England and vowed to myself to explore them further. Although I had access to plenty of comparative material for France, a comment of a former professor, James Allen Vann, came to mind, reminding me that few working historians get a chance to expand their horizons much beyond the limits of inquiry set out in their dissertations unless they try their hand at something new shortly thereafter. In my work on London I had come across many references to interchanges between the English and the Dutch, and I vaguely knew of the importance of the cross-channel connections between the two nations. Deciding to put off the comparison with France for another day, then, I set myself the task of learning Dutch and the history of the early modern Dutch Republic. One early result was a study of a well-educated Dutch physician who moved to London and had initial success only later to be charged with malpractice for the use of a new remedy he developed.

In studying the past of yet another country, many of my assumptions about the patterns of early modern history were altered. The early modern Netherlands was an unusual place. The polity was self-evidently unable to live from its own resources, as some of its neighbors imagined themselves able to do. Everyone could clearly see that Dutch power depended on shipping and exchange in ways that were less obvious elsewhere. My encounters with what was then called the "history of discovery" and in more recent years with global history and the economic history that underpins it, changed my thinking even further. Yet the old ambition of composing a work of history not torn apart by dividing the world into a conceptual part and all the rest remains. One of the key terms for seeing how such a history might work is what early modern people themselves called the "passions," which they considered to be the forces that created change, not only in minds but in bodies, and not only in individual persons but in all things.

I can only hope that the results produced from this mixed set of historical agendas is helpful to others. And I hope that even those who disagree with the arguments of the book will find some of the descriptive material on which it is based to be of interest. When I began the project, most of the work on which it depended was written in Dutch, and although over the past two decades an enormous amount of the best current work has been published in English, there may be some things here to which few English readers will otherwise have access.

Perhaps the reader will indulge me in one last personal comment. Much of the following interpretation of the period rests on trying to understand how passions for goods shaped collective behavior and belief even more than the moral pursuit of the good. But this does not mean that I consider everything people are driven to do or to think are the result of economic interest. Far from it. My intention is to stress how certain ways of life shaped the cultural values of certain kinds of people, which in turn gave shape to the meaningful questions they asked; once the (ever-changing) rules of their game were in place, people used them creatively, not only for material or political advancement but also for entertainment, edification, and pleasure, not only for utilitarian advantage but also for exploring the unknown and finding things out. I am not trying to overdetermine why people acted or wrote or spoke or thought as they did. Yet the material constraints on our mortal condition are real, the goads and shackles of economic life not least among them. Despite the fall of the Berlin Wall two decades ago, then, the materialist aspects of this study will tempt many into considering it an argument about how economic forces determined what people thought, but it is a danger I am willing to risk.

The book has been a long time in the making: more than twenty years, in fact. Much of the delay can be explained by having to learn a new language and new histories to carry it out; other reasons lie in having spent much of that period preoccupied with university service and academic administration, first in Madison, Wisconsin, and then in London, England, which kept me from libraries and archives far more than I would have liked. At the same time, such work gave me much greater experience of the world, which has greatly affected how the arguments of this book are presented.

It is a collective project in ways I could not have imagined when I first embarked on it, being influenced by profound and sometimes deeply disturbing changes in the world but especially by a number of chance but friendly encounters with people from many different places who were generous in sharing their knowledge, which cannot but sustain some faith in human sociability and wisdom. Above all, I received enormous help and encouragement from Dutch friends and colleagues, most especially Antonie Luyendijk-Elshout; Harm, Gemmic, and Saskia Beukers; and Toon and Tip Kerkhoff, who all patiently taught me an enormous amount about their country and its history while making me and my wife, Faye, feel entirely welcome. Hisa Kuriyama further helped me understand some things about early modern Japan, while many more recent friends have introduced me to other parts of Asia and Europe. My colleagues and students in Madison, Wisconsin, and the dean of the University of Wisconsin School of Medicine and Public Health, Phil Farrell, offered encouragement for the project. Rob Howell and Jolanda Taylor began the work of teaching me Dutch, while during the academic year 1989–90 I lived in the Netherlands and carried out research thanks to the combined support of the Fulbright Commission and the Netherlands America Foundation for Educational Exchange, the National Endowment for the Humanities, the National Library of Medicine, and the Graduate School of the University of Wisconsin–Madison. In developing my thoughts, I have had the chance to get the advice and criticism of different academic audiences, to whom I extend my thanks; some of these lectures became published articles: I therefore also thank the editors, publishers, and referees of my published articles since 1990 (included in the Bibliography) for their help.

Since my arrival in London, colleagues and associates in The Wellcome Trust Centre for the History of Medicine at UCL have been constant sources of conversation and stimulus. I am especially grateful to Michael Neve, who read and commented on early drafts, Andrew Wear, who responded in detail to one of the chapters, the members of the Early Medicine Reading Group for remarks on still another, Christopher Lawrence and Karen Buckle, both of whom read through and responded to the penultimate draft, and to others, especially Patrick Wallis, who carefully read two chapters, and the anonymous referees for Yale University Press. I also thank Sharon Messenger and Caroline Overy for their knowledgeable help in searching out literature and illustrations and arranging for the permissions, as well as for assistance in checking the Bibliography. Anne Hardy, Alan Shiel, and Debra Scallan have patiently supported me in my administrative role when I have been preoccupied with other things. Lisa Jardine has been a constant source of good cheer and counsel. The staff of the Wellcome Library for the History and Understanding of Medicine are, as always, unfailingly helpful; the support of UCL and the Wellcome Trust has made it possible to finish the book. Jean Thomson Black's ever-supportive editorial guidance saw the results through the Press, and Laura Jones Dooley's copyediting improved my expression manyfold. And to Gabe, who in taking me for regular walks has reminded me of other worlds, and most especially to Faye, who has sacrificed so much for this while keeping us going, I offer my sincerest gratitude.

1

Worldly Goods and the Transformations of Objectivity

It is a great mistake to conceive of the rise of modern science as an appeal to reason. On the contrary, it was through and through an anti-intellectualist movement.

—ALFRED NORTH WHITEHEAD, *Science and the Modern World*

Like most works of history, this book offers an account of how we got to be the way we are. It seeks to share an understanding of how some ways of knowing the world that are too often taken for granted, or seem simple-minded—particularly the detailed and exacting description of natural objects—grew into what came to be called "the new philosophy" of the sixteenth and seventeenth centuries. The aspect of the modern world explored most fully is, then, sometimes called "the rise of modern science," or "the scientific revolution." It took place during the first age of global commerce. To give the study some geographical focus, most of the examples are taken from the region of northwestern Europe loosely termed the Dutch Republic. They in turn emphasize the foundational significance of what we would today call the life sciences and medicine rather than physics and mathematics, although a more complete account would have included more on what is commonly called technology than could be offered here. In other words, this book points to some of the means by which those who were deeply invested in the materially "real" came to think that their own criteria for judging truth had wider applicability; for the growing dominion of a certain kind of knowledge economy had consequences for the content of science.[1] By looking at the rise of science in this way, we can clearly see that, as many contemporaries themselves acknowledged, the new philosophy arose not from disembodied minds but from the passions and interests of mind and body united. And by examining how natural knowledge was changed in one place we can glimpse the larger canvas on which the human experience of our own age still exhibits elements of the terrors and tragedies, desires and successes, of the first global age.

Between the voyages to the Indies by Christopher Columbus and Vasco da Gama in 1490s, and the consolidation of the Amsterdam-London commercial axis at the end of the 1690s, long-distance seaborne trade throughout the world fell largely under the control of merchants along the European Atlantic coast. They in turn financed the apparatus of their respective nation-states to guarantee their security, and in some places, such as the Dutch Republic, even waged wars abroad in their own right. As they and their associates gained a dominant position in their own worlds, the matters they valued most, whether material goods, social manners, cultural symbols, or intellectual pursuits, also came to dominate the lives of other people. The culture of the exchange economy was not the only culture vying for influence, even in the home countries of the merchants. But it had enormous consequences for how most people of affairs, even university professors, came to understand the ways of nature, for it changed the terms of reference for intellectual investigation. Like commerce, science arose not from liberating the mind from the world but from keenly interested engagement with it.

Among the figures who appear below are some of the most eminent natural investigators of their age, although today they are not well known outside the Dutch-speaking world: botanists like Bernardus Paludanus, Carolus Clusius, Paulus Hermann, and Isaac and Jan Commelin; naturalists and physicians working in the East Indies like Jacobus Bontius, Georgius Everhardus Rumphius, Hendrik Adriaan van Reede, and Willem ten Rhijne, or those resident in The Netherlands like the "tea doctor" Cornelis Bontekoe; medical professors such as Pieter Pauw, François dele Boë Sylvius, and Herman Boerhaave; philosophers like Justus Lipsius, René Descartes, and Benedict Spinoza; anatomists such as Drs. Nicolaes Tulp, Regnier de Graaf, Jan Swammerdam, and Frederik Ruysch. Yet the discovery of the world did not take place in libraries or lecture halls alone, nor only among people we would today call "intellectuals." Ship captains, officers, sailors, and surgeons reported on their experiences and systematically collected information and objects. Diplomats, merchants, and travelers in foreign lands took careful note of what they saw and sent back specimens. Apothecaries eagerly accumulated information about the goods that passed through their shops, displayed strange bits of nature to draw in customers and demonstrate their command of information about the world, and dug gardens for the cultivation of plants. Herb wives and wise women occasionally deigned to explain a bit of their knowledge to grand men who asked them for it. Medical practitioners of many kinds made and sold medicines, sometimes chemically prepared, or offered advice about health and sickness rooted in their knowledge of nature. Gardening enthusiasts sought out beautiful and exotic botanicals, substantial people assembled collections of strange and interesting objects in their homes, civic leaders attended public anatomy demonstrations, the poor entered hospitals in which they might receive experimental treatment or surgery, and almost everyone seemed to be consuming larger amounts of fine spices such as nutmeg, cinnamon, and sugar, or new medicinal imports that quickly became common items of consumption, such as tobacco and chocolate from the New World, and coffee and tea from Southern Arabia and the Far East.

As these last examples indicate, in order to understand what was occurring in The Netherlands it is crucial to see what was happening in the Dutch world more generally. By the middle of the seventeenth century, the Dutch seaborne empire had become the most extensive in the world, while the Dutch had at the same time become acknowledged leaders in many areas of medicine and natural science, partly because of their contacts in Asia. With their commercial enterprises backed up by naval power, the Dutch rapidly took over most of the Portuguese places in Asia, and in many parts of Africa and South America, while adding yet other new ports of call to the list of their possessions from North America to Southeast Asia. At the same time, to accomplish their ends, they necessarily relied on local knowledge, meaning that people throughout the world also contributed to the growth of what we now call science, sometimes through mutually beneficial interactions, sometimes by appropriation or coercion. Unfortunately, Dutch history has been given little attention in the English-speaking world despite some wonderfully dramatic and captivating narrative histories of the Dutch Revolt in our own language, above all John Lothrop Motley's grand studies of over one hundred and fifty years ago.[2] To help accommodate this strangeness, the plan of the work below approaches history from the lives of the people involved in some of the most significant events while recognizing that their lives are illustrative rather than "typical." One of the consequences is that the emphasis falls on those persons who are well-documented when one of the arguments of the book is that many people, of all ages and social classes, and from many parts of the world, contributed to the developments investigated here; but any other approach would have created either additional narrative complexity or lifeless generalizations. This work is, therefore, admittedly episodic, supposing that movements in the larger world are most clearly seen from the vantage point of particular persons, who like everyone else made a difference for good or ill. Despite the cross-cutting complexities of life, moreover, a narrative framework has been used as much as possible in order to keep confusion at bay.

This introductory chapter sets out the main ways in which the economic transformations of the first age of global commerce placed a high value on careful descriptive information about objects, and how such values shaped priorities for knowing about nature; a second chapter further explores the intellectual values of the Dutch merchants' information economy. They are followed by an account of why invoking religion or the Reformation to explain the rise of science—as is often done—is inadequate, by sketching in the innovative work of naturalists such as Clusius and Paludanus, and the founding of the university of Leiden and its medical faculty and botanical garden. The ways in which commerce affected medical practice and knowledge, even anatomical investigations, with special reference to Amsterdam, follows. A subsequent chapter on some important studies of medicine and natural history in the Dutch East and West Indies, centered on the work of the physician Bontius, illustrate how matters of fact were gathered, transformed, and exchanged as part of a worldwide commercial network. But new methods of accounting for the human body and its attributes certainly gave rise to new ideas: the sixth chapter therefore goes over the ways in which the views of the "philosopher" Descartes were changed by his long residence in the Dutch Republic, during which he studied anatomy at great length and turned his attentions to some of its consequences for an analysis of the passions; some examples of how others, such as the brothers De la Court and Spinoza, drew on Descartes's work for their own analyses of politics and ethics illustrates some of the consequences of such studies in the most fraught intellectual debates of the period. The following chapter shows just how much the new understanding of animal and human bodies depended on exacting methods for manipulating objects that were imported from other technical activities. The powerful and continuing interest in the medicine and natural history of the Indies came to be manifest later in the seventeenth century in both magnificent books and in specimens cultivated on grand garden estates, the subject of the eighth chapter. The next gives an account of how a Dutch physician, Willem ten Rhijne, worked with colleagues in Japan to produce one of the earliest European analyses of acupuncture and moxibustion, and a translation of some works on Chinese knowledge of how to examine the pulse; it emphasizes the difficulties of transmitting some kinds of knowledge across cultural barriers whereas other kinds of information could flow easily. The penultimate chapter, focusing mainly on the work of Herman Boerhaave and Bernard Mandeville, concentrates on how objectivity helped to further the study of medicine and natural history but also provided the intellectual foundations for philosophical materialism, which deeply threatened the political and religious establishment. Some thoughts of a comparative kind, showing how the Dutch examples can be generalized into an account of the rise of the new science in Europe, end the story.

Underlying the connections between the history of early modern commerce and science are echoes of an old-fashioned historical literature, in which the maritime ventures and intellectual transformations of the sixteenth and seventeenth centuries were sometimes treated together as part of the common "discovery" of the world.[3] For instance, one of the founders of the history of science, George Sarton, defined the history of science as the "discovery of objective truth." The more recent view of an eminent Dutch historian of science, Reijer Hooykaas, also pointed to the moment when, "not incidentally but in principle and in practice, the scientists definitively recognized the priority of Experience. The change of attitude caused by the voyages of discovery is a landmark affecting not only geography and cartography, but the whole of 'natural history.' It led to the a reform of all scientific disciplines—(not only of the mathematical-physical)—because it influenced the *method* of all the sciences, however much their mathematization might be delayed."[4]

In retrospect, it is apparent that Sarton and his like (often called "positivists") took it for granted that discovery was a more or less self-evident process, so that if one went to the trouble of looking closely, new information about the world would readily become apparent. It was simply a question of going out and doing it. For various good reasons, such statements have often been put aside in recent generations by historians who have given much more attention to the social, linguistic, and cultural causes of the construction of scientific knowledge. By the middle of the twentieth century, gestalt psychology, art history, linguistics and ethnography, and many other disciplines, were clearly demonstrating that human beings usually interpret the world in expected ways, which are in turn imbibed from the other persons around us rather than directly from nature: in other words, many prominent figures were arguing that concepts and culture have large effects on how we interact with the world around us. Soon historians and philosophers were reading this modernist aesthetic back into the origins of the scientific revolution of the sixteenth and seventeenth centuries, emphasizing a mental revolution arising from pure thought, sometimes termed a "metaphysical" shift.[5] Since the famous work of Thomas Kuhn and others in the late 1960s, such views have been amended by a variety of arguments adapted from sociology, anthropology, and cultural studies, which have been advanced by historians to explain why one competing theory is chosen over another, giving birth to a large literature that now sometimes goes under the rubric of "social construction."[6] The aim of most such work remains to explore the origin of ideas and concepts.[7]

This book does argue that "discovery" lies at the heart of the scientific revolution: or more precisely, that determined investigations into matters of fact laid the groundwork for generalizations about nature. But it does not assume that discovery is self-evident. More importantly, its underlying theme is that the search for wisdom became a search for knowledge, or rather the kind of wisdom rooted in understanding *why* nature is as it is became subordinate to the kind of wisdom rooted in understanding *how* natural things are. To put it more technically, proponents of the new science rejected teleology—arguing that nothing would come from trying to understand the purposes intended by nature or God—instead, they valued the search for exact description of natural things as they could be grasped by the senses, allowing comparison, alteration, and use for material betterment. Knowledge of concepts became "speculation"; certainty, or at least highly probabilistic knowledge, was tied to real objects and specimens. What this book takes up that Sarton and his like did not, then, are two underlying questions. Why did an enormous amount of personal time and effort, and economic and other resources, come to be devoted to seeking out and acquiring precise and accurate descriptive information about natural things? The activities of commerce, including the trading ventures once called voyages of discovery, have a very important place in the answer to this question. And how did these investigations come to be at the center of "natural philosophy"? Here, the kinds of values embedded in commercial ways of life can offer additional explanations. Even the university professors had to start paying attention to the implications of this shift in values.

CONSUMPTION AND TASTE

The low countries or "netherlands" comprised the northern remnant of the old Middle Kingdom carved out of Charlemagne's empire. The name came from a landscape of few promontories and many wide patches of soggy earth and bog, a region in many places subject to periodic flooding despite local systems of intensive water management that included dredging, ditching, and diking. These lands formed a great wedge running from north of the mouth of the Somme, east over marsh and pasture to the hills and woods of the Argonne, then north to about the mouth of the Ems, and back southwest along the dunes and sand beaches of the coast—that is, it ran from what is today northern France into what is modern Germany, including Belgium, The Netherlands, and part of Luxembourg. Two great rivers ran through these low countries from the southeast to the northwest, the Rhine and the Maas (or Meuse), draining into the North Sea the deep waters produced by frequent rain and the heavy snowmelt of the distant Alps.

The region was not without some advantages. In southern provinces such as Flanders there was much rich arable land as well as fine pasture for war- and draft-horses. To the northwest, the low river deltas of sand and clay composed a land of countless lakes, ponds, and marsh, but here more than enough peat could

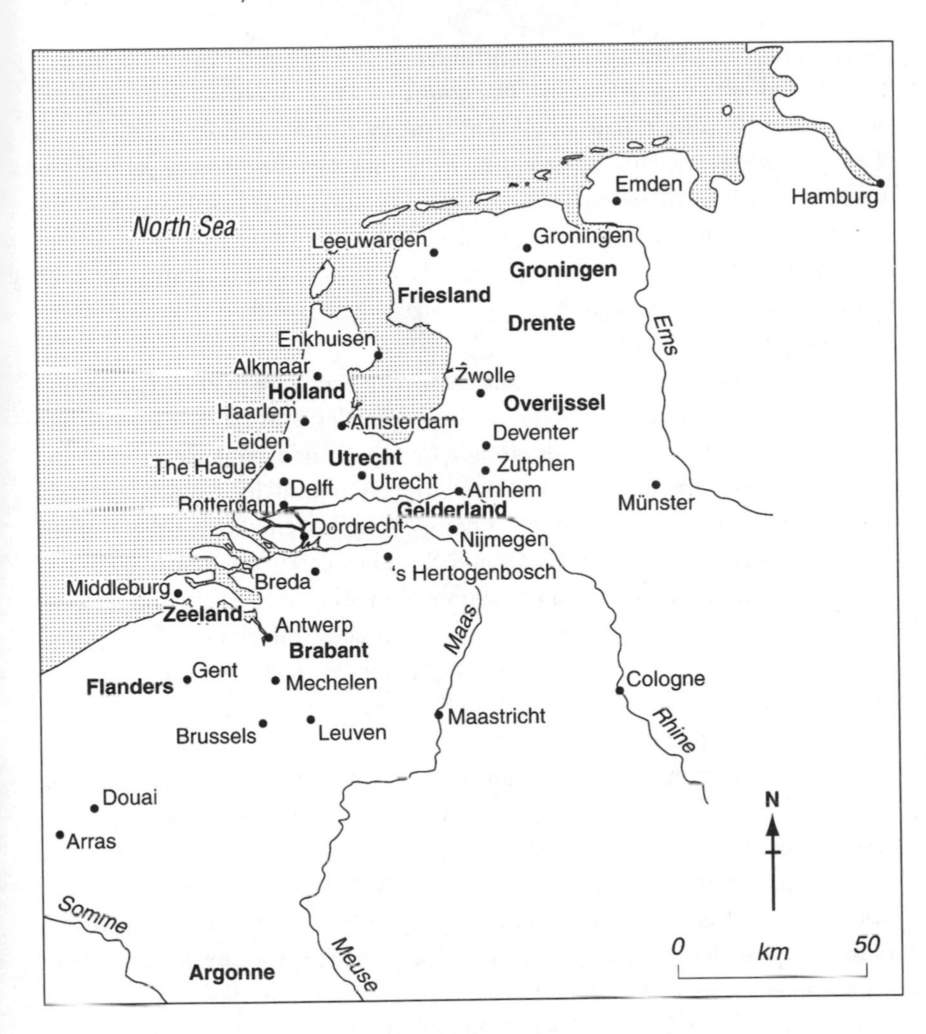

The low countries

be cut and dried for domestic and commercial heat, while water-laden pastures harbored cattle, enabling a large net export of meat and cheese. The soils even allowed for growing oats and barley in quantity (although not much wheat and rye), with the surplus exported most efficiently in barrels after the conversion of the grain into beer and ale. Along with the Basques, the coastal netherlanders had also been early pioneers in deep-sea fishing in the later Middle Ages, going for herring, cod, and haddock. They even began to hunt whales in the North Sea and around Iceland.[8] Indeed, an early-seventeenth-century English adventurer, Captain John Smith, argued bluntly that the "poore Hollanders" had become hardy, industrious, and wealthy by fishing in all weathers in "this their Myne," the "silvered streames" of seafish.[9] The politico-legal legacy of living in lands not easily subject to manorial production also meant that a relatively small number of people owed service to lords, allowing ready diversification of occupation. The southern cities, particularly, became famed for the quantity and quality of the cloth and tapestries made there from wool imported from Spain and the British Isles. By a variety of means, then, it became one of the most heavily populated regions of Europe, a region as highly urbanized and dependent on commerce as northern Italy.

It was especially because of commerce that its people prospered. Goods sent from the Mediterranean to the North Sea or the Baltic often traveled through this region. Near the coast, goods brought by river could be put aboard seagoing ships to reach the British Isles, Scandinavia, the Baltic, and the Iberian Peninsula. In turn, the "mother-trade" of the economy brought grain, tar, timber, leather, and other essentials from the Baltic and Scandinavia for use and re-export to other places, in return for wine and salt brought mainly from Iberia, and for local cheese, beer, and salted fish, as well as luxury goods transshipped from everywhere. To move their commodities the netherlanders had developed the *fluyt*, a broad-bottomed and shallow draft ship that could carry large amounts of bulk goods while requiring little manpower due to the deployment of new pulley systems in the rigging. All this gave regional merchants advantages in the carrying trades, and they drove that trade hard, ending up as the major shippers of northwestern Europe.

They also became beneficiaries of the direct trade with Asia. That began after Vasco da Gama rounded Africa in 1497–98 and Pedro Álvares Cabral followed up with a great fleet returning to Lisbon loaded with spices, resulting in spectacular financial success despite the loss of half his thirteen ships. For centuries, Europeans knew only that spices originated in far-off lands to the east, south of Cathay, the fabled widespread region known generally as "India," which according to some accounts was ruled by the wondrous and powerful Christian prince Prester John. Here lived strange people, animals, and plants: priests who stood in hot sands all day looking straight at the sun as well as forest-dwellers who lived from the scent of flowers alone.[10] The islands further to the southeast, "the Indies," were a region of even more wonders and of fabulous wealth. Marco Polo never got there, although he reported that reliable pilots and seamen said that 7,448 islands were to be found in the Indies, "And I assure you that in all these islands there is no tree that does not give off a powerful and agreeable fragrance and serve some useful purpose." Moreover, "there are, in addition, many precious spices of various sorts," including pepper both white and black.[11]

A long list of roots, seeds, flowers, saps, barks, fruits, and woods made up the category of aromatic and tasteful substances known as spices, but the most sought-after came from Southeast Asia: pepper, ginger, cinnamon, cloves, nutmeg, and mace. Black pepper was from the immature dried berry of a climbing vine that grew on the Malabar coast (now Kerala) of the Indian subcontinent, on the large island of Sumatra, and at a few other places in South and Southeast Asia; white pepper was the seed within the ripened berry. Ginger came from the rhizome of a plant that was sometimes thought to be the root of the pepper vine, since it came from the same general region of India; cinnamon was from the bark of a shrub that grew mainly in Ceylon (now Sri Lanka). (Cinnamon is still often confused with cassia—the most common form of "cinnamon" imported to North America—which originates in southeast China.) The three remaining exotic spices on which European medicine and cooking depended came from even farther away: the spiky one, named after the Latin word *clavus* (nail) and metamorphosed to the English "clove" and the more literal Dutch "kruidnagel," is the immature dried flower of a tree found only on some of the "spice islands" of the Moluccas (Malukus), in what is today Indonesia; nutmeg and mace came from a fruit tree that grew only on six small islands in the tiny archipelago of Banda, the southernmost of the Spice Islands. (Nutmeg was the seed of the fruit, mace the dried red flesh covering the seed.) These spices were shipped in stages from Southeast Asia to South Asia, and then overland or by ship via the Persian Gulf and Euphrates to Aleppo and Beirut, or via the Red Sea to Cairo and Alexandria; at cities on the eastern fringes of the Mediterranean, Catalan, Genoese, and especially Venetian merchants acquired the spices and sold them on to others in Europe.[12] These merchants were known as "grocers" (for dealing with quantities by the gross), "apothecaries" (from *apotheke*, or storehouse), or most commonly and simply, *spezieri*: "spicers" in English, or "specyer" or "kruidenier" in Dutch.

At first, the spicer-apothecaries handled another comestible from Asia, too: sugar. Sugarcane may have been first domesticated from crosses of wild varieties in the ancient southwest Pacific and carried throughout the region by Polynesians, who loved the taste of its sap. By the second and first centuries BCE it was being cultivated in China and India where, in the Indus Valley, followers of Alexander the Great came across it.[13] For the ancient Roman encyclopedist Pliny the Elder, sugar was rarer than pepper, and consequently was virtually excluded from ordinary cooking. Cane sugar (or simply sugar) "is a kind of honey that collects in reeds, white like gum and brittle to the teeth. . . . It is only used as a medicine."[14] The reeds of which Pliny spoke—the sugarcanes—were introduced to Mediterranean cultivation after the Islamic conquests of the seventh and eighth

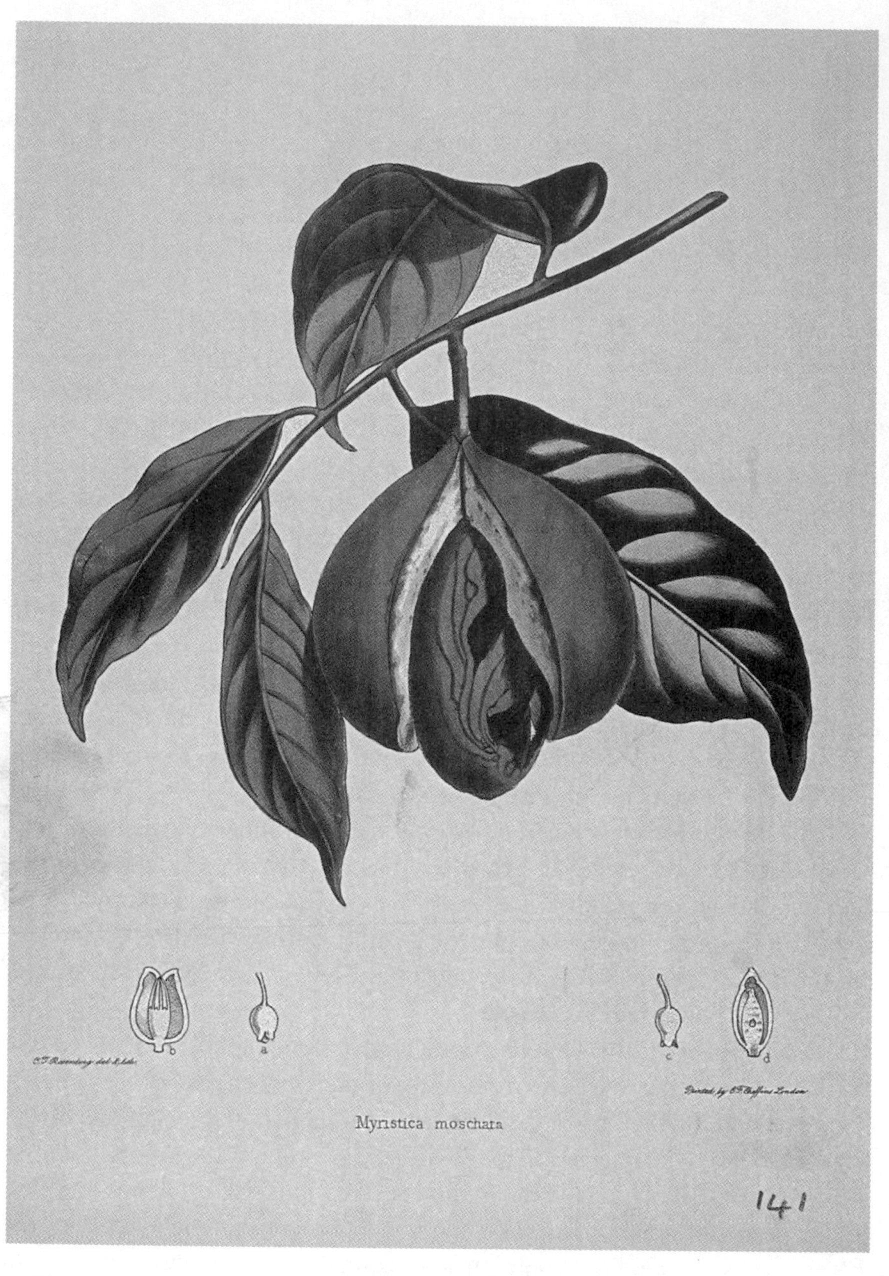

The nutmeg fruit, showing the seed within. Lithograph by C. Rosenberg, c. 1850.
Courtesy of the Wellcome Library, London

centuries CE. Crusaders of the twelfth century found the cane grown and its sap refined in Syria and Palestine, and quickly profited by exporting sugar to Venice and other western ports. When the Christian fortresses of the Holy Land were retaken late in the thirteenth century, sugar production for the European market shifted to Cyprus, Crete, and Sicily, with a smaller amount in Spain and Portugal. About a century later, sugarcanes were carried by the Genoese to the tropical Atlantic of Madeira and the Canary Islands, being cultivated there on large estates.[15] The Sicilian invention of a new kind of sugar press in 1449 made production even more profitable. By the end of the fifteenth century, an even more efficient system developed by which raw cane was imported to Venice and refined in huge boiling factories into cone-shaped sugar loaves with syrup as a by-product, both of which were exported along with other spices, bringing additional wealth to the city.

Many peoples of the Old World craved the taste of spices. Their passions had helped to foster long-distance commerce for many centuries. When the Portuguese first arrived in "India," then, they entered a region where long-established trade routes flowed, not only to Europe, but by sea from East Africa to South Asia and on to the Spice Islands, China, and Japan, connecting places like Zanzibar and Basra to the Coromandel coast, Java, the Banda Islands, Indochina, and the east China seas. Vasco da Gama depended on North African merchants to act as his intermediaries, but the embarrassing quality of the gifts he presented to the Samorin of Calicut (twelve lengths of striped cloth, twelve coats with hoods, two barrels of oil and honey) opened him to ridicule by them, "who observed with glee that the presents were 'scarcely worthy of the poorest merchant in Mecca.'"[16] The Portuguese nevertheless soon grew rich after forcibly gaining entry into the intra-Asian trade, disrupting established relationships and co-opting others while also diverting a portion of the products of Asia back to Europe.[17] Their intervention in the spice trade also threatened to cut off the supply of spices via the Levant and the Mediterranean, especially via Alexandria and Aleppo to Venice. As the Venetian Girolamo Priuli worried immediately upon receiving news of the event in 1501, "now that this new route is found, the King of Portugal will bring all the spices to Lisbon."[18] The Portuguese could purchase spices in India at a fraction of the cost that the Venetians obtained them in Alexandria, and consequently they had only to undercut slightly the usual selling prices to make huge profits. Of course, the king also had to protect his spice traders in Asia and to force access to markets that were forbidden, which meant reinvesting great sums in a sizable and widespread military establishment in Asia necessary for the task. But the king took a large share of goods for himself and taxed the merchants he licensed to cover the costs of their protection, thereby profiting handsomely.[19] Although the direct Portuguese route to Europe did not eliminate the markets relying on eastern Mediterranean outlets—that would come only after the Dutch monopolized the trade more than a century later—Priuli was right to see it deeply threatening the customary European spice trade, as it also worried the traditional carriers in the Indian Ocean.[20]

As it happens, however, the development of a major outlet for the spice trade via the Portuguese had enormous ramifications for northern Europeans, too, especially for Antwerp, and eventually for Amsterdam. Again Priuli put his finger on the matter: "There is no doubt that the Hungarians, Germans, Flemish, and French, who formerly came to Venice to spend their money on spices, will all turn towards Lisbon, for it is nearer to them, and easier to reach."[21] These northern customers soon influenced events further by moving the main market for the spice trade to Antwerp. For after some initial experiments in how best to sell his spices, the king of Portugal found it best to rely on a syndicate of merchants to disburse the Asian goods brought to Lisbon. They in turn conveyed the goods to Antwerp, where they could be sold in such a way that the syndicate could control the prices. Antwerp had already begun to be a hub for commercial exchanges in northwestern Europe, but it especially benefited from the growth of Portuguese commerce.[22] It also became a place of refuge for many of the Jewish and "new-Christian" merchants expelled from Spain and then Portugal, who brought their knowledge, capital, and personal networks with them. The Portuguese syndicate consolidated Antwerp's place as the most important meeting place for merchants from the lands around the Baltic, North Sea, Atlantic, and Mediterranean.[23] The election in 1515 of Charles V as Holy Roman Emperor confirmed the importance of the new system, since it was accomplished only with the financial backing of the great German mercantile houses, themselves deeply involved in the new trade. The emperor in turn further promoted the use of Antwerp as the major trading center of his northern territories.

Moreover, in trying to get to the Spice Islands by sailing west, Columbus had famously bumped into a new hemisphere, which also produced new wealth for places like Antwerp. Although some of the newly encountered species of the region, such as tobacco, would eventually become great bulk commodities themselves, it was first the sugar trade that benefited most. Sugar plantations had already spread to the islands off Africa's western coast and now took root throughout the Caribbean: cane was brought to Santo Domingo as early as Columbus's second voyage.[24] Not many years after, in 1515, a surgeon there named Gonzalo de Vellosa imported sugar masters from the Canary Islands to set up a refinery.[25] Within a few decades, attempts to use the local inhabitants in a forced-labor plantation economy that included sugar growing resulted in the extermination of whole peoples and the decimation of others, followed by a system of harsh chattel slavery based on imported labor from Africa. One result was that sugar became plentiful in Europe, being used not only as a medicine and taken as a sweet but as a substance that could be molded into all kinds of shapes, like clay. At the marriage of Alexander Farnese and Princess Maria of Portugal, celebrated in Brussels in 1565, more than three thousand pieces were made from the finest sugar not only to serve as the very plates and utensils but as decorative statuettes, some representing horses and men weighing up to nine or ten pounds each: the total amount of sugar used for this one feast alone may have amounted to over six thousand pounds in weight, costing perhaps three thousand ducats.[26] Another consequence of sugar production shifting to the Atlantic and Caribbean was to benefit Antwerp again, which became the main center of sugar refining, with as many as nineteen factories in the city in 1556.[27]

With such commerce, the sixteenth-century low countries produced a remarkably rich and diverse culture. Whereas the Renaissance of the fifteenth century belonged to northern Italy, the sixteenth century's belonged to the low countries. No other region in Europe aside from northern Italy contained as many towns and cities. By the mid-sixteenth century Antwerp had developed into the second most important mercantile city in Europe after Venice, with a huge population of about one hundred thousand.[28] Its printers, artists, and many others working in the luxury trades were internationally recognized for their work. So, too, the leaders of the city could boast that Antwerp had become, "not only the first and principle commercial city of all Europe, but also the source, origin and storehouse of all goods, riches and merchandise, and a refuge and nurse of . . . virtues."[29]

TASTE AND OBJECTIVITY

It is important to note the association made in this boast between "a storehouse of goods" and a "nurse of virtues." It underlines fundamental values that were coming to the surface during the period called the Renaissance. Throughout the growing networks of urban commerce that linked the most populated regions of western Eurasia monied contemporaries experienced an expansion in disposable income—what is sometimes termed a consumer revolution.[30] A large proportion of wealth was, of course, spent on literal consumption, with people ingesting enormous quantities of sugar and spices imported from abroad, in turn fueling the contemporary overseas enterprises of the Portuguese and the Spaniards. It has been calculated that during the three hundred years from the time of Columbus to the end of the eighteenth century, about one pound of goods for every person in Europe was imported from Asia alone each year—mostly, it should be added, in the form of comestibles.[31] Other sorts of wealth were also on display. By the end of the fifteenth century the interior household spaces of the well-to-do were filling up with luxury goods. Lords and merchants were not only underwriting the cost of fabulous altarpieces and other public displays of religious piety, not only purchasing great horses, fine armor, luxurious clothes, and beautiful tapestries, but they were also acquiring well-crafted furniture, linens, antiquities, painting and sculpture, books and manuscripts, strange and lovely items of nature, and other rare and beautiful objects. Many of the wealthiest Italians, particularly, modeled themselves on their Roman ancestors, whose remains were all around them, collecting antiquities (especially coins and medals), works of ancient art and literature and offering positions in their households to students of antiquity who could explain and interpret these things.[32] At the height of the Renaissance in Italy, the erudite Leon Battista Alberti argued that possessions, including impressive buildings and beautiful books, were important for the happiness of any family, just as wealth was the source of friendship and praise, fame and authority, even being necessary for the prosperity of the state. Valued objects had become "goods" alongside personal virtues. As the historian of art and society Richard Goldthwaite has put it, "possessions become an objectification of self," perhaps "for the first time."[33]

The new goods were therefore imbued with moral qualities—virtues—and could be discussed as examples of human betterment. The possession of objects recognized by others as good "demonstrated taste more conspicuously than wealth."[34] In the use of the word "taste," it is notable how attributes previously personal and bodily became displaced onto luxury objects, whether consumed or accumulated, for having discriminating taste grew to mean possessing fine personal judgment about material things.[35] It arose from instances such as demonstrating the ability to discern correctly on the palate the spices in a dish or the region from which a wine originated. But the ability became generalized to mean knowledgeable discrimination of any sense impressions, so that one could tell quickly the quality of, say, the weave of a piece of cloth, the use of the brush in a painting, or the interlacing of parts in music. People of good taste in turn signaled their ability to make these fine discriminations by nuanced word and gesture, in the process marking themselves off from those who could not. Some were supposed to be born to it, like the young woman in the fairy tale who proved she was a princess by being discomfited by a pea beneath several mattresses. Most learned by example. Still others were acknowledged to have acquired good taste from education: the so-called humanists assumed that the right sort of people could be further refined by tutoring and knowledge of the world. But around the common sensitivities marked by good taste, men and women of both the warrior nobility and mercantile wealth, lords of church and city, and their servants and learned associates, could find common ground. As Hans-Georg Gadamer put it, "*The concept of taste* was originally more a *moral* than an aesthetic idea."[36]

But what one knows via the senses is grasped directly. "Taste is therefore something like a sense. In its operation it has no knowledge of reasons. If taste registers a negative reaction to something, it is not able to say why. But it experiences it with the greatest certainty."[37] The knowledge that comes with taste is, then, very real, but the knower cannot explain *why* he or she knows, only *that* he or she knows. Comparisons and analogy can make the knowing articulate, but they cannot inculcate it in someone who has not had the experience of it. For instance, someone with a fine taste for food can coach another to note this or that aspect of flavor, which is "like" something else: "a hint of plum" or the aroma of cinnamon, or even a "coloration" in a musical line. This kind of knowledge is passed on by experience, example, or imitation; it cannot be learned from the giving of reasons. Many modern European languages other than English have words to indicate this kind of knowledge, words like German and Dutch *kennen* rather than *wissen* or *weten*, or French *connaître* (from which the word "connoisseur" moved into English in the mid-eighteenth century) rather than *savoir*, all of which indicate knowing by acquaintance rather than by reasoning. In other words, what one knows from the senses as filtered by taste and experience has to do with particular instances rather than universal propositions. You have encountered *this* person or *this* sound before, and you therefore know something about it, even if the knowing is not based on reasoned explanation. Tasteful collectors therefore imparted a knowledge of sense impressions that, once acquired, could be asserted, demonstrated, and perhaps even put to use, but not predicted or explained from first principles. So the goods of commerce embodied not only particular moral attributes but particular kinds of knowledge, giving pride of place to the knowledge of the tangible world.

The boast of Antwerp had, therefore, actually been a bit longer, claiming that it was "a refuge and nurse of all arts and sciences" as well as virtues. Yet the kind of arts and sciences developing in Antwerp, like the kind of moral virtues rooted in goods rather than "the good," presented grave challenges to the kind of knowledge traditionally inculcated in the schools (which was termed *scientia* in Latin, or weten and similar words in the vernacular languages). Universities were the preserve of the professors who had studied and disputed for long years and then passed on their knowledge to students by lecturing and debating. They valued demonstrative certainty above all else, wishing to draw conclusions that could be shown to follow from necessity. Such demonstrative certainty came from reasoning by clear and certain steps from premises known to be true. The classic examples are demonstration by dialectic, in which a proposition (thesis) is contradicted (antithesis) and resolved by a proposition containing truths from both (synthesis) or by the use of syllogism, in which a proposition known to contain true and universal assertions is linked to another that refers to a new premise, yielding new truths conclusively (as in "All men are mortal, Socrates is a man, ergo Socrates is mortal"). These methods could be clearly explained and could be tested not only by writing things down but by debating with an opponent. Because such methods yielded demonstrative certainty, knowledge of this kind could be built into philosophical systems of great range and power capable of being passed on to others by explanation. Above all, they had the capacity to reason about the causes of change, probing for why matters were as they were.[38]

The word associated with the new ways of knowing, however, was "curiosity," which had nothing to do with understanding the reasoned causes of things. It is a complicated word, having been associated with sin but gaining in positive connotations.[39] Like taste, virtue, and other such words, it was too closely linked to the experience of the world to have a precise definition. A recent study of its use in early modern Germany has shown "that people talked and wrote about curiosity for conflicting purposes, and that they could do so because there was an enduring lack of consensus about what exactly curiosity was." The term "often came to encompass not only people's desire to know or possess something but also *what* they desired to know or possess"; and exactly because it was linked to desire it was not a concept but an aspect of life given expression in ordinary language. In other words, the word "was often a 'plot summary,' an aggregate of examples linked by family resemblances that could be pointed out but not actually explained." The author of this sentence, Neil Kenny, concludes that "if the case of early modern curiosity is anything to go by, then 'concepts,' in the usual senses of the term, do not exist within the flow of history. . . . Once one stops trying to discover concepts and their 'empty coherence' in the past, then the inventiveness, dynamism, and irreducible particularism of early modern uses of curiosity becomes more visible."[40] Put another way, the *curiosity* of the period points not to something about "the minds" of people alone but to how their experiences involved them in finding out about the myriad things around them.

Curiosity about particular things even gave rise to a new word: "fact," or "a matter of fact," a term (as we now know from the work of Barbara Shapiro) that had been borrowed from law. In legal terminology, the word "fact" meant a deed, since the Latin *factum* meant a thing done, and from it Dutch adopted *feit*, French took *fait*, and English derived both *feat* and *fact*. The word usually had a legal application as a term for "what had happened," a true account of which needed to be established in order to begin the process of rendering a judgment about it. Generally speaking, only late in the sixteenth century did the word start to be applied in a way that meant something that had "really occurred . . . hence a particular truth known by actual observation or authentic testimony, as opposed to what is merely inferred."[41] Matters of fact served well to convey the kind of information acquired in the forge of experience, but they yielded only probabilistic knowledge, just as in a law case, where the facts can be known with a high level of probability, perhaps even with a moral certainty, but never with an absolute certainty.[42] For many kinds of worldly decisions, such as determining a legal case, deciding on whether a financial decision has been a good one or not, giving an account of a treatment for a particular kind of fever, or judging the quality of a wine, factual knowledge—knowledge by acquaintance with the thing—is not just adequate but necessary. The many people benefiting from the new economies of exchange therefore not only valued material goods, they also valued matters of fact.

One certainly sees the rising power of an appreciation for the knowledge of things in early modern painting, which richly depicted the details of life. Such "realistic" art became one of the glories of the low countries in sixteenth and the seventeenth centuries. These were pictures "from life" (*naer 't leven*) rather than "from the imagination" (*uyt de gheest*), as one of the best contemporary commentators, Karel van Mander, put it.[43] The greatest skill was applied to creating representations of the straightforward appearances of things in all their meticulous detail. From the medieval period, and especially from the fifteenth century, representational art had developed dramatically in conformity with the late medieval interest in the details of the creatures of nature that showed itself in sculpture and manuscript illumination.[44] Illuminators added illustrations of a variety of recognizable plants and insects to the borders of manuscripts such as books of hours (meant for contemplation) and breviaries (books of prayers). Netherlandish painters such as Jan van Eyck increasingly included naturalistic details in the background of their works, tricking the eye into imagining that it was seeing nature itself; later painters were further affected by the example of the German artist Albrecht Dürer, one of the first to devote himself to compositions centered entirely on natural subjects, as in his famous sketch of a hare, drawing of a stag beetle, or portrayal of "The great piece of turf" of 1503.[45] By the sixteenth century, the fine detail of facial expression, draped fabric, architectural form, landscape background, tools, flowers, and other items, had moved from incidental detail to the focus of attention throughout Europe.[46]

It is therefore possible to speak about *objectivity* as a kind of knowledge being cultivated in the early modern period: a knowledge appertaining to a detailed acquaintance with objects. While some have argued that objectivity is known only

Saint Katharina, with marginal butterfly and flowers (including rose, violet, daisy, and columbine). From the Mayer van den Bergh Breviary, Brugge, c. 1510. Museum Meyer van der Burgh, Antwerpen © collectiebeleid. Copyright IRPA-KIK, Brussels

Albrecht Dürer, *The Great Piece of Turf*, 1503.
By permission of the Albertina, Vienna

in contrast to "subjectivity," the same authors agree that of course the "notion of getting a true picture of nature" existed long before 1800.[47] The appropriate use of the word was indicated by Paolo Sarpi in the early seventeenth century when writing about religious relics: "a worship did belong unto them besides the adoration due unto the Saint worshipped in them, calling this worship Relative, and the other Objective."[48] Even if the word was not often used in the period, we ourselves can use "objectivity" as a term referring to matters that pertain to the knowledge of objects without reference to intuition or innate knowledge, the corporeal knowledge of things that can be experienced by the bodily senses, information from which can be exchanged.[49] "Objectivity" therefore implies that knowledge is emergent from bodily experience, not something apart from it. For

Hans Boulenger, *Tulips in a Vase*, 1639.
By permission of the Rijksmuseum, Amsterdam

the sake of simplicity, then, it is possible to say that many early modern people considered the highest kind of knowledge to be the result of study of the objects (*res*) of nature.

COLLECTING OBJECTS AND SPECIMENS

The high value placed on the knowledge that came from acquaintance with objects (*kennen*) rather than discourse (weten) began to dominate natural philosophy from the Renaissance. One of the most important problems solved by kennen was the decoding of the handwriting and words in ancient manuscripts. Antique manuscripts were one of the first kinds of possessions to be collected in bulk. Simply touching something that had been handled by the ancients could convey a thrill of connection with their spirit, while examining their words seemed to transmit meanings more or less directly from their age to the reader's. Many well-educated, well-to-do, and politically active men therefore found collecting works by ancient authors exciting. But trying to decipher what the books contained was certainly no easy or straightforward task. The collectors of manuscripts consequently also collected scholars in their households who could make out something of the meanings of these writings. Learning how to read the handwriting (*paleography*) was the first great difficulty. Deciding what the words meant once they were deciphered (*philology*) was often just as difficult. As with other matters of taste and fact, getting a sense for the meaning of a word, whether a new word or a familiar one used in an unfamiliar way, can be achieved only by noting examples, studying how a word was used in a variety of contexts, and then applying discriminating judgment to make appropriate inferences. It was a question of finding a fit rather than seeing an implication in a proposition. Philology and paleography presented the problems of particulars rather than general reasons, puzzles to be figured out by trying, and trying again, until an appropriate meaning was found.

The difficulty of knowing what particular words meant was especially great when it came to the huge vocabularies associated with all the variety of natural things (*res naturae*), a subject called "natural history." The term borrowed from the Greek word *historia*, which meant "a learning by inquiry" or "the knowledge or information so acquired."[50] Natural history was, therefore, an account of nature based on information acquired by the investigation of natural things. The best-known text on the subject was the first-century Roman encyclopedia of Pliny the Elder, his *Historia naturalis* ("Natural history"). It was a huge, sprawling work of many volumes, emphatically conveying a view of nature that considered the multitudinous expressions of existing things rather than their underlying unity, a collection of the "nature of things" rather than a single "nature," no doubt because Pliny was a polytheist. Pliny therefore included in his work a description of all the things of heaven and earth, covering cosmography, astronomy, geography, chorography, mineralogy, and meteorology, and the living things of this earth and their purposes, including fishes, birds, insects, and other animals, herbs, shrubs, and trees, their economic and medical uses to humanity, and the investigation of the tools, costumes, customs, and beliefs of strange people.[51] Pliny's remained a text much mined by authors of medieval encyclopedias, bestiaries, herbals, and other works that touched on natural things, with a few authors adding their own new investigations to it.[52] By the mid-fifteenth century, about two hundred manuscript versions of Pliny's *Natural History* existed. It was first printed in 1469, and over the course of the next twenty years, twenty-two versions appeared, with the more than forty commentaries on it giving it a degree of attention "rarely equalled." Indeed, commenting on Pliny became "a characteristic scientific genre of the Renaissance."[53]

With both the development of philology and the collection of multiple manuscript versions of Pliny's work, attempts were made to reconstruct his original text and to understand his original meaning. Because of the rise of Greek studies at the end of the fifteenth century, in particular, scholars such as Giorgio Valla, Alessandro Benedetti, and Niccolò Leoniceno were examining the sources Pliny himself had relied on in order to figure out what things he meant to indicate when he used certain words.[54] The most important of these sources was Dioscorides, a contemporary of Pliny (living around 65 CE) who had traveled in the lands of the eastern Mediterranean and composed a work in Greek best known as the *Materia medica* (the "Matters of medicine"), describing plants, animals, and minerals and their medical uses. The famous printer of Venice Aldus Manutius brought out a Greek edition of Dioscorides in 1499, and this was followed by a reliable Latin translation by the French medical humanist Jean de la Ruelle, with a commentary expanding and correcting Dioscorides. The Spaniard Elio Antonio de Lebrija (or Nebrija), who had become a committed humanist during his peregrinations in northern Italy, developed an important commentary based on Ruelle's edition, adding an alphabetical lexicon of about two thousand Spanish terms for the names that occurred in Dioscorides' text. By the middle of the sixteenth century, editions and commentaries on Dioscorides by Pietro Andrea Gregorio Mattioli, who spent much of his working life in the region around Trent in the Italian Alps, were the most important, although rival experts continued to disagree vociferously about the details of his identifications. This kind of work on Pliny, Dioscorides, and other sources indicated not only that many of Pliny's manuscripts had become corrupted over the years but that Pliny himself had sometimes erred in identifying some of the natural things he mentioned.[55]

But if ancient authors themselves did not use words correctly in all cases, comparing versions of a text would not necessarily establish exactly what they meant or should have meant. In other words, it was important to compare not only words with words but words with objects. The search for this kind of information about the world often went by the Latin term *venatio*, "to hunt."[56] The hunt was no mere leisurely academic examination of old books but an urgent investigation of natural things themselves. Apothecaries and physicians, for instance, hoped to rediscover real examples of the powerful drugs used by the ancients. One "simple" much used by them had been rhubarb root. Careful inquiry showed that true rhubarb came from somewhere in "China," being marketed to Europe mainly through Muscovy (Russia) and the Ottoman Empire. Investigators were able to distinguish it from the common rhubarb obtained readily throughout the Near East partly by the different effects of the two kinds. True rhubarb was said to work by purifying the humors via a gentle purge, having astringent as well as cathartic effects, whereas common rhubarb was neither as gentle nor as effective. Yet despite hunting in markets and making inquiries of merchants, an accurate description of the living plant, as well as its place of origin in Asia, would remain uncertain well into the nineteenth century.[57] Similar investigations went into New World plants. A Spanish practitioner in the Caribbean found a balsam on the island of Santo Domingo that he asserted to be the equal of, or even better than, the greatly beneficial balsam known to classical medicine. He certainly wished to market it as such. The Spanish crown itself became deeply involved in trying to assess his claims.[58]

Some of the implications of this approach toward knowledge can be seen well in the work of François Rabelais. Although best known for his satirical *Gargantua and Pantagruel*, in his professional capacity Rabelais had issued one of the first accurate Latin translations of the Hippocratic aphorisms and is said to have been the first to lecture with a Greek text of Hippocrates in front of him.[59] When read in this light, his *Gargantua* can be seen to take pleasure in exploding earnest dogmas and challenging musty old academics who are tripped up by the unexpected. The world was not built by the human mind and could not be known by reason alone. Rabelais therefore located truth in the simple and strange in all its variety rather than in its supposedly hidden and unitary meanings. While reveling in the endless complexities and unexpected events of life, Rabelais avoided the formlessness of madness; for him, multifarious events and things spoke more truthfully about what is "really" going on beneath the surface than could words alone. As one of his most penetrating commentators has written: "I consider it a mistake to probe Rabelais' hidden meaning . . . for some definite and clearly outlined doctrine; the thing which lies concealed in his work, yet is conveyed in a thousand ways, is an intellectual attitude, which he himself calls Pantagruelism; a grasp of life which comprehends the spiritual and sensual simultaneously, which allows none of life's possibilities to escape."[60]

The same exacting, descriptive attention given to botany and other subjects in natural history was therefore also being called for when physicians described diseases. Of course, physicians had always had to pay attention to the symptoms they saw in their patients and clients. But medieval physicians sought to incorporate these into discourses for their patients, sometimes written out, in which they discussed what they saw in light of what they knew about the causes of the patient's troubles, and further, giving advice about how they should change their habits and diet in order to live well and long. These *consilia* were similar to the advice offered by a clergyman or lawyer after consultation about one's state of affairs, in which the descriptive information was less important than analysis of the causes and the recommendations for change to which it led.[61] By the middle of the sixteenth century, however, medical humanists were recovering the "true" Hippocrates from the Greek texts, finding in him a careful clinician who (especially in the text called *Epidemics*) described the signs and symptoms of disease very carefully, sometimes including a prognosis about what would come next but not commenting on causes.[62] To gain a true knowledge of disease, then, one had also to go directly to the things themselves for thorough and exact description, proceeding like the famous Hippocrates.

As a result, physicians came to write detailed descriptive notes about the conditions they saw (case "histories"), often sending them to one another in letters, and sometimes collecting them together and publishing them as *Observations*. It even became common to collect the observations together in groups of one hundred, published as *Centuries*. The German physician Guilhelmus Fabricius Hildanus, for instance, was only one of the many who published "Centuriae" of observations in the late sixteenth and seventeenth centuries, a practice he probably picked up when he was a young man in the early 1580s from one of his older humanist colleagues, Reinerus Soleander (Sondermann).[63] From the nearby low countries, Pieter van Foreest (Forestus) became widely known throughout Europe for the number and quality of his published *Observationes*.[64] As one historian has remarked, "Collections of curationes and observationes were the primary medium for the circulation of medical information in the early modern period."[65]

Compound medicines also came in for investigation. An example is theriac, one of the most complex mixtures, a dark and sticky substance thought to be a wonderfully effective antidote to all poisons and a general preservative against most diseases. Many ancient authors recommended it, and a few listed its ingredients, from true rhubarb to serpent skins, bezoar stones, bits of mummy, amber, and so forth, some receipts recording up to eighty-one simples that were to be included in this one medicine. Theriac consequently required enormous time and expense to compound. In great cities like Venice and Florence, an annual ceremony for the official mixing of a huge batch of the medicine took place with great pomp and dignity, presided over by physicians and magistrates, after which it was certified and doled out to apothecaries and merchants for sale locally and abroad.[66] Venetian theriac became particularly well regarded for its superior quality and despite its high price was often specified by name in prescriptions written by physicians throughout Europe. Scholars therefore took care to discover the proportions of the ingredients for theriac specified in the texts and to identify accurately the correct simples that went into it. This active hunt for the authentic ancient remedies helped to bring about "a quiet revolution in simples."[67] None of it could have been done by reflection and discussion alone, only by actively searching things out.

The same was true when it came to the knowledge produced by another of the enthusiasms of the consumer revolution: the private garden. Even in built-up cities like Venice, pleasure gardens flourished.[68] The Latin word commonly used for such spaces, *hortus*, had originally applied to the kitchen gardens of Roman farms, which were walled or fenced to deter vermin and shelter tender shoots.[69] But under the late republic and the empire, Romans had begun to imitate their neighbors much further to the east in constructing pleasure gardens. One of the seven wonders of the ancient world had been the "hanging gardens" of Babylon, built on artificial hillsides terraced by Nebuchadnezzar II for his wife, who had herself come from more verdant lands to the east, the region that became central Persia, whose cities contained some of the loveliest gardens of the known world. It was from this region that a local word came into the vocabulary of a people living on the fringes of Persia: the Greek language adopted the word *paradise*. "Paradises" were spaces other than living quarters, enclosed by walls and pierced by sturdy gates to keep out the unwanted, where precious fruits and flowers could be kept safe from sudden winds and thieving hands, and where hosts could entertain guests over food and drink. To such gardens, streams and fountains brought water and palms brought shade, making them comparatively cool even at midday; here business and pleasure could be conducted away from interfering ears and eyes; and here countless precious fruits and flowers could grow to delight the mouth, nose, and eye. The most remarkable gardens of the medieval period—so much imitated in European romances—remained those of the Islamic world, at Persian cities like Isfahan or Iberian ones such as Cordova and Granada or, after the conquest of Constantinople, in the Ottoman Empire. Sultan Mehmed II built the "abode of bliss," Topkapi Sarayi, on one of the seven hills of Constantinople, a fantastic place that included twelve gardens staffed by hundreds of *bostangi* (who also served as some of the sultan's most trusted bodyguards—and executioners).[70]

Trying to keep up with their Islamic neighbors at the same time that they were developing a taste for things Greek and Roman, Renaissance Italians took a keen interest in having a classically styled garden as part of a great house. The founder of Italian humanism, Petrarch, made a point of situating many literary conversations in gardens, while Giovanni Boccaccio set his *Decameron* in the aristocratic garden of the Villa Palmieri, where the young people played in the morning, passed the afternoons quietly, and told stories in the evening. Alberti wrote of grottoes and gardens as part of his famous book on architecture of 1452, and shortly thereafter the Medici family of Florence redesigned its villa at Careggi along the lines of an antique estate rather than the medieval fortress it had been, constructing in it a garden overlooked by a double loggia, a feature that soon had many imitators. Other architectural treatises that contained garden designs followed, all insisting on symmetry and geometrically figured parterres within walled or fenced enclosures, containing water, trees, flowers, and herbs.[71]

But although in Italy gardens continued to concentrate on designs of shaped mass greenery combined with rock and water, a keen interest in gardens showing off the form and color of individual plants was also clearly evident by the early sixteenth century, especially further north. For ages, only a few flowers had seemed worthy of mention in poetry and prose: the ubiquitous rose and lily, violets, columbine, and occasionally (white) iris, heliotrope, and mandrake. Carnations were added after Crusaders brought them back to Europe. By the beginning of the sixteenth century, however, the number of different plants named in gardens and depicted in books was multiplying rapidly. The growing array of new plants being brought back from faraway places, especially from the Ottoman lands of the Near East, contributed much to this trend, as gardeners started to cultivate exotics.[72]

Given the growing appreciation for direct investigation of botanicals for use or pleasure, medical faculties in universities began to appoint professors of materia medica and to construct botanical gardens for their teaching. The first such professor seems to have been appointed at the papal university in Rome in 1514, and he also had the use of a garden for demonstrating medicinal plants.[73] Some years later, Luca Ghini became the first professor of simples at Bologna and first director of the botanical garden at Pisa, which was laid out around 1544. A botanical garden for the university of Padua soon followed,[74] Bologna constructed one in 1567, and other medical faculties throughout Europe followed suit. For study, a kind of botanical collection that supplemented the gardens was popularized, if not invented by, Luca Ghini in the 1530s and 1540s: the *herbarius*, sometimes also called a "dry garden" (*hortus siccus*). Ghini found that leaves, flowers, and other thin parts of plants could be placed between sheets of paper, which were in turn pressed by weights while allowing them to dry.[75] The properly dried remains were a more or less permanent record of the living plant, giving clear evidence of shape and texture, although the colors tended to fade. Notes could be added to the page to which the specimen was attached for later reference.

Yet the case of Padua's famous garden clearly shows how much was owed not only to medical utility but to the patrician interest in pleasure gardening. The Venetians constructed the Paduan garden at enormous expense, on a grand

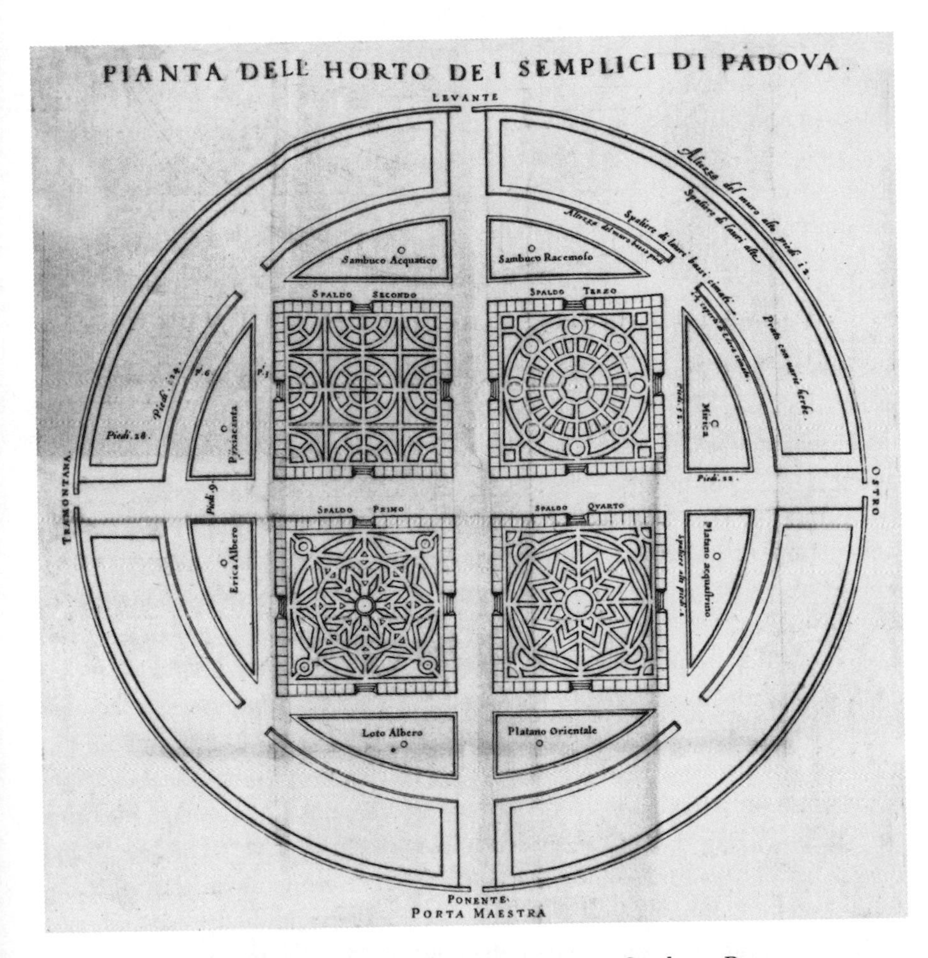

Plan of the botanical garden at Padua. From Girolamo Porro, *L'horto de i semplici di Padoua*, 1591. By permission of the British Library

scale, for the propagation of exotics as well as the teaching of medical botany. A great circular earthen rampart, built according to the latest military construction methods and penetrated by four tunnels, closed off the garden from the rest of the (irregularly shaped) grounds while also allowing visitors to walk along its top and look down into the design of the beds. The enclosed parterres, divided into four square quarters, were designed as complex geometrically figured plots, planned by the learned Venetian cleric Daniele Barbaro according to the best architectural and mathematical principles of the 1540s. Barbaro used circles, squares, and triangles while playing with the problem of squaring the circle and the use of the "magic number" found in Vitruvius. Naturally, the wonderfully impressive layout of the garden demonstrated the exquisite good taste of the rulers of the Serene Republic. But it was not especially practical for teaching. The most valuable specimens disappeared, necessitating the removal of the rampart around 1552 in favor of a more traditional brick wall, while the very elaborate beds must have broken up any simple ordering scheme for the plants, making it harder for the students to memorize their simples. In the early 1590s, therefore, as the use of the garden for teaching medical botany began to take precedence, the new prefect, Giacomo Antonio Cortuso, introduced ambitious plans to redesign the layout along more utilitarian lines.[76]

The new aristocracy of taste collected other specimens of nature, too. Some of these objects were interesting or beautiful, but many could not be identified from ancient sources, hence falling into the category of the strange, exotic, or "curious."[77] As tokens of the exotic lands they had visited, travelers brought back strange and wonderful things both made by human artifice and gathered from nature, which collectors in turn purchased.[78] In most cases, such curiosities would have been placed in libraries or mixed among objects of religious veneration or fine art.[79] By the 1560s, however, efforts to situate the growing amount of accumulated *naturalia* in rooms of their own ("cabinets") apart from other goods can be clearly discerned.[80] One of the first known cabinets was assembled by Francesco I de' Medici, Grand Duke of Tuscany, who set aside a room full of all sorts of special natural and man-made objects, ordered to reflect the glory of the prince by a display of the marvelous and imperial.[81] About the same time, Hans Jacob Fugger of Augsburg built a similar *wunderkammer*.[82] The Medici and Fugger families had both developed huge family-run merchant firms that depended on long-distance trade (the Medici with the established fabric and spice trade of the eastern Mediterranean, the Fuggers with the new worldwide silver and spice trade of the Spanish and Portuguese), and each had aspirations to impose themselves as local political rulers (the Medici became grand dukes while a branch of the Fugger family became counts). The display of exotic materials demonstrated their connections with, and knowledge of, the wide world. They were soon imitated by princes, such as the nearby dukes of Bavaria (whose seat was near Augsburg) and William IV of Hessen-Kassel.[83]

But as with libraries or gardens, collecting naturalia, too, required expertise. The ability to make discriminations about whether the natural objects on hand or on offer were rare or common, unimportant or unusual, was critical to giving a cabinet of curiosities significance. To advise him on his assembly of naturalia, then, Hans Jacob Fugger employed the physician Samuel Quickelberg (or Quiccheberg), originally from Antwerp and educated at Basel.[84] Medical people themselves even began their own collections. When the Zurich physician

The museum of Francesco Calzolari. From Benedetto Ceruti, *Musaeum Francisci Calceolarii junioris Veronensis*, 1622. By permission of Plymouth City Museum and Art Gallery: Cottonian Collection (29)

Conrad Gesner was offered a post as family tutor and librarian by Johann Jakob Fugger in 1548, Gesner paid Fugger a visit to look over the situation, and while in the end he turned down the offer in order to remain in Zurich, Fugger's collections inspired him to begin to assemble his own assembly of naturalia.[85] Other famous collections were being put together by the physician and botanist Michele Mercati in Rome, the physician and botanist Ulisse Aldrovandi in Bologna, the apothecary Ferrante Imperato in Naples, and the apothecary Francesco Calceolari (or Calzolari) in Verona.[86] A catalogue of Calceolari's three-room collection makes it clear that he began assembling it by setting aside the exotic herbs sent to him that were no longer living and so not suitable for his garden.[87] The association between apothecaries and natural history became legendary, with their shops often depicted with stuffed crocodiles and other bits of naturalia among the objects in the foreground. In Calceolari's case, for instance, his collection no doubt increased his reputation as an expert in the exotic materials he sold, including the medicinals on which his business depended. At least it drew many distinguished gentlemen and scholars to his shop, who signed a ledger recording their interest, making a visit to it of even greater significance for others.

A steady trade in naturalia developed, with a few brokers even buying up objects at dockside and reselling them to collectors. On most ships, occupants could engage in a limited amount of "private trade" according to the space allotted. Common sailors and soldiers would bring back anything they could stuff, along with their bedding and clothing, in their duffel bags. People of higher rank, like surgeons, could bring along chests.[88] Captains and senior merchants had the use of a small room and whatever could fit in it. Clearly, small and light items of high value would bring the most profit. Senior employees might have the means to trade in noble metals and jewelry, but even sailors might be able to bring back a few exotic items of naturalia, or a small amount of medicine and spice, for a tidy little profit.

People visited these collections in part because of the meanings with which they were imbued. But as with other matters of taste, the meanings to be found in the specimens on display were not discursive or reasoned but arose instead from relationships between particulars. For instance, flowers like the rose had long symbolized true love and other qualities. Or rather, the rose did not symbolize love but presented it to us. All created things were considered to be interrelated, so that any one object was connected to others by tangled webs of sympathy or antipathy. Works like the occult illustrated romance of the *Hypnerotomachia poliphili* of 1499 suggested that the way gardens themselves were laid out could make them places of great power. The macrocosm of the universe could be present in the microcosm of the garden, so that by assembling certain things in particular ways, gardens could embody some of the powers of the universe. The objective knowledge accumulated from cabinets of naturalia, too, was laden with significations. Quickelberg's printed description of Fugger's cabinet in 1565, the first known catalog of a collection, suggests this clearly. He showed how the assembly of the Fugger cabinet implied that each of the objects had a relationship to the macrocosm, which taken together represented the universe. Various kinds of sacred and human inscriptions made up the first of five groupings, followed by a second group of objects produced using natural things (such as fantastic pieces made from precious metals, glass, pottery, wood, or stone, and so forth), which in turn were followed by three groups of things produced by nature itself, divided into whether they came from the realms of earth, water, or air.[89] A description of

Calceolari's shop suggests something similar: he displayed a range of simple and compound medicines, diverse minerals and precious stones, remains of rare animals and fish, various kinds of earths, and roots. It contained, in short, all those things that are "most beautiful, rare, and good," assembled into a one of the most exquisite and singular "Universal Theaters" of the age.[90]

It should come as no surprise that Calceolari and Imperato were also expert at handling naturalia, for as apothecaries they were merchants who dealt in expensive produce from exotic lands. In handling a range of expensive and exotic, sometimes potent and always sought-after commodities, the apothecaries ranked among the wealthiest and most influential groups in Italian cities. With the rapid growth in the quantity of imported sugar shifting its use toward the table in addition to the sickroom, "confectioners" who specialized in sugar and the dishes made from it split off from apothecaries.[91] The apothecaries in turn divided into two general groups, some remaining long-distance importers, others selling their wares directly to the public. A growing number of the latter were turning into dispensing pharmacists, who in turn began to specialize in handling natural products only, from medicines to pigments, giving up the modeling of wax votive objects and similar activities, for instance.[92] Increasingly limited in their dealings to medicinals and other natural products of high value, keepers of apothecary shops required knowledge about identification and usage. They became expert in knowing the exact details of the plants, animals, and minerals in which they dealt and their uses and preparation, often including new chemical methods. They also collected other exotic items that came their way as well.

Many apothecaries also became expert gardeners. In their gardens they could grow common or acclimatized plants for medical use, but as with their collections of naturalia, they also grew interested in many other kinds of plants. Moreover, as esteem for a knowledge of plants grew, physicians were also stimulated to try to keep ahead of apothecaries and others. As the eminent Parisian physician Jean Fernel put it in the mid-sixteenth century: "The knowledge, collection, choice, culling, preservation, preparation, correction, and task of mixing of simples all pertain to apothecaries; yet it is especially necessary for the physician to be expert and skilled in these things. If, in fact, he wishes to maintain and safeguard his dignity and authority among the servants of the art, he should teach *them* these things."[93] Valerius Cordus took a similar view in composing the Nuremberg pharmacopoeia the *Dispensatorium* in order to limit the remedies sold in the shops to ones approved by the physicians—his was one of the first and perhaps the most copied work of its kind.[94]

By the mid-sixteenth century, therefore, an enthusiasm for natural history and medicine can be observed everywhere. In botany, for instance, one of the first vernacular herbals to appear in print was printed at Louvain in 1484 (Johan Veldener's *Herbarius in Dietsche*) and was frequently imitated in following decades.[95] In the Germanic lands, a group of new and richly illustrated books were published on botany: Otho Brunfels's *Herbarum vivae icones* ("Illustrations of plants done from nature") printed at Strasbourg from 1530 to 1536, and the even more impressive illustrated work of Leonhard Fuchs, *De historia stirpium* ("The natural history of plants") which had appeared in 1542, with several subsequent editions, several in a small-size format for easy consultation in the field. Also in 1542, the Zurich physician Conrad Gesner published his *Catalogus plantarum* ("Catalog of plants"), and soon, inspired by a manuscript of Claudius Aelianus on natural history he encountered in the library of Johann Jakob Fugger in 1548, Gesner turned to compiling encyclopedias on animals and minerals to complement his work on plants.[96] In France, Guillaume Rondelet, one of the greatest early naturalists, contributed much to building up the botanical garden of the university of Montpellier, pursued anatomy with vigor, and wrote works on medical diagnosis and materia medica, although he is most famous for his publication on fish (*Libri de piscibus marinis*, 1554–55). In England, William Turner published a well-known work of botany in the middle of the century, while at the beginning of the next, John Tradescant the Elder and the Younger became famous both as gardeners and as collectors of naturalia.[97] The natural history work of Spaniards Nicolás Monardes and Francisco Hernández was superb, even if that of Hernández was never fully published; the Portuguese book of Garcia da Orta on the medicinals of the East Indies became fundamental;[98] from the low countries, three botanists became especially well known in the mid-sixteenth century, Dodonaeus (Rembert Dodoens), Lobelius (Matthias de l'Obel), and Carolus Clusius (Charles de l'Escluse); the Italians remained almost too numerous to notice.

Just beneath the surface of this excitement about the diversity of the world lay theological problems, but they had mainly been tamed. Simply reveling in the particularities of nature echoed the classical and pantheistic outlook of Pliny and his like, in which all aspects of nature were redolent with innate powers (*virtues*).[99] To a good Christian, however, there were not many forces in nature but one God, who had created nature rather than being embodied in it and whose powers lay concealed rather than evident. "The heavens declare the glory of God; and the firmament sheweth his handywork," began Psalm 19 (King James Version), but the apostle Paul wrote against the pantheistic pagans who "changed the glory of the uncorruptible God into an image made like to corruptible man, and to birds, and fourfooted beasts, and creeping things" (Rom. 1:23). Yet, to counter heresy and unbelief, including pantheism, the fourth-century bishop of

Hippo Augustine had found it helpful to write of the two books from which evidence of God could be drawn: the book of revelation, in which God revealed himself by the word (*logos*), and the book of nature, in which God revealed himself through what he created. In arguing against the Manicheans, for instance, he wrote about how the book of scripture could be heard by the learned, who could read its words, while the book of nature could be seen by anyone.[100] Created things therefore expressed the nature of God quite as much as did his word, and read correctly they revealed not so much the multitude of things and powers as his deeper presence.[101]

Some medieval university professors had therefore felt empowered to investigate natural things on the path to knowledge of the divine, incorporating such information in their encyclopedias, bestiaries, herbals, and other such works.[102] By the end of the fifteenth century, the phrase "natural theology" had appeared as a way of suggesting that a defense of Christian belief could be based on evidence of God the Creator alone. An argument rooted in the things he created seemed useful in converting Jews and Muslims when arguments based on reason alone had failed, as they were recognized to have done by the later fourteenth century.[103] For instance, an early printed book titled *Theologia naturalis* (1480), originally a work of 1436 written by a professor at Toulouse, Raimundo Sibiuda—better known as Raymond Sebond—had been written for such a purpose, as well as to hearten his brethren in Christ who did not believe as firmly as they should. Originally titling his manuscript Liber creaturarum (seu Naturae) seu Liber de homine ("Book of created things, or nature, or book of man"), Sebond took the view that the books of nature and revelation were one and the same. Following his predecessor Raymond Lull, he explained the ladder of creation from minerals and inferior living things up to humanity, and from humanity to God, all as evidence of the truth of Christianity. Sebond's argument rested on how the knowledge of created things, coupled with faith, gave a more secure foundation than rational argument. (In this way he was also fighting Averroist arguments for "double truth," in which the truths of nature and of religion must be evaluated according to different standards.) The book and its argument proved to be very popular in the sixteenth century. Michel de Montaigne's translation of it into French (the *Théologie naturelle* of 1569) is the best-known version because of the later fame of the translator, but Sebond had many other editors and translators, too.[104] In such works, the book of nature took its place alongside the book of revelation without danger of pantheism. It was a book open to all faiths, and its interpretations had an ancient and respectable pedigree common to all Christian theologians of the period.

OBJECTIVITY IN ANATOMY AND MEDICINE

And yet, Montaigne saw all too well that arguments about natural theology ran into philosophical trouble because they relied on the evidence of the senses, which could be misleading. His *Apologie* for Sebond's natural theology, composed mainly in 1576, is the longest and one of the most elaborate of his *Essays*. He agreed that Christianity depends on faith and grace rather than reason, and he gave much evidence about the hypocrisy of so-called Christians who lack such faith. Humankind is no better than other animals, our knowledge makes us neither happy nor good, and in any case we mortals have no real knowledge of how things really are. "There cannot be first principles for men, unless the Divinity has revealed them; all the rest—beginning, middle, and end—is nothing but dreams and smoke."[105] But if the mind is inadequate for acquiring certainty, so are the senses: "To judge the appearances that we receive of objects, we would need a judicatory instrument; to verify this instrument, we need a demonstration; to verify the demonstration, an instrument: there we are in a circle. Since the senses cannot decide our dispute, being themselves full of uncertainty, it must be reason that does so. No reason can be established without another reason: there we go retreating back to infinity."[106] For both Montaigne and Sebond, then, human understanding was nothing unless buttressed by the mysteries of faith and grace, while Montaigne was also doubtful that, given the inadequacy of our senses, we can learn anything certain from the book of nature.

But that was more muted in his later writings. Elsewhere, Montaigne suggested a way out: simple things can speak truth. In his famous essay on cannibals, for instance, he wrote of how he got most of his information from a servant who had spent ten or twelve years in Brazil: "This man I had was a simple, crude fellow—a character fit to bear true witness; for clever people observe more things and more curiously, but they interpret them; and to lend weight and conviction to their interpretation, they cannot help altering history a little. . . . We need a man very honest, or so simple that he has not the stuff to build up false inventions and give them plausibility; and wedded to no theory. Such was my man."[107] Elsewhere, like Rabelais before him, he wrote of how the wisdom of Socrates lay in his common touch, the simple language of cobblers and shepherds. Even before both Rabelais and Montaigne, Erasmus had famously written of how truth is so strange that it can best be known through the mouths of fools, while those who lived according to the tenets they were taught by "serious" people knew nothing important.[108] In such statements, these authors were clearly drawing on a genre of Christian piety that placed the best hope of salvation in the poor and ordinary people to whom Christ spoke rather than in the politically and socially powerful.

The terrible suffering and destruction of the wars that raged throughout Europe in the sixteenth century, many prompted by disagreement over philosophical and theological principles, were surely evidence enough for doubting whether mortals could really understand the truth. But, they were saying, if doctrine were laid aside, simple facts might almost speak for themselves and their creator.

At the same time, public anatomy lessons had become exciting events in many cities and universities, teaching that the human frame could reveal itself to the eyewitness. Well-placed people, even influential religious authorities, had certainly believed for some time that important truths could be found in bodies. Magistrates and princes who had the power of life and death frequently wanted to get to the bottom of things before making decisions—this was especially true of the growing number of civic authorities trained in the law. From at least the thirteenth century, at some places and in some cases, postmortem dissections had been carried out to determine cause of death, especially to decide if violence—including poisoning—had been involved.[109] The medical practitioners charged with such things looked for unusual signs in the dead body, such as discoloration or putrefaction of organs, pools of congealed blood or strange objects (such as masses in the heart or lungs). By the sixteenth century, it had become almost customary for the bodies of kings, queens, other high-ranking nobles, and influential religious figures to be inspected after death by royal physicians and surgeons before embalming, so that assurances about the state of the body upon death could be issued. It had also become common to open the bodies of deceased religious leaders who were suspected of sainthood, for certain signs in the body (including the presence of a sweet smell instead of putrefaction) might confirm the sacred character of a person.[110] Moreover, as a concern for naturalism in sculpture and painting grew, artists such as Leonardo began to use their personal networks to gain access to the bodies of the dead to study the body's musculature; by the middle of the sixteenth century, painters' guilds were frequently inviting physicians to give them anatomy lessons.[111] Probably because surgeons were called in for such purposes, their guild charters often included the annual right to dissect in public one or more bodies of executed criminals.

Ordinary people in the early modern world may not have felt as much revulsion for cutting open the body of a recently deceased human being as we might think. There were of course religious strictures, partly due to worries about the wholeness of a person and related concerns about the resurrection. To put the point in Christian terms, a human being was a living body infused with an active spark (*anima*, or "soul") capable of apprehending God. Humans were not, then, merely embodied souls waiting to escape their earthly confines. Ancient Neoplatonists, Gnostics, Manicheans, and other philosophical and religious schools might speak of the soul as divine and the body as corrupt or even evil, so that the good soul strove to leave the body, but not orthodox Christian theologians. Indeed, souls would not only be reunited with their bodies at the resurrection, they had to be. A person could not exist without being an animated body. Even souls in heaven awaiting Judgment Day were therefore incomplete until the bodily resurrection. Moreover, the most critical belief of Christians was that God himself took on bodily form in the person of Jesus Christ to experience fully his creation and the human condition, deciding in the end to redeem fallen humanity through the sacrifice of the crucifixion and the promise of a future life.[112] Or as Lorenzo Valla put it in his *De voluptate* of 1431, by becoming embodied God had expressed his love for the world (*caritas*), teaching humanity to value the miracle of life; if humans would equally value it, we could live in harmony with one another and the rest of creation.[113] The body was sacred, and treating it contemptuously was a desecration. Most anatomy regulations therefore suggested that demonstrations be performed only on the bodies of executed criminals of low birth who came from outside the city, so that relatives who were citizens would not be offended.[114] Indeed, qualities of aggression and brutality seemed necessary for anyone who cut up animal bodies, much less human ones. Anatomists consequently sometimes earned the common epithet of "butcher" or worse.[115]

But matters were not so simple. Almost everyone would have witnessed death and even have handled corpses, since dying was much less separated from the living than it has now become. Moreover, dead bodies had the power to heal. Many remains of bodies and objects associated with saints—relics—had healing properties, at least for followers of the old religion. Yet even leaving aside powers of the holy, matter associated with death sometimes had great potency for good. For instance, many early modern medical receipt books included the use of powdered moss that had grown on a human skull.[116] The touch of a recently executed person's hand on the sores of scrofula, on the heads of children suffering from epilepsy, and in other conditions was thought to heal, as could items they had worn on the occasion.[117] Because of their associations with such things, perhaps, executioners themselves were thought to possess particularly good abilities to treat various illnesses.[118]

Therefore, as with the relationship between the enthusiasm for gardening and the study of botany, mid-sixteenth-century physicians mobilized their influence to pursue anatomical studies in an academic manner both to keep abreast of developments that were moving rapidly ahead outside the schools and to further that knowledge. Physicians and their students needed to be trained in anatomy if they were to be as capable as the surgeons. Moreover, many parts of classical medical literature had made the case for anatomy as a part of intellectual discourse. Famous authors like Galen argued that the ordinary functioning of the human body, as well as its abnormal or diseased expressions, were rooted in various aspects of the body, making it important to know the parts in order to know about normal and abnormal physiology. Anatomy made it possible to understand the functions of the observable parts, to assess their relationships with each other and, supplemented by reason, to probe into the causes of things. It did not have direct practical use in treating disease (except to a few surgeons); its usefulness was indirect, helping to explain why bodily events happened as they did. But teaching by cutting (*anatomia*) had become available in some medical faculties from the fourteenth century.[119] If, then, physicians were to be known for exploring the truth rigorously, wherever it took them, they would have to examine the body, even if it engaged them in a potentially cruel practice.

Such arrangements were not merely utilitarian: they were spurred on by great and growing public excitement about new anatomical studies. Far and away the most famous of them was that of Andreas Vesalius, son of a Brussels court apothecary, who in 1543 published *De fabrica humani corporis* ("On the fabric of the human body"). Like other humanists who showed that the ancient authors needed correcting, Vesalius demonstrated that even the great Galen himself had incorrectly passed off information taken from animal anatomies as information about the human body.[120] After beginning his education at Louvain, he traveled to Paris, where he immersed himself in the new Hippocratism. The connection between the Hippocratic appeal to the eyes and anatomy is still found in modern languages, as in the English "autopsy," which had become commonly applied to postmortem anatomies by the mid-seventeenth century; but in his day the Latin *autopsia* (derived from the Greek word for eyewitness) was a term of rhetoric in which one appealed to authority based on being present at an event. The appeal inevitably used the first-person singular: "I saw" or a similar construction.[121] Such reshaped personal experiences were becoming legitimate sources of knowledge: in the works of contemporary physicians like Geronimo Cardano the boundary between "impersonal academic or scientific discussion and personal history" began to be breached "regularly and persistently."[122] Witnesses could then adopt the bureaucratic language of objective administration to turn accounts of their experiences from personal impressions into universalizable descriptions.[123] Vesalius was therefore sometimes taken for a Hippocratic.[124]

The emphasis on exact description of the body helped make curiosity into a positive passion, in anatomy as well as other subjects. In *De fabrica*, Vesalius recounted for his readers—as if they would admire him for it—how he had once waited at nightfall outside the closed gates of Louvain so that he could quietly cut down and hide the skeleton and ligaments of a criminal who had been hung

Engraved title page of Andreas Vesalius, *De humani corporis fabrica*, 1555.
Courtesy of the Wellcome Library, London

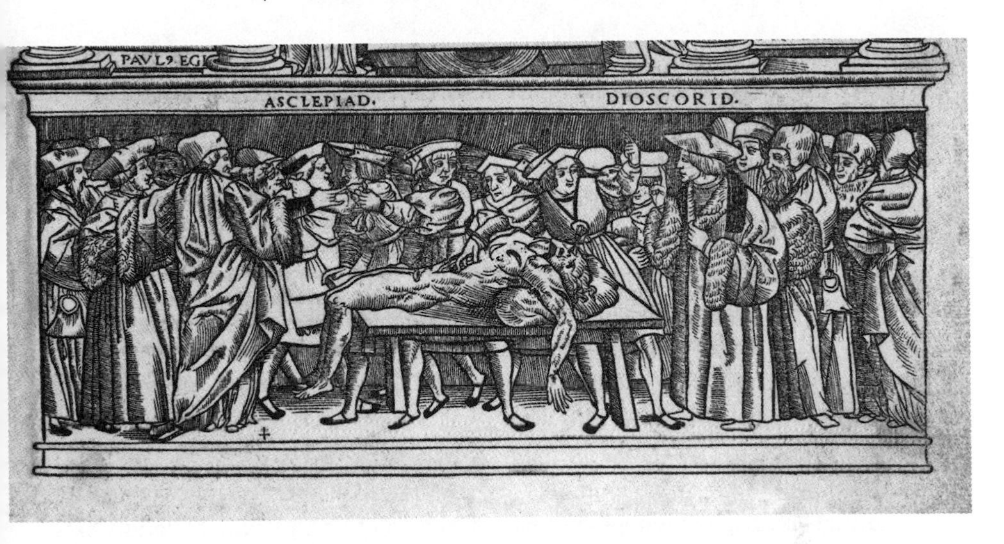

Bottom panel of the title page of Johann Winter von Andernach, *De anatomicis administronibus*, 1531. Courtesy of the Wellcome Library, London

in a cage exposed to the elements, smuggling it back to his rooms the next day piece by piece for study.[125] (It should be added, however, that his behavior was not without precedent among anatomy students: as early as 1319 anatomists and their students in Bologna were accused of surreptitiously digging up recently buried corpses for private study.)[126] Andrea Carlino has noticed that a similar curious excitement about the body is depicted in a woodcut from about the time that Vesalius studied in Paris: the title page of a newly edited version of a Galenic work on anatomy published in 1531 by Johann Winther van Andernach shows a heated discussion around an anatomy table. The illustration makes no distinction (as was customary during formal anatomies) between the lecturer (*lector*), dissector (*sector*), or *demonstrator*. Instead, "the right hand of a youthful student . . . is immersed in the viscera of the cadaver, and the student's left hand is raised as if to reinforce a statement made to a person in solemn professorial vestments. . . . Also in the centre, slightly to the left, another youth holds up the entrails in the direction of a severe figure in a toga."[127] Even young students were excited to get their hands dirty in digging deep into the bowels of truth; with publications like Vesalius's making a reputation, everyone seemed to want a look in.

Placing a high value on what can be known about the world objectively arose, then, from the same movements that gave rise to the revolution in consumption of the Renaissance. Certain ways of life encouraged people to pay attention to the things that could be known through the bodily senses. This tasteful objectivity began with the descriptive facts, the credibility of which was guaranteed by personal credit, the sharing of information, and collective decision-making based on plain and precise descriptive language. Such knowledge could also lead to generalizations, even those of number, in order to establish patterns among the things themselves, which held the best hope for a materially better tomorrow. As cities and the financial capital they produced became ever more important for the larger political systems of which they were a part, the values of the urban merchants, including their intellectual values, were increasingly well regarded throughout society. Objectivity had for ages been a common way of knowing about the world among those who handled or made objects, or used them to coerce others; philosophers, on the other hand, had been concerned with the study of the general principles of nature by which the changing and unchanging could be analyzed. In parts of early modern Europe, however, even academic philosophers became increasingly interested in descriptive knowledge of natural objects, including their material construction and spatial relationships to other objects. In medicine, objectivity facilitated the development of the study of anatomy, materia medica, and case histories. Many things could emerge from this analysis but not an understanding of the ultimate causes of things. What people knew was better called kennen than weten, although hope of deriving the good from an understanding of goods kept the other vocabulary in use. The intellectual activities we call science emerged from the ways of knowing valued most highly by the merchant-rulers of urban Europe.

Sir Francis Bacon was therefore only the most famous of those who based natural knowledge on the high value given to objectivity. Elevated to the position of lord chancellor of England, and fed up with both the "unfruitful" philosophy of the schools and thoughtless empiricism, he advocated a middle and generative way, which he called the way of the bee. The bee "gathers its material from the flowers of the garden and field, but then transforms and digests it by a power of its own. And the true business of philosophy is much the same, for it does not rely only or chiefly on the powers of the mind, nor does it store the material supplied by natural history and practical experiments untouched in its memory, but lays it up in the understanding changed and refined. Thus from a closer and purer alliance of the two faculties—the experimental and the rational, such as has never yet been made—we have good reason for hope."[128] The hope of intellectual honey therefore came not from brainpower alone, or even for the most part: it "does not rely only or chiefly on the powers of the mind." That path leads only to intellectual pride, and vain imaginings passed off as truths. Bacon claimed instead that real knowledge arose from the going out and gathering in of things and information about them, followed by pondering them, checking them, refining them, and going out for yet more matters of fact.

As people became more accustomed to making judgments about natural things according to the sensitivities of objectivity, the probabilism first associated with this kind of knowledge almost disappeared. As the English naturalist and later president of the Royal Society and London College of Physicians Hans Sloane put it around 1700: "Observations of Matters of Fact, is more certain than most Others, and in my slender Opinion, less subject to Mistakes than *Reasonings, Hypotheses,* and *Deductions* are. . . . These are things we are sure of, so far as our Senses are not fallible; and which, in probability, have been ever since the Creation, and will remain to the End of the World, in the same Condition we now find them."[129] Sloane therefore considered the matters of fact to be things that held true no matter the qualities of the persons exchanging them, nor the difference of place from which they came or to which they were going. Or to put the case in more modern philosophical language, there are some kinds of recognizable natural kinds, such as gold, that are not "constructed" in the same way as the knowledge of trace compounds in the body.[130] In other words, some types of scientific facts are accepted concepts; others are descriptions of objects and the events with which they are associated. Or to revert to early modern categories, certainty could be founded on the things themselves, whereas the meaningful statements generated about them did not share the same kind of firmness: that was the fundamental distinction between facts and speculation. It is the former, the plain matters of fact, that gained increased attention in the early modern period. Of course, even the most careful observation is discriminating and selective, but such discrimination is not necessarily a distraction from the truth, since many things can be brought into view only by focused attention.

Thus, the material details of the world as perceived by the senses became the foundation for a new approach to knowledge. In turn, it owed much to the acquisition and transmission of accurate information. It was from encounters with nature and its goods, and with like-minded persons, more than from attempts to rise above life on earth, from desire and interest more than intellectual distance, from warm hope of material progress and gain more than otherworldly aestheticism, that natural knowledge was transformed. Countless people were involved in the production, accumulation, and exchange of the natural knowledge upon which commerce depended, and the high value they placed on accurate description of the created world—those "matters of fact" that would be true in any circumstance—became a measuring stick according to which they could judge other forms of knowing. Valuing natural knowledge of such a kind suited activities based on exchanging goods rather than on seeking the Good, for deep in the roots of this kind of knowledge economy lay a moral economy weighted according to bodily pleasures and pains. It somehow does not seem so long ago.

2

An Information Economy

In this matter again many Philosophers insist that whomever wishes to go through life prudently and to obtain a knowledge of things should begin with traveling afar.

—CAROLUS CLUSIUS, *Aromatum, et simplicium aliquot medicamentorum*

The ways of knowing that surrounded objectivity established highly probable facts despite a world in constant change. This appreciation for things and their descriptions gave substance to the culture of taste and objectivity and was in turn built on the sociable interactions of exchange. Like the values of taste and consumption, exchange began from the precise knowledge of things that came via personal experience, but it also included the ability to transform one value into another. Methods of exchange, then, also had fundamental implications for establishing the value of certain kinds of knowing, turning information into knowledge.

EXCHANGE AND TRAVEL AS TRANSFORMATIVE EXPERIENCE

Objects themselves might be said to be stable and enduring, at least for as long as the processes of decay can be forestalled. But they can produce transformations in the people who exchange them. One of the social analysts of a century ago, Georg Simmel, underlined the changes induced by the process of exchange. All human relationships involve exchange of one kind or another, he noted, for even in ordinary conversation something is exchanged. In exchange one gains something by giving something, altering one's state in the process. When an exchange of word, gesture, or object takes place, the two parties involved are changed in terms of what they know, experience, or possess. Due to the exchange, each party has given something but also has had something added, which is the source of *value.* "When we subsume the two acts or changes of condition" that occur between the parties to an exchange, "it is tempting to think that with the exchange something has happened in addition to or beyond that which took place in each of the contracting parties. This is like being misled by the substantive concept of 'the kiss' . . . into thinking that a kiss is something that lies outside the two pairs of lips, outside their movements and sensations." In other words, value is something that expresses a state within people rather than something abstract, something over and above, or separate from and in addition to, them. We nevertheless commonly reify the outcome by associating values with objects. Or put another way, "the value of the gain is not, so to speak, brought with it, ready-made, but accrues to the desired object, in part or even entirely through the measure of the sacrifice demanded in acquiring it."[1] Value is read into objects but is rooted in a change of personal state, a consequence of interpersonal exchange.

Simmel was rejecting the classical understanding in which exchange is a mere necessity allowing something gathered or produced to reach the person or place where it can be put to use.[2] Early political economists developed theories that rooted monetary value in labor, utility, or scarcity rather than in exchange itself, seeing price, for example, as the reification of all the social relationships that go into producing things rather than a value placed on exchange. On such accounts, the "middlemen" who engage in exchange should merely add the costs of transportation and handling to the natural price of the things transported, and perhaps add a small profit for organizing the venture. Anything else often appears to be unfair or immoral, an exploitation of what should be a simple relationship between producer and consumer. Aristotle himself argued something similar, while theologians held that the prices charged by middlemen were usurious and hence sinful; ordinary people sometimes rioted against the prices charged by middlemen as not befitting a moral economy; and analysts such as Karl Marx could write that the "immorality of trade" was obvious.[3]

Simmel, however, not only rejected the view that trade was immoral, he went so far as to locate in exchange the roots of the highest ascetic sacrifices and aesthetic goods: "Exchange is just as productive, as creative of values, as is so-called production." Expanding on the point, he noted that "value and exchange constitute the foundation of our practical life." Or, developing the point more universally, he exclaimed, "Exchange is the purest and most concentrated form of all human interactions in which serious interests are at stake."[4] Others have gone further by emphasizing that "the conceptual foundations upon which both economics and culture rest have to do with notions of value,"[5] which is especially apparent where issues such as aesthetic appreciation, taste, social cachet, pleasure, and playfulness affect what value is placed on objects and their associations.[6] Perhaps exchange is even what makes humans different from other primates: primates may make and use tools, communicate, live in societies, and so on, but only humans exchange things.[7] Moreover, in this view, demand is as important as production. As Arjun Appaduari put it, demand is not "a mysterious emanation of human needs [nor] a mechanical response to social manipulation" but rather "emerges as a function of a variety of social practices and classifications."[8] Or, as an account of the modern artist J. S. G. Boggs put it, after finding he could live by trading images of money, "Faced with so many successful transactions, and such interesting ones, he began to sense how the transactions themselves, beyond the simple drawings, were the true aesthetic objects."[9]

The transformations wrought by exchange can therefore be life-enhancing in many ways. But only if they are between more-or-less equal partners, as Simmel's analysis of the exchange of a kiss suggests. When something else takes place between two parties, such as a forced kiss or worse, or a bad bargain made because of deception or coercion, or when appropriation, theft, or even destruction occurs, personal change is also a result, although far from a life-enhancing change. Bad interactions might not be exchange relationships in the full meaning of the term, but they certainly involve interchanges shaped by the movement of objects, or the absence of such movement. It is manifest that the acquisition of objects and even of accurate information about them sometimes—too often, in fact—occurred in ways that were unequal and destructive, as well as through equal and mutually beneficial exchanges.

The causes of change therefore lie within persons, causes that early modern people termed the "passions."[10] They are not the watered-down "emotions" of the modern world, which in today's conversation implicitly emerge from mental states. For early modern analysts, the passions were movements arising from body as well as mind, expressions of life united rather than divided, powers that moved in and through one's whole being. They prompted action and thought together. Derived from the Greek *pathos*, meaning something that has befallen one, what one has experienced or suffered, and moving into Latin as *passio*, the term carried three meanings: bodily suffering, as when someone was termed a "patient," or as in the "passion of Christ";[11] being acted upon, suggesting passivity or allowance; and—following on this—a movement of the *anima* caused by some force (as in "Elias was a man subject to like passions as we are").[12] The passions could even be taken to be the forces behind all natural actions: Sir Francis Bacon wrote that "the principles, fountains, causes, and forms of motion, that is, the appetites and passions of every kind of matter, are the proper objects of philosophy."[13] "Suffering from passion" therefore suggested allowing oneself to be controlled by faculties other than reason. These were, in turn, the result of movements that occurred especially in the sensitive soul or (as Plato termed it) the middle soul. The movements of the sensitive soul were in turn bound to the *spiritus*, faculties, and organs of the body.[14] Consequently, the relations between one's soul and one's body were intimate, complex, and dynamic. The famous ancient dictum to "know thyself" (*nosce teipsum*) was directed toward precisely these relationships.[15]

Related to passions are interests. Curiously, however, even recent histories that question many established assumptions continue to make the case for the disinterested nature of scientific knowledge. As Steven Shapin put it in his short account of the scientific revolution, natural knowledge was, or at least was meant to be perceived as, "benign, powerful, and above all *disinterested*."[16] To make a similar point, Lorraine Daston wrote about the "moral economy of science." It is a phrase introduced by the great historian of the English working classes, E. P. Thompson, which he used to speak about the uncommoditized, interpersonal bonds that existed in local communities, which often united them against those who placed the highest premium on money.[17] Daston adapted the phrase to describe the "web of affect-saturated values that stand and function in well-defined relationship to one another," which refer, "not to money, markets, labor, production, and distribution of material resources, but rather to an organized system that displays certain regularities."[18] This she used to criticize the concepts of the "norms" of scientific communities described by the sociologist Robert K. Merton: like him she wished to explain the values of the scientific community, including why it is consensual, but she also wanted to allow for conflict and change. To do so, she properly introduced a concern for affect and emotion. Yet the passion she considered to have most affected the development of early modern science was "wonder," which, she declared, was described as "remarkable . . . for its disinterestedness."[19] Such views are, on this point at least, redolent of Plato and Immanuel Kant and his successors, who taught that entangling the mind in worldly pursuits was a distraction from the cultivation of intellectual virtue (*bildung*), a form of life that should have no goals outside of itself.[20]

But if one examines those who were involved in the close searching out of natural objects and events, it is noticeable that they were deeply interested. Indeed, when giving an account of the rise of the modern world, one historian, Albert O. Hirschman, noted how words like "interest" were coming to be used in a positive way by the eighteenth century.[21] The English word "interest" is enormously diverse in its connotations (as are the Dutch words related to *interesse*). It is a powerful and fruitful word, sometimes even linking personal attributes such as curiosity to social and economic relationships. Indeed, according to the *Oxford*

English Dictionary, "there is much that is obscure in the history of this word." From the late fifteenth century onward, as a noun the word was used to indicate such matters as the "relation of being objectively concerned in something, by having a right or title to, a claim upon, or a share in" it, as in legal, spiritual, or financial concerns in something; "the relation of being concerned or affected in respect to advantage or detriment"; and twelve other major meanings with multiple nuances. Someone can exercise one's "interest" to get another a job, to earn money by lending to a bank or other party, to pursue one's selfish desires, to have a share in a business, and so on. The last of five meanings of the verb yield the modern most common usage: "To affect with a feeling of concern; to stimulate to sympathetic feeling; to excite the curiosity or attention of." In other words, there are countless ways in which people had an interest in nature. Consequently, those who had the most to gain or lose from knowing about the world—those who were most interested—often had the best claim to speak credibly about it.

If knowledge of the world depended on the transformations wrought by exchanges rooted in the passions and interests, it also depended on moving about. Not only did the lives of early modern Europeans bob about on a rough sea of events, thoughts, and passions, they were also very often in the process of trying to get to someplace else. Most men and women took to the road because they had to, going from high pasture to shelter as the weather dictated or moving in the search for employment and enjoyment, fleeing disease and famine, joining armies or avoiding them, searching for love or crashing through seas in search of foreign goods. Masses migrated from one rural region to another according to the seasonality of work, or from rural regions to cities, where they took work as laborers or sailors, or settled for a period as servants and apprentices before returning home, or instead staying on to raise families, or simply added their lives to the high death rates of urban environments.[22] Others felt compulsions, visiting distant relatives, setting out with bands of pilgrims, or taking up distant posts on behalf of church or sovereign.[23] In some parts of Europe, journeymen were even expected to move from place to place, working with different masters to improve their skills in a craft. The *perigrinatio academica* of students from one university to another in search of intellectual masters was similar. And merchants had to travel to acquire and trade their goods.

Of course everyone knew that setting foot on a road often altered the voyager. It mattered not whether people traveled for necessity, education, or the acquisition of wealth: travelers broke old habits of life in the process of encountering new people, places, and customs. Archaic legends throughout the world told of life-altering journeys. A famous example in the so-called Western tradition is one of the first works to be recorded in an alphabetic script, Homer's *Odyssey*, telling of the transformation of Odysseus from a merely good warrior-prince into one of the greatest of men—a model of patience, prudence, ingenuity, and supple physical and mental toughness—through the experiences he encountered on his long way home from battle. Later Greeks and Romans also looked to the more historical example of Alexander the Great, who went wherever he liked, to the edges of the known world, sending back knowledge of people, places, and things to his former teacher, Aristotle. Fantastic stories about travelers still abounded in early modern Europe, whether as figures of romance, like the chivalric adventurer Sir John Mandeville or the knights of the Round Table, or as historical legends such as Marco Polo. The importance of transformation also expressed itself in the enormous popularity of Ovid's *Metamorphosis* in early modern Europe. New stories of strange peoples and places from both the East and the West created additional excitement. While many travelers must have been as transformed by their voyages as Odysseus, even the stay-at-homes learned things from the tales they heard and the people who passed by.

Because much of this traveling about removed people from systems of governance that presumed rootedness, it also often created social disorder and subversion. By the sixteenth century, religious authorities often did their best to persuade ordinary people against going on pilgrimages, since they saw them more as an occasion for indulging in worldly passions and vanities than as acts of piety. They also often attributed the rise of moral relativism and "libertinism," even atheism, to travel.[24] It encouraged people to consider the various religions they encountered to be merely customs of a locale that, like other customs, were different but equal. Two early-seventeenth-century examples illustrate the point. The famous French aristocrat and author René Descartes, who traveled widely in Europe in his youth and lived outside France for most of his adult life, wrote, "I have recognized through my travels that those with views quite contrary to ours are not on that account barbarians or savages, but that many of them make use of reason as much or more than we do."[25] Similarly, only after he visited lands to France's north, the low countries, did his correspondent Marin Mersenne come to think for the first time that philosophers of a different faith from his own might be both moral and worth listening to.[26]

To encourage the benefits of travel but to minimize the problems, many people of affairs urged young men to undertake properly disciplined travel—in imitation of the travels of members of respectable merchant houses—usually under the guidance of a tutor.[27] The accomplished Sir Thomas Elyot, for instance, recommended travel as part of a young English gentleman's education in his *Boke Called the Governour* (1531). Some decades later, Lord Walsingham, who had spent time on the Continent during the reign of Queen Mary, wrote to a nephew that books alone "are but dead letters, it is the voice and conference of men that giveth them life and shall engender in you true knowledge," for which he should travel.[28] In a similar vein, Bacon declared, "Travel in the younger sort is part of education; in the elder a part of experience."[29] Indeed, by the Elizabethan period in England, travel had become "an 'art,' to be practiced by a properly taught young man in order to complete his education."[30] In France, too, the great iconoclastic philosopher Pierre de la Ramée, better known as Petrus Ramus, in his writings of the 1560s and 1570s made travel a necessity for the discovery of truth, while around 1579 the famous Michel de Montaigne recommended travel for shaping *jugment* and *entendement*.[31] In the low countries, Philip Marnix, one of the well-educated political leaders of the low countries, recommended strongly that after studying languages and acquiring a sound religious and physical training, a young man should travel. He was echoed by the famous scholar Justus Lipsius. Lipsius wrote a letter to Philippe de Lannoy in 1578 (published shortly afterward) that explicitly drew on the example of Homer's Odysseus to claim that travel enriched one's insight, knowledge, and character by bringing one into contact with new people who had different rites, manners and civil customs. People in the northern netherlands often referred to Lipsius's ideals about travel. The diary of a late-sixteenth-century resident of Utrecht who took several journeys opened with a string of mottoes from Lipsius and other humanists on the importance of travel for broadening the mind and expanding knowledge. Not surprisingly, perhaps, its author held the view that his soul's time here on earth was a "sojourn," several decades before John Bunyan developed that commonplace for English readers in his *Pilgrim's Progress*.[32] By the middle of the seventeenth century, traveling the usual routes for this kind of edification would be called the Grand Tour.[33] Parochialism—staying put in one's own parish—became a word with negative connotations.

It is no surprise, then, to note the turn of phrase about intellectual "movements," for both the content and the framework of knowledge could be reshaped in the encounters with strangers. A model developed by sociolinguists is helpful here: they have noted that new words and information are often introduced to strongly knit social groups via loose acquaintances rather than close friends or relatives. Core members of a group develop their own ways of doing things and tend to imitate one another, or at least their leaders. But people who are not core members, such as the children of friends, tend to be the ones to introduce innovations like new linguistic expressions or fashions in dress to the insiders. Drawing on a study of "hysterical contagion" in a textile plant, Mark Granovetter generalized that the most influential innovators tended to be "individuals with many weak ties," since they were "best placed to diffuse" their ideas. Put another way, "Weak ties are more likely to link members of *different* small groups than are strong ones, which tend to be concentrated within particular groups." Paradoxically, then, "weak ties, often denounced as generative of alienation . . . are here seen as indispensable to individuals' opportunities and to their integration into communities; strong ties, breeding social cohesion, lead to overall fragmentation."[34] Subsequent sociolinguistic findings in both Philadelphia and Northern Ireland "emphasize the need for acknowledging the importance of loose knit network ties in facilitating linguistic innovations."[35] Putting this concept into historical language, it might be said that intellectual movements, like linguistic changes, were rooted in the travels of people who did not know one another particularly well but who met and conversed, exchanging words and concepts, or even misunderstandings, in a creative way. Following their meeting, people sometimes decided to stay in touch in the future, creating networks of literally distant acquaintances who fostered the exchange of letters and books as well as gifts and objects. Webs of connections first established through personal meetings therefore lie behind abstractions like "the Renaissance," "the Scientific Revolution," "the Enlightenment," and so on. Travel proved critical for creating the loose communities who led intellectual movements.[36]

COMMERCIAL AND SCIENTIFIC PRACTICE

Moving about in the world and exchanging things with others were the basic practices of merchants. Commercial practices of the early modern period began by bringing people and goods together in one vicinity. Medieval merchants had traded their goods more or less one-for-one: they put things they deemed valuable aboard ship and traveled to places where they might be exchanged for other goods or money, and they either returned with their new goods or exchanged them in turn at another place for what they valued more highly, doing so at city fairs that might occur once, twice, or in rare cases three times or more during the year in any one location. In a very few places, merchants could come together regularly to trade. By the 1530s, Antwerp was attracting so many merchants from so many locales with so many different goods to trade that they could simply ship their goods there to be stored, knowing that an appropriate buyer would easily be found. To facilitate this kind of trading, in 1531–32 Antwerp built the Nieuwe Beurs, a covered place for the merchants to meet that was open every day but Sundays and important religious holidays. (In the early seventeenth century, Amsterdam followed Antwerp's example by building its own Beurs.)[37] In other words, Antwerp became a permanent staple market, the major entrepôt of northwestern Europe.

The Amsterdam Exchange (Beurs) in the seventeenth century.
Unattributed engraving after an old woodcut in the Nuremberg Museum.
By permission of Mary Evans Picture Library

Once inventory had been accumulated in one place, however, it could be transformed into something with universal value, usually embodied in the strange thing called "money."[38] From at least the thirteenth century, most of urban Europe had been monetized: that is, people commonly used currency to pay taxes or settle accounts. But with monetization came other problems, such as how to convert the value of one set of coins into another. Coins were minted at a variety of sovereign territories, with the content of noble metal in them also varying with time. One might call it the problem of commensurability: How does one find the common denominator among diverse coins, allowing comparison?[39] Specialized money changers and bankers arose who dealt with such problems by developing methods of conversion, enabling people to use currency in their possession to settle accounts in another currency.[40] Perhaps their methods prompted philosophers of the period to try to quantify the similar qualities in diverse things, such as the hot or cold qualities of different medicinal herbs, as a way of making them commensurable, too.[41] Metaphors from agriculture and small crafts could not help one understand what was occurring here: one was not planting seeds to get "growth," or even turning labor and materials into a product. Rather, people trading in money became wealthy by finding the lowest common denominator among diverse items and exchanging them, making things that seemed to have little in common commensurable and taking a profit from recognizing the common denominator. Once brokers found methods to allow them to become a medium for the exchange of relative value, coins themselves even increased in value. From at least the end of the fifteenth century Girolamo Butigella and other jurists realized that the value of a coin could be greater than its metallic content because money possessed some sort of added value by virtue of its easy exchangeability. By the mid-sixteenth-century, a French jurist, Charles Dumoulin, could explicitly declare the value of a coin to be simply what others assess it to be, that is, its value in exchange (a particular problem in probate when former debts needed to be paid in money that had changed its value in the meantime).[42] The value of money was therefore a cipher, a denominator of value that allowed very different things to be made comparable through a system of negotiation, and it was therefore valuable in itself. And through it, otherwise incomparable values embodied in things themselves could be turned into number, and counted.

The Exchange was, then, first and foremost a meeting place for the exchange of information—and a place where the accuracy of information was highly valued. As a public forum, parties trading on the Exchange knew the prices at which the goods had recently been going for: indeed, recent prices were written down for the use of other brokers, and traders circulated such information to one another. The collection of commodity prices and exchange rates had been a necessary part of late medieval Italian banking practice, and perhaps the Italians also were the first to begin to print such information; it is certain, however, that a commodity price "current" was being published in Antwerp by about 1540, with an exchange rate current appearing at roughly the same time. The publication of such information was a service provided for the general community of merchants rather than for an individual house or firm, allowing for long-distance public circulation of information about commodities and exchange rates. Indeed, the mutual dependence of merchants on the accuracy of the published lists and of the publishers on the accuracy of the information reported to them made the data published in the currents very reliable.[43] Along with the distribution of information about prices went information about events that might affect business—they became the first printed newspapers. At the Exchange, further discussions could take place about the exact quality of commodities to be traded, where they had been purchased, for what price, and what they were likely to fetch when sold in another place. The Exchange was therefore "a mustering field not only for the coincidental surplus production . . . but also for information" about commodities and exchanges worldwide, helping to stimulate collective decisions by merchants on the allocation of capital.[44]

Access to precise and up-to-date information remained fundamental to business practice.[45] Amsterdam set up a direct postal exchange with Antwerp in 1568 to facilitate the exchange of business information and within a few decades had established four postal centers, one for correspondence with the southern netherlands, France, Spain, and Portugal, one for Hamburg and the Baltic, one for all correspondence "beyond Zwolle" (to Cologne, the upper Rhine and the German "Rijkspost"), and one inland, for other Dutch cities. Throughout this system, certain writers would collect information and undertake to send out newsletters; in 1592, the States General itself even contracted with Hendrik van Bilderbeeke in Cologne to collect and supply regular news at a salary of two hundred pounds, later raised to three hundred, with his news from there sent to The Hague and copied to other Dutch cities. By the 1590s and early seventeenth century, brokers can be found who made their living from collecting and sending specified newsletters to their clients, with regularly printed and numbered newspapers appearing in the second decade of the century.[46] The post offices, newsletters, and newspapers at first depended on merchants for their main clientele, focusing their content on information about prices, goods, the events of war and peace, and other facts that might affect business.

But business depended on exchanging not only commodities and information but promises. One kind was a sort of paper money, recognized first by the Antwerp law courts in 1507 (and throughout the low countries by 1537): a promissory note made out to the bearer. Merchants had previously written out IOUs as a form of credit, the borrower certifying that he would pay the lender a certain amount on a certain date. The promissory note, however, could be traded like any other commodity. That is, payment was made not necessarily to the lender but to whoever possessed the note. The notes themselves could therefore be exchanged in lieu of credit or sold to raise ready cash: the first evidence of discounting such a note (that is, selling a mature instrument for less than its face value, with the new bearer taking the risk of being paid his due from the payee) is from 1536. Something similar soon allowed not only money but goods to be traded in a similar way, as bills of exchange developed. By 1541 these had developed into legally enforceable promises to hand over specified goods to anyone possessing the bill. Both promissory notes and bills of exchange were traded on the Exchange, too. Soon merchants could also raise capital on the market by trading paper "stocks," representing a share of "interest" in a business relationship. Paper instruments themselves had become commodities. In other words, by the mid-sixteenth century, merchants could bring goods, money, or simply written promises to Antwerp and there trade them with anyone, all year round, allowing the merchants to deal in any and all commodities and to put excess capital or credit to work by trading pieces of paper.[47] This freed many more merchants from the necessity of traveling about with their goods or with cash in hand in order to conduct business, allowing them to concentrate on finance.

One of the most critical assessments lying behind commerce, then, was another characteristic considered essential for the development of scientific knowledge: dependable honesty and related attributes, such as "credibility" and "credit," from the Latin *credo* ("I believe"). For a great merchant, honest dealing and creditworthiness in all that he said were the foundation of his way of life. "His word is his bond" applies even better to the businessman than to the gentlemen. Of course, personal credit lay behind most material exchanges in the period, even in small villages.[48] But whereas sectarians might trust only others of their common faith, or villagers only people of long acquaintance, merchants had to find other ways to weigh trust among diverse and sometimes brief relationships. In the world of the Exchange, then, from trust "came recommendations, guarantees and credit. Trust was perhaps even more important than capital, the main function of which indeed was—and is—to generate trust, and thereby credit." In other words, one of the essential ingredients in the making of money from money was credibility: money generates credit; credit gathers money.[49]

The grandest merchants therefore possessed—and had to possess—sound personal reputations earned from years of consistent honesty and the meeting of obligations. Participants in their world were keen to display signs of honesty in their gesture, word, and dress. Public modesty and consistency of word and deed counted for far more than codes of arms or extravagant behavior, which might signal an egoist or spendthrift who would use money for excessive personal pleasure and public display rather than for reinvestment (which would be of benefit to other merchants). Cities were certainly not places for fine horses, hunting, or dueling. Merchants also tended to frown on gambling and whoring. Instead, they valued more domestic pleasures, getting together in the halls of the militias, guilds, or other civic associations, or gathering to celebrate the conclusion of a contract or a personal anniversary. There they might collectively indulge in oceans of good drink and mountains of fine food, cultivating a discriminating taste for wines, spices, and tobacco. They also displayed a sense of humor in bringing out trick glasses and goblets that spilled all over unwary drinkers and became well known for their farces, although many of these practices gradually came to be associated with lower-class rowdyism instead of well-to-do laughter.[50] Moreover, rather than dressing like peacocks in a wide palate of colors with gold and silver thread and gemstone highlights—as courtiers did to draw attention to themselves—merchants and their wives mainly dressed in dark cloth with white accents of bleached and starched linen. On closer inspection, the fabrics might be expensive silk or beautifully woven brocade, but one needed to look closely at textures rather than from afar at sparkling jewels to see the wealth invested in their clothing (although sometimes the sons and daughters of the rich could not forebear the chance to show off in brightly colored garments).

The leading men and women of city life therefore liked to think that real value emerged not from haughty princely authority but from the common outlook of more widely distributed, less eminent but important opinion-makers, people like themselves. The most loved play of the Dutch playwright G. A. Bredero—a social farce, *The Spanish Brabanter*, first performed in 1617—included many likable characters based on honest, materialistic Amsterdammers while making fun of the polite pretensions of southern aristocrats. The noble fop from Brabant, Jerolimo, appears on stage well dressed but without a penny to his name, exclaiming, "This city's *magnafique*, but what a grubby folk! / In Brabant we're all quite exquisite / In dress and bearing—in the Spanish mode— / Like lesser kings, gods visible on earth." But of course appearances deceive. Jerolimo is so proud and dissimulating that for a moment he convinces even street-savvy prostitutes that he possesses wealth, whereas instead the apparently poorly dressed Amsterdammers have quietly amassed the real thing.[51] Max Weber's analysis of the Dutch burgers as displaying an unworldly asceticism rather misses the point; Simon Schama's sense that they were "embarrassed" by their riches—in the sense that they did not flaunt their wealth—gets closer to the bone.[52] The demand for consistent modest public behavior to show that they were not high risk-takers in their personal lives nor careless with their purses was a necessary component of being trustworthy and dependable, which in turn helped to establish their creditworthiness.

Similarly, scientists are also among those who seek to give or take "credit" where it is due. Bruno Latour and Steve Woolgar have therefore discussed how one sense of credit among scientists is as a kind of commodity to be exchanged. Such observations led them to comment that "there is no ultimate objective to scientific investment other than the continual redeployment of accumulated resources. It is in this sense that we liken scientists' credibility to a cycle of capital investment." Their arguments about how modern science is done therefore sought to show how *credibility* was "materialized" into facts.[53] In a similar way, Pamela Smith suggested that the early modern alchemist and projector Johann Joachim Becher turned symbols into money, largely through the use of *Kredit*, which could be taken in different and sometimes conflicting ways as a term either for aristocratic honor or for financial reputation. Becher also seemed to be aware of the "paradox" of regeneration through consumption, which was a rich vein of alchemical analysis.[54] Steven Shapin offered a different view, trying to boil down the essence of credibility to trust, which he thought in turn depended on social authority, with the only people who had enough of it to arbitrate early modern claims of natural knowledge, in his view, being gentlemen.[55] Others have seen princes as the arbiters of truth.[56] But in the Dutch world, at least, gentlemen, aristocrats, and princes had a reputation for being notoriously fickle in their behavior. They might feel bound to discharge debts of honor while at the same time feeling no similar obligation with regard to their financial debts.

Moreover, promises were not based on personal trust alone. They were reinforced by various powerful rituals backed up by coercion: in other words, they were rooted in the power of contract. Contracts—written agreements promising one thing in return for another, in the exact language of notaries and lawyers, with enforceable penalties for nonperformance or noncompliance—became the bedrock of society in the low countries. Investment in enterprises other than landholding, especially, depended on more than custom, relying on written and enforceable agreements to back up promises. For instance, the preamble to the Perpetual Edict of 1540, issued by the Hapsburg emperor Charles V, spoke not only of checking heresy but also of other aspects of necessary good order and justice, such as preventing people from fleeing their creditors. Absconding from debt would be treated like common theft, with wives or anyone else who aided and abetted such fugitives becoming liable for the debts, while if the thieves were caught they would be summarily dealt with and hanged. The same edict allowed the payment of interest on debts while prohibiting monopolies because they threatened the ability to make money from freely contracted enterprises.[57] All kinds of economic activity, including collaborative undertakings such as trading ventures, could flourish only when "property rights" had in such ways been abstracted beyond real property (that is, land and buildings) and specified by law and enforced by local or national bodies.[58] By the 1570s, contractual obligations were so deeply embedded in urban society in the low countries as to be assumed to apply to almost any relationship. The Dutch Revolt of the time was itself often portrayed as standing up against a party who was in breach of contract: the king was said to be acting without consideration for ancient liberties as guaranteed by contractual charters. Even divine relationships could be viewed similarly: the period witnessed a shift from an older concern with the seven deadly sins to a newer set of anxieties about the Ten Commandments, while "covenant theology" developed the theory that God and humankind were bound by mutual promises and obligations and was around long before receiving its clearest definition from the Leiden professor of theology Johannes Cocceius in the 1640s.[59]

It could also be said that the whole commercial system depended on something like a religious hope for a better future. It did not, of course, lie in the hope of being taken up after death into the eternal and timeless world, but anticipated a "secular" future in the world of time (from *saeculum*, an age). Most historical discussions of changing concepts of time have focused on the development of a sense of its uniformity. It was famously the invention of mechanical timekeepers that conveyed the view that time is uniform. The sense that time changes with the seasons, with one's age, with peace or war, and during moments of stress or bliss accords with human felt experience. But mechanical clocks moved steadily, invariably (aside from mechanical inconsistency), dividing the day into equal hours. Now it was the night or day that changed according to the hour, not the hour that altered according to the light. Clocks quickly appeared in the towers of guildhalls and other municipal buildings, striking the hours to regulate commerce and other activities of large numbers of people: workers in the low countries had been complaining of "working to the clock" from at least the fourteenth century.[60] By the lifetime of Galileo Galilei and Descartes, both natural philosophers and musicians could take the uniform nature of time for granted, "timing" events according to regularized beats.[61] As Norbert Elias put it, "The significance of the emergence of the concept of 'physical time' from the matrix of 'social time' can hardly be overrated."[62] In the same period, however, another sense of the relationship between time and human life developed; there was a growing sense that new methods of using time could bring material goods. Capitalist forms of economy depended not only on drawing attention to the rapid passage of time or on making work more regularized. The Dutch financial world also depended on new methods of commerce that extended time: long-term arrangements, which required personal commitment to behave at an appointed date in the future as specified: something called "investment." Indeed, as economic historians have pointed out, "the essence of capital is time."[63] Commerce thrived on secularization.

In the world of commercial credit and contract, living honestly according to the future consequences of one's promises came to be one of the chief marks of credibility; one of its hallmarks was the associated clarity of speech. Lawyers and notaries might use jargon, but for those who understood it, this technical language made for greater precision and less ambiguity. Again, the contrast with the European upper classes is illuminating: whereas noblemen and women might speak clearly to their inferiors when ordering them about, when they spoke with their peers and superiors complex power relationships encouraged the shadow-worlds of allusive and metaphorical courtly speech. In commerce, however, being clear and consistent was a sign of truth and credibility. In establishing one's credibility, then, the cultivation of exact and unadorned speech, or "plain dealing," went down well. If the commercial world of the netherlands placed a high value on plain speaking, so did the new science. Plain speech was noted at the time as being very important to clear descriptive expression and analysis. The physician Cornelis Bontekoe, for example, introduced his book on the benefits of tea-drinking with the following common sentiment: "I am accustomed to pay more attention to the subject, and to the truth of what I say than to the fair choice of words and eloquence of style: all the more since I believe I am eloquent enough if I can make myself understood, since the only standard of speaking and writing is that of being understood."[64] Previous historians have sought the origins of this aspect of the new philosophy in the plain style of Puritan preaching.[65] More recently, it has been associated with the rhetorical techniques of "virtual witnessing" that also marked the emergence of the new philosophy.[66] But the ways in which urban commerce might have encouraged plain speech have not been explored.

In short, then, a number of values were shared by both merchants and those we would now call scientists, including: travel, seeing things afresh, exchange, commensurability, credibility, the hope of a better material future through worldly activity, and a preference for plain and precise language. Above all, among the values shared by science and commerce were a certain kind of interested engagement with objective knowledge and an attentive appreciation for collective generalizations based on exacting information about the objects in with which they dealt. Exchange values, openly based on both passion and calculation, placed certain forms of knowing about objects, even living objects, front and center. When such values began to reorient natural philosophy, something recognizably like modern science emerged.

A NATION OF MERCHANTS

The commercial methods pioneered in places like Antwerp were built into the foundations of the new nation-state emerging in the northern provinces of the low countries in the later sixteenth century. As the wars of the Dutch Revolt devastated the southern provinces, the center of mercantile exchange for northwestern Europe shifted north from Antwerp to Middelburg, Delft, Rotterdam, Enkhuizen, and especially Amsterdam.[67] Like other northern cities, Amsterdam benefited enormously from the wealth and knowledge brought to it by refugees from Antwerp and other cities of the south. After the establishment of the seven United Provinces in the mid-1580s—the seven northernmost provinces who combined in a close alliance against Hapsburg impositions—wealthy urban merchants with both capital and information about global trade had flocked to the northern netherlands from their places of refuge, bringing their political outlook, social values, and business skills with them and making the Dutch Republic the center of what some have termed the "first modern economy."[68] Amsterdam, for instance, experienced a threefold rise in the number of its merchant community between 1585 and 1620 (from about five hundred to about fifteen hundred), with the refugees from Antwerp alone increasing the city's total capital stock by about 50 percent.[69] Many of the immigrants had experience in organizing some of the carrying trade from the Baltic and the Levant, a few had experience in venturing as far as West Africa and the Caribbean, and there were even immigrants who had recently expanded into the Muscovy trade, sending ships north around Norway to Archangel laded with silver, spices, silks, Mediterranean goods, herring, wine, and salt, from whence they brought back furs, caviar, rhubarb, and other expensive products, many from Central Asia.[70] Former Antwerpers also brought sugar refineries in large number to Dutch cities, especially Amsterdam.[71] With their collective involvement, the re-export of Portuguese spices to other parts of Europe via Amsterdam had become so great by 1594–97 that the city virtually controlled the European trade in colonial goods.[72]

The United Provinces was more or less run by these urban merchant oligarchs, becoming the greatest contemporary exception to the rule that nation-states were monarchies. The state ordinarily had neither a strong center nor an authoritative prince, nor even a powerful civil service; it did not even have a well-developed sense of collective nationhood.[73] There was neither a national supreme court nor a unified treasury: each of the seven provinces had its own set of courts and fiscal arrangements, each contributing a proportional share to the finances of the union. Sovereignty therefore remained with the several provinces individually, within each of which various groups—especially the leading citizens of the numerous cities—jockeyed for power and kept an eye on one another. With representatives from a wide spectrum of interests watching their every move, the civil service remained small, and always subordinate to the offices they served. The most influential denizens were therefore the great merchants and their wives who, when they also held political office, were called the *regenten* ("regents"). They imagined the civic polity as a large family, in which the fathers and mothers had duties toward those less able to care for themselves. Indeed, in the early seventeenth century, the Dutch use of the word *regent* was reserved to the men and women who acted as a father or mother in their civic community, holding office in the city government (a male-only option) or on one of the boards of the civic charities (where female regenten often governed the institutions inhabited by girls or women).[74] Whatever power the regenten possessed therefore came not from their individual but from their collective persons as forged in meetings and committees. This enormously complex political system functioned only because, as with business, interests were negotiable and because they found that in the end some interests were common to all.

The interest that most firmly linked the people of the United Provinces was a bit of economic magic called the national debt. The debt gave concrete expression to Dutch collective confidence, paying for the troops, weapons, fortifications, and ships that protected the country, while it also bound together everyone with even a small income into a system of credit and interest. Or to put it another way, a certain share of the livelihoods of a great many people was invested in collective goods that smoothed over social problems and reduced collective risks, which helped to make profit-making activity more secure and less expensive. The new methods of warfare required expensive new technologies in shipbuilding, fortification, gunnery, munitions, and the endless drilling of regular (and regularly paid) troops in the use of new weapons and formations, all of which required mountains of money.[75] But certain kinds of fiscal arrangements originating in the province of Holland, which on the heels of the Revolt reached out to bind all the seven provinces, managed not only to maintain the defense of the state but to spread prosperity in doing so.[76] Cities, provinces, and even the whole union paid for necessary expenditures in ready money, raised mainly from low-interest loans, which were in turn paid back in small but regular amounts over long periods from tax revenues.

Not only the large merchants but many small investors participated in this method of transferring money to the state in return for a steady income—almost anyone who came into a bit of money could in effect lend it to the government in return for guaranteed, long-term income (which was much less risky, if less lucrative, than speculating in business ventures). Indeed, many of the loans were paid back in the form of annuities, some of which lasted for the lender's lifetime, others of which lasted for a fixed number of years and could be inherited or sold—with the consequence of secondary markets developing in these instruments.[77] (The calculations for public lotteries and annuities performed by leading regenten concerned with public finance also provided the basis for the development of mathematical probability.)[78] Lenders also were investing social capital in that they expected and supported the continued solvency of the state throughout the duration of the loan. Because so many persons invested social and financial capital in this way, and directly benefited from state payments, they were also amenable to living with very high levels of taxation compared to other Europeans countries—with grumbling but without rebellion—because they felt they were getting as much or more value than they gave. Because the loans were paid back in small amounts over a long term from tax revenue, which in turn came from several sources, especially excise taxes, the state avoided the debt crises and defaults that so plagued monarchical governments. The experience of personal trust being rewarded with a steady income boosted confidence in the fiscal responsibility of the managers of the state, which meant that they in turn could borrow at low interest rates, averaging around only 4 percent—several times lower than, say, the king of Spain. Although wages remained steady in the first half of the seventeenth century, there was real growth in per capita income, meaning that the growth in income was coming not from wage labor but from investments, with a substantial proportion of commercial profit taken out of riskier partnerships and more safely invested in government, benefiting many small investors.[79] The secret to the success of the Dutch Revolt, then, lay in a widely shared, well-managed national debt.

It helped that few individuals gained exemption from taxation. Even the capitalists, as they were called (perhaps for the first time), were taxed: in 1621, in Holland, certain extraordinary taxes were levied on land, houses, obligations, manors, tithes, and offices, with those owning more than two thousand gilders' worth of such property called "capitalists" ("half-capitalists" were defined in 1625 as owning one thousand to two thousand gilders).[80] A steady flow of revenue from taxes also meant that expectations about the state's income from year to year developed, which in turn provided a talking point for negotiations among the various parties: a proto-state budget. All this meant that the fiscal steadiness of the Republic proved remarkable, allowing competing groups to strike compromises and find common cause, and making it possible for a relatively small nation to mobilize enough resources to stand up to a huge power like Spain or, later, France. By encouraging the pursuit of individual wealth within a managed polity full of checks and balances, the regenten could pay for their collective security while enjoying material luxuries. It was an alliance of financial and political knowledge that could be seen to be about collective values, something like what Adam Smith would later call "the wealth of nations."

By the 1630s, Dutch merchants had virtually created a state resembling an aggressive commercial firm, the "Republic, Inc." They had combined commercial finance and state policy to such an extent that they were able to shoulder their way into most of the important commercial markets around the world, even managing to monopolize most of the trade with Asia, from source to market. The most obvious instrument for their ability to create states and wage war in the interest of grasping as much commerce as possible was the Dutch East India Company: the Verenigde Oostindische Compagnie, known as the VOC, or simply "the Company." It was a new kind of corporation in world history.[81] In their early days sailing to Asia, the Dutch felt weak in entering a region where other merchants had long been established in trade. A century before, the Portuguese had forcibly disrupted some of the long-established relationships of the region and co-opted others in order to profit from the intra-Asian trade.[82] The Dutch at first tried to avoid the more powerful Portuguese fleets, which tended to head northeast from the Cape of Good Hope toward their great rendezvous at Goa. The Dutch therefore instead sailed south, not to the Indian subcontinent but directly to the Southeast Asian archipelago. They also had to avoid the Portuguese-controlled city of Malacca, which controlled the Straits of Malacca and provided a base to which all sorts of people from eastern and western Asia came to exchange goods.

The region they had entered was peopled by ethnically diverse and complex cultures inhabiting the coastal cities throughout Southeast Asia, often going by the rubric of "Javanese" no matter what island they came from. Most were commercially sophisticated, had a social hierarchy based on wealth derived from trade, possessed slaves, often worshipped according to Islamic teachings, used the Malay language in commercial interactions, built sometimes very large freight ships (the word associated with Chinese ships, "junk," derives from the Javanese *jong*, a word dating to at least the ninth century), and produced other items requiring skilled craftsmanship. Inland, in the often heavily forested and mountainous interior regions of the sometimes very large islands, could be found states ruled by charismatic leaders, often religiously Hindu or Buddhist. The spice trade provided the basis for the intra-Asian commerce upon which many of these various states depended for their wealth and power. Geographically, the most critical areas lay near the Sunda Straits, the gap between Sumatra and Java. Through the straits passed much of the shipping between China and south Asia, and between the Spice Islands and the western destinations of India, Ceylon, Arabia, Africa, and Europe. India itself consumed twice as much of the spice produced in Southeast Asia as Europe, and China took perhaps three-quarters of all the pepper production of Sumatra and other parts of the region.[83] In exchange, cotton textiles and silver jewelry flowed to the Southeast Asian archipelago from the Indian subcontinent, together with silk and other luxuries from China.[84]

During their first ventures, Dutch merchants and captains depended on fair trading practices and good political relations to obtain the spices they desired. When they first arrived in the region, the Dutch were often welcomed as competitors and rivals of the Portuguese. Many local princes therefore seized the opportunity of playing one European group against another for both profit and power. When the prince of Atjeh (in the north of Sumatra) declared war on the Portuguese factories in his region in 1600, for instance, he looked for and used Dutch allies.

After the first successful trading ventures of the late 1590s, however, various problems of organization were recognized. Because every voyage was undertaken by a different partnership (*partenrederijen*), each Dutch fleet was in competition with every other one trading in Asia, including those sent out by compatriots.

Such temporary companies acted according to a familiar contractual arrangement: the partners invested in a voyage or group of voyages, took their profits on the return of the ships and sale of the goods, and then individually decided to end their association or continue in the group for another expedition. Out in the Indies, each venture bid against the others for produce (increasing investment costs), while on return each had to vend their goods at a price lower than the others (reducing income), with consequent reduced profit for the investors. Moreover, the rivalry among ventures made it very difficult for the competing merchants and captains to coordinate their power against common threats. A single managerial system for the Eastern trade would, however, create new advantages, making the trading companies not only more profitable but powerful enough militarily to take on the Portuguese directly. The States General therefore compelled the various Eastern trading companies to negotiate with one another about a common strategy. Between 1600 and 1602, fraught discussions took place under the watchful eye of Johan van Oldenbarnevelt, the advocate of the States of Holland, and with the personal intervention of the stadholder, Prince Maurits. Finally, in March 1602, the Verenigde Oostindische Compagnie was set up. The companies that had previously been involved in Eastern ventures were turned into "chambers" that met in the cities that had been their headquarters (Amsterdam, Middelburg, Hoorn, Enkhuizen, Delft, and Rotterdam), and each chamber was allotted a certain number of delegates for the general board of directors, who were called the Heren XVII ("Gentlemen Seventeen" or "Lords Seventeen"); as in other Dutch assemblies, the largest chamber, Amsterdam's, with eight members, dominated the board but did not hold a voting majority. The officers of the VOC had to swear an oath of allegiance to the States General and promise to keep the states informed of events in Asia, while during periods of wartime emergency they had to transfer money, ships, manpower, and equipment to the Republic on easy terms. In return they were given the power to negotiate treaties, build fortifications, enlist soldiers, and otherwise act as a sovereign power east of the Cape of Good Hope.[85] The VOC was, therefore, a profit-making, semi-independent arm of the Dutch state.

With the founding of the VOC and their new ability to coordinate their efforts in Asia, the Dutch could take independent action more effectively.[86] In December 1603, the VOC sent out a fleet of a dozen heavily armed ships to attack the Portuguese forts in Mozambique and Goa, but it and subsequent expeditions had little success against the centers of Portuguese power in India or Malacca. The Heren XVII therefore reorganized their efforts, forgoing direct assault on the main Portuguese strongholds in favor of getting the sources of spice production into their own hands. In 1605 a VOC fleet managed to capture the Portuguese castle on Ambon, which had been established to maintain their position against depredations from the sultan of Ternate. During the sixteenth century, the island had seen large-scale planting of cloves, which local growers traded for rice, textiles, musical instruments, gold and silver jewelry, and other imports. Portuguese attempts to force this trade into their own hands had, however, led to friction with the majority population of the island, who increasingly turned to Islam against the Catholic Portuguese and the minority population (*uli siva*). When Steven van der Haghen arrived at the head of a fleet in 1605, the Dutch were therefore welcomed by the majority of the local population as enemies of the Portuguese, and they threatened a bloody insurrection in support of a Dutch assault on the fortress. In the face of this threat, the Portuguese governor surrendered the castle, its eighty-three cannon, and its garrison of six hundred without a shot being fired, thereby ceding effective control of the trade in cloves. The Dutch renamed it Fort Victoria. In 1607 a monopoly contract was also drawn up with the powers on nearby Ternate for cloves in return for protection against other European powers.

The VOC began trading directly with the Indian subcontinent in the same year Fort Victoria surrendered, when a ship arrived on the east (Coromandel) coast where Portuguese influence was weak. VOC operations in the area came to be directed from Fort Geldria, in the city of Pulicat. Attempts to trade with the western (Malabar) coast of India were blocked for some years by the Portuguese, but in 1616 Pieter van den Broecke managed to obtain privileges from the Mogul Empire allowing the VOC to operate out of Surat. From Surat the VOC entered Persia, setting up a factory at Gamron (now Bandar Abbas) in 1624, enabling access to the southern parts of the Levantine trade. A station established at about the same time at Mocca, at the entrance to the Red Sea, completed the VOC's ports of call in what came to be called the "western quarter," from Surat to Mocca.[87] The establishment of the English at Masulipatnam in 1611 and Armagon in 1620, and of Danish East India Company factories at Tranquebar and Serampore in 1616, posed no immediate threat.[88] The Coromandel coast and western quarter of VOC activities supplied cotton textiles and other goods that were traded for spices and other products within Asia. Controlling large amounts of the seaborne intra-Asian trade became the key to the VOC's long-term profitability. Seeking access to the silver produced abundantly in Japan, the VOC was also able to establish a factory there in 1609, competing directly with the Portuguese and Spaniards.[89]

Many of the VOC's early successes came because its internal organization—unique for the time—gave it new abilities to carry out middle- and long-term planning. As a joint-stock company, it most unusually (compared to partnerships and regulated companies) accumulated a permanent capital fund to pay for its operations.[90] That is, no longer was money invested in a ship or a voyage paid back on completion along with any profits. Rather, people invested sums of money in the VOC chambers to be used as the VOC saw fit, for which they received a proportional share in the company. With the support of the States General, instead of paying off the investors after making immediate profits, the Heren XVII reinvested the income from all the voyaging back into the company to build up capital reserves, to construct facilities in The Netherlands and in Asia, to finance military operations, and to trade on a continuing basis. Not until 1610 did shareholders receive a dividend, which then turned out to be 75 percent, although issued in kind, in the form of mace; other payments in kind followed, and finally some cash was paid to shareholders in 1612. On rare occasions, the books were closed and the shareholders were offered an opportunity to cash out. But new shares were seldom offered to anyone beyond the original circle of investors, which meant that when old investors sold shares they went for a high price, so almost all members of the VOC stayed invested in the Company in return for whatever dividends the governors saw fit to pay. The largest investors also generally ran the chambers and got elected to the Heren XVII, while many others of them also sat in municipal, provincial, or national assemblies, combining business and political power in their persons. In other words, perhaps for the first time in history, a publicly chartered trading company came into being that gave the directors the ability to accumulate a permanent capital stock and a related material infrastructure to use as they saw necessary, and with their many other activities in other businesses and civic duties, the investors in the VOC could count on their interests being well represented throughout the Republic. Within the VOC, information circulated to committee, where it was used to assess current situations and future activities. With coordinated management and a growing permanent body of human and material resources, the Heren XVII could make plans to expand VOC operations as opportunity and determination allowed and in coordination with the state.

It was no easy matter to do, however: the Heren XVII had to plan for the building, fitting-out, provisioning, and manning of ships sent out each year, the sale of goods brought back to The Netherlands by the return ships, and the various needs of their factories in the East for manpower, weapons, fortifications, ships, merchants, and clerical staff. Given the time it took to traverse the distances with broad-bottomed ships and fickle winds, they had to make plans according to a two-year cycle: fitting out a ship already built and getting it to its Asian destination took almost a year of intensive labor, and almost another year was necessary for it to load cargo in the appropriate places in Asia and return to The Netherlands; the construction of new ships and other necessities needed to be planned far further in advance. Official letters and other documents might make the return more rapidly—in under a year—since they could be passed on by the most rapid means, but to learn of developments in the East, assemble the necessary personnel and equipment to respond, and transport it to its destination usually took a couple of years of exhausting work.[91] To carry out such a complicated business, the governors of the local chambers met regularly, and their representatives, the Heren XVII, met three times a year at one of the chambers, on a rotating basis, to coordinate their activities. They were assisted at first by the secretary of the Amsterdam chamber, after 1614 by a permanent counsel (*advocaat*), and after 1621 by two *advocaten*; after 1606 they also relied on a number of standing committees appointed by the chambers. Out in Asia itself, beginning in 1609, the Heren XVII appointed a governor-general to carry out their orders and administer daily activities, and a local council arose to aid and advise him.[92]

A pattern of business soon developed. The main meeting of the year lasted three to four weeks in the autumn after the yearly return of ships from Asia. At this meeting the Heren XVII decided on the equipping of the next season's outbound ships, the amount of precious metal to be sent to Asia for purchases, the goods to be ordered from Asia, and the manner of auctioning off the imports to get the highest prices. In the spring, financial plans were firmed up based on the outcome of the last auctions. The summer meeting reviewed the correspondence from Asia and drew up the lengthy orders of the Heren XVII, which were sent out with the first ships. At every meeting, the gentlemen also reviewed the financial situation of each chamber. Each chamber also undertook the task of building ships for their use. The shipyards of the chambers—especially in Amsterdam—became huge hives of activity, building, repairing, and outfitting ships of various sizes. The largest were the East Indiamen, designed to be *retourships*, or ships for sailing out to the Indies and back many times on the high seas: they were high, square-sterned, full-hulled, and capable of carrying cannon, similar to the warships of the Republic's fleet but not as fast or as fully armed. Other smaller ships were intended for a few years of return service before spending their last years in the intra-Asian trade, whereas some ships were built for Asian service alone.[93]

The Heren XVII also took stock of what they had on hand in The Netherlands, noted current and anticipated prices, and placed orders for new goods with their merchants in Asia. Although to modern eyes the spices attract attention (and are usually thought of only for use in cooking), the order placed in 1617 is revealing for the amount of imports used in medicine as well:

pepper, 70,000–100,000 "bales" (sacks; used in medicine and cookery)
cloves, "as much as possible" (used in medicine and cookery)

nutmeg, 1,000 "barrels" (*bhaar*; used in medicine and cookery)

mace, 300 barrels (used in medicine and cookery)

long pepper, 5,000 pounds (used in medicine and cookery)

galingale (*galleguen*; a rhizome with a hot, ginger-peppery flavor; used in medicine and cookery), 6,000 pounds

ginger and cinnamon, "as much as there was space available" (used in medicine and cookery)

lignum aloes (a scented resin), 6,000 pounds of the best kind (used in medicine and cookery)

India rubber (*gommelack*), 30,000 pounds

camphor from Borneo, 6,000 pounds (used in medicine and cookery)

China root (a medicine), 30,000 pounds, but "fresh and scentless"

benjamin (a gum from *styrax benzoïn*, used in medicines), 20,000 pounds of the best that can be found

musk, "none"

dragon's blood (a red resin from an ancient tree found on Sumatra and nearby places, often used as a color in varnishes and as a medicine), "none until further order"

wax, 200,000 pounds

wood of cassia fistula (a less valuable member of the cinnamon family; used in medicine and cookery), 3,000 pounds

spikenard (a well-regarded oil from India; used in medicine and cookery), 5,000 ounces

cubebe (another form of long pepper; used in medicine and cookery), "a good amount"

raw borax (used in soaps, enamels, and ceramics), 5000 pounds[94]

Other contemporary orders also ask for cardamom and sugar (used in medicine and cookery), amber, indigo, and bezoar stones (concretions from the stomachs of certain goats, used as an antidote against poisons and powerful diseases). In 1622, the VOC set aside three hundred thousand gilders for trade in "cloth from the Coromandel coast, all sorts of rarities, drugs, and porcelain." And there were always significant amounts of other items such as medicinal and cooking herbs and roots, special medicines, and various exotics.[95]

Other products gradually also became important for sale in Europe. Tea (from Japan and China) turns up at auction sales in 1651–52, and coffee (*cauwa de Mocha*) in 1661–62, about a decade after they first started to be brought back

The four continents making offerings to Amsterdam, who holds a caduceus and a shield bearing the arms of the city. Title page from *Historische beschryvinghe van Amsterdam* (Jacob van Meurs, 1663). By permission of the Rijksmuseum, Amsterdam

in smaller quantities.[96] Both were first introduced for their beneficial medicinal effects, and only gradually, as they became common foods, did they become massive items of trade. Saltpeter from the Coromandel coast, sugar and cotton textiles from the Coromandel and Malabar coasts, copper from Japan, raw silk from Persia, Bengal, and China, carpets from Persia and other places, and fine porcelain from China all also came to play an important role in VOC sales, none of them being a monopoly of the Company. The wealth tied up in such imports can be gathered from a passage in the diary of an official with the English admiralty after he toured a Dutch East-Indies ship captured in the Second Anglo-Dutch War. On 16 November 1665, Samuel Pepys wrote that Lord Brouncker and Sir Edmond Pooly "carried me down into the Hold of the India Shipp, and there did show me the greatest wealth lie in confusion that a man can see in the world—pepper scatter[ed] through every chink, you trod up it; and in cloves and nutmegs, I walked above the knees—whole rooms full—and silk in bales, and boxes of Copperplate, one of which I saw opened." Pepys was noting only the main items stored onboard, while silver, jewels, and other items of highest value would have been removed at once, under guard. Nevertheless, it was "as noble a sight as ever I saw in my life."[97] Another East India ship, from 1697, contained the following inventory: low in the holds were placed four hundred chests of Japanese copper, 134 pieces of Siamese tin, twenty-five tons of sappanwood, 580,281 pounds of black pepper, fourteen hundred bags of saltpeter; higher up were packed "candied ginger, nutmeg, cloves, cardamom, ginseng, white pepper, benzoin, cotton cloth and yarn, raw silk, drugs, various textiles including Bengal, Persian, and Chinese silk, cinnamon, indigo, civet, tea, and two small cases of birds' nests" (for soup).[98] One recent authoritative estimate finds that European shipping via the Cape (of which the bulk was carried by the VOC) had completely undermined the overland caravan trade by the 1620s.[99]

WISDOM FROM SELF-INTEREST

The values inherent in the world of commerce were explicitly and self-consciously recognized to be at the root of the new science by contemporaries. One of the most thoughtful analyses to make the point came from Casparus Barlaeus, acclaimed as one of the best minds of his generation. In early 1632 he gave an inaugural address on the foundation of an advanced school in Amsterdam. A sign of Amsterdam's civic pride, the new Athenaeum was to be a university in all but name, and Casparus Barlaeus had been appointed one of its two first professors. Given his new position, he would of course wish to praise learning. Barlaeus could also take it for granted that the search for wisdom was considered important for any magistrate. It was already enough of a commonplace that Cornelis Pietersz. Hooft, speaking as one of the most influential of the libertine Amsterdammers of a generation before, could simply declare that only people of both education and wealth should hold political power.[100] The very fact that the city was establishing the Athenaeum made the continuance of such values plain. The city had also recently combined and strengthened some of the already excellent teaching available at its advanced Latin schools. Because something like 7 percent of late-teenage Amsterdam boys went on to university intending careers in church, law, medicine, or public affairs—a very high rate for the period—there were also practical reasons for founding the new school in Amsterdam: their expenses would be considerably lightened if they studied at home.[101]

It was in full awareness of the material benefits of the Dutch world that Barleaus took up the subject of the union of wisdom with commerce to mark the occasion of the foundation of a new school. He did so in a speech to be delivered before the social, economic, and political leaders of the wealthiest and most powerful city of the Dutch Republic. For his theme, he chose a modern twist on a classical work that would be well known to all the former schoolboys and schoolgirls in the audience, Martianus Capella's *Marriage of Mercury and Philology*. In the original version, the mortal Philology—a young woman who studies books all night—acquires immortality by her marriage to Mercury, the god of commerce. Barlaeus substituted the activity of Mercatura (trade) for the god Mercury, and the efforts of Sapientia (wisdom) for the person of Philology, remaking the myth into a metaphor that spoke for the proud new world of Amsterdam.[102]

He nevertheless faced a difficult challenge in trying to reassure some other people of learning that creating wealth would not harm but instead increase "ruminations of the mind." To people who devoted their lives to the pursuit of learning or the study of God, giving attention to worldly matters was often considered to be a distraction at best, a sin at worst. Various passages in the New Testament, for instance, indicated that wealth was inimical to salvation. Perhaps the best known is the Gospel According to Mark, which commented, "It is easier for a camel to go through the eye of a needle, than for a rich man to enter into the kingdom of God" (10:25). Some of the most famous authors of classical philosophy had also argued something similar. The eighth book of Plato's *Republic* had opined that "in proportion as riches and rich men are honored in the State, virtue and the virtuous are dishonored."[103] Aristotle's *Politics* tempered this a bit, proposing that the getting of wealth is a natural and important part of household management. But for him, gathering goods was compatible with virtue only as long as it was limited to the provision of necessities, for otherwise wealth corrupted.[104] In other words, many of the best authorities had set at odds the values of *negotium* and *otium*, business and the peaceful life, work and contemplation.

Barlaeus was therefore bold in his intention to refute the common assumption that commerce stood in opposition to virtue and the pursuit of wisdom. Some learned Dutch commentators had already moved part way toward such a reconciliation. Dirk Volckertsz. Coornhert, for instance, an advocate of toleration and religious liberty and a constant thorn in the side of the dogmatists and their fellow travelers, wrote a Dutch dialogue of 1580, *Coopman* ("The merchant"), in which he explored how a merchant could live as a good Christian. He thought that if the search for gain was undertaken not with the intention of piling up riches but for the general good, so that the virtuous merchant would generously give away his gains to good causes (such as charity, or supporting the Revolt), then wealth and virtue were compatible.[105] In this sense, Coornhert took a line followed by the Reformed church more generally. On the problem of usury, for example, they argued that public banks and money exchanges were allowable as long as their profits went to charitable causes. Even private lending, which technically debarred a person from the communion table, was generally tolerated because it was necessary for the creation of the wealth that supported the struggle against Spain.[106]

But there was a fundamental problem at the heart of Coornhert's strategy. One of his interlocutors put it this way: "How can the merchant seek wealth but yet not desire to possess it?"[107] That is, if merchants could be moral only by giving away their wealth, why would they seek it? This problem had, of course, been forced on the mind of others, too. For instance, the Reformed minister and early advocate of taking the war to the Spaniards in the West Indies, Willem Usselincx, laid many of the troubles of the early West India Company (WIC) at the feet of greedy merchants. He wanted people to support the WIC for the glory of God, for the damage it could do to their common enemy, Spain, and for the benefits it would bring to the Republic. But in the end, he had to admit that "the principle and most powerful inducement [for attracting investors] will be the profit that each can make for himself."[108] Even for a man of the cloth, then, the common ship of state moved toward the good when the quest for private gain drove men to ply its oars, even if this had nothing to do with individual salvation. The general good could result from the acquisitive spirit or even from immoral promptings like greed.

But Barlaeus went further. He did not simply overlook personal fault in order to see the general good. Of course, there were general goods that came from such personal strivings: trying to make a profit could indeed solve grave problems, such as inducing merchants to ship grain from storage to where it could be sold for a high price and thus breaking a famine, an example employed by Cicero. But for Barlaeus, the merchant could do good even when keeping the benefit of his activities rather than only by giving away his wealth; commerce could itself be among the best pursuits of human life. Others were arguing for something similar. In France, for instance, a powerful new current of moral writing focused on *amour-propre* (self-love) as the stimulus for most human behavior.[109] Similarly the Dutch jurist Hugo Grotius, with whom Barlaeus shared many interests, argued that the most elemental law of nature was self-preservation, and the next most fundamental right was therefore self-interest. On this argument about natural law rather than on the special rights of governments, which were often said to be derived from divine law, he had based his work on free trade, *Mare liberum* ("Freedom of the seas," 1609)—one of the founding documents of international law. The outcome was a view of civil society that, as Richard Tuck has put it, "is a construct by individuals wielding rights or bundles of property," with governments only possessing rights held by individual persons.[110] That the pursuit of such self-interest is right and natural is also implied by Barlaeus. He, like Grotius, also held as a corollary that self-interest made people sociable, since by engaging with one another they obtained what they needed. Commerce therefore also brought people together in ways where they learned from one another. By engaging fully in the part set for each of us in the theater of the world by an omnipotent and omniscient God, his will would be done. God was therefore like a merchant; indeed, a contemporary Dutch translation of Barlaeus's address even used the term "the great Factor" (meaning head of a trading station) to refer to God.[111]

In discussing the relationship between commerce and wisdom, then, Barlaeus began with comments on the general good, showing that periods of great learning and great wealth went together, to which he added a distinct note of local pride.[112] A quick look around showed anyone a city of harbors and docks, canals, locks, and other waterworks, to which fleets of ships from all over the world sailed richly laden with merchandise. The result could be seen in the splendid edifices seen on all sides and in the general affluence of Amsterdam's citizens. In addition, their actions showed them to be people of prudence, obedience, modesty, reverence, and lawfulness, appreciating above all good public order. In the wise and noble leaders of the city, Mercury and Pluto (the gods of intelligence and wealth) found a home, and here in this center of commerce wisdom abided.

In this part of his argument, about how virtue and magnificence came from the union of learning and worldly activity, Barlaeus was building on two lines of argument. One was the well-established position that the active life (*vita activa*) was more virtuous than the contemplative life (*vita contemplativa*). The civic humanism of the Renaissance Italian city-states had long ago turned the duties of male citizens to participate in political, military, and economic affairs into new virtues (or more properly, *virtú*).[113] Barlaeus had no difficulty in turning the moral qualities of the man of action into those of the man of commerce, since commerce was so bound up with courageous activity in the world. The second part of this aspect of Barlaeus's argument drew on new interpretations of history. Poets, playwrights, and philosophers in many places were asserting that contemporary material progress supported an extraordinary flourishing of arts and sciences even greater than in Rome.[114] For instance, Louis Le Roy, professor of Greek at the Collège Royal in Paris, had written a universal history in 1576 showing that "the conjunction of Power and Wisdom," or might of arms and of letters, created the greatest nations. But while God's Providence had in previous instances laid even these mighty nations low, he thought that the wheel of fortune could be reforged into a line of constant progress if the learned worked hard to "carefully preserve the arts and sciences, as also all other things necessary for life."[115] In England, just a few years before Barlaeus spoke, Sir Francis Bacon had been advancing similar arguments about the union of philosophical and worldly utility in constructing a just and powerful state. It might be expected, then, that Barlaeus's audience needed little persuasion to agree that the union of worldly enterprise and learning in their own city made this place and this moment one of the most noble in history, one that would persist for ages.

But Barlaeus took this line of argument a large step further in showing not just that knowledge and commerce were in conjunction but that they arose from the same source in the human spirit—a kind of love that produced them both. He discoursed on antiquity to show this, introducing examples of moral conduct. To gain their true ends, both the sage and the merchant had to act according to the dictates of natural virtue: to moderate their desires, to cultivate honest conduct in all things, and to value all matters in helping them to their ends. Precisely these connections between virtue and knowledge were being underlined in a set of vernacular terms new in Barlaeus's day. The Italian word for a person who combined both tasteful knowledge and natural virtue was *virtuoso*.[116] The English adopted it for themselves. In Dutch, *liefhebber* (or German *liebhaber*), indicated something similar: someone who had a sincere and discriminating love of things that indicated inner virtue.[117] With similar aims, the French later coined *amateur*, from the Latin *amare* ("to love"). Indeed, this language about the deep associations between personal and material goods quickly came to mean someone who collected objects, whether made by fine craftworkers and artists or by nature. Such people spoke about how precious objects were exemplifications of the best part of the human spirit or God's creation, feeling themselves uplifted by them. As patrons and collectors, they had the ability to identify and to bring forth enduring examples of the good and the beautiful despite human sin and mortality. Barleaus seems to have had these virtuous collectors of information very much in mind as the best embodiments of modern wisdom, for in going on to show that commerce encouraged the study of "speculative philosophy" he gave a telling list of its parts. It included nothing on theology or traditional philosophy, not even natural philosophy, certainly nothing of the occult philosophy. Instead, he drew the attention of his listeners to geography, natural history, astronomy, languages, and the study of the various characteristics of different peoples. To our eyes, he was describing nothing like speculative philosophy, but information-based subjects about nature and humankind.

Finally, in his conclusion Barlaeus returned to showing by example that ancient cities clearly displayed how great wealth and philosophical excellence developed together, the one supporting the other. By implication, philosophers who promoted contemplation rather than action were mistaken. So were those who tried to instruct communities to live according to God's plan as set out by reasoned theories: they would only divert merchants and philosophers from exploring their authentic paths through the world. It was not from doctrine but from the interactions found in buying and selling, and in the search for knowledge that was another aspect of exchange, that modesty, honesty, and natural truths emerged. The ways of life on which the Dutch political economy depended were thus reflected in a certain kind of objective investigation of nature, the kind referred to by Barlaeus as "speculative philosophy" and which we would call natural history.

It was therefore not only bulk commodities that the regenten valued. They also highly prized individual specimens of particularly rare or beautiful appearance and the knowledge rooted in them. Many examples could be introduced from the high culture of fashionable clothes, furnishings, painting, and books, but perhaps the clearest example from the Dutch world is that of the so-called tulip mania, which peaked just three years after Barlaeus's address.

Many new flowering plants arrived in the low countries from Ottoman lands. The Turks planted tulips by the thousands in the sultan's gardens in Constantinople, frequently stuck them into their turbans (from which the European name "tulip" derived), and often depicted them in the naturalistic art that established itself in the Ottoman world in the second half of the sixteenth century.[118] So it was that the tulip had been first described for Europeans by Pierre Belon during his travels in the Levant in the late 1540s.[119] It had a further notice from one of the most influential seekers of new garden plants, Ogier Ghiselin de Busbecq. Originally from Flanders, and an able diplomat, Busbecq took up the very sensitive posting of the Holy Roman Emperor's ambassador to Constantinople for the purpose of negotiating a peace treaty with Suleiman the Magnificent. To do so, he resided in Ottoman lands for most of the period from 1555 to 1562.[120] He took with him on his embassy a physician born in the low countries who was also a fine botanist, Willem Quackelbeen.[121] Busbecq and Quackelbeen identified many plants unknown in Europe or known only through the ancient descriptions of Dioscorides, and they sent many seeds and cuttings to the well-known Italian botanist Pierandrea Mattioli and others, for which Mattioli offered sincere and

Turkish (Iznik) pottery dish, painted with tulips, hyacinths, and carnations, c. 1560–1565. By permission of the Courtauld Institute of Art Gallery, London

profuse thanks. It was plants from Ottoman gardens that revolutionized European horticulture, and Busbecq and Quackelbeen introduced many of them.[122] He and Quackelbeen are, for example, credited with introducing the horse chestnut, at least one kind of gladiolus, "true" *Hermodactylus* (probably "Dutch Iris"), sweet flag, spiny broom, lilac, the plane tree, grape hyacinth, and other species to Europe.[123] (He also convinced the emperor to purchase from a Jewish physician at the Ottoman court an ancient and beautifully illustrated manuscript of Dioscorides, now called the "Codex Vindobonensis," which remains in Vienna.)[124] In the letters he wrote from Turkey and afterward published, Busbecq noted that the Turks were "passionately fond of flowers, and though parsimonious in other matters they do not hesitate to give several *aspres* for a choice blossom." He also commented that in traveling from Adrianople to Constantinople "we were presented with large nosegays of flowers, the narcissus, the hyacinth, and the tulipan (as the Turks call this last). . . . The tulip has little or no smell; its recommendation is the variety and beauty of the colouring."[125]

The first record of a tulip growing in Europe is from 1559, when the natural historian Conrad Gesner visited the garden of the Augsburg councilor Johannes Heinrich Herwart and saw one there.[126] Probably only a few years later, the soon-to-be famous naturalist Carolus Clusius saw one for himself. In 1569, Clusius also obtained tulip seeds from Busbecq, acquired during Busbecq's residence in Turkey.[127] Clusius was enthralled with the diversity of form and color of the flowers exhibited by this species, and he cultivated many varieties while spreading enthusiasm and specimens among his acquaintances. The tulip became one of the most widely coveted of the flowering plants introduced to European gardens in the late sixteenth century, and the Dutch became particularly associated with its cultivation.

One of the attractions of the tulip was that it came in an astounding assortment of shapes and sizes, with new varieties continually emerging by accident or design. Tulips ranged from the small spiky flowers of those from the wild mountains of eastern Asia to the tall, rounded blossoms of the cultivated garden plants, some with petals overlapping to make the flower look like a deep cup, others just touching shoulders to flatten out like saucers, almost all standing proudly upright toward the heavens. They bloomed variously from early spring to early summer. The colors ranged through the spectrum from white to yellow to red, even to dark purple, missing only true black. Most petals appeared in a solid color from stem to edge, but some were "broken," meaning that they had colored streaks on a solid background. These breaks seemed to occur by chance, with a bright red tulip, for example, blooming the following year with yellow streaks. Once the change occurred, however, it usually remained in succeeding years, and it sometimes spread to others nearby, suggesting that broken tulips expressed something unusual in the soil where the break occurred. (In the twentieth century the breaks would be explained as the result of an infection by a virus.) Breaks among yellow and red varieties appeared most often, while such occurrences among whites were much harder to come by.

The most prized tulip had it all: a white petal on a base hinting of blue, with broken flames of vivid crimson shooting up to the top. One of the appraisers of tulips in 1625, Nicolas Wassenaer, had no doubt that Semper Augustus was the best. In 1624 its bulbs were valued at twelve hundred Dutch florins each (about four times the annual wages of a skilled laborer), but in the following year their owner was offered more than double that for each of them, even though the

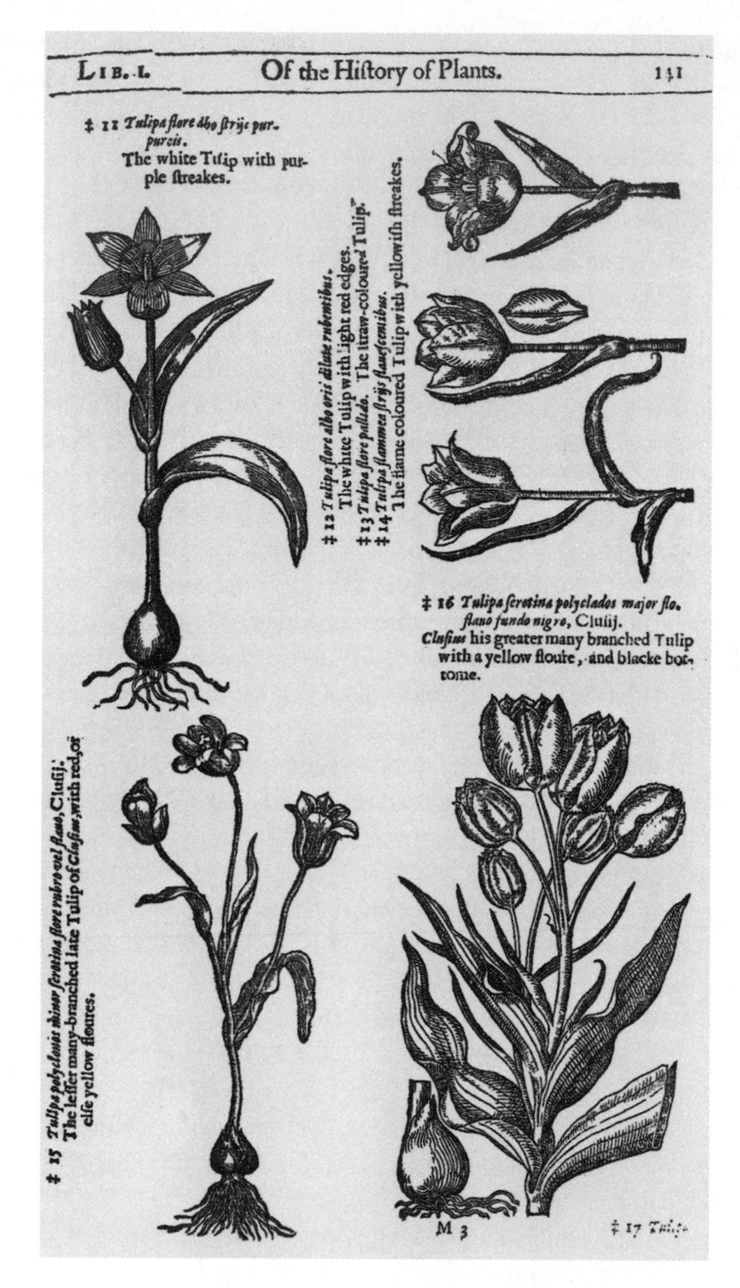

Tulips, from John Gerard's *Herball* (1633); that in the upper left is labeled "The white Tulip with purple streakes." Courtesy of the Wellcome Library, London

bulbs themselves had multiplied. The owner sold one to a close friend for two thousand florins, but with the restriction that the buyer could not pass on any of the bulbs to anyone else without the seller's permission. By 1636 and 1637, when the "tulip craze" was reaching its height, it was said that Semper Augustus bulbs were worth ten thousand florins each.[128]

The man who owned this most valuable of tulips was Adriaan Pauw, son of a powerful and conservative Calvinist family of merchants and politicians from Amsterdam (some of whom had been among the first investors in the VOC), himself a distinguished moderate Calvinist servant of his country and first cousin to Pieter Pauw, professor of botany at Leiden and colleague of Clusius.[129] It was in the gardens of his estate at Heemstede that he planted Semper Augustus. Like other members of the urban magistrates of his generation, he turned a working estate into a country retreat, and it would have been incomplete without a pleasure garden.[130] Pauw had concluded negotiations to purchase the estate in 1621 for thirty-six thousand gilders: it was situated near the great inland lake called the Harleemermeer—which was being converted into arable land through a huge draining project, in which Pauw probably invested—and the purchase brought him the noble title of Heer van Heemstede.[131] Twelve bulbs of Semper Augustus were known to exist in 1624 (worth about one-third the price of the Heemstede estate at then-current prices); all were in Pauw's hands. To multiply the effect of the dozen spring blossoms and give the impression that he had even more, Pauw had a gazebo covered with mirrors constructed. By devoting land and property, employees, personal energy, and other valuables to the cultivation of gardens, Pauw was expressing an appreciation for shaping nature in ways that could be shared with others of like feeling and judgment, a moral appreciation associated with knowledge and taste.[132]

Pauw's Semper Augustus was valuable, then, because when people saw it they experienced, or could be taught to experience, or could be taught to pretend to have, certain pleasurable sensations; the sight elicited conversation as well as contemplation. Or more accurately, fanciers were able to share with other people their appreciation of the flower by exchanges of gesture and word, which were in turn associated with other cultural assumptions, such as the goodness of natural beauty. Fanciers with access to many gardens, and dealers who handled many varieties, produced ranked lists of tulip flowers, in effect serving as experts on good taste and discernment. Some of the lists expanded to illustrated booklets, with the flowers handpainted on sheets of paper or, for more mass audiences, as woodcuts (to which coloring was often added by hand for an extra price). Emanuel Sweerts published the first such illustrated guide, the *Florilegium*, in 1612, followed by Chrispijn van der Passe's *Hortus floridus*, with a host of others soon seeing print.[133] The only woman known to have made her living by painting at the time, Judith Leyster, became involved in producing illustrations for the tulip trade, as did Rembrandt's master, Jacob van Swanenburch.[134] The distribution of such works helped create a consensus about the relative beauty of one variety in comparison to others. Some people might disagree with the high value placed on this or that variety, but in the ability to discriminate between varieties and assess subtle differences of opinion also lay chances for exchange.

Thus like others of its kind, the value of Semper Augustus lay not only in the loveliness of its blossom but in its ready potential for commerce. The lovely but ephemeral flowers of the tulip were embodied in their hardy bulbs, the enduring tokens of particularly constituted varietal expression, ready to burst forth in predictable glory the following spring. Like onions, which once dug up could easily be stored for long periods, tulip bulbs can easily be accumulated and handled. While fertile seeds took roughly seven years to produce a mature bulb and flower, and with results that were unpredictable, varieties bred true from bulbs. Flowers cross-pollinate with others of this widely varied species, and moreover, their seeds do not convey the streaking that made the "broken" varieties so lovely and valuable. The only way to be certain of having either the same or a broken blossom in the following year was to propagate new specimens from bulbs. To do this, the gardener cut the flower heads after the petals fell, leaving the tulip's leaves to wither away gradually on their own, which produced a relatively large bulb with smaller offsets that was dug out of the soil ("lifted") in the summer and stored in a cool, dry place until autumn planting. Bulbs had the advantage over leafy plants of not needing much attention, and once dug were readily transported, making them easy to give, sell, or steal; yet they would faithfully produce flowers the next season just like the blooms they had produced the year before, while their offsets grew into new bulbs. Tulip bulbs were therefore a bit like cowry shells or even coins, whose material presence could count as a representation of value that could be exchanged for other things.[135] In other words, tulips are a good example of how shared values are often abstracted into the realm of "currency," things that have an agreed current value exchangeable for other things. In the case of tulips, then, because of the materiality of the bulb from which the flower's beauty arose, their moral and aesthetic value could readily be converted into price. Demand for tulip bulbs arose more generally from the high cultural value placed on pleasure gardens and flowering plants, together with the unusual properties of the particular tulip plants themselves.

The chance for profit followed from the desire for exchange. As a person growing up in a world that valued certain kinds of experiences of nature, including the cultivated pleasures that plants could bring to the eye, and as someone with an education that gave him the tools by which he could interpret botanical forms along with the most expert, Pauw had a sense of the value of Semper Augustus and acquired it. Yet not only wealthy fanciers valued tulips. The Netherlands was remarkable for the fact that even ordinary craftworkers might have paintings hung in their workshops and homes; just so, ordinary people could also learn something about the rules according to which the beauty of flowers was appreciated by those of a higher station. Many people could participate in growing and exchanging tulips for money, even if they were not particularly wealthy. As the variety of early introductions became more numerous, whether through gift, sale, or theft, the bulbs came into the hands of many people beyond the small group of original cultivators. As regard for gardens was affirmed and broadened, the market for garden plants also expanded rapidly, not only in the low countries but in France, Germany, and many other regions; tulips became one of the early staples with growers and dealers. The bulbs could be named, measured, weighed, and stored. In this way, buyers and sellers both knew what they were handling. For varieties that were agreed to be particularly lovely and unusual, bulb dealers could get very high prices, since everyone seemed to want some. Because no special gardening experience was needed to cultivate the flower, even people of relatively modest means and a bit of soil outside the back could grow and propagate tulip bulbs, hoping for a break and the wealth that would follow. As demand rose, so did prices. And as prices rose, increasing numbers of people got into the market for bulbs not for personal or social pleasure but because they saw them as a way to invest. Even dormant brown bulbs could be associated with a colored picture of the flower they would produce, could be compared to other flowers, and could be weighed and assessed for maturity and number of offsets; moreover, because the bulbs were often exchanged at public auctions, a knowledgeable investor could compare past prices again current ones, giving a reasonable hope of measuring the consensus about which types were valued most highly and perhaps even predicting future value according to the movements of taste indicated by the auction prices.

A speculative market took off, based on a sense that the market for bulbs would always rise. The ingenious financial methods developed in the early modern period also allowed investors to profit who did not have a personal interest in growing bulbs in their gardens; indeed, they did not even need to touch one. As with other expensive commodities, bulbs could be traded on paper: the owner of a bulb would take a sum of money from someone and write down a promise of delivering the bulb in the future, such as when it could be safely lifted. But delivery of the bulb on the appointed date would be to whomever possessed the piece of paper on which the promise was written, which meant that the promissory notes themselves, not just bulbs, could be bought and sold. This meant that if people in the market noticed demand rising for a certain kind of tulip, the promissory notes would be traded for higher prices, and conversely someone with notes for varieties with weakening demand could unload them before they dropped further. The original owner got his money and the person taking delivery in the end also made money if the bulb's value had risen above what he paid for the promissory note; in the meantime, many other people might have handled the note and made a profit by selling it on.

The bulb market was not, then, very different from other kinds of transactions occurring among the greatest merchants of the day at places like the Amsterdam stock exchange (Beurs). The major difference was that when it came to bulbs, it took much less capital to get into the game, at least at first, for ordinary kinds of tulips. Trading in bulbs, contracts, and speculative futures drew in large and growing numbers of people, even being done in taverns and inns or wherever else gatherings of interested investors took place, giving the impression of endless growth. As the investments of ordinary people in the trade grew ever greater, critics started to call it the *windhandel* ("trading in wind" or, in a less savory connotation, flatulence). The frequent satires about the business, which often relied on variants of old expressions about fools and their money, might even have been appreciated by those inside the business as long as they kept getting richer—at least on paper. Prices shot up, rising to unheard of levels during 1634–37. The average price for a single bulb at an auction in Alkmaar on 5 February 1637, arranged to benefit an orphanage, rose to sixteen thousand stuivers (eight hundred gilders), the amount of money a junior merchant might make in a year if employed by the VOC in the rich Asian markets; at that level, stately houses in Amsterdam were not much more expensive than a handful of the most desirable bulbs. But the alchemy of ever-increasing demand for bulbs ultimately failed. When for some reason demand slackened, probably because the market was getting too expensive for even wealthy investors, the bubble burst. In the middle of February, only a few days after the huge prices paid at Alkmaar, an auction failed because there were too few buyers at the prices being touted. The bottom quickly fell out of the market, with prices plummeting, causing large numbers of ordinary people to lose most or all of their investments. The lawyers and magistrates had years of work ahead trying to sort out the transactions and ensuring that the bulb growers got at least a fraction of what they had been promised.[136]

Some of the most advanced forms of social and economic transaction, therefore, underpinned the bulb market. They included trading in a futures market based on promissory notes. That is, people contracted to purchase a bulb at an appointed future date at a set price, hoping that by then it would be more valuable but risking a fall in value—but no money changed hands between the grower and the investor until the future date. This not only allowed buying and selling of goods without having to transport the goods to a particular place for physical exchange but also permitted financing according to expectations of future developments. The grower, too, also had to risk the potential failure of the investor before the due date while gambling that the best time to commit to a sale was the present. That is, if one can gain resources now and pay later based on the income earned from using the resources, or pay low rates for future goods that will be worth more when in hand, both borrower and lender will be better off. As long as the demand for tulips continued to rise, everyone made money, including the notaries who wrote up the contracts. It all depended on having confidence in one's expectations for the future and—at least for those at the heart of the market—a firm understanding of the attributes of the objects in which one dealt.

The so-called scientific revolution resulted from movements in the world and in persons, leading to countless efforts to find out matters of fact about natural things and to ascertaining whether that information was accurate and commensurable. This discovery of the world—its geography, peoples, plants and animals, and astrological and alchemical associations; the accumulation of specimens of it, the cataloging of its variety, and the detailing of its structure—created extraordinary public excitement, as well as bringing about unanticipated consequences. "Matters of fact," like objects, traveled with people, who moved about exchanging goods and information. In the process, local knowledge was often transformed into universalizable truths. Trust and credibility rooted in modesty and work, supported by plain speech and the rule of law, oriented toward finding out and accumulating a knowledge of the exact details of the material world and exchanging them commensurably: these constituted the values of the hardheaded merchant and his fellow travelers just as much as they did the values of the naturalists and physicians. Indeed, these values supported the very fabric of the Dutch Republic, making it a bastion of safety against their enemies and a place of refuge for those who lived from knowing about how best to transform worldly things into valued specimens of consumer taste and personal good. Objectivity had the power to whet the appetites, even to alter perceptions, concepts, and moral strictures. It did not float above the world but was deeply involved with it.

3

Reformations Tempered

In Pursuit of Natural Facts

Between ourselves, these are two things that I have always observed to be in singular accord: supercelestial thoughts and subterranean conduct.
—MONTAIGNE, "Of Experience," *Essays*

The terrible tragedies of the Reformation and Counter-Reformation were doubly fierce because of theologians and philosophers who were convinced that they knew the ways of God and what he wanted from humanity; many of them also intensely disapproved of the concern for worldly things. Theological rationalists of many kinds, whether Protestant or Catholic, were organizing, turning out more people like themselves, and persuading political leaders to support their agendas. On the other hand, to people like Desiderius Erasmus, Rabelais, Montaigne, and many others inculcated in the new ways of knowing of the Renaissance, the theologians' convictions merely proved their pride of intellect. At best, mortals could know the surface of God's creation, and perhaps catch occasional sight of the shadows of his hand at work. As for obtaining demonstrative certainty about what he intended either for humankind or for individual persons, or the nuances of his moral laws, only someone of overweening pride could think that they really knew, let alone think that their view should dictate the behavior of others. Such coolness to dogma in the name of liberty found refuge in a new polity in the northern low countries and in a new university in one of the region's largest cities, Leiden. The clerics almost managed to take over this land and this university, too, but freedom of inquiry survived, just.

Although some historians have been tempted to see the rise of the new science in the United Provinces as due to Protestantism, in particular to the Calvinism that dominated the public face of the Dutch Republic, it is clear that what was far more important was the ability to escape the intellectual constraints that religious figures of many kinds wished to place on everyone. At the time of the condemnation of Galileo, for instance, Calvinist clerics were as opposed to the proposition that the Earth moves around the Sun as any member of the pope's entourage might be.[1] When considering the emergence of the new science in their own country, then, Dutch historians themselves have tended to emphasize the fundamental importance of learned criticism—often associated with Erasmian humanism—rather than any religious doctrine. They have sometimes considered the combination of humanism with an undogmatic Calvinism to have been helpful in fostering natural science.[2] But even those Dutch historians most sympathetic to religious impulses in science have argued that other changes, such as the growing emphasis on factual information and the erasing of the classical distinction between the natural and the artificial, were more important.[3]

In the English-speaking world, however, the so-called Weber thesis about the "Protestant ethic" as the source of the modern world's rationalized sociopolitical system has often been used to explain the rise of modern science, as well.[4] To oversimplify, Max Weber argued that the shift from the cozy communitarianism of the older world to the "disenchanted," striving, calculating, and bureaucratized modern world was prompted in large part by the rise of ascetic Protestantism (in the English variants often associated with "Puritanism"). This was already an old trope when Weber was writing, with Karl Marx and Friedrich Engels themselves, for instance, having associated the work ethic and capital accumulation with Calvinism.[5] Weber thought that he had a way for identifying how Calvinist doctrine did its work: it taught that only a few elect among the mass of humanity would be saved and that they had been predestined by God for this from the beginning of time, which, Weber argued, caused people to probe their life and conscience anxiously in search of signs of salvation. The best signs of salvation came from material success, suggesting that God's grace had helped one on the way, which in turn prompted Calvinists to set aside ordinary pleasures in favor of hard work and accumulation. A powerful Protestant—more particularly Calvinist and Puritan—distaste for bodily pleasures therefore provided the belief system that turned material interests into social values. In the 1930s, Dorothy Stimson used Weber's theory to account for the rise of science.[6] Shortly thereafter followed the most influential work on how early modern Protestantism and capitalism laid the foundations for modern science, in the Harvard dissertation of Robert K. Merton, who later became a powerful voice in American academe. He took Weber's Protestant ethic and coupled it with British Marxism to demonstrate that modern science arose from the empirical pursuit of material advantage, which he thought to have been prompted by Puritan asceticism.[7] Following in his wake, a great deal of work has been devoted to the "Puritanism and science" argument. The most important revision was offered by Charles Webster, who shifted the ground from Puritan asceticism to eschatology, arguing that Puritans devoted themselves to utilitarian investigations of the natural world as a way of contributing to Christ's Second Coming.[8] Most historians, however, have forgotten or ignored the materialist aspects of Merton's argument in favor of the religious ones alone. The question has been reframed as "What religious view best encouraged the rise of science?" There have been arguments to emphasize Anglicanism or, within it, latitudinarianism, or Lutheranism; still others have pointed out Catholic contributions to science, especially those of the Jesuits. An even more general position is that Christianity or monotheism gave rise to science.[9]

But as many have pointed out, the Weber thesis itself is fatally flawed. To take but one example, a Dutch economic historian, J. H. van Stuijvenberg, showed that Weber's argument about religious doctrine and economic attitudes is tentative, inconsistent, historically confused in mixing later Calvinist interest in predestination with the teachings of Calvin himself and early Calvinism, sociologically unable to show whether ethical teachings from the pulpit affected the lives of their auditors, and, most important, theologically unsupportable, since there are no contemporary Calvinist theological writings that accord with Weber's sense of the Calvinist ethic until the eighteenth century or thereabouts.[10] Others have argued instead that the "hedonist-libertarian" ethic, or consumerism, or other personal and social values fit the case for accumulation better.[11] If the Weber thesis is flawed, so are many of the arguments rooted in it. Bluntly put: there is no reason to think that Protestant clerics were any more favorably disposed to capitalism or science than clerics of any other faith. Religious persons saw an intense concentration on worldly things as a potentially dangerous distraction from what they considered far more important, the study and worship of God and his ways. The development of natural knowledge in the Dutch Republic, then, depended not only on the new values of commerce but on the ability of those pursuing the facts to escape the domination of rational dogmas of many kinds.

LEARNED HUMANISM AND NATURAL HISTORY

How naturalists escaped the constraints of religious doctrine and its consequences can be seen in the life of one of the most famous naturalists of the low countries, Charles de l'Escluse, better known by the Latinization of his name, Carolus Clusius. Raised in the old religion while dying in mainly Calvinist Leiden, he was always chary of being forced to make public declarations of faith.

Benefiting from an excellent education, opportunities to investigate nature and to publish about it, and networks of fellow enthusiasts, in his long life Clusius exemplified the new generations of naturalists. Born in 1526 at Arras, in the province of Artois, a bit south of Antwerp, Clusius came from a highly literate family. His father, Michel de l'Escluse, first held the post of clerk and receiver for the great Abbey of Saint-Vaast that dominated Artois and later purchased an estate near Armentières that brought a minor title of nobility; Clusius's mother, Guillemette Quincault, had a brother who was prior of the abbey (in rank second only to the abbot himself).[12] Probably the eldest of the couple's children, Clusius had at least three brothers and three sisters. At age fourteen, following several years of preliminary education in religion, literature, and Latin, he attended the well-regarded school at his uncle's abbey.

The low countries had long been replete with many outstanding schools. In the early fourteenth century, the mystical writings of Meister Eckhardt carried the spiritualist message forcefully, while a few decades later the "Modern Devotion" arose around Geerte Groote, a citizen of Deventer, teaching that all people, not just monks, nuns, and clergymen, should be humble, industrious, and constant in remaining devoted to God.[13] A follower of the new teachings, Thomas à Kempis wrote a work early in the fifteenth century considered one of the great pieces of Christian devotional literature, the *Imitatio Christi* ("Imitation of Christ"). As a sign of this popular Christian mysticism, throughout the low countries communities of beguines and begards had come into being—the former made up of women, the latter of men—who lived in separate houses grouped in walled communities within the cities. They did not take religious vows but devoted themselves to good works and a chaste life of service to God and humanity. Because their religious views were not controlled by the institutions of the church, however, they often became centers of heterodox opinion. The interior piety that was the true way to God needed to be nourished by reading scripture and other parts of sacred literature. Associated with these spiritualist movements was the founding of grammar schools. For instance, a group calling itself the Brethren of the Common Life, who lived in common but without taking religious vows or separating themselves from the world, established many schools, a number of which became noted for the very high quality of their teaching. The most famous personification of these values was Erasmus of Rotterdam, who was originally educated by the Brethren of the Common Life.[14] Local parishes, dioceses, and municipalities also promoted schooling. Even relatively poor boys and girls were often sent to elementary schools for a few years before they were put to work, learning to sing, read, and write in the common tongue. Grammar schools offering Latin abounded for those of about age eight or older who wished to study further.[15] After his uncle's death in 1543, then, Clusius was sent away to the Latin school van Houchaert (Eucharius) in Gent, a place led by someone who had studied philology in Italy before returning to make the school into a place with a regional reputation for excellence in the classical languages. By the mid-sixteenth century, many larger cities in the low countries merged their several grammar schools into "great schools" or otherwise found ways to encourage the new language studies that accompanied humanism.

Urban adults often continued to find enjoyable classical edification in the so-called chambers of rhetoric. They had long existed in most cities and many large towns of the low countries as places where learning outside the religious establishments continued to be furthered. The chambers of rhetoric brought together men and sometimes women of many occupations and backgrounds, from shoemakers and craftworkers to great merchants. The chambers had begun in the late medieval period as fraternities for the performance of Christian mystery plays during fairs, carnival, and the end of Lent, organizing themselves like guilds with a "king" at their head. The members gathered regularly, often accompanied by good-hearted eating and drinking. By the sixteenth century they were writing and performing their own plays and verses centered mainly on classical allegories or —especially at the height of the Reformation—New and Old Testament stories. With the advent of printing, they also began to publish their verses. Some of the most famous ones of the sixteenth century were written by Anna Bins, who took her place in an Antwerp chamber.[16] Their more civic tasks were to provide Latin mottoes and addresses for formal occasions, such as the entrance of a great dignitary to their city. Municipal governments often granted them annual subsidies or other support, partly for their public functions and partly for their value as sources of edifying entertainment and even tourism.[17] By the sixteenth century, chambers of rhetoric also engaged in public competitions with one another: a certain chamber would circulate one or more questions and host a gathering for a kind of rhetorical tournament to see who could best answer, with one chamber emerging the winner (and leading the drinking bouts that concluded the events). It was said that the competition in Antwerp in 1561 brought together 1,893 rhetoricians, who spent more than one hundred thousand English pounds in drink and other entertainment.[18] Some of the plays put on by chambers of rhetoric betrayed evangelical sympathies, causing them to be targets of suspicion, although the Calvinist clergy often grew to dislike them, too, as seminaries of free thought.[19] They continued to flourish in the early seventeenth century, serving as important sources of education, furthering their members' ability for stylish expression, and further acquainting them (and their audiences) with classical and biblical themes.[20]

After three years of polishing his ability with Latin and Greek and his skills in disputation, at the age of twenty Clusius went on to the Collegium Trilingue at Louvain (or Leuven), the regional university, where the teaching of Hebrew was also available. During the period he spent there, from 1546 to 1548, he clearly took advantage of the fine Latin instruction of Petrus Nannius and the Greek of Hadrianus Amerotius, developing a fluent classical style. His focus, however, in accordance with his father's wishes, was on the study of law. Students of law had been early beneficiaries of the humanist movement, and jurists remained one of the most important constituencies for the new learning.[21] Indeed, the humanist revival of Roman law left a deep impression on legal methods in the low countries. Clusius obtained his law licentiate in his early twenties, two years after entering Louvain.

Before entering on his professional work, however, he and a friend, Johannes Edingus, set out on a study trip to the Lutheran university of Marburg. Young people had before them not only the precedents of legendary travelers like Odysseus and Alexander the Great but the more prosaic *peregrinatio academica*, the traveling for education that was the medieval equivalent of the classical search for teachers of wisdom. Clusius and Edingus set off with the goal of sitting at the feet of one the most famous German jurists of the day, Johannes Oldendorp. During that autumn of 1548, however, the professor was absent at Augsburg for a treaty conference acting on behalf of his prince, Philip the Bold, following the first major war of the Reformation within the Holy Roman Empire. (The Lutheran princes of the Schmalkaldic League had been fighting against the Catholic emperor, Charles V.) In Marburg, despite his apparently orthodox Catholic upbringing, Clusius lodged with one of the local professors, a Lutheran theologian, Andreas Hyperius.

It was still possible for a young scholar raised in abbey schools and a Catholic university to study at a Lutheran university without difficulty. The leading local authorities of the urban low countries showed a certain generosity of spirit in allowing for a variety of religious views as long as it did not lead to disturbing the peace. The clearest statement of their outlook was penned by Thomas Erastus in 1568 (his book was published posthumously in 1589), in which he argued that the civil authority should have supremacy over ecclesiastical questions. This created serious conflict with superior political authorities who wished to have their religious views imposed on all people. Even when in the 1540s the government of Charles V proclaimed stricter measures for the suppression of religious heresies, many magistrates resisted enforcing them against people who dissented from Rome but kept the peace. They were happy to be members of the broad and catholic Christian church but disliked the Inquisition and dogma imposed from

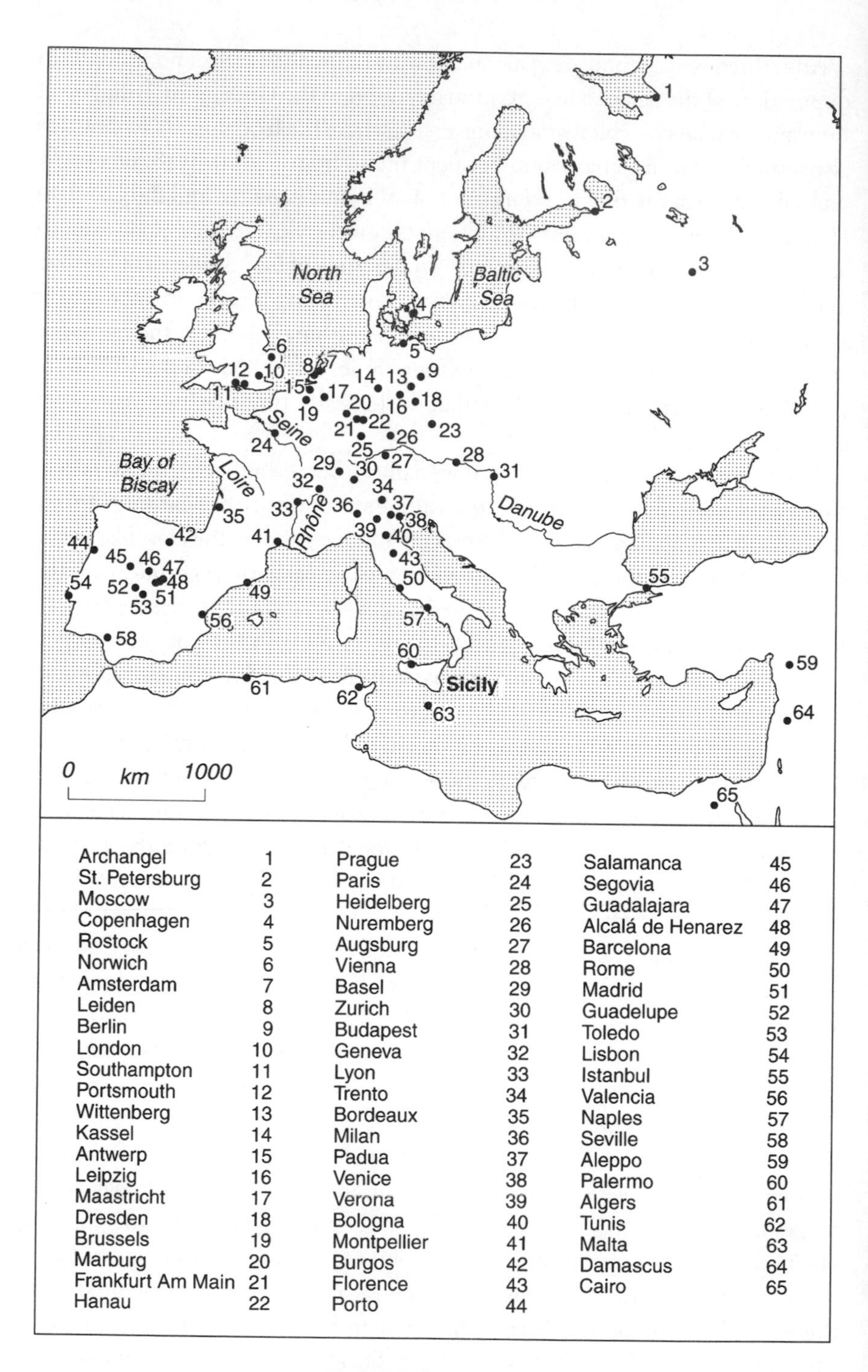

Early Modern Europe

Rome and viewed with suspicion the great numbers of monks begging in the streets, spreading "superstitious" dogma among ordinary people.[22]

But although the magistrates would readily allow discussion and debate on religious as well as other matters, they would not brook subversions of their juridical supremacy. They often disliked innovators in religion who, following the confrontation with Martin Luther at the Diet of Worms in 1521, wished to make their own ideas about theological doctrine supreme. The first public burnings of heretics who threatened the establishment occurred in Brussels in the summer of 1523, and at least thirteen hundred more would be arrested over the next three decades or more, many of them dying by judicial execution, often in the dark of night so as not to cause public disorder; the actual numbers might be several times as high. The most notorious group, the Anabaptists, openly advocated tearing down the contemporary system of worldly government to make way for a holy order: they observed adult baptism, taught that property should be held in common, believed that men and women were equal, sometimes engaged in lovemaking outside the legal bonds of marriage, worshipped God by speaking as the spirit moved them, and believed that Judgment Day was imminent. They had much success in spreading their message and in organizing underground churches in cities and towns of the low countries and nearby, briefly took over the city of Münster in 1534–35, and almost managed a similar revolution in Amsterdam.[23] Where they appeared openly, Anabaptists caused magistrates to react brutally against their radical version of the Reformation. Aside from the well-organized Anabaptists, there were few large groups who advocated insurrection to build the City of God on earth, but individuals and small groups who did so were confronted. For instance, the religious views of a learned humanist named John Calvin, a native of the southern low countries, at first seemed no threat to public order. Calvin set out powerful new ideas about God and salvation that were at odds with the doctrines of both Catholicism and Lutheranism. His views on matters such as the Mass spread widely in nearby regions, and because his works were based on the latest linguistic and textual studies, they were often held in high regard by the well educated. From 1541 he headed a theocratic regime in Geneva, making the city the bastion of his teachings but also a model of public order. It would not be until the 1560s that religious radicals in the low countries took up the banner of Calvin in their public confrontations with the Catholic establishment and not until then that "Calvinism" became a threat. The efforts of the magistrates were, then, directed mainly toward upholding the civil order over which they presided, not toward policing consciences so long as outward civilities were observed.

Clusius therefore apparently went along happily to the lectures of his housemaster Hyperius. He also went for walks with him examining the local flora. Indeed, the first certain evidence of his botanizing comes from a later note in one of his books about having collected a certain plant in the woods of Hesse when out with Hyperius, which would place the event during his period in Lutheran Marburg, about 1548.[24] Some months later, in the summer of 1549, Clusius and Edingus traveled on to more distant Wittenburg, where Clusius became part of the circle around the famous Lutheran humanist Philipp Melanchthon. There he would have been directly confronted with the view that only firsthand investigation of God's creation—not reasoning about it—could truly show his providence, for at about the time that Clusius arrived, Melanchthon had just published his lengthy textbook arguing this point, *Initia doctrinae physicae* ("Introduction to theories of nature," 1549). Like many other contemporary humanist natural philosophers, Melanchthon acknowledged medicine and natural history to be at the heart of new discussions in natural philosophy.[25]

Because the first clear evidence about Clusius exhibiting an interest in natural history and botany comes from the last years of the 1540s, during his period in Lutheran Marburg and Wittenburg, some have considered that German Lutheranism caused him to take an interest in nature. The keen enthusiasm for botany in the German-speaking world is often placed in a Protestant context, for the new and richly illustrated books on the subject by Otho Brunfels, Leonhard Fuchs, and Conrad Gesner were written by Protestants: Brunfels and Fuchs had converted to Lutheranism, while Gesner followed the religious reformer Huldrych Zwingli. Augustinian natural theology and an excitement about natural history certainly tinted Luther's views: he wrote that the flowers of the field were creatures of God and reflected their source, serving as lessons for humans, and he said that "we are now beginning to have the knowledge of the creatures which we lost in Adam's fall."[26] But in remarks such as this Luther was probably incorporating the current excitement about natural history into his devotional framework rather than expressing a theological opinion that was in any way at odds with other faiths. For similar reasons, texts by the Lutheran Melanchthon that were not explicitly theological might be used readily at Catholic German universities, such as Ingolstadt.[27] Consequently, although the first evidence of Clusius's active botanizing comes from Lutheran Marburg, we must be careful not to leap to conclusions, ascribing it to a theological position. He may have been prompted to pay attention to botany by natural piety, but that could be found in many places and among many creeds. If circumstances were otherwise, he might equally well have been captivated by the subject in Catholic Italy, France, or Spain.[28]

After a time at Wittenburg, Clusius and Edingus soon decided to embark on medical studies in France, Clusius setting out for Montpellier (apparently on

The illustrators of Leonhard Fuchs's *De historia stirpium* (1542), depicted by themselves at the end of the volume. Courtesy of the Wellcome Library, London

the advice of Melanchthon), where he arrived after a leisurely year and a half of travel, botanizing along the way. When he arrived there in the autumn of 1551, the medical faculty at Montpellier was among those renowned for humanist initiatives and natural historical investigations. To modern eyes, its most famous member was someone who had completely absorbed the new excitement about matters of fact, François Rabelais. Clusius resided in Montpellier with one of Rabelais's colleagues, Guillaume Rondelet, who became one of the most celebrated natural historians of the day. Rondelet seems to have been open to all kinds of religious ideas, since he burned his theological books in 1552 when his friend, Bishop Pellicier, was imprisoned on suspicion about his opinions. Clusius became one of Rondelet's chief assistants during the writing of his grand work on fishes of the sea, *Libri de piscibus marinis.* The professor enlisted the enthusiasm of local fishermen, who would bring him interesting specimens dead and alive; for the living fish, he built large tanks fed by piped water so that he could observe them. Clusius put his talents to work first in helping Rondelet to collect material and information and then by helping to arrange the information in a

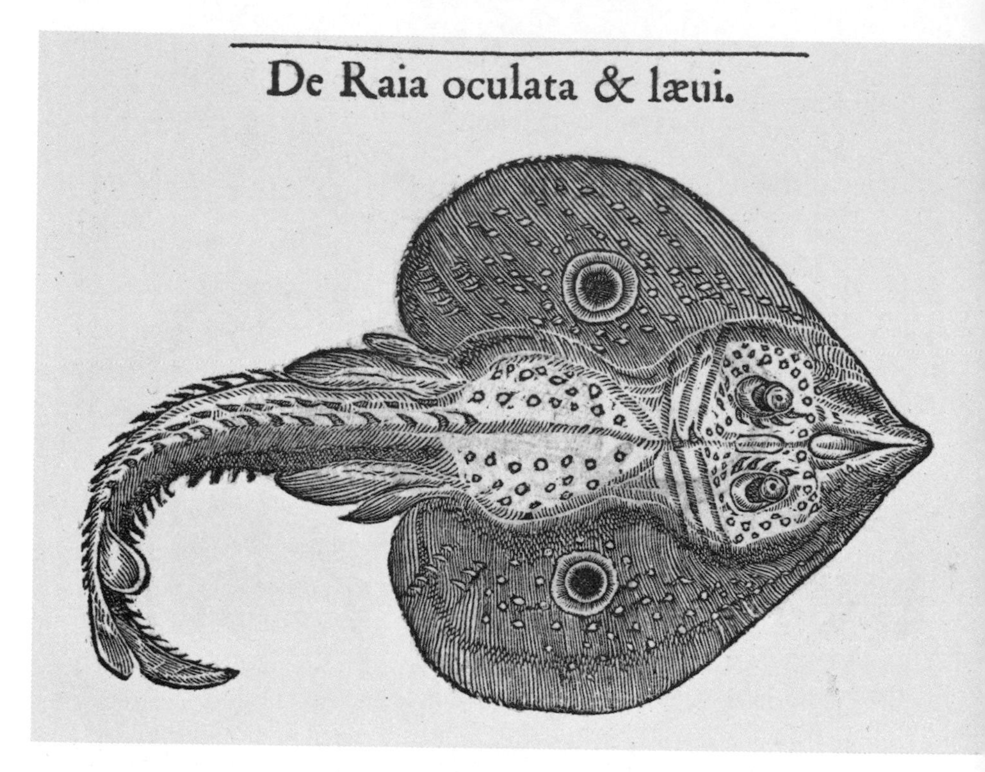

A ray, from Guillaume Rondelet, *Libri de piscibus marinis* (1554–55).
Courtesy of the Wellcome Library, London

sensible order and turning the results into polished Latin. Rondelet thanked Clusius in print when the book was finally published in the spring of 1555.[29] Clusius also traveled about the region of southern France, studying plants as he went. His time in nominally Catholic but increasingly Calvinist Montpellier clearly confirmed in Clusius his lifelong commitment to the exciting new study of natural history.

At the beginning of 1554, with the last proofs for Rondelet's soon-to-be famous book delivered to the printer, Clusius started on a long peregrination back to the low countries, spending the next five years in and around Antwerp with a short trip back to Louvain. His return may well have been prompted by the war between the ruler of the low countries, Charles V, and the king of France, Henri II. When Charles abdicated in 1555, his son and heir, Philip, returned to the low countries from England to take over the reins of government and oversee the war. The war between the Hapsburgs and Valois was finally brought to an end in April 1559, and in the same year Philip departed for Madrid to take up the reins of power there, leaving his illegitimate half-sister, Margaret of Parma, in charge of the low countries. Some months after the conclusion of the war, in 1560, Clusius moved to Paris, where he again studied medicine, perhaps also encountering famous intellectuals like the philosopher Petrus Ramus and the poet Pierre de Ronsard. After hiring himself out as a tutor and guide, he traveled to Catholic Cologne and Augsburg, where he arranged to tutor the son of the enormously wealthy (and Catholic) merchant family of Fugger, taking him to Spain, where the family firm had offices. He and the young Jacob Fugger traveled for more than a year in the Iberian Peninsula. He met up with the young Fugger in Antwerp in early 1564, then left for Paris and traveled on down the west coast of France to Spain, moving on to Burgos, Valladolid, Salamanca, Ávila, and Madrid. They then turned north to Segovia and back via Guadarrama, then northeast to Alcalá de Henares and Guadalajara before returning to Madrid by the beginning of September; then they set out again from Madrid, this time west to Toledo, Guadalupe, Badajoz, and Lisbon, which they reached by the end of the month. They ventured on further travels from there, not long after the beginning of 1565 arriving in Seville, after which they toured the southeast coast of Spain before returning to Madrid in mid-April, where Clusius dropped off his pupil. He then made his way back to Paris and Antwerp, arriving in the first half of June. As always, he met many people with whom he had conversations and opportunities to inspect their libraries, gardens, and collections of curiosities; he put much effort into personally investigating the botany of the places through which they traveled; and he collected newly published books on the exotic botany of Asia and the New World. The fact that he traveled in Catholic lands went with out comment, despite the growing religious divisions in the low countries.

RELIGIOUS WAR, BOTANICAL SUCCESSES

On his return to the low countries, however, Clusius found religious and political tensions boiling over, and for the next several years he became involved with the political opposition to Hapsburg policies that rigorously compelled their subjects to adopt stringent new Catholic doctrines. The long-running Council of Trent had concluded deliberations in the summer of 1563 with firm and strict definitions of Catholic orthodoxy (while Calvinists at the same time codified their principles in the Heidelberg catechism), and consequent Hapsburg efforts to undergird the Catholic Church in the netherlands by installing new bishops and the Inquisition while also reinforcing its institutional and fiscal base evoked powerful opposition, which mainly fell on Margaret of Parma's chief minister, Antoine Perrenot, lord of Granvelle. He became so unpopular with the nobility of the region that Philip was forced to recall him.[30] But resentment at the centralizing political and religious innovations of their Hapsburg rulers continued. In protest, a large number of noblemen, gentlemen, and magistrates boycotted the regent's court. Many petitioned Margaret about their grievances in April 1566. In writing to a friend about this so-called Request, Clusius used the phrase *de nostris rebus* ("of our affairs"), identifying himself with the movement; he also translated the petition into Latin for circulation among friends in Germany. Such sentiments are also in evidence in a missive of May, in which he wrote about having received someone's letter *in mediis nostris motibus* ("in the midst of our efforts"). About the same time, he received a severe wound to his right leg, cut to the bone, suggestive of a knife or sword thrust, and was forced to remain in bed for June and July, although by mid-August he was hobbling about again.[31] By then, mobs of iconoclasts were smashing up churches and cathedrals throughout the low countries. As the end of 1566 approached, Clusius had become pessimistic that developments were spiraling out of control; he worried about the mass field conventicles and the spread of Anabaptism, ending a letter by saying that one had to trust in and pray to God and hope that the voice of the Evangelists and Jesus Christ would be preserved.[32]

To Clusius's early-twentieth-century Dutch biographer, F. W. T. Hunger, Clusius's behavior shows him working for the Calvinist interest, and he therefore interpreted many of Clusius's movements in light of his "anti-Catholicism."[33] More likely, however, Clusius was a constitutional conservative, opposed to the policies of the Hapsburg rulers and possibly to the new doctrines of Trent but not to the traditional and broad church in which he had been raised. An even more interesting, related possibility exists: to a more recent historian of the political and religious troubles of the period, Clusius's anti-inquisitorial actions and personal sentiments may place him among the members of the Family of Love.[34] The Family of Love (Huis des Liefde, literally "House of the Loving") was a group interested in inner piety rather than outward show, anticipating an imminent cleansing of the world by God but in the meantime allowing members to confess any religion required by the authorities. The Familists were inspired by the spiritualist teachings of Hendrick Niclaes and (after 1573) Hendrik Jansen van Barrefelt (alias Hiël), who in turn had been deeply moved by late medieval works of piety such as the *Imitation of Christ* and more contemporary hopes and fears about the Second Coming. For Niclaes and his followers, being led by the spirit was as important as learning from the Bible and the Scriptures; with the Second Coming, no particular church would be supreme, but the saved of all faiths would live in harmony. Many Familists were wealthy and well-educated merchants who knew full well the negative consequences of religious conflict for business; many others supported the Erasmian tradition of scholarship, morality, and hope. It was, therefore, a movement trying to preserve a certain spiritual tradition within the catholic and universal church against the theological precisionism of Trent and the intolerance of the Inquisition.

But its members opposed Reformed Protestantism even more strongly. For instance, Niclaes taught with God's help, people could contribute to their own salvation, even to strive for perfection on earth, against the Protestant doctrine of justification by faith alone, and especially against the severe Calvinist views about grace and predestination that condemned all but a select few to damnation. For such reasons, "the main enemies of the Family of Love were Protestants," while Niclaes himself "abhorred the Reformation, considering Protestants as objectionable sectarians and schismatics, whose only effect had been the replacing of Catholic ceremonies by far worse ones."[35] In the mid-1560s, most known Familists were in Antwerp, and like Clusius, they were deeply involved in attempts to resolve the uprising peacefully. For instance, the final draft of the petition of the autumn of 1566, demanding that the Brussels government grant religious liberty in return for a levy of three million gold florins, was completed in the house of the brother of the wealthy Familist Luís Pérez (who was also a *marrano*, that is, of a formerly openly Jewish family).[36] When the Calvinist uprising was succeeded by a hardening of positions and the calling up of the Spanish army from Italy, many Familists (like others) left for Cologne and other refuges abroad. The Antwerp Familists might then be said to represent a broad-church Catholicism working for the reunion of humanity.

Because their views were so closely guarded, it is hard to be certain about whether people like Clusius were Familists, but the likelihood is great. The Antwerp publisher who issued his works was the Familist best known to history, Cristophe Plantin; Clusius also worked closely with the Familist mapmaker Abraham Ortelius (whose published maps of southern France and the Iberian Peninsula came from information provided by Clusius); and one of Clusius's most famous later friends and correspondents was the philosopher Justus Lipsius, who was also deeply involved with the group. The Antwerp Familists also had sympathizers at the Brussels court, most notably the Spaniard Benito Arias Montano, another of Clusius's friends. Montano arrived in Antwerp in 1568 to direct the production of the Polyglot Bible (which printed the various books in the main five original languages), a project Plantin had proposed and for which he did the printing; by the end of his stay in Antwerp (1575), Montano himself had become a Familist.[37] Late in his life, when he wrote to Clusius from Seville in a moment of homesickness for old Antwerp, he called Clusius his "best and most sincere friend."[38] Moreover, the Antwerp circle around Plantin and Montano, which included

Clusius, was in close touch with the universalist scholar Guillaume Postel, who advocated the unity and redemption of all humanity through reconciliation.[39] Even Clusius's later period in Vienna might be viewed as in keeping with the Familist program, for until later in the reign of Rudolph II, the Austrian Hapsburgs supported the search for universalism on the assumption that through it they would fulfill their destiny of reuniting Christendom and perhaps even laying the foundations for becoming universal monarchs over the whole world.[40] (Other contemporary monarchs held similar opinions about their own historical destinies.) At the same time, Clusius was in touch with people associated with the Czech Brethren, another well-educated irenic group.[41] In any case, he certainly belonged to the large constituency, perhaps the majority of people, whom historians of the low countries have come to recognize as preferring not to take sides in the religio-political conflict unless forced to do so, being more interested in things other than precision of doctrine (and so deeply disliking the Inquisition).

Tellingly, Clusius's reputation as an exacting botanist was secured during this period of severe political conflict. In 1567, he published an edition of a book on the exotic plants of Asia first written in Portuguese by Garcia da Orta, the *Colóquios dos simples e drogas e cousas mediçinais da India* ("Colloquies on the simples and drugs of India") of 1563.[42] Orta had gathered information on the botany of the East Indies from observation supplemented by all kinds of reading and conversations with local people and visitors. He also employed others to send him information, plants, and seeds from further away. He had come from a family of New Christians who settled in Portugal after the expulsion of the Jews from Spain in 1492 (his father had apparently been forcibly baptized in 1497). Probably beginning around 1515, Orta traveled to Spain to study arts, philosophy, and medicine at the universities of Salamanca and Alcalá de Henares. He returned to his hometown in Portugal in 1523 before taking up residence in Lisbon three years later, where from 1530 he managed slowly to work his way into the professoriate. When the establishment of the Portuguese Inquisition began to make life increasingly uncomfortable for New Christians, however, Orta took ship for Goa in 1534 along with his friend and patron Martim Affonso de Sousa, captain-major of the Indian Ocean (afterward governor-general of Portuguese Asia). Orta accompanied Sousa on several military campaigns along the west coast of India and settled in the burgeoning capital city of Goa rather than returning with Sousa to Portugal in 1538. During this period, Orta not only served as personal physician to many of the viceroys and governors of Goa but became the personal physician and friend of the sultan of the powerful neighboring state of Ahmandnagar, Burhān Nizām Shāh, and from various sources and merchant ventures—including the collection and transportation to Europe of Asian spices and materia medica—he became quite wealthy. Because of threats from the Portuguese Inquisition, two of his sisters with their families and his mother also came to settle in Goa. Until his death in 1568, Orta acted publicly as an orthodox Catholic and cultivated friendships with Franciscans, Dominicans, and Jesuits, but it was probably his well-placed friends who protected him. Almost immediately after his death, inquisitors in Goa forced one of his brothers-in-law to testify that Orta had remained as faithful to Jewish law and custom as possible, then arrested, tortured, and persecuted several other relatives of his, even burning one of his sisters at the stake in 1569 "as an impenitent Jewess" and exhuming Orta's body in 1580 and burning it in an auto-da-fé.[43]

In Goa, five years before his death, Orta published the *Colóquios*, dedicating it to his friend and patron Sousa and including in it the first poem in print by his acquaintance Luis de Camões (whose "Luciads" are considered one of the greatest examples of lyric poetry of the sixteenth century). The work's method is ingenious in presenting the best of the two schools of thought in which Orta had been educated: the philological and the factual. It takes the form of a dialogue between Orta and his fictionalized alter ego, one Dr. Ruano, who is said to have studied with Orta at Salamanca and Alcalá. The character's name shows Orta's humanist upbringing, for it was a play on words suggesting the bettering of the famous ancient Greek *Materia medica* of Dioscorides: Orta had studied at Salamanca and Alcalá with Elio Antonio de Nebrija (or Lebrija), who had in turn become a committed humanist during his peregrinations in northern Italy in the later fifteenth century and introduced to Spain the Latin and Greek humanist program in mathematics, astronomy, geography, and natural history. Nebrija had taken the new humanist edition of Dioscorides by Jean de la Ruelle and added a commentary and an alphabetical lexicon of about two thousand Spanish terms for the names that occurred in Dioscorides' text. Ruelle's name meant "of the narrow street" in French, while "Ruano" means "related to the street" in Spanish.[44] Orta's character Dr. Ruano therefore is undoubtedly a personification of the modern editors of Dioscorides, whom Orta spent his work correcting.

In Orta's dialogue, the character Ruano begins by asking to know the names of all the medicinal plants in all languages, as well as their local uses, and in each subsequent colloquy Ruano questions Orta and advances quotations from classical and Islamic sources. Orta's character plays the part of the new natural historian: he gives answers based on knowledge of the things themselves. For instance, in the chapter on cardamom, Ruano begins by posing a knotty problem: "How can it be explained that the use of *Cardamomo, mayor y menor*, as we now use it in Europe, is not in conformity with the teaching of Galen, Pliny, nor Dioscorides?" Orta replies that he will let Ruano "see it as clear as the light of noon." He refers to the Arabic writings on cardamom, lists the names for it in Malabar, Ceylon, Bengal, Gujerat, and the Deccan, explains that neither the ancient Greeks nor Latins knew it, and says that the famous medieval translator Gerard of Cremona confused things further when converting some of the Arabic passages into Latin. To show how little Pliny knew, "I will ask you to look at one," then asking a "boy" to pick a black cardamom from the garden in which the conversation was set. The discussion goes on with each authority put and contradicted by Orta's evidence.[45] Although the element of this discussion that is most apparent is Orta's use of his own and contemporaries' experiences with things themselves, it is also clear that he drew heavily on texts as well, especially ones about the drugs and plants of South Asia. Such works had been produced in many languages, including from at least the ninth century in Arabic, and Orta knew many of the Arabic texts from their translation in Latin. He had also taken the opportunity while in Goa to learn some Arabic, since his transcriptions of many words show his contact with native speakers.[46] More generally, his judgments privilege descriptive detail. He preferred to write of things he had seen, although when that was not possible he had to resort to conversations with other people as credible sources of information. He had many opportunities to meet a variety of knowledgeable people of different backgrounds, places, and experiences, including learned Hindus and Muslims. In other words, Orta used direct experience as a potent method of contradicting other "experts," although in building a case for positive knowledge he also had to rely on others he considered credible. It was an impressive work of botany and materia medica.

Because Orta's book had been printed in Goa it was scarce in Europe, but within nine months of its publication, Clusius had obtained a copy during his travels in Portugal when tutoring Jacob Fugger.[47] On returning to the low countries, during the growing troubles of the winter of 1565–66, Clusius set out to translate and edit it. At the end of each translated chapter he added his own comment, which he clearly distinguished from the original with an asterisk and a different typeface. In his annotations, Clusius added his own descriptive detail, including where he had observed the object himself, and used and cited all other printed accounts that he could find about the item.[48] It was a remarkably able demonstration of his humanist skills in linguistic facility, editorial discipline, collation of sources, and careful observation. After bureaucratic delays caused both by the contemporary political upheavals and by his severe leg wound, he finally obtained a license to print the book in September 1566; the drawings of the objects he commissioned from Peter van der Borcht, and the woodcuts made from the drawings by Arnolt Nicolay were complete by October. Clusius dated his dedication of the book to his former student Jacob Fugger on 13 December, and the book appeared at the Frankfurt book fair in early April 1567—within four years of its appearance in Goa—securing Clusius's reputation.[49] It went through several editions and was quickly translated into other languages. Clusius's edition meant that other authors, such as the Spaniard Christovão da Costa could base their own work on Orta's even when they had no access to the original.[50] Clusius later issued another version of Orta, the *Exoticorum libri decem* (1605), to which he added a Latin version of Costa's book as well as the work of Nicolás Monardes on the medicines of the New World, which he had encountered on a trip to England after having previously met Monardes in Spain[51] (Joseph Justus Scaliger corrected the Arabic in this edition).[52] Throughout, even when relying on other authors, Clusius was investigating nature through true descriptions of phenomena rather than speculations about their causes.

About the time the first edition was printed, however, the Duke of Alva arrived in the north. In addition to the iconoclastic fury, further petitions in 1566 and 1567 had demanded that the Brussels government grant religious liberty in return for tax revenue, which the Hapsburgs considered blatant subversion. Margaret managed to weather the storms in office, but political maneuvering at the Spanish court caused Philip II to decide to end all opposition with a demonstration of overwhelming force. He sent his most accomplished general, Alva, and some of his best troops from Italy to the low countries, where they arrived in August 1567. The Iron Duke waged a brutal campaign against groups and persons opposed to Margaret and Philip, sometimes massacring whole towns that held out in order to cow anyone even considering disloyalty. Clusius found a protector in his friend Jean de Brancion, a member of the Hapsburg court, who had a house and garden in Mechelen (Malines), between Antwerp and Brussels. Since 1455, Mechelen had been the seat of the supreme court of the low countries, not only serving as a court of last appeal from all the provinces but seeing to administrative and fiscal affairs—like the Parlement de Paris—making it quite possible that Clusius was continuing to act as a lawyer there.

But his period in Mechelen would once more prove to be fortunate for Clusius. Many gardeners of the low countries exhibited an enthusiasm for exotic plants, establishing ornamental gardens that impressed by the variety and rarity of specimens on display. As elsewhere, among the first to grow exotics were the apothecaries: in Antwerp, for instance, the apothecary Pieter van Coudenberghe became famous for his garden, which contained hundreds of different plants around 1548; by 1558 his garden was reported to contain four hundred exotic species, even reaching six hundred by 1568.[53] When Philip II set out to construct the lawns, avenues, and flowerbeds for his palace at Aranjuez in 1561, then, he imported twenty-four Flemish gardeners and brought in crates of plants from the

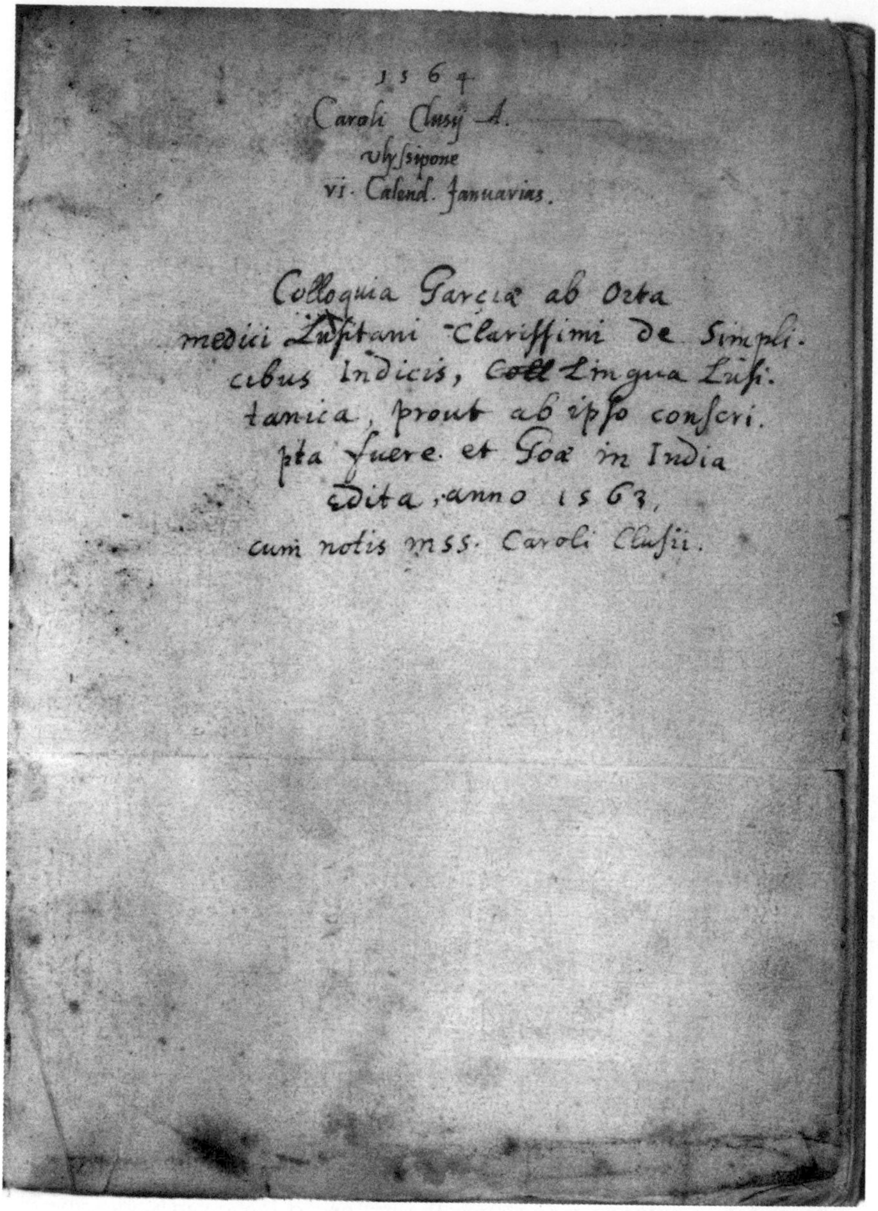
1564
Caroli Clusij A.
Ulyssipone
VI. Calend. Januarius.

Colloquia Garciæ ab Orta
medici Lusitani Clarissimi de Simpli-
cibus Indicis, ~~Coll~~ Lingua Lusi-
tanica, prout ab ipso conscri-
pta fuere et Goæ in India
edita, anno 1563,
cum notis mss. Caroli Clusii.

Inscription in the hand of Clusius, inside the front cover of his copy of Garcia da Orta, *Colóquios dos simples* (1563). By permission of Cambridge University Library

low countries.[54] In England, too, a great deal of information about new plants for gardens and agriculture arrived from Flanders and elsewhere in the low countries.[55] As it happened, the local city physician of Mechelen was Rembert Dodoens. His popular vernacular herbal, the *Cruydeboeck* of 1554, had been dedicated to Mary of Hungary, governor of the low countries on behalf of Charles V. He may have further encouraged Clusius's interest in natural history.[56] More significantly, it was in Mechelen that Clusius apparently first encountered a specimen of a flowering plant with which his name would ever after be associated: the tulip. In the early 1560s, a sack of bulbs arrived in Antwerp. They had been added to a shipment of cloth from Istanbul and were mistaken for onions and mostly eaten, although a few ended up in a vegetable garden, from which some were saved by a merchant of Mechelen, Joris Reye, who had a keen interest in ornamental plants. It was Clusius, then living in Mechelen, who told that story, and it is therefore likely that he or Brancion possessed some bulbs from Reye's stock: a feathered tulip was later named after Brancion.[57]

Clusius may have continued to be politically active during this time. Alva's violent actions were superficially successful in suppressing most open opposition to the government at Brussels, but underneath they created bitter resentment, which coalesced around a band nobles, soldiers, sailors, and rabble proudly calling themselves "Beggars" (from the earlier petitions) led by William the Silent, Prince of Orange and titular leader (stadholder) of the province of Holland.[58] By the end of the 1560s, they had been forced to the margins of the North Sea, including ports in England, remaining beyond the Iron Duke's reach. In 1571, Clusius made his second trip across the Channel (he had traveled to England as a tutor to some young gentlemen in 1561). There he met various apothecaries and botanists and acquired a copy of the just-published second volume of Nicolás Monardes's work on the medicinal plants of New Spain. He also acquired the (sweet) potato, which was among the exotics acquired by Sir Francis Drake—Clusius would become a great advocate of this new food. When Clusius returned from England via France, however, he was arrested crossing into Hapsburg territory for trying to bring across a great deal of concealed gold and silver money. He gained release following the confiscation of his goods and the payment of a fine.[59]

It would not be long before he fled the low countries. By the time of Clusius's return, new taxes had been introduced, causing many businesses to shut down and even forcing Clusius to set aside the printing of the flora of Spain on which he had been working. Widespread violence soon broke out again, too. In 1572, Queen Elizabeth of England was compelled to expel the armed ships of the Beggars from English ports; they in turn captured the city of Brielle in Zeeland, and from there the followers of the Prince of Orange began reclaiming the northwestern provinces of Zeeland and Holland. Many city militias went over to the militants, and religious exiles poured back into the provinces to bolster the opposition. Mechelen was one of the first cities south of the Maas to go over to the Orangists. But at the beginning of October the Duke of Alva appeared at its doors. Although the Orangists fled in advance and the rest of the citizens threw open the gates to Alva as a show of loyalty, the duke decided to make an example of the place: he allowed his troops to sack the city and massacre its inhabitants as a warning to other towns that entertained the Orangists.[60] Clusius managed to survive these troubles, but with the death of his father early in 1573 releasing him from the obligation to stay nearby and with the local military situation deteriorating rapidly, Clusius arranged through Busbecq to obtain a passport to the (Catholic) emperor's court in Vienna, where he arrived in early November.[61]

Shortly after arriving in Vienna, Clusius became a member of the staff of Maximilian II, charged with establishing an imperial medicinal and botanical garden. In the spring of 1574, during an uncertain period of peace between the Ottoman and Hapsburg Empires, he also traveled in lower Austria and Hungary, exploring the botany of the region. With the help of Baron Balthasar de Bathyány, he was able to employ an artist to paint the mushrooms of Pannonia.[62] He was also able to obtain many more varieties of tulip from Ottoman contacts, including a pasha in Budapest.[63] Otherwise, he remained for many years in Vienna except for another trip to England in 1579 and short trips to the low countries in 1580 and 1581.

In Vienna, Clusius lived among many scholars and artists who self-consciously dealt in the Mannerist outlook, which emphasized material detail signifying occult powers.[64] The label Mannerism is meant to characterize an attitude imbued with carefully detailed naturalism coupled with irregularity, drama, suggestion, intellectuality, and virtuosity. The emblem book was perhaps its most representative product.[65] Inspired by Egyptian hieroglyphics, which were in turn thought to reveal powers and meanings directly to the mind, the emblem book printed images, usually accompanied by allusive and explanatory texts, which revealed hidden correspondences to those who had the proper insight. One of the most notable emblem books was written by the Hungarian Joannes Sambucus (János Zsámboky), who obtained a place at the court of the Holy Roman Emperor in Prague and became one of Clusius's acquaintances.[66] It should be noted that in Mannerist emblems, objects and representations of them took precedence over the word—the word helped to elaborate ineffable knowledge that was contained in the objects themselves but did not replace their bodily multiplicity of meaning. For instance, an exquisite work of calligraphy that had been made by Georg Bocskay in 1561–62 for Emperor Ferdinand I was amended by the illuminator and miniaturist Joris Hoefnagel in the early 1590s for Emperor Rudolph II:

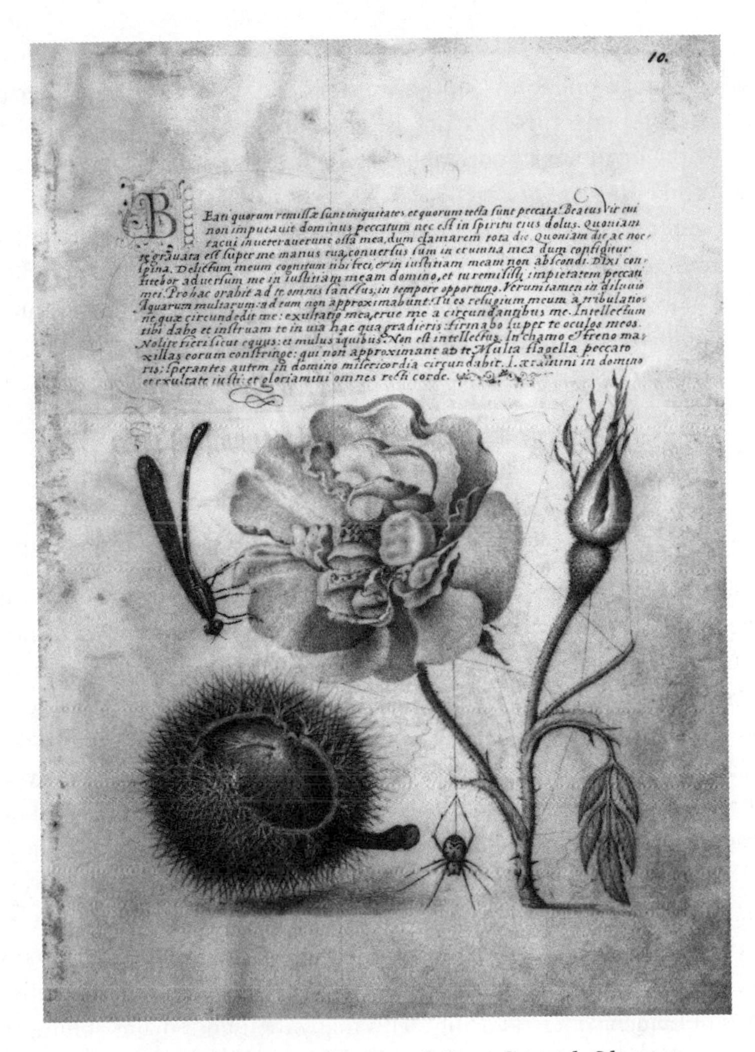

Joris Hoefnagel, *Damselfly, French Rose, Spanish Chestnut, and Spider*, painted 1591–96 in a work of calligraphy by Georg Bocskay, Mira Calligraphiae Monumenta of 1561–62. The J. Paul Getty Museum, Los Angeles

he added superb depictions of flowers, insects, and other *naturalia* "to demonstrate the superior power of images over written words."[67] William Ashworth has therefore stressed the "emblematic" aspects of contemporary natural history.[68] Such connections are further evidence for how humanity's highest quest was becoming "the elevation of things into a spiritual realm."[69] Many of Clusius's English acquaintances, such as Philip Sidney, Sir Walter Raleigh, and John Dee, possessed a similar outlook.[70] But in his public expressions, at least, Clusius remained far more concerned with precise description than with hidden forces. His original works, including important books on the botany of the Iberian Peninsula (in 1576) and Hungary (in 1583), gave exact reports of such things as the pistils, stamens, and even pollen of the flowers, using precise language and accompanied by fine illustrations, without explicit comment on sympathetic powers or any other causes of things—not even comments on the strength of the four qualities (hot, cold, wet, dry) by which the medicinal virtues of herbs had been classified for many centuries. Like other botanists of his generation, Clusius found more than enough interest in objectivity, in the exact description of the diverse creatures found in nature. The rest is silence.

A NEW UNIVERSITY AND THE STUDY OF NATURE

Clusius left for Vienna just about the time the Prince of Orange was turning the tide against the Hapsburgs. When Clusius returned to the low countries, it would be only after another twenty years, at the invitation of his gardener friends, to take up a position in the medical faculty of the newly founded university in Leiden. The university was a reward to a great and loyal city. At the end of 1573, as the Hapsburg troops were successfully retaking the provinces of the low countries, having just recaptured nearby Haarlem after a prolonged siege, they dug in outside Leiden, the second largest city in the province of Holland. They were forced to move away in the spring of 1574 to counter the arrival in the eastern provinces of German and French forces supporting William of Orange. But as soon as the Hapsburgs had seen off their opponents, they returned to lay siege to the city again. During this second siege, lasting from April until 3 October, the residents of Leiden suffered terribly, with many thousands dying of hunger. Because of resolute leadership the city nevertheless held out for William until his troops broke the dikes and sailed in over flooded fields on a wind that seemed to many to be providential.[71] By the time William's forces arrived, there was hardly a person left with enough strength to stand and greet them. But the relief of Leiden struck a major blow to Hapsburg power in Holland and Zeeland, forcing the Hapsburgs to peace negotiations at Breda. To strengthen his hand in negotiations further, William of Orange and the States of Holland and Zeeland decided to set up a new university.

The establishment of a university was an act of sovereignty, demonstrating to the world the newfound independence of the two provinces.[72] Moreover, the nearby ancient universities of Cologne and Louvain, as well as the university at

Douai, which had been founded by Philip in 1562, were bastions of Counter-Reformation Catholicism in now-alien territories where it might be dangerous for the sons of the Orangists to study. There was a need for a university that could inculcate in the local youth the values of free humanistic inquiry; it would succeed even more if it could attract the sons of others, too. After considering various options, it was decided to settle the new university in Leiden, partly as a reward for its heroic loyalty in the Orangist cause: the only larger city, Amsterdam, still held out for the Brussels government. (The often-told story that William offered Leiden a choice between exemption from taxes or a new university, and that the city chose the university, is the invention of the seventeenth-century poet and historian P. C. Hooft.)[73]

In keeping with the general concerns of William's council, the original ambitions of the university were not—as is sometimes said—simply to train Calvinist clergymen; rather, they were to educate a new generation of leaders for the developing nation while preserving religious liberty. Although many of those who took up arms for William were militant Calvinists, the Prince of Orange had led the fight in the name of liberty of conscience and the preservation of local rights and privileges against a powerful and centralizing government.[74] William's policies struck a compromise between people who opposed Hapsburg policies yet wished to remain members of a nondoctrinaire Catholic or Protestant church without hounding people of different faiths and people who were committed to spreading the Calvinist gospel and rooting out Catholicism. The former were widespread among the municipal councils of the low countries, but often silently obeyed the government in Brussels when faced with news that Alva was on his way; the latter were so determined in their opposition that they were far more willing to risk life and limb in taking up arms against the duke. The divisions between the two factions could be bitter. For instance, when a synod was convened in 1571 among the Dutch Reformed who were exiled in Emden, one of their leaders, Laurens Jacobsz. Reael, opposed the militants for trying to impose "a new popery" of dogma on them all.[75] For a time, at least, the old ideals of opposition to any strict ideology governed the policies of the Orangists. Their views would be embodied in the new university they founded at Leiden.

The Prince of Orange's intentions are clear from his letter to the States of Holland and Zeeland of 28 December 1574 putting forward his reasons for establishing a school: it would be "a firm support and sustenance of freedom and good legal administration of the country, not only in matters of religion, but also with regard to the general welfare of the people." The students would be trained "in both the right knowledge of God and all sorts of good, honourable and liberal arts and sciences, serving the legal administration" of the provinces.[76] He also wanted the university to become distinguished for its teaching in order to attract students from throughout the region and thus spread his ideals concerning liberty of conscience. In response, the States of Holland and Zeeland issued a charter for the university in the name of King Philip II on 6 January 1575. In the early years, the leader of the university's governors was Jan van der Does (or Dousa, lord of nearby Nordwijk), who had played a critical role in saving the city from the Hapsburgs after studying in the intellectually vital Paris of the late 1550s and early 1560s.[77] He was one of those who spoke stoutly for the Revolt as a defense of liberty. Liberty "was presented as the political value *par excellence*, the 'daughter of the Netherlands,' the source of prosperity and justice," which needed defense from Philip II. Like many others of his generation, too, Dousa held that liberty included not only freedom of "body and goods" but freedom of conscience and (sometimes) speech.[78] In a volume of poems published to celebrate the university's opening, Dousa praised the prince for having left the academy intellectually free for the cultivation of all the muses, for which reason students would come from everywhere and make it the chief place of learning of the age.[79]

The militants made some early attempts to make the new school conform to Calvinist doctrine, but they were resisted. After all, once freed from the yoke of one religious bureaucracy a great many people—more than half the population in quite a few cities—declined to become members of any particular confession.[80] For instance, despite suffering terribly from its siege by Hapsburg troops, Leiden's municipal government itself was not dogmatically Calvinist: the city's liberal Calvinist preacher, Caspar Coolhaes, once remarked that "it was not Calvin who died for our sins."[81] The militants complained mightily about such libertines, but for the moment the liberals had the upper hand.[82] Consequently, when in the university's first year a group of visiting Calvinist theologians proposed placing the academy under the jurisdiction of the local religious council (the consistory), the city government called the plans "inquisitorial" and fought back.[83] Moreover, given the prince's policy of favoring freedom of conscience, when in June 1575 he finally approved a governing structure for the university, he did not give control to the religious professionals. Instead, the new academy would be governed by three curators appointed for life by the States of Holland and Zeeland,[84] together with four burgomasters of Leiden who each served a four-year term. After 1578 there was explicitly no religious test for entry.

Teaching was organized according to the customs of the region. Seven years of education in Latin, language, and literature were expected before matriculation to the university. Students continued their studies in the "arts" faculty—usually called the "philosophical" faculty at Leiden—which gave more advanced instruction in language and literature as well as other subjects such as philosophy and mathematics.[85] When students completed their work, they received a baccalaureate degree, which in turn provided a basis for advanced study in the three higher faculties of theology, law, and medicine, which could in turn award doctoral degrees.[86] Students were free to follow courses in any faculty, although they needed to meet the requirements of the one in which they enrolled in order to receive a degree. Aside from the divinity students, who were supported on scholarships and therefore lived apart in the so-called States College, students rented rooms in town rather than living in residential colleges, so there were no religious or political house rules to govern them, either. Quite a few early students were Catholics, and university officials protected their right to study. Dousa and his fellow curators were intent on making their new school into a place of intellectual freedom and rigor rather than a school for sectarians. Nevertheless, it would be many decades before a degree from Leiden was recognized outside the northern Netherlands, since neither Catholic nor Lutheran sovereigns had an interest in acknowledging it. Philip II accused the school of promoting the Calvinist sect and in 1582 banned anyone from studying there.[87] The Catholic Church refused to acknowledge the university, and in 1603 Pope Clement VIII went so far as to excommunicate anyone who enrolled as a student in Leiden. In 1597, Huguenot king Henri IV recognized Leiden's degree in philosophy and civil law as valid in his kingdom, and his successor, Louis XIII, did the same for Leiden's medical doctors in 1624. But not until after the Peace of Münster of 1648, which formally recognized the Dutch Republic as an independent nation, was a Leiden degree recognized in most places in Europe. Until then, many Dutch students who could afford it continued to take their degrees at older, foreign universities, often after first studying in The Netherlands.[88]

The curators made strenuous efforts in the first few years of the university's founding to attract famous professors to the school without religious prejudice, and they had some successes. The first was Justus Lipsius, who joined the law faculty in 1578, from where he also taught history; he became deeply involved in developing a strong arts faculty.[89] Lipsius felt the turmoil of his own lifetime keenly, leaving his home in Overijse in 1576 for Louvain because of the mutinies among the unpaid Spanish troops that year, then fleeing further when the army under Don Juan advanced on Louvain (which was sacked), finally escaping from Antwerp to Leiden, which by then was north of the fighting. Lipsius had become well known for his 1574 edition of the work of the Roman historian Tacitus, and the interpretative works he produced while in Leiden inspired by Tacitean stoicism made him truly famous. He had formerly moved in the Lutheran and then Catholic worlds, and he was associated with members of the Family of Love, although he nominally became a Calvinist when he came to Leiden. Two other scholars well known in their own day but somewhat less eminent than Lipsius were also offered professorships: Bonaventure Vulcanus in Greek and Latin in 1581, and Thomas Sosius in Roman law in 1584. Neither man had to give up his Catholicism.[90] In 1585, the former Antwerp printer (and printer of Clusius's works), Christophe Plantin, came to Leiden to found a university press and wrote with pride to an old friend that members of the university did not have to convert to Calvinism, only obey the civil authorities.[91] After Lipsius left, another famous humanist—by reputation perhaps the greatest Protestant intellectual of his age—was recruited to the law faculty in his place: Joseph Justus Scaliger. Scaliger also taught Latin, classical studies, and history.[92] Although he had broken with the Catholic Church and vigorously opposed the new Jesuit order, Scaliger also remained undogmatic on theological matters.

Following the assassination of the Prince of Orange in 1584 by a Catholic fanatic, however, the ability of the students and faculty at Leiden to develop their views as they considered best came under attack by militant Calvinists and their political allies. While the nondoctrinaire libertines were coming to recognize that an independent state in the north had considerable benefits, the goal of many exiles and militant Calvinists remained to reclaim the whole of the Hapsburg Netherlands by force of arms. By 1579 the hope for a reunion of all the provinces had evaporated. The Union of Utrecht in that year combined the seven northern provinces, which officially became the United Provinces or, more colloquially, Holland, named after the largest and wealthiest one among them. In the southern low countries, antagonism to Hapsburg policies remained strong, but since most of the urban magistrates and militias remained loyal to the old faith, the Spaniards gradually retook most of them. William's assassination galvanized new energies for action, with the consequent rise in power of the ideologically committed.[93] Queen Elizabeth of England now agreed to support the Dutch cause—which soon led to war between England and Spain—and she sent her favorite, the Earl of Leicester, with English troops to lend a hand. The states believed that their cause needed a princely figurehead and accepted Leicester as governor-general of The Netherlands.

Leicester was initially hailed as a protector of Dutch liberties and a patron of letters. But he quickly decided that the merchant-princes of Holland were not interested in taking orders from him, nor were they as serious as the militant Calvinists about waging war to retake the southern provinces. Ensuing frictions developed between him and the States of Holland, with the Calvinist activists lending him their support. Leicester implicitly supported the rising of anti-Catholic guild and militiamen in the city of Utrecht in 1586, which made Utrecht a bulwark of militant Calvinism. At Leiden, some of the Calvinist theologians disliked the nondoctrinaire atmosphere of Dousa's early university, and plotted revolution. Chief among the Leiden militants was Adrianus Saravia, whom Leicester had reappointed rector of the university in 1586. Saravia wanted Leiden to follow Utrecht into the camp of the committed Calvinists. During the summer of 1586 a rumor circulated that he, as rector, would move the university to Utrecht, which was prevented when the city of Leiden responded with a polite but firm remonstrance. A year later, Leicester began planning a coup against the States of Holland that would have seized the two largest cities of Amsterdam and Leiden, and Saravia entered deep into his plots. The opening move in Leiden failed, although Leicester nevertheless tried to take Amsterdam before again being foiled. After this failure Leicester left for England, bringing this unseemly period to an end and allowing the libertines to maintain control for the moment.[94] Saravia also managed to flee to England and thereby to escape beheading. The curators and professors of the university who valued freedom of conscience no doubt breathed a sigh of relief, although for a decade Leiden remained a place sometimes unfriendly to English visitors.[95]

Such events, however, had created an atmosphere of crisis. The celebrated Lipsius threatened to leave Leiden and The Netherlands for places where doctrinal conformity did not threaten freedom of inquiry.[96] His book of 1584 on the consolations of Stoic philosophy, *De constantia*, had been written in the tradition of ancient authors like Boethius as well as in sympathy with such contemporaries as Montaigne, but the religiously committed had taken him to task for setting out a view that spoke of the antique Fortuna with hardly a word for the Christian God.[97] The failed coup determined him to leave, although negotiations with the curators and a few months of sick leave managed temporarily to smooth things over.[98] Lipsius's next book, *Politica*, was aimed not at personal conscience but at governing the behavior of rulers and leaders, and was equally rooted in antique philosophy rather than Christian religion. He drew heavily on the Roman Stoic Seneca, to show that subjects should be obedient and politically passive while pursuing their own inner life of contemplation according to the guidance of reason; and he drew on Tacitus to argue that rulers in turn needed to use power effectively to bring about the highest public goods, those of peace and security. There was nothing republican about his views.[99] Nevertheless, when it was published in 1589, *Politica* created tremendous controversy, both for its political passivism and for its dependence on classical sources at the expense of Christian ones. The States of Holland soon felt pressured to establish a committee to make recommendations on the university, and one recommendation returned in 1591 was that all professors subscribe to "the true Christian religion," that is, Calvinism.[100] Although the states did not adopt this recommendation, Lipsius left for

Catholic (and monarchical) Louvain: passive Stoicism apparently flourished at least as well in the south as in the north, perhaps even better.[101]

GARDENS AND ANATOMIES

One place that remained a bastion for investigations of nature without concern for religious conformity was the medical faculty. When the ceremonial inauguration of the university had taken place on 8 February 1575, the allegorical procession through the streets included a mounted representation of medicine, accompanied by four figures on foot identified as Hippocrates, Galen, Dioscorides, and Theophrastus; it may even be that Paracelsus was also meant to be included.[102] Certainly, the plan of study penned by Guillaume de Feugeray (Feugeraeus)—on leave from the Reformed university of Rouen to give advice on setting up the university—contained a section on medicine, stating that "the physician, as a most faithful servant of nature, must be shaped and instructed somewhat differently" from the two other higher faculties of theology and law. Although physicians must, like others, listen to lectures and spend time in formal speaking and disputing, after their first year they should move on to frequently "examining, dissecting, dissolving and transmuting the bodies of animals, vegetables and minerals." They also had to learn medical practice by following an honorable and learned doctor and examining and imitating his diagnoses and treatments of the sick. Only then would they truly deserve the title of *Doctor*, with "the dignity of preserving and restoring of health" through serving and investigating nature.[103] In other words, if Leiden were to live up to the highest ideals of medical education, students would have to attend not only the classroom, disputation hall, and library but also the anatomical theater, botanical garden, chemical laboratory, and clinic.

When the university opened, it had two medical professors on its lists, Pieter van Foreest (Forestus) from Delft, and the local Geraerdt de Bont (Bontius). There was little to do in the first years, however, since the first student in medicine (an Englishman entered in the register as "Jacobus Jaimes")[104] did not enroll until September 1578; about a month later, the first Dutch student matriculated, Rudolph Snell, who soon went over to teaching mathematics.[105] Forestus apparently never lectured. Bontius gained a full-time colleague in 1581 when Johannes van Heurne (Heurnius) was appointed. Despite the politico-religious crisis in Leiden of the later 1580s, it was then that clear evidence of regular medical lecturing occurs. How Heurnius and Bontius divided up the teaching at first is not certain, but during the winter of 1587 (shortly before Leicester's attempted coup) there is written evidence that Heurnius was lecturing on medical practice and

Bontius was teaching anatomy and prognosis according to Hippocrates, while five years later Heurnius was lecturing on Hippocrates and Bontius on practice according to Paul of Aegina.[106]

The Leiden school remained resolutely practical and descriptive in its academic orientation. In 1592 Heurnius published a textbook, *Institutiones medicinae* ("Medical institutes"), among other works, while Bontius published little and ordered his papers to be burned on his death, but both clearly showed their commitment to the new Hippocratism, which stressed careful observation and a reluctance to jump to conclusions about the causes of disease. In that way they were demonstrating their currency with the latest medical thought: by the 1570s, editions of Hippocrates were being published far more commonly than editions of the other great classical author, Galen.[107] The Leiden medical faculty remained committed to the Hippocratic approach for many decades, a sign of which was that students trying for the medical doctorate had to be examined on two of the Hippocratic aphorisms: the aphorisms would be assigned in the morning, and the intending doctor would then have to explain them in an hour-long disputation in the afternoon.[108] The practical and observational orientation of Hippocrates also encouraged the development of natural history as a part of medical study at Leiden.

The curators of the university therefore managed to promote one of their pet projects under the umbrella of the faculty of medicine: the establishment of a botanical garden. In the process, they also acquired an anatomy theater. As one might expect of well-to-do people, many of the curators and their associates were avid gardeners with an interest in naturalia. Perhaps some were even inspired by their professor Lipsius, who began to cultivate a garden immediately on arriving in Leiden and set the second book of *De constantia* in a garden, transforming the usual Epicurean theme of pleasure gardens into a Stoic view of the garden as a place of retreat and study.[109] For Lipsius and his like, who considered solitude necessary for developing the intellect, the garden, along with the "closet," was a place one could be alone with one's thoughts and books or where a few friends could gather for learned conversation.[110] He praised gardens as places of refuge from "cities and troublesom assembliees of people."[111] Botany might be a subject critical to the interpretation of nature, but it was also one of the great interests of well-to-do people and objective enough to escape the concerns of the dogmatists.

The curators therefore made appointing someone to head up a university garden a priority. In 1582 they brought on board someone truly famous: Rembert Dodoens. He had published on botany, including the vernacular *Cruydeboeck*, but had also produced many other studies, including a work on transubstantiation that showed him to be a good Catholic, and he had served as court physician to Holy Roman Emperor Maximilian II and his successor, Rudolph II, before returning to Antwerp in 1580. Because of their mutual interest in gardening, Dodoens had been on friendly terms with Lipsius for some years.[112] When he came to Leiden, there was no suggestion that he had to change his faith. But hopes that the sixty-six-year-old Dodoens would build up botanical teaching and add his fame to the young faculty ended with his death early in 1585. Dodoens's death was followed by the troubled period of Leicester and Saravia. Yet in April 1587, just before Leicester's attempted coup d'état, the curators took a major step by obtaining from the city a plot of open ground that adjoined the new academy building they had just acquired for lecturing (the former convent of the White Nuns). In the same month, the medical professor Bontius was given an increase in his salary to lecture on medicinal herbs in the summer and human anatomy in the winter. Bontius did little with the garden, perhaps because of the political upheavals that followed Leicester's departure but more probably because he had too many other obligations. A couple of months later, Lipsius approached an acquaintance and fellow Familist about becoming head of the hortus, but the renowned Clusius for the moment declined on grounds of being too elderly.[113] In 1589, Pieter Pauw (Pavius), a nephew of the powerful Amsterdam regent Reinier Pauw, therefore joined the faculty as a temporary ("extraordinary") botanical professor. One of Bontius's first pupils, Pauw had traveled and studied in the Baltic and Italy before returning to Leiden and had large intellectual ambitions and wide associations.[114]

Not only did Pauw bend his energies to developing the garden, but he also made the first sustained efforts to bring anatomical teaching to the university, which had also been envisaged from the beginning. Public interest in anatomy lessons was certainly high: the chancellor of the city of Arnhem wrote to the stadholder in 1566 that a "Doctor in medicine and surgery" had come to the city asking the authorities to supply him with a "corpse or body," in which case he would demonstrate the art of anatomy to all for use and edification.[115] By the middle of the sixteenth century, many surgeons' guilds also required those aspiring to be a master to learn human anatomy. In 1555, Philip II granted the Amsterdam surgeons the right to obtain the body of one executed criminal per year for dissection, a ceremonial occasion that first took place in the Sint Ursulaklooster (cloister of Saint Ursula). It followed an instance in 1550 in which the Amsterdam magistrates presented the body of an executed thief to the surgeons for dissection, in the course of which the victim's skin was carefully removed and tanned to hang permanently in the city guildhall as a warning to other lawbreakers, which led the Amsterdam surgeons to be called "people skinners."[116] The surgeons of Leiden (like the surgeons in Amsterdam after 1619) conducted their public dissections over the Waag, the main building in which market officials presided over the official weighing and measuring of goods; in Delft, Forestus dissected bodies of those who died in the hospital in which he held an appointment, and regular public anatomical lessons occurred under the auspices of the surgeons' guild.[117] And the work of Vesalius at Padua had become well known throughout the learned world.

Having become friendly with one of Vesalius's successors at Padua, Fabricius ab Aquapendente, and being eager to bring the latest medical learning to the northern netherlands, Pauw began giving public demonstrations of the fabric of the human body immediately after his arrival, in the winter of 1589, at the same time that the garden was being organized. (In an age without chemical fixatives, the bodies putrefied, often making the practice of dissection hard on the nose and stomach, although decay was slowed in the cold winter months. Additionally, the depth of winter was a period associated with the pre-Lenten Carnival, when transgressive behavior tended to be tolerated more than usually.)[118] Pauw's public demonstrations were conducted in a side chapel of the former church that in 1577 had been the original location of the university, the Faliede Bagijnenkerk (Church of the Faille-Mantled Beguines). The church had proved too small for the university, and in 1581 the curators obtained the new academy building across the Rapenburg canal and down a few doors, next to which the garden was established, but the curators kept the old building, planning to set up a library on its top floor (which would have to wait on other priorities).[119] For a while, half its ground floor was left to the students for their fencing exercises and the other half was turned into an Anglican chapel for some of the foreigners. Pauw found that the former side chapel above not only had available space but windows on three sides, making the light suitable for his purposes, and this he took over for his anatomy lessons. At the end of the chapel, Pauw built a wall to set it apart from the rest of the building and started demonstrating each winter.

After becoming a regular professor in medicine in 1592, Pauw became an even more energetic anatomist, working to establish a better, theater-style construction for the lessons, which were very popular. He probably learned from his friends and acquaintances in Padua about the plans there of Hieronymous Fabricius to erect a permanent anatomy theater at his own expense, which was built in 1594.[120] In 1593 Pauw advanced plans for the erection of Leiden's own permanent theater in the former chapel, which was finally completed at the end of 1597. In the same year the States of Holland ordered one criminal executed during the winter months to be turned over to Leiden for anatomical study. In the permanent theater, a revolving table stood on one post in the middle of the room with six circular risers surrounding it, each of which had a railing so that the elevated onlookers could lean in and get a relatively close look without falling on

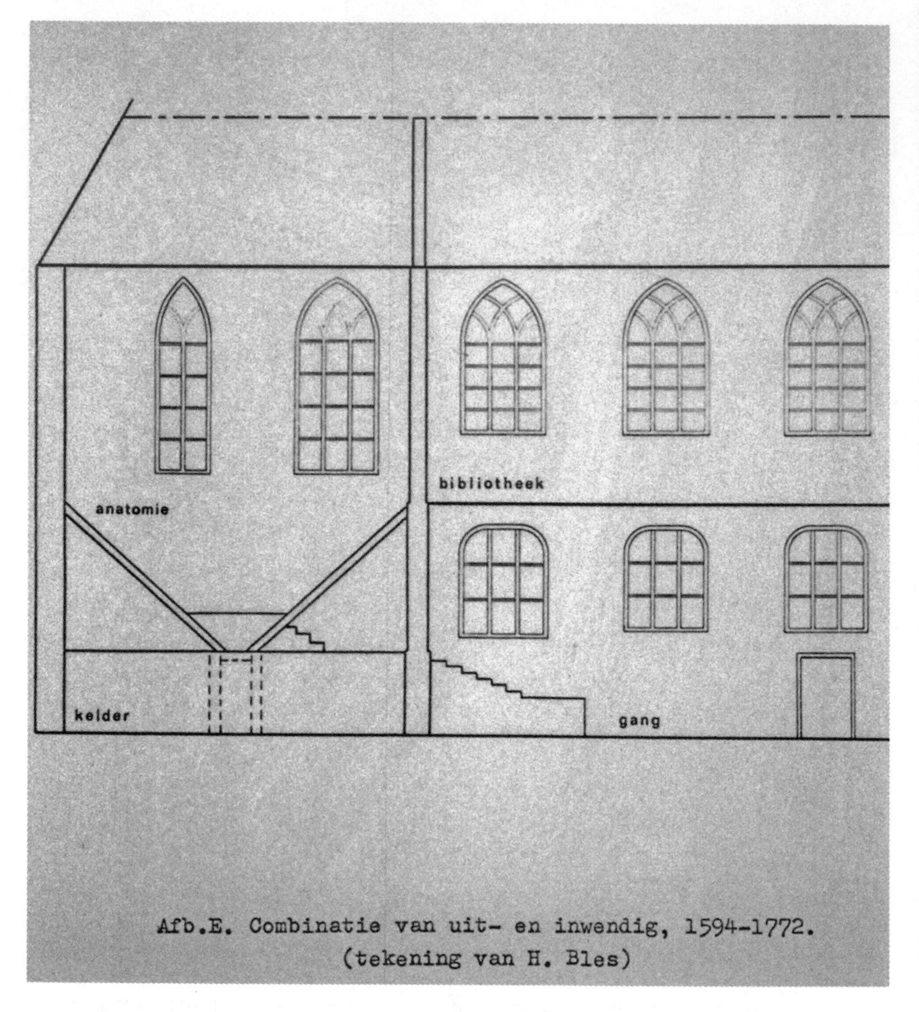

Reconstruction of the interior of Faliede Bagijnenkerk, including Pauw's anatomy theater on the left, library above right, and entry hall below right. By A. J. F. Gogelein, courtesy of the Rijksuniversiteit Leiden

those in front. When the public dissections occurred, the first and closest ring would be reserved for professors, curators, burgomasters, and other dignitaries; the next two would be reserved for the barber-surgeons and medical students; the three highest ones would be available for other people, including students in other faculties. Over the entrance was situated a cabinet in which surgical and medical instruments were displayed. Pauw had assistance in organizing the public anatomies from one of the members of the garden staff; an entrance fee was collected from viewers and the money was placed at his disposal in order to defray the costs.[121]

The curators considered Pauw's anatomical demonstrations to be so important that a bell was rung to announce them and other classes were suspended.[122] He trained a generation of students throughout the university in the complexities of nature through his anatomical demonstrations. Moreover, Pauw wrote a commentary on the first four books of the classical medical encyclopedist Celsus, reconsidered the opinion of Hippocrates on wounds, issued a revised edition of Vesalius's *Epitome*, and set out his own more complete study of the human skeleton, in which he made some especially important observations on the skull and its development.[123] He claimed that over nineteen years he dissected more than sixty human bodies, several of them children. As was the custom at Padua, where he had studied, Pauw also both dissected and vivisected animals and the fetuses of dogs.[124] In his correspondence with such contemporaries as Guilielmus Fabricius of Cologne, Geneva, and Lausanne, and Thomas Bartolinus of Copenhagen, he also reported various observations that they considered important enough to publish and to attribute to him.[125]

The curators did not give up their search for a famous, permanent professor for their planned botanical garden, however. In 1591 they tried to attract Berent ten Broecke, better known by his learned pen-name, Bernardus Paludanus, from Enkhuizen. Apparently born in Steenwyck in 1550, he may have attended the nearby renowned humanist Latin school in Zwolle. Perhaps, like so many others, he was caught up in the back draft of the Revolt of the later 1560s; in any case he took to the road in the later 1570s for further education and edification, moving through Germany, Poland, and Lithuania—areas associated with netherlandish trade in the Baltic—before heading south to the Republic of Venice and its University of Padua, where he enrolled in April 1578. There he was able to become immersed in the latest studies in anatomy, natural history, and medicine. In July he took ship to the East along the Venetian spice routes, to Syria, and then through Palestine to Egypt, returning by early 1579. Sometime that year he visited Rome and Naples, returning to Padua; in December he set out again for Naples and then Malta, stopping on his way back at Sicily in April 1580 and Rome in early May before returning again to Padua. In July he took his doctorate of philosophy and medicine. Traveling north along the trade routes to Innsbruck and Augsburg, then to Nuremberg, by the autumn he was in Leipzig and Waldenburg, where he served several months as physician-in-ordinary to the family of Prince Von Schönburg. He visited Dresden and other localities and thought of traveling to Santo Domingo in the Americas, where the Fuggers had business interests; but in February 1581 he left the court of Waldenburg and traveled to the Frankfurt fair, down to Strasbourg and then Stuttgart, Heidelberg, Cologne, Cassel, and Wolfenbüttel before finally heading for Hamburg and from there sailing back to The Netherlands.[126] Among the many naturalists he conversed with along the way was Clusius, whom he met in Vienna in 1577, on his way to Italy; they kept in close contact thereafter, exchanging specimens and information.[127] Paludanus first settled in the old trading center of Zwolle as city physician (being paid by the city to help the sick poor and advise it on all matters related to medicine and medical practice)[128] before moving to the growing port of Enkhuizen at the end of 1585, where in February 1586 he again secured the post of city physician.

On his travels, like Clusius and so many others before him, Paludanus made careful observations and assiduously collected acquaintances, information, and books. He also visited several cabinets of curiosity and sought out the kinds of objects worth collecting for himself, eventually becoming well known as one of the foremost collectors in the northern Netherlands. An inventory of Paludanus's early collection exists from 1592, compiled by one of the visitors who accompanied Friedrich, later Duke of Würtemberg-Teck, on a visit, which was partly published in 1604. In various tables, the inventory lists minerals and coins (by far the largest section), followed by dried plants and animal parts, and several hundred examples of human handiwork (*artificialia*).[129] Among the objects listed are various types of earths (such as earth of Damas, which had the color of flesh and was reputed to have originated from the body of Adam), rocks and marbles, precious stones, medals, and tokens, various preserved fruits and grains, precious woods, dried fish and reptiles, many insects, the horns of a variety of animals, shells and corals, clothes and other objects used by "savages" and foreigners, objects of art done in ivory, mummies and funerary furnishings from Egypt, and jewelry, crowns, and other beautiful and precious handcrafted objects, including weapons.[130] The collection was obviously large and varied enough to attract the interest of visitors like the future duke. The impression of the importance of the collection is reinforced by the fact that in November 1593, the young Philipp Ludwig II of Hanau-Münzenberg, then studying in Leiden, made a trip to North Holland, visiting Paludanus and his collection. Afterward, he commented that his cabinet had made Paludanus "famous throughout all Europe and the world."[131]

The curators therefore wanted to attract Paludanus to their medical faculty, expecting him to bring them his collections of naturalia and artificialia, which would enhance the reputation of the university. In addition to being a notable collector of naturalia, Paludanus also had a good reputation as a botanist. During the negotiations with the Leiden curators, he sent them two ground plans depicting the grand botanical garden in Padua from his time there as a student, perhaps as a model for what he would want for Leiden.[132] Nevertheless, Paludanus declined the curators' invitation and remained in Enkhuizen, perhaps—as Paludanus said in excusing himself—because his wife preferred not to move.[133] More likely he found that becoming a professor, especially after the religious conflicts that had caused Lipsius's departure, could not hold a candle to remaining the city physician of a great commercial center. (As it happens, just one year later Jan Huygen van Linschoten returned to Enkhuizen from Portuguese Asia with a mass of new information, and Paludanus would become the virtual coauthor of the famous resulting book.)

Although they had failed to attract Paludanus, the curators made the best of local resources as they continued their search: they promoted Pauw to a regular professorship in botany and anatomy and hired a director of the hortus from nearby Delft, Dirk Outgers Cluyt (Clutius), friend and kinsman of the Delft physician and nominal Leiden professor Forestus. Cluyt was one of many excellent naturalist apothecaries, keeping a large shop on the Wijnhaven, behind which was a large and much-frequented garden where he raised costly plants. He had great energy as well as an excellent knowledge of materia medica, and he began plans for a comprehensive medical garden.[134]

The curators' next choice was Clusius. Like Paludanus, Clusius had earlier been reluctant to become a professor. Rather than take up Lipsius's offer to join the faculty in Leiden when he left Vienna for Frankfurt in 1588, aged sixty-two, Clusius had gratefully received an annual subsidy from the Landgrave of Hessen-Kassel, Duke William. But the secretary to the board of curators in Leiden, Johan van Hogelande, had entered into correspondence with Clusius about their common passion of gardening in earlier years.[135] For the same reason, Clusius also knew another influential person involved with the university: Marie de Brimeu, Princess of Chimay, who had been in correspondence with him since 1571.[136] Born to considerable wealth, Brimeu had been welcomed to The Netherlands as an eminent political refugee (her husband continued to support the Hapsburgs), and because her husband retained all her possessions, she received financial support from the States General. They even granted her a house in The Hague in 1593. She was on good terms with both Clusius and Paludanus because of their mutual interest in gardening, and she became closely involved in university affairs.[137] Hogelande had kept Clusius informed about the curators' plans for a botanical garden, including their offer to Paludanus, and he apparently also told Clusius of Paludanus's refusal. In December 1591 he offered Clusius the position. Immediately following was a letter from Brimeu strongly urging Clusius to come to Leiden.[138]

Clusius was emphatically a lawyer and botanist, however, not a physician, and the position of head of the botanical garden lay in the medical faculty and required the candidate to teach. He was obviously reluctant to join the medical faculty, and only at their meeting on 24 June 1592 did the curators have a reply from him with proposals they could discuss. He could not possibly come to Leiden until the autumn of 1593, he wrote, because the season for digging his bulbs and other plants had passed, and he demanded an honorarium of three hundred *rijksdaalders* (twelve hundred gilders) per year plus travel expenses and no teaching duties. These were heavy requests that, if met, would secure for the university a famous man only as honorary professor of botany. On the other hand, Clusius had developed a botanical garden for the emperor in Vienna, would share his enormous knowledge with local garden enthusiasts and the staff, and would bring along his valuable collection of plants. After considering the offer for a couple of months, the curators accepted his conditions, and Marie de Brimeu wrote him the same day urging him to come. With such persuasions and, shortly after, the death of Duke William (whose son and heir, Moritz, stopped his honorarium), Clusius decided to accept and prepared to move to Leiden the next spring. Before he could move, however, he suffered a terrible fall on Easter Sunday, 25 April, which kept him bedridden for ten weeks; it also left him ever after with a limp and an inability to dig or tend a garden. Determined nonetheless to help Clusius make the transition to Leiden, Brimeu took a house with a garden in the city and offered it to him for his private use. He and his specimens finally arrived in mid-October 1593, almost two years after negotiations had opened.[139]

Over the preceding summer, Cluyt had begun to lay out the public garden under Pauw's supervision, but Clusius was in no condition to do much after his arrival, and by spring 1594 it was clear that he would be unable to work in the new hortus himself. Moreover, probably because the famous professor was not teaching them, the students fomented a minor revolt. Hogelande had to write Clusius a letter demanding that he teach, but Clusius refused based on his understanding of his contract; Pauw was apparently also upset about Clusius's behavior, since Clusius described the younger man as arrogant and jealous.[140] Yet after his arrival in Leiden Clusius continued to be very busy dealing with botany enthusiasts, being flooded with requests from other gardeners for information and for specimens, and he in turn collected assiduously. He clearly gave most of his energies to cultivating his botanical friends and private garden and to writing. He also worked with the famous artist Jacques de Gheyn II, who made a series of beautiful and botanically accurate watercolors of individual flowers under Clusius's direction, stimulating the development of Dutch flower painting.[141]

It was, then, the knowledgeable apothecary Cluyt rather than the famous pro-

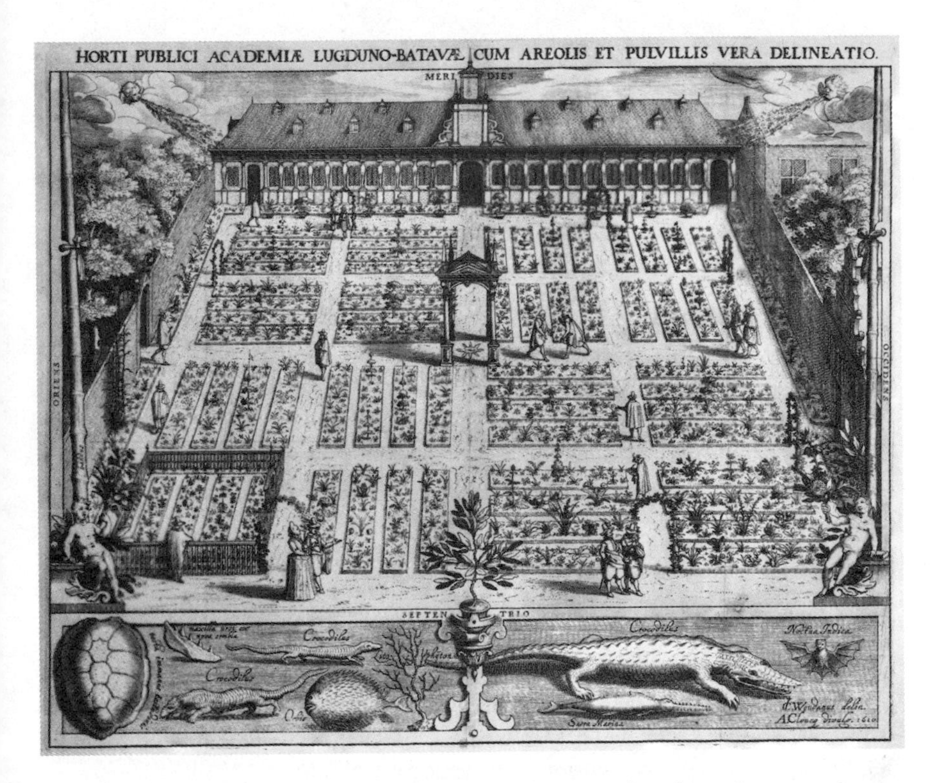

Illustration of the botanical garden of the University of Leiden, 1610; note the illustrations of specimens of natural history on display in the ambulacrum at the back of the garden. By permission of the Rijksmuseum

fessor who laid out and planted the hortus with a team of gardeners under Clusius's nominal direction in the summer of 1594.[142] Given Clusius's distractions, the curators had appointed Cluyt as Clusius's assistant on 2 May 1594 to do the work of developing the garden; six days later they made Cluyt the same offer they had made to Paludanus: he would be paid four hundred gilders per year to demonstrate the plants in the garden to the students in the summer and to teach them minerals and other naturalia in the winter.[143] By the end of September, Cluyt had finished the garden. It was divided into quarters, with three of the quarters containing sixteen beds and the other one twelve, with eighteen, twenty-six, or thirty-two plantings per bed (depending on the size of the specimens), for a possible total of fourteen hundred plantings; the inventory listed about 1,060 species. In keeping with the interests of the day, only about a third of the specimens had medical interest, the rest being exotics or ornamentals.[144] Although Clusius had been consulted for the original design, Pauw had apparently kept instruction in mind and persuaded Cluyt to narrow the beds, altering them from Clusius's sense of proper proportion but making them easier to view by the students.[145]

The garden also made possible teaching in natural history. When the seasons ended instruction in the garden, teaching continued indoors. Although the university had not acquired the spectacular collection it would have had Paludanus come, it did collect specimens ranging from minerals and tortoise shells to Egyptian mummies and crocodile hides. In 1599, a gallery with large glass windows was added to the south side of the hortus, allowing the wintering-over of delicate botanicals and space for the collection of naturalia. The professors of botany no doubt also made sure that a start was made toward assembling a herbarium of dried plants. Cluyt and his son Outgaert also gave lessons using Cluyt's remarkable private collection of perhaps four thousand botanical illustrations.[146] After Cluyt's death in 1598, seventeen students petitioned the curators to have his son Augerius replace his father as head of the hortus, but the university senate decided instead to require Pauw and Bontius to instruct the students in the garden; the demonstrating apparently fell mainly to Pauw, while Bontius lectured on Dioscorides and his best modern interpreter, Mattioli. One of Bontius's sons later called Pauw "the greatest botanist of our age."[147] Clusius himself spent his last years working on additional editions of botanical books and helping his friends, although he also had to contend with mice who ate his bulbs and thieves who broke into his private garden and carried off plants.

NATURAL HISTORY FROM THE EAST INDIES

Clusius nevertheless eagerly took account of the new information arriving from Asia via Dutch channels. Many of the merchants who fled north in the later 1580s brought with them an intense hatred of King Philip, whom they saw as the chief supporter of the Inquisition. The Iberian spice trade with the Far East was an especially tempting target. If the Dutch could shoulder their way into the spice trade directly, instead of being mere re-exporters, they could not only make greater profits for themselves but remove one of the main sources of the enemy's wealth, strengthening themselves and dealing the Hapsburgs a great blow. It was in Enkhuizen—the place from which Paludanus could not be enticed for the professorship of botany—where the chief pieces of information necessary to begin the long-distance voyages to Asia were accumulated. At the end of the sixteenth century, Enkhuizen was one of the most important commercial hubs of northwestern Europe. It provided a safe harbor north of Amsterdam and the other great cities of Holland, lying on the edge of a vast province of water called the Zuiderzee (the "southern sea") that gave ready access to the

North Sea and routes to the Baltic, and to the chief cities of the northern low countries. In the "mother trade" with the Baltic, its sailors and ships were many, and well regarded, including one of the former sea captains of the city, Lucas Jansz. Waghenaer, who in the later sixteenth century developed improved methods for charting coasts and seas that became so widely used by coastal pilots that the English simply named them "wagoners" after him.[148]

It was to Enkhuizen, too, that Jan Huygen van Linschoten returned with information and experience enough to open the purses for funding the first Dutch attempts to sail directly to the Spice Islands. To people who pay attention to the "history of discovery," Linschoten is one of the most famous men of his day. But he did not start out to be a "discoverer," nor to be a patriot. He was born in Haarlem about 1563, and his family moved to Enkhuizen during the bitter warfare around their hometown in the early 1570s. The family's move to Orangist Enkhuizen did not, however, mean that they were politically or religiously motivated, only that they wished to escape the war. Indeed, when it came time to make their way in the world, Linschoten's two older brothers traveled to Seville, a major commercial and administrative center and the hub for trade with the New World. Jan joined his brothers there at the end of 1579, and all three found service with high-ranking men of the Catholic world. Jan and his brother Willem found themselves in the chief Portuguese port of Lisbon not long after Philip of Spain also became king of Portugal in 1580. From Lisbon, Willem took employment as a clerk on a ship bound for the Indies, at the same time securing for Jan a position in the suite of Vincente de Fonseca, the new archbishop of Goa—capital of Portuguese India—who was outbound on the same ship. The fleet left on 8 April 1583, stopped at Mozambique in August, and reached Goa on 21 September. Linschoten became one of the archbishop's most trusted staff members, so there could have been no hint that he was anything but accepting of the old religion. When the archbishop returned in January 1587 to report to King Philip, Linschoten was placed in charge of his palace and affairs. When news of the archbishop's death arrived in Goa in September 1588, however, the household was left without a master. Linschoten worked his way back to Europe, taking service as a merchant with the firm of Fugger. He left Goa in January 1589 in a fleet that on 22 July reached the safety of the Azores, where he stayed for two years helping to organize the salvage of wrecks as English sea wolves brazenly patrolled the area looking for targets in the wake of the failure of the great Armada. He returned to Lisbon at the beginning of January 1592. Somewhere along the way, however, he had decided to go home. He reached Enkhuizen on 3 September, after almost thirteen years away.[149] He brought with him an enormous fund of information about the Portuguese Indies, as well as two feathered skins of birds of paradise, a sample of Chinese writing on Chinese paper, other texts written on palm leaves, seeds from the *arbor triste* of western south Asia, and other bits of naturalia.[150]

The world in which Linschoten set foot again had changed. When he left, Enkhuizen had been a safe place on the fringes of a bitter civil conflict; when he returned, it was an important city in a polity calling itself the United Provinces that was waging a war for independence against Spain and Portugal. Linschoten put his knowledge at the service of the leaders of his new country. It was the perfect moment: the city's merchants must have been buzzing with news of the recent English capture of a Portuguese-Spanish carrack off the Azores, from where Linschoten had just returned. The *Madre de Dios* had been sailed into Dartmouth escorted by an English fleet just a few days before Linschoten returned to Enkhuizen, and after the looting by locals was controlled by Sir Walter Raleigh (released from the Tower of London for the purpose), it was found to contain over 475 "tons" (large barrels) of pepper, over 50 tons of cloves, almost 40 tons of cinnamon, about 3.5 tons of nutmeg and an equal amount of mace, 15 tons of ebony, yards and yards of silk, calicoes, and rugs, chests full of musk, pearls, amber, silver, and gold, and great quantities of drugs, dyes, jewelry, and other items—to name only the richest products aboard this one ship. One senior English official calculated it to be worth the amazing sum of over five hundred thousand pounds sterling, more than a king's ransom.[151] With his knowledge of the Portuguese enterprises in Asia, including information about sailing routes in the Indian Ocean, Linschoten's timing could not have been better for those who hoped to acquire similar riches. Not long after, the brothers Cornelis and Frederick de Houtman also returned successfully from a period of commercial espionage in Lisbon.[152] Eager to rob Spain of some of the Asian wealth that supplied its war machine and generate a stream of gold they could tap, the Prince of Orange and the government of the United Provinces themselves debriefed the returned traveler.

Among the merchants who had fled north and taken an interest in the advantages of a direct trade in spices was Balthasar de Moucheron, who had built his fortune in Antwerp and was now heavily involved with the Muscovy trade. He had connections with another newcomer, Petrus Plancius, a committed Calvinist minister who settled in Amsterdam and devoted his energies to furthering new and more accurate methods of navigation and mapping, and to teaching navigation to a broad audience in a local church. Plancius brought with him techniques of mapmaking that had been pioneered in Antwerp by Gemma Frisius and his best-known pupil, Gerard Kramer, or Mercator, who became the preeminent cosmographer of the period, developing a mathematical method for projecting the curved surface of the globe onto flat paper that accurately showed the relationship among landmasses; the Mercator projection is still often used today.[153]

Plancius's theoretical methods sometimes proved useless for navigating at sea.[154] They did, however, prove to be fundamental for keeping track of any location between known points and hence for mapping, which made geographical details visible in spatial relationships, helping merchants and civil servants keep track of knowledge and plan new ventures. Plancius consulted with the States of Holland and the States General, the Prince of Orange, and merchants about plans to travel to the Indies directly. Plancius's partner, Cornelis Claesz., set up a large Amsterdam firm dealing in accurate maps (printed and hand-drawn), charts, rutters, atlases, and travel accounts, which contained not only information for the curious but explicit details for navigators and merchants.[155] Information from Linschoten and Plancius suggested that the seas on the China coast could be reached from the north. With Linschoten in hand, and with support from the States General and the Prince of Orange, Moucheron therefore organized a fleet to look for a passage to China beyond the Archangel route he already knew.

In June 1594, three ships set sail for the Northeast Passage, one fitted out in Enkhuizen with Linschoten aboard. The voyagers found openings in the land and ice far to the northeast, seeming to indicate the existence of a passage, although they were forced to return before they could go further. On returning, Linschoten reported orally and in writing on the results to the states and the prince. In July 1595 another seven vessels sailed north, with Linschoten as one of the two chief commissioners in charge, but they did not get as far as the previous year before being forced back. Yet a third attempt was made in 1596, although in a famous tragedy the crew, commanded by Willem Barentsz, was forced to winter over on bleak Nova Zembla, where most of the officers and men died—many battling polar bears.[156] Fortunately for Linschoten, he had stayed at home during that voyage (having been made treasurer of Enkhuizen).

At the same time that the search for the Northeast Passage was being attempted, Linschoten's information was also helping in the organization of another enterprise to get to the Spice Islands by the known way taken by the Portuguese, that of sailing around Africa. In 1594 a group of nine wealthy and powerful Amsterdammers organized themselves into a Compagnie van Verre ("Long-distance company") to fit out a fleet. Again, several of them were immigrants from the southern Netherlands who had experience in arranging long-distance ventures in the Muscovy trade. Others were members of the Amsterdam elite who were able to secure the cooperation of the States of Holland and Zeeland for the provision of cannon and smaller weapons for the enterprise as well as encouragement in the form of an exemption from import and export duties. Because of good intelligence from Linschoten about the sea-lanes and winds used by the Portuguese, they hoped to avoid open confrontation by sailing directly to the Spice Islands south of the Portuguese bases in the Indian Ocean. Their four ships, led by Cornelis de Houtman, set out in the spring of 1595. They carried an early version of Linschoten's book, including his sailing directions for India, the eastern seas, and the American coasts, translated by him from the manuscripts of Portuguese and Spanish pilots.

Just as these first Dutch attempts to reach the Indies began, Linschoten published a vernacular book about the East, further raising public interest. Titled *Itinerario ofte Reijs-boeck van Ian Hughen van Linschoten naer oost ofte Portugaels Indien* ("Itinerario, or the travel-book of Jan Huygen van Linschoten to the East or Portuguese Indies"), it was published in early 1596, and portions were quickly translated into several other languages, including English, German, French, and Latin.[157] Linschoten had begun to write his account after his return from Goa and the Azores, and at his request the States General issued a patent (*octrooi*) for his text on 8 October 1594. The book consisted of three parts: an account of his voyages; sailing directions for India, the eastern seas, and the American coasts translated from Spanish and Portuguese manuscripts (a version of which had been sent along with Houtman's first voyage); and a description of America and the east and west coasts of Africa taken from other authors. The first part held the greatest interest for most people. It is full of incident and exotic description, and richly illustrated.[158] The places mentioned range through the Portuguese Indies, running from Mozambique to Hormuz, the west coast of the Indian subcontinent, the Maldive Islands, Ceylon, up the east coast of the subcontinent to Bengal, then along to Siam, down to Malacca, Sumatra, Java, the Moluccas, over to Borneo and the Philippines, up to China, and across to Japan; then (in chapter 27) Linschoten begins describing the people and habits of India and other places. In the second half of part 1 (from chapter 45) he describes the plants and animals he saw or of which he had good accounts.

In writing his book, Linschoten had the energetic assistance of a fellow citizen of Enkhuizen, to whom he dedicated the work: Paludanus. One can well imagine how eager a collector and naturalist like Paludanus must have been to become acquainted with Linschoten when Linschoten returned to Enkhuizen bringing not only information about, but objects from, the Indies. With his own experience of travel, enthusiasm for careful and accurate observation, and enormous knowledge of natural history, Paludanus was the perfect partner for teaming up with Linschoten to produce a book on the Indies. Indeed, he helped so much that he was virtually co-author of Linschoten's *Itinerario*, turning the work from a personal account of travels into a description of the East Indies in line with the best new work on natural history. One kind of help he gave was to introduce Linschoten to recent books on which he could model his account. Paluda-

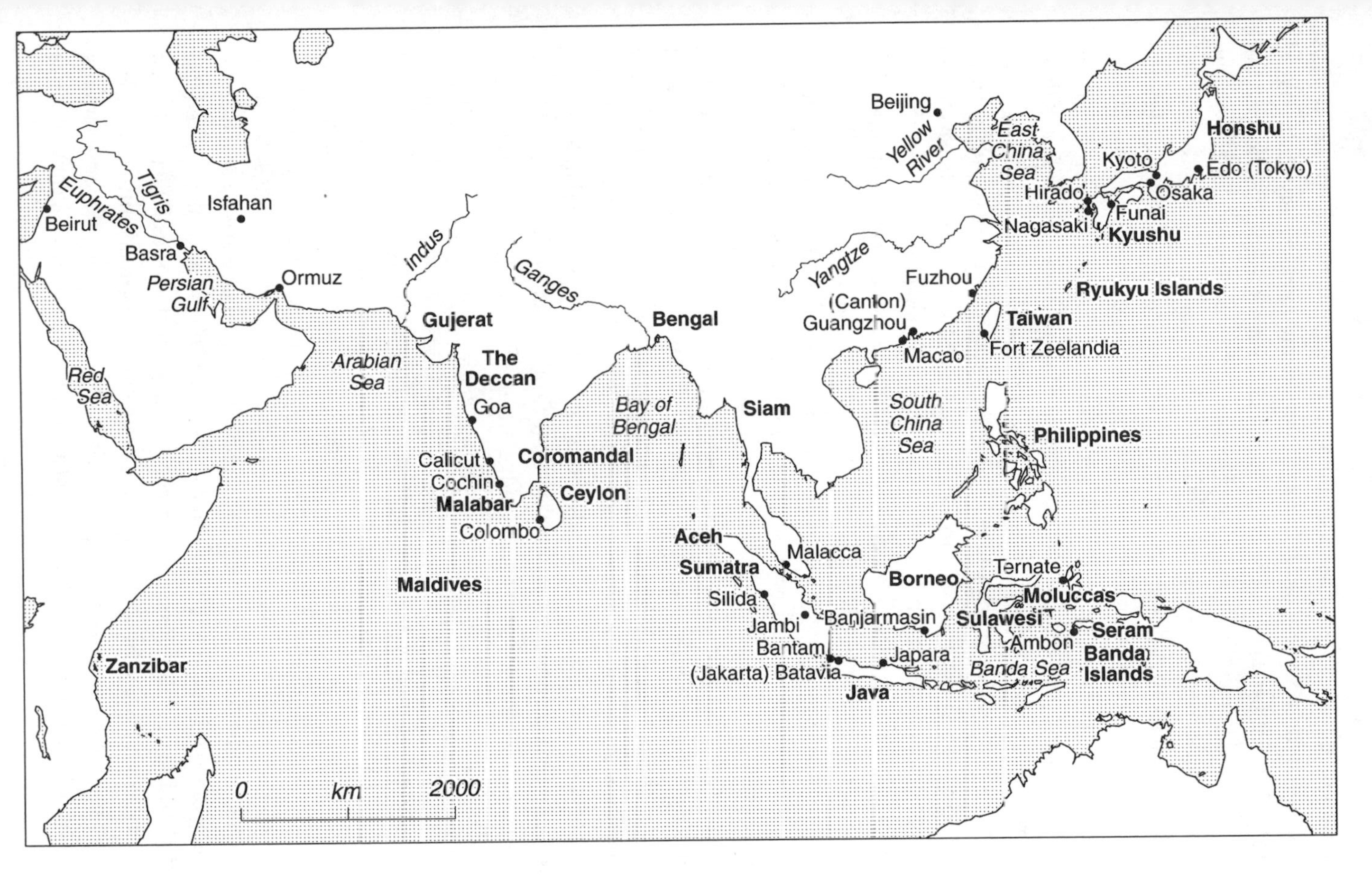

The Dutch in Asia

nus showed Linschoten other accounts of the Indies, such as Joseph de Acosta's *Historia natural y moral de las Indias* ("Natural and moral history of the [West] Indies") published at Seville in 1590. Acosta's work provided the basis for part 3 of Linschoten's *Itinerario.* It also piqued his interest deeply enough for Linschoten to undertake a translation of it into Dutch, which appeared two years later. The information about Africa in Linschoten's book came from an Italian publication of 1591, supplemented by the account of a Dutch captain who in 1593 was the first to visit the Gold Coast of Africa, Barent Ericksz. of Enkhuizen (who sold many of the objects he collected on his voyage to Paludanus):[159] the first of these works was probably supplied by Paludanus, the second certainly was.

Paludanus also carefully edited Linschoten's account in light of published information on the Indies, carefully identifying his commentary in extensive annotations. As Clusius did in editing Orta, Paludanus added his words at the end of Linschoten's chapter, in a different typeface and marked with his initials. For example, in chapter 23 Linschoten tells of China and along the way describes porcelain and what he knows about how it is made (which Europeans were unable to duplicate for many decades to come). At that point Paludanus adds a paragraph expanding on Linschoten's information. The addition summarizes the account on the making of porcelain taken from the humanist Joseph Caesar Scaliger's *Book of Subtleties,* which explains that it is made from a special kind of powdered shell and then buried for one hundred years, before concluding that Linschoten's own version "appears to be true," since he says that porcelain is made of an earth, just like pots and other containers in The Netherlands, making Linschoten's account more probable than Scaliger's.[160] In the second half of part 1, on the natural history of the Indies, the number and length of Paludanus's addenda increase greatly, almost doubling the size of many of the descriptions, particularly the botanical ones. In the section on nutmeg and mace (chapter 66), for instance, Linschoten describes the appearance of the tree, where it is found and whence the fruit is sent, what the fruit is like to the sight, touch, and taste, how nutmeg is preserved, the way the tree grows, the reputedly unhealthy climate of the region where it grows, and the local names for nutmeg and mace. Paludanus adds that nutmeg is good medically as a remedy for pain in the head, uterus, or muscles, gives a closer description of the fruit—repeating much of the sense of what Linschoten had written—and describes the physiological benefits derived from eating nutmeg and mace.[161] (The journal of one of the first Dutch voyages to the Spice Islands, that of Jacob van Neck, described nutmeg's use less discriminatingly as strengthening the nerves, sharpening the memory, warming the stomach, and stopping diarrhea and as an all-around remedy against illnesses

Bernardus Paludanus, holding a sample of a pepper plant. Engraving by H. Barij after G. van Nieuwenhuysen. Courtesy of the Wellcome Library, London

having a cold origin.)[162] For most of his additions, Paludanus relied on books by earlier physicians who had a knowledge of Asian materia medica, many of whom had written in Arabic and been translated into Latin, such as Avicenna, Serapion, and Rhazis. He turned most often, however, to Clusius's edition of Orta to make Linschoten's account both more accurate and inclusive.

Two years after they sailed, three of the four ships of this first Dutch voyage to Asia returned, with only a third of the original crew still alive, but yielding a small profit and much experience. An anonymous, first-person account appeared in print less than two months after the survivors returned, with a second edition shortly thereafter; accounts of the voyages to find the Northeast Passage also appeared quickly.[163] In all these narratives, descriptions of exotic nature reinforced in readers the sense of sharing the experience with the author. They exhibit a mingling of geographical, commercial, and natural information much like Linschoten's book, although description of events took precedent over close description of the animals and plants. A survivor of the first voyage, Willem Lodewycksz., wrote an account called *D'eerste boeck* ("The first book," 1598), whose mimetic retelling of events included many passages about exotic plants and animals as well as places and people. Some of the accounts are invented—a Madagascan "salamander," for instance, is presented as living in fire, although the association with that element is entirely mythological and no salamanders exist on Madagascar—while much of his account of Java is taken from Linschoten's book and Portuguese sources. But he also gave a somewhat melancholy account of the cruelties of sailors who tied empty barrels to the tails of sharks, of the warm dreams of basking sea turtles being cut short by boat hooks that pull them aboard to be eaten, and of flying fish that try to escape larger pursuers in the water only to be caught and swallowed by birds or, falling on deck, consumed by sailors.[164]

Following the success of the first voyage, there was a rush of new investment accompanied by a rush to acquire new information. The original members of the Compagnie van Verre merged with another trading company of mainly southern immigrants to form the Oude Compagnie ("Old Company"), which sent out a fleet of eight ships in 1598. With the tremendous publicity of the first venture and the reimposition of the Iberian embargo on Dutch trade in the same year, two smaller fleets from Zeeland and two from Rotterdam had also set out for Asia in 1598.[165] The two fleets from Rotterdam took Magellan and Drake's route around South America, one failing to reach the Indies, the other losing most of the ships—although one of them, the *Leifde*, was blown onto the coast of Japan in April 1600, depositing a few survivors there, including William Adams, the famous English pilot who would open up Japan to Dutch commerce.[166] The two fleets from Zeeland went around Africa and made profits, while the fleet organized by the Oude Compagnie turned a remarkable 400 percent profit.[167] Given such precedents, other ventures using the route around Africa quickly followed.

Naturalists, too, were eager for news and objects of the East Indies. Professors Clusius and Pauw wrote the investors in the 1599 Vierde Schipvaart ("Fourth voyage"), asking them to collect and describe the exotic plants encountered, and they agreed to charge someone with these duties, a surgeon named Nicolaas Coolmans.[168] When the ships returned in 1601 without Coolmans, who had died on the trip, Clusius, undeterred, wrote a letter to the governors of the newly organizing VOC again asking them to collect information and specimens. Again, they responded positively, conveying to the apothecaries and surgeons of one of their first fleets orders to bring back several kinds of items: samples of plants (by collecting small branches with their leaves, fruits, and flowers, all pressed between sheets of paper), examples of species of economic interest, "all kinds of strange trees" together with drawings and descriptions when possible and information about their uses, and even strange sea fish and plants.[169]

At this time Paludanus sold his first collection of naturalia and started another: a collection of overseas exotica. When Duke Friedrich purchased Paludanus's first collection around 1600, at least one acquaintance worried this would leave Paludanus dispirited.[170] But he remained eager to expand his knowledge with information and material brought back from more distant parts. Several of the items returning with Linschoten seem to have ended up in Paludanus's hands, and these were quickly supplemented by other material tokens from the Dutch voyages to the Spice Islands: for instance, a survivor of the first voyage, Franck van der Does, managed to bring home a Hottentot spear and other artifacts, which ended up in collections, perhaps in Paludanus's.[171] Indeed, within a few more years Paludanus had built up an even more impressive collection than his first, filled with specimens from Asia, Africa, and even the New World, including the dried skin and feathers of two birds of paradise from New Guinea (possibly the specimens Linschoten had brought back). The collection brought him fame. A local history of Enkhuizen written in 1603 noted that many people came to the city to see Paludanus's collection, assembled from items the world over: it was said by an anonymous Italian traveler in 1622 that his collection even superseded that of the famous Ferrante Imperato of Naples.[172] Travel guides referred to his cabinet, and the accounts of travelers mention visits to it. Quite how the collection was displayed is not known, although at least part of it was kept in a new type of furniture: cabinets containing drawers.[173] Paludanus even received a visit from the whole court of the so-called Winter King, Frederik of the Palatinate and his wife, Elizabeth, accompanied by the court of Maurits, Prince of Orange. Other names appear in the visitors' book, which contains about nineteen hundred names, many readily identifiable as noblemen and women, diplomats, and learned men (including Clusius, Dousa, Jan van Hout, Scaliger, Nicolas-Claude Fabri de Peiresc, and Ole Worm).[174] After his death the cabinet, first in the hands of his wife and then passing to others, continued to be visited in Enkhuizen. The largest part of it was sold to the ducal court of Schleswig in Gottorp, which was eventually absorbed into the royal collections of Denmark—some of Paludanus's specimens are still identifiable in Copenhagen.[175]

A decade before Clusius's death in 1609, then, information and specimens from a raft of the new trading ventures to far-off lands was flowing into Leiden and other cities. The first engraving of the Leiden botanical garden, made just before the end of the century, gives prominent place to images of a turtle shell, a stuffed crocodile, and the jawbone of a polar bear, all of which could be viewed in the ambulacrum in the university's garden where the naturalia were kept. The remains of a large turtle would have been easy enough to come by in the Indian Ocean, the crocodile in Egypt or Asia, and the polar bear in the northern seas, through which the Northeast Passage was being sought. By the time of Pauw's death in 1617, Leiden's collection of naturalia had expanded to include medicinal simples in boxes, a swordfish, bamboo, various eggs, different fruits, plants and fishes from the East Indies, corals, shells, hoofs and shin bones of the elk, a parrot, a barnacle goose, an armadillo, an anteater, and animal skins, birds of paradise, a string of teeth and a hammock from the West Indies, paper with Chinese characters and depictions of plants in ink, clothing from Russia and Japan, and many other curiosities.[176] The people who valued the knowledge represented by the cabinets of Paludanus and the university of Leiden included not only students and intellectuals, noblemen and noblewomen, and other members of the wellborn but people from the merchant elite, magistrates, guild masters, and all sorts of ordinary folk.

When Clusius died at eighty-three he was famous for his extensive and detailed knowledge, much of it accumulated during his travels. He had published exacting works of descriptive botany accompanied by fine illustrations. He was the first to give an account of fungi (207 species listed in a supplement to his *Rariorum plantarum historia* of 1601), and he paid close attention to botanical information from both the East and West Indies. He translated a variety of botanical works into Latin and Dodoens's Dutch herbal of the low countries into French. While editing them he added many relevant matters of fact he had found for himself. Clusius translated into Latin accounts of voyages, including those of Pierre Belon's travels to Greece, the Near East, and Egypt, Thomas Harriot's to Virginia, and Gerrit de Veer's to the Arctic. Because of his broad mind and ease of manner, opportunity and energy, position and courtesy, Clusius was known to and respected by almost every important botanist and gardener in Europe, and many other people besides. His excellent command of Latin and ability in Greek brought him high regard as a scholar, but he also spoke the vernacular languages of the French north and south, Spanish, Portuguese, a variety of Germanic tongues found in Holland, the Rhineland, Wittenburg, and Vienna, and even some English. He was therefore able to gather much of his information on the names and uses of plants from conversations with the local people who collected them. He received countless letters from friends and acquaintances and wrote huge numbers in turn, often accepting and sending enclosed seeds and specimens, thus building an international horticultural network of personal obligation through a web of mutual exchanges. He also invested much energy into acclimatizing exotic plants for European gardens, becoming a clearinghouse for new species. In the process, he paid as much attention to the beautiful as to the useful. He became the chief propagandist for the potato, but he also gets the major credit for introducing flowering plants such as new varieties of irises and lilies, and anemones, narcissi, and hyacinths, while he became most famous as an advocate for tulips. Perhaps when we add his ability with languages to his work in naming plants, we would do well to think of Clusius as a veritable Adam, a conceit that would probably have pleased him as long as it was tongue-in-cheek. Clusius epitomizes, then, both the best botanical work of the period and the well-traveled humanist man of the world, moving about relatively freely among the well to do and well placed while also communing with herb-wives, gardeners, and landladies.

Clusius the traveler and networker, Clusius the linguist, Clusius the exact observer, Clusius the seeker after hidden mysteries, Clusius the devout but undoctrinaire: all this fit well with contemporary views of what was necessary for someone to learn about the created world. A variety of people who advocated what came to be called the new natural philosophy held that wide experience, linguistic skills, a good memory, a clear head, and the cultivation of virtue, when coupled with exhaustive and sometimes exhausting investigations of things themselves, led to real and solid knowledge of nature. The values represented in Clusius' work were shared by many others in his day and originated not from a change in worldview so much as from paying close attention to the things of the world, which were in turn associated with the worlds of commerce and liberty. For people like Clusius, as for his friends, it was the particulars that counted, not arguments about general principles.

In the low countries, then, the Renaissance interest in objectivity remained strong despite religious and political conflict. People carried on worldly studies despite brutal war and religious bigotry, although it was no easy task. The religious and philosophical battles of the day threatened life and liberty at worst and threw up distractions at best. Yet a culture of religious piety rather than doctrine and of excellent educational institutions cheek by jowl with intense commercial activity had long sustained the magistrates' sense of propriety and developed in them great sympathy for liberty of conscience as long as it did not lead to law-breaking. Clusius was merely one of the most famous of those building the study of nature on this humanist legacy, who worked to transform scholarship from a matter of debating hidden causes to accumulating erudition about things, in the process sidestepping the most intense religious disputes. The attempts of the clerics to make everyone subscribe to one view or another of God's injunctions were resisted in part by the examination of God's creation instead, a move supported by travel and commerce, the love of gardens and naturalia, an excitement for anatomy, and a practical concern for clinical description, all arising in the culture of taste and objectivity.

4

Commerce and Medicine in Amsterdam

Moreover, even in the body, though it dies like that of the beasts, and is in many ways weaker than theirs, what goodness of God, what providence of the great Creator, is apparent!

—AUGUSTINE, *The City of God*, book 22

In January 1632, a well-to-do physician and magistrate of the city of Amsterdam commissioned a large portrait showing himself and several surgeons expounding the moral lessons of anatomy. The result was one of the most famous paintings of the seventeenth century: *The Anatomy Lesson of Dr. Nicolaes Tulp*, painted by a young artist who had just moved to the city, Rembrandt van Rijn. The painter broke convention in the way he portrayed the lesson, which was performed in late January and early February. He did not pose his subjects standing at ease with a skeleton or body in their midst simply announcing human mortality, as painters had depicted Tulp's predecessors. Rather, he invites the viewer to see Tulp and his companions frozen in a moment, stopped in action. The surgeons—Jacob Block, Hartman Hartmansz., Adriaan Slabberaen, Jacob de Witt, Mathys Calkoen, Jacob Koolvelt, and Frans van Loenen—look on with attention, or seem to have just caught the thought. The body they are exploring, that of a thief who had been punished for acts of murderous violence, Adriaen Kint, looks still fresh from the gallows.[1] Tulp seems caught in the midst of uttering a word, no doubt something about how wonderfully God made man.[2] His left hand is raised, perhaps as a tribute to the surgeons' guild for which he was the anatomical lecturer: the Greek *cheirourgia* meant "working with the hand," especially "the practice of surgery."[3] His raised hand counterpoises finger and thumb while in the other hand he holds an instrument in which he has gathered up the muscles and ligaments of the arm that make such a motion possible. The great-

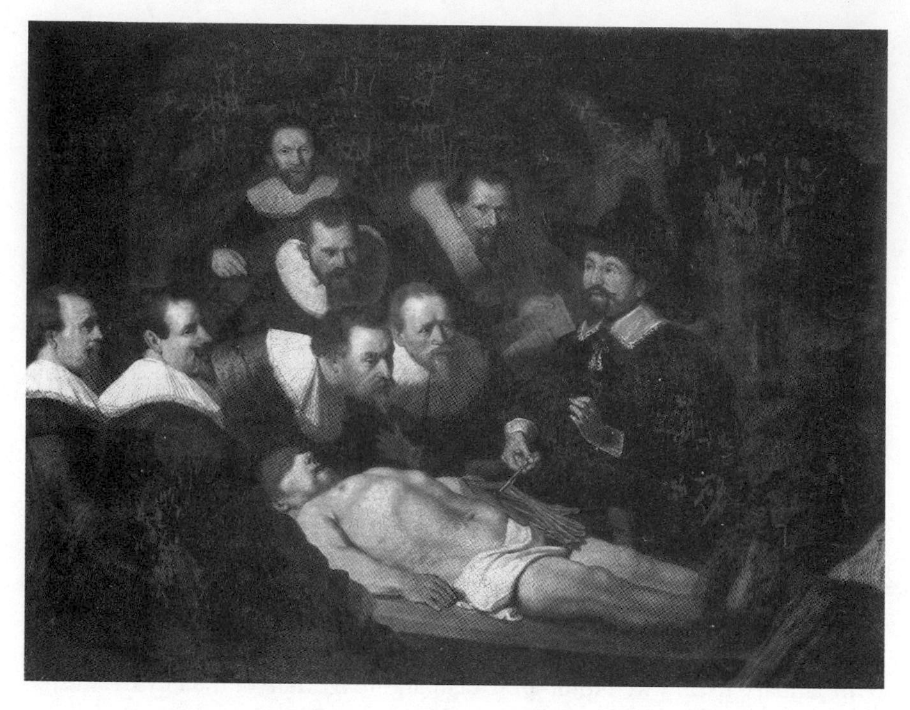

Oil painting, copy of Rembrandt van Rijn's *The Anatomy Lesson of Dr. Nicolaes Tulp*. Courtesy of the Wellcome Library, London

est miracle of creation lies spread out before them for intimate, penetrating inspection, ready to give up its secrets to the careful investigator whose movements of inspection are made using the very same ligaments they are inspecting. The anatomist anatomizing himself; bodily wonder wondering.

Although he was not yet forty when he commissioned his most famous portrait from Rembrandt, Tulp was a leading figure in his city, one of the regenten: he not only served for fifty years as one of the two dozen members on the city council but gained great wealth—indeed, he was the first person in Amsterdam to ride in a horse and coach.[4] In years to come he would use his influence to bring in new municipal statutes to regulate medical practice and practitioners, and through the marriage of his daughter he united his family with one of the most powerful merchant houses in the country, the Six family. But Tulp may have commissioned this painting to make up for disappointment: he had not been appointed one of Barlaeus's colleagues. Tulp had obtained the highly visible post of anatomical demonstrator (*praelector anatomiae*) to the surgeons' guild in 1629, but when the Athenaeum was established in 1632, it did not include, despite earlier proposals, a professorship of medicine or botany, an absence that may have wounded Tulp since he might well have expected it to come his way.

However powerful he might have become, then, even a Tulp represented only one part of a complex and highly competitive commercial and intellectual environment. He did not always get his way. He was a physician in a municipality full of many kinds of medical practitioners, a strict Calvinist moralist during a period when the "libertines" were often in power—the new Athenaeum, and Barlaeus himself, avoided being tied to Tulp's religious party. Both Barlaeus and Tulp agreed that humankind's full attention should be directed to the details of nature. As intellectuals, they sometimes stood shoulder to shoulder against popular ignorance and disorder. When in 1639 the city opened a new anatomical theater over which Tulp presided, Barlaeus wrote the inaugural poem, part of which was painted in gold letters over the balcony so that the audience could absorb its moral message, although that message suggested something less centered on predestination than Tulp may have liked. Between them—one a wealthy magistrate and physician who worked hard to give institutional form to the medical community of Amsterdam, the other the chief public intellectual of the chief commercial city of the country—lay important differences of interpretation. But the high regard they both had for the moral virtues of investigating matters of fact—the more precise the better—provided the common ground over which they and many others argued about interpretation. Even their differences, then, made it clear that neither the rules of commerce nor nature itself provided a simple and clear set of instructions for how to order human lives.

MEDICAL COMMERCE AND MATTERS OF FACT

The associations between the knowledge of useful matters of fact and commercial interests were very clear in the medical marketplace. Even patients with small resources had an enormous range of practitioners from whom to choose. In addition to self-diagnosis and self-help, and the advice and assistance of family, friends, and neighbors, local cunning folk (also called "wisewomen" and "wisemen") and midwives were widespread, helping with matters of fertility and marriage, conception and pregnancy, childbirth, baptism, motherhood and childrearing, and death and the dead.[5] They were usually considered helpful to members of their communities and so were not, contrary to some accounts, as frequently accused of witchcraft as many historians have assumed (although persecutions for witchcraft were not common in the low countries in any case).[6] More visible in the historical record, and probably more visible to the early modern Dutch themselves, were the many medical practitioners who sold their services in villages, towns, and cities for money, sometimes moving from place to place to do so. Given the robust state of the Dutch economy, more than enough people had sufficient disposable income to pay for medicines and medical care when needed, and a few had the means for such care whenever it was desired. The medical marketplace therefore grew rapidly.

Dutch cities had long allowed people selling medicines such as herbs, oils, and ointments to participate in the markets periodically set up for the exchange of goods. Some were local people who offered their wares on a regular basis; others came for a time and moved on. By the later sixteenth century, the market regulators, who were ultimately responsible to the city councils, had usually moved medical salespeople away from the center of the markets to their edges. The names sometimes applied to them by respectable burgers—*empirici* ("empiric") or, worse, *kwakzalver* ("quacksalver," or "quack")—were taking on negative connotations, but as long as these people obeyed the rules of the market like everyone else, they were not hindered. Nor was the fact that they commonly offered their own medical "secrets" for sale a particular problem, since even the most respectable physicians of the day might have their secrets. In addition, medical itinerants would regularly come through cities and villages to couch for cataracts, cut for bladder stones, excise hernias, and otherwise treat difficult problems that local surgeons considered too risky. By the end of the sixteenth century these "free masters" (*vrije meesters*) usually needed a certificate from the town councils to offer their services in the municipalities, ordinarily only after an examination by a committee of the surgeons' guild, although the magistrates could overrule suspicious surgeons if a vrije meester had enough support from local petitioners who wanted help. Much more worrisome for the magistrates were people they considered to be harmful, such as those who diagnosed the illnesses of strangers only by examining their urine (*piskijkers*, or "piss inspectors"), or the vrije meesters who falsely claimed degrees or princely licenses from foreign lands.[7]

Such practitioners marketed their medical expertise on the basis of its utilitarian benefits. A sense of their claims can be found in the advertisements some of them had printed, which could be circulated by hand, posted at busy spots, or placed in newspapers (after these last came into being in the early seventeenth century).[8] A few sixteenth-century medical pamphlets and broadsides survive from what must then have been a common medium. One of the earliest survivals, from the early sixteenth century, was an advertisement for a "noble" oil prepared and sold by Karel de Minne, a surgeon and apothecary from Antwerp; another from the sixteenth century was a "delightful" (*costelic*) laxative sold by Peter of Dort.[9] The printing presses were also turning out a large number of medical handbooks meant for ordinary readers. They contained advice on how to treat ill-

Engraving by W. French after G. Dou's *The Quack Doctor.*
Courtesy of the Wellcome Library, London

Prudentia warns an alchemist of the abuse of fire. Sixteenth-century engraving by C. van de Passe after M. de Vos. Courtesy of the Wellcome Library, London

nesses and accidents, including recipes for making the necessary medicines. One of the most popular—appearing in at least seven editions—was the *Medicijnboec* ("Medicine book"), intended for "common people" (*ghemeyne man*), originally written in German by Christoph Wirtsung, first appearing in Dutch in 1589 in a translation by Carel Baten, a well-educated and much-traveled physician from Antwerp who moved north to Dordrecht, and later Amsterdam, after the Revolt. Baten also translated practical works on surgery.[10] In addition, many almanacs and other cheap pamphlets contained bits of medical advice, such as how to coordinate venesection with the phases of the Moon.[11] Medical advice also filled many of the pages of the so-called books of secrets or, as they were called in the Germanic languages, books for the head of household: *huisvater* books. They became a publisher's staple.[12] The general subject of which they formed a part was therefore sometimes called "natural magic."[13] Giambattista della Porta's famous book by that title, translated into most European languages, including Dutch, contained recipes for all kinds of household activities, from cookery to perfumery, from embarrassing your enemies to changing your hair or eye color.[14]

Another group of practitioners emphasized their expertise in providing remedies made using medical alchemy ("iatrochemistry"). They used various methods to apply heat to dissolve and alter substances, breaking them down into their component parts—hence the Dutch name for chemistry, *scheikunde*, the "art of separation." One of the most common alchemical processes was to extract the essence of substances through distillation. Distillation apparatus had become powerful enough by the end of the thirteenth century for a powerful cordial, *aqua vita* ("water of life"), to be extracted from wine; by the sixteenth century a tremendous variety of these liquors (or "liqueurs") containing alcoholic essences were on offer, from mint to licorice. Using distillation and similar methods of preparation, medicines could be made in large quantities relatively cheaply yet last for long periods without obvious loss of potency. At the same time, the method of chemical preparation itself disguised the ingredients, helping to keep the secret remedies private. As a result, many people who marketed medicines of their own making found iatrochemistry to be fundamental to their business.

There is evidence of a long tradition of iatrochemical investigations in the low countries. For instance, there are the shadowy fifteenth-century figures of Isaac and Johan Isaac Hollandus, most likely father and son, who probably originated in the province from which they took their name (although some think they came from Flanders). They apparently developed an improved alchemical furnace and experimented with metals, stones, vegetables, and fluids of the human body, such as urine and blood. They may also have been the originators of the important theory that natural things were composed of three states: the fluid (mercurial), the combustible (sulfurous), and the fixed (salt), as in a crude but common example of the application of heat to wood causing it to separate into smoke (the fluid state), fire (combustible), and ash (fixed).[15] (Only in the seventeenth century did manuscripts of Isaac and Johan Isaac Hollandus get into print; how widely their manuscripts circulated beforehand is unclear.)

The ultimate alchemical goal for some of these "artists" was, however, more ambitious: to work with substances in such a way that the essential stuff that lay behind all things could be materialized in its pure form, which was called the philosopher's stone, commonly thought to look like a yellowish or reddish powder. The philosopher's stone contained powers of life and transformation. It could in turn be combined with other matter to yield gold, but it could also yield the elixir of life, a panacea against all illnesses including aging, and the liquor alkahest, a universal solvent from which other potent medicines could be produced.[16] Such iatrochemists usually claimed to do everything by the operations of nature: the powers of nature might be hidden ("occult"), but by working with these occult powers they were simply exploring nature's potentials.

The best-known author of works on iatrochemistry was Aureolis Theophrastus Bombastus von Hohenheim, better known by his humanist pen name, Paracelsus (meaning "better than Celsus"). Paracelsus came from Einsiedeln, near Zurich, but had followers in the low countries, as elsewhere. Paracelsus announced ideas about surgical and medicinal healing in works that combined the lore and legend of miners and other ordinary people with the high-flown Neoplatonism of the intellectuals of his generation.[17] After various travels and almost getting hanged in Salzburg for his involvement with the Peasants' Revolt of 1525, he obtained an appointment as city physician of Basel in 1527, which gave him the right to teach at the university. A year later, the local physicians and apothecaries had forced him to depart, but Basel would continue to be one of the few places where chemical medicine was taught at a university. The last years of his life were spent moving from place to place, writing voluminous tracts (mostly published after his death in 1541), and experimenting with new remedies, especially ones prepared chemically. He was a major promoter of the idea that the essences of things give rise to their superficial appearances, that the work of these essences can be discussed in terms of the three principles (mercury, sulfur, and salt), and that the creative power of nature, the *archeus,* imbues all things so that our bodies and organs have their own archeus, too, to carry out natural functions. Other powers can interfere with these archeii, however, causing disease; the cure for each disease, therefore, depends on using medicines that have archeii to counteract the particular disease-causing archeus.[18]

Many of Paracelsus's works found their way into Dutch. For instance, a surgeon from Delft, Pieter Volck Holst, translated a book on surgery by Paracelsus (as *Die groote chirurgie,* "The Great Book of Surgery," 1555), to which he affixed a preface avidly supporting iatrochemistry. Paracelsus's smaller work on surgery and his work on hospitals was translated by Martin Everaerts as *Die cleyne chirurgerie ende tgasthuys boeck* ("The little book of surgery and hospital practice," 1568); some of his works on medicines and the three principles were translated by Jan Pauwelsz. as *Theophrastus des ervarenen vorsten alier medicyns, van den eersten dry principiis overgheest* ("Theophrastus [Paracelsus], the experienced prince of doctors, from the first three spiritual principles," 1580).[19] By the later sixteenth century, even princes throughout Europe actively furthered the studies of iatrochemists, sometimes even working in the laboratories themselves. While the excitement about medical chemistry has been interpreted as a mainly Protestant enthusiasm, it was in fact quite widespread, being encouraged not only by Queen Elizabeth of England and the son of Clusius's patron, Prince Moritz of Hessen-Kassel, but by Catholic princes such as Emperor Rudolph II, King Philip of Spain, and the prince-bishop of Liege and archbishop-elector of Cologne, Ernst of Bavaria.[20]

The expertise of apothecaries (*apothekers*) was somewhat different, resting on their claim to know about the often exotic substances in which they dealt. As elsewhere in Europe, they had once been long-distance merchants but were by the later sixteenth century owners of retail shops that sold medicines to the public, including the prescriptions written by physicians. (The wholesalers were now designated as *kruideniers*, "grocers.")[21] It had been common in the low countries generally, as in Amsterdam, for both the apothecaries and the physicians to belong to the guild of Saint Lucas, to which artists like the painters also belonged, although in Amsterdam after 1579 the apothecaries seem to have been gradually moved to the *kramersgilde* ("retailers' guild").[22] The number of substances in which they dealt was great and, given Dutch commerce in the East and West Indies, growing rapidly. One list of the cost per pound of various drugs on sale in Amsterdam from 1609 to 1637 includes rhubarb, scamony, opium, various kinds of turbit, manna, zedoarc, fine aloe, fine China root, collaquindida, senna leaf, mustard seed, cultivated sarsaparilla, wild sarsaparilla, myrrh, refined borax, benzoin, gum tragacanth, cinnabar, petroleum, ambergris (per ounce), civet, mace in the husk, pure mace, ground pearls, mercury, sandalwood, sandalwood branches, white incense, galengal, cassia fistula, Greek camphor, iris root, spikenard, spica Romana, oil of clove, eastern bezoar stone, western bezoar stone, long pepper, dragon's blood, cubebe pepper, ammoniac, refined sal ammoniac, Venetian turpentine, galbanum, cardamom, bitter almond, aged coculus, opoponac, euphorbe, mecho(a)can, cucumber, cassia ligna, new English saffron, ginger from Porto Rico, ginger from San Domingo, ginger from Brazil, Chinese ginger, anise from Bari, Venetian anise, anise from Alicante, anise from Malta, Polish cumin, Sicilian cumin, capers from Majorca, capers from Toulon, and sugar from Saint Thomas.[23] The up-front investment in inventory and shop supplies, including weights and measures, and various kinds of highly decorated pots and jars to contain the materials, must have been enormous.[24]

Like their brethren elsewhere, some Amsterdam apothecaries acquired large collections of naturalia. The most famous was undoubtedly that of Jan Jacobsz. Swammerdam (father of the more famous naturalist). He kept an apothecary shop in Amsterdam called *De Star* on the Oude Schans near the Montelbaenstoren—the last sight many sailors glimpsed on their way out of Amsterdam's harbor—thus right by the main quays of the East and West India Companies. He was well placed to purchase specimens and objects from sailors who had just put into port, and he collected a huge cabinet of curiosities, situated on the first floor above the shop. The elder Swammerdam became especially well known for his collection of old Chinese porcelain. But he had mounds of other materials, too: coins and medals; a strange, small Chinese image stamped in silver; a gold Japanese idol; a gold representation of Gustavus Adolphus (the Swedish ruler and general who for a while overran the Catholics armies in the Thirty Years' War); an artificial mouse with copper wheels and iron springs that could walk about; a Turkish almanac with colored letters; a Chinese almanac; silver from Mexico; bloodstones; pumice from Iceland; a sample of a "miraculous earth" called "fixed milk of the Virgin Mary" (*Miraculosa terra, seu lac Virginis Mariae concretum*); three eagle-stones (said to be collected from eagle's nests, and of great medical efficacy); seventy corals; a branch of a tree called Rose of Jericho; two birds of paradise ("with feet," contradicting the common legend that they did not have any); edible swallows' nests; seven crabs from the Moluccas; a stone from Portugal that cured fevers (*lapis antifebrilis*); an Indian millipede; a sea star; some insects; nineteen hundred shells; a unicorn horn six feet, three inches long; and more and more. Among the people who visited his cabinet in the early 1660s were the Danish scholar Ole Borch (visiting twice), the German Knorr von Rosenroth, and the Frenchman Balthasar de Monconys.[25] He must have purchased many of the items for his collection, as well as many of the imported medicinals that he sold from his shop, from sailors, soldiers, merchants, and captains returned from far-off places. After his death in 1678 his sons and daughter argued about how best to dispose of the collection, for which they published a catalog drawn up by his naturalist son Jan, which listed the objects on 142 pages of two closely printed columns, organized roughly into stones and minerals, plants, animals, and *artificialia* (*konstwercke*).[26] Jan had argued that it should be sold as three "complete" cabinets, or the individual items auctioned off, although in the end the children agreed to try to sell it all to one buyer. This made it harder to sell, however: the cabinet was valued at sixty thousand gilders, but it sold for just ten thousand—still a substantial sum.[27]

Another group of organized practitioners were the surgeons. By the mid-sixteenth century they were growing distinct enough from the barber-surgeons (who continued to cut hair, perform common venesections, and deal with blemishes and other superficial ills) to organize their own corporations, making them subject to closer regulation by their gilds and the municipal magistrates, although the details varied from city to city.[28] In the sixteenth century Dutch ordinances often dealt with two kinds of surgeons, those who treated illnesses that appeared on the surface of the body, such as pox (syphilis), cancer, scrofula, excrescences,

Apothecary shop, with distilling apparatus in the lower left. Title page from Daniel Sennert, *Institutionum medicinae*, 1644. Courtesy of the Wellcome Library, London

ulcers, and so on, and those who used instruments to remove bladder stones, repair hernias, and extract teeth.[29] By the later seventeenth century, however, the regulations of surgeons spoke more of ranks than of tasks. The kinds of assistance they offered can be seen in the statutes of the surgeons' guild of Utrecht (there called the Collegium medico-chirurgicum), which in 1676 adopted two sets of fees for various common procedures, presumably distinguishing between masters and journeymen.[30] They included:

> For each bandaging of a complex wound or ulceration done under a doctor's supervision who was called in by the patient, 10 or 6 stuivers; without a doctor, six or four stuivers; if there are two surgeons present, the amount shall be the same as if a doctor was present.
>
> For every bandaging of a simple wound or ulceration at the house of a patient, six or four stuivers; done at the house of the surgeon, four or two stuivers.
>
> For taking care of complicated fractures or dislocations of other parts, such as a clavicle, one ducat or a rijksdollar [about six or four gilders], and for simple fractures or dislocations of the same, two gilders ten stuivers or one gilder ten stuivers respectively, and for complete costs of the bandaging of a complicated fracture or dislocation as above, fifteen or eight stuivers respectively, and for binding up simple ones, twelve or six stuivers.
>
> For visiting patients and changing their bandages, with a doctor present, six stuivers, without a doctor, four stuivers.
>
> For trepanation, seven or five gilders.
>
> For setting a seton [to make a small, weeping ulcer to drain away bad humors] one gilder ten stuivers or fifteen stuivers.
>
> For cutting out cancers in the breast, fifteen gilders or seven gilders 10 stuivers.
>
> For draining the chest or stomach by paracentesis, five gilders or a rijksdollar [worth four gilders].
>
> For amputating an arm or a leg, ten gilders or seven gilders ten stuivers.
>
> For cutting out any fleshy growths or swellings, one, two, or more gilders in proportion to the size, and in case of a difference about the fee, the town physician to make the final determination.
>
> For making a blister, twelve stuivers, the medicine and the first bandaging included.
>
> For a venesection on the arm, ten stuivers or six to four stuivers; in the foot, twelve stuivers or six.

These surgeons also dealt with prolapsed intestines and wombs, removing dead or nearly dead infants stuck in the mother's birth canal, setting broken bones, and various other procedures. Major operations, such as removing a bladder stone, posed dangers to the patient that were very serious but manageable so that only a few of the most expert master surgeons engaged in such practices, while the risk of death from operating on the abdomen, thorax, or head were so great as to be undertaken only in the rarest of cases.

Urban environments allowed physicians to flourish, too. They by definition had a medical doctorate obtained after a university education in their subject: in The Hague, for instance, the magistrates passed an ordinance in 1622 prohibiting any "doctors" from giving out drinks, waters, or other medicines or from visiting the sick unless they first presented their degree to the magistrates.[31] Even the word *physician* implied one who had studied *physic*, a word derived in turn from the Greek word for nature (*phusis*). From his knowledge of nature, a physician could advise on how best to retain or recover health by working in accordance with nature's dictates. As learned givers of advice, physicians were therefore like the clergy and lawyers. But the practical orientation of much of the university teaching in the Dutch medical faculties was carried over into the vernacular, which pointed to the activity of the persons rather than their qualifications: the word for healing, *genees*, was expanded to an honorific *geneesheer*, literally "sir healer," while the practice of physic became *geneeskunde*, the "art of healing." In this, Dutch was less like English and more like French, in which physicians were called *médecin*, a word derived from the Latin *medico*, originally meaning to treat or to dye cloth with tinctures.[32]

The early modern urban environment of the low countries supported many physicians. They could obtain a degree from Leiden, of course, as well as from Franeker (1585), Harderwijk (1600), Groningen (1614), and Utrecht (1636) or any number of universities nearby The Netherlands.[33] The Leiden curriculum might be considered typical: as in the other faculties, the medical professors lectured four days a week, with Wednesdays and Saturdays reserved for formal disputations (organized wrangles over medical principles and particulars), examinations, promotion exercises, book sales, and other academically related activities; Sunday was a day of rest.[34] Students acquired a knowledge of the foundations of nature, then turning to the particular application of natural principles in matters of health and disease by examining the "naturals, non-naturals, and contranaturals." Studying the naturals meant examining the makeup of the healthy body, from the organs, humors, and faculties to the "temperament" or "constitution" of the whole body to the three fundamental changes of generation, growth, and nutrition. In the mid-sixteenth century, these "naturals" had been renamed "physiology" by the Parisian physician Jean Fernel. The contra-naturals were things that operated against the ordinary course of nature, such as birth defects or defective limbs. The non-naturals—six in all—were the activities that affected the natural human body, keeping it in health or bringing on disease: air, food and drink, exertion and rest, sleep and waking, retentions and evacuations, and passions of the mind.[35] Following an examination of this first and most basic part of medicine, students would go on to learn the principles behind the causes and signs of disease (diagnosis), the course of health or disease (prognosis), the preservation of health and prevention of disease (hygiene), and methods by which to restore a sick body to health (treatment). These five general subjects (physiology, diagnosis, prognosis, hygiene, and treatment) could be studied according to the best modern knowledge but together composed the medical "institutes," or the elements of the subject, in the same way that theology and law had their own foundational institutes. Depending on the resources available, additional instruction could be had in the botanical garden, in the anatomy theater, and even (after the later 1630s in Utrecht and Leiden) in a local hospital ward.

Dutch university medicine never favored the teaching of medical astrology. Astrology had been one of the chief methods by which physicians determined the effects of the larger world on the health and welfare of particular people, and because it involved the study of astronomy, medical astrology was the main reason why medical professors provided much of the advanced teaching in mathematics in late medieval and sixteenth-century universities.[36] At mid-sixteenth-century Louvain, for instance, astrology constituted an important part of the curriculum, with the medical professor there, Gemma Frisius, becoming one of the chief mathematicians and cosmographers of his generation.[37] Because of the connection between medical astrology and mathematics, it is not surprising that the first Dutch student in medicine at Leiden, Rudolph Snell, also became its first professor of mathematics in 1579. Yet despite the connections of Leiden physicians with mathematics and astronomy, astrology never became a regular part of medical teaching there or elsewhere in the northern provinces. Among contemporaries there were certainly intellectual movements that had begun to question the legitimacy of astrology. The influential Lipsius, for example, rejected astrological causes of fortune, agreeing with a critic that the "stars and their positions in heaven" were among the most important "instruments of fate" but not its cause.[38] The involvement of the medical faculties in these debates is not yet well understood, however. Some physicians mounted a defense of astrology, yet others were beginning to raise public questions about the subject.[39] Early in his life the noted Parisian professor Fernel had expressed the common interest in astrology as a critical adjunct to medicine, but by 1545 he was indicating serious doubts about its utility for medicine. His doubts may have arisen from checking the agreement between anticipated astrological influences and bodily expressions in his patients or from beginning to doubt more generally the powers associated with natural magic.[40] At the same time, it should be noted that in the writings attributed to Hippocrates, the only passage referring positively to "astrology" is in the often-read "Airs, Waters, Places," where in paragraph two it is said that knowing the future pattern of the weather can help a physician maintain health and predict the course of diseases.[41] This is, however, more a general comment about meteorology than astrology and a far cry from the complex methods of prognostication based on mathematical calculations of the movements of the heavens with regard to particular people, which is what constituted late medieval and early modern medical astrology. Even at Louvain, serious doubts were being cast on astrology by the 1580s both because of a lack of experiential verification of its practical outcomes and because of internal contradictions in its theories.[42] It may be, then, that the Hippocratic commitments of Heurnius and Bontius and most other Dutch medical practitioners, which emphasized knowledge derived from experience, are at least partly responsible for their lack of concern about astrological instruction.

Another part of medicine also remained unavailable in most universities for several decades: iatrochemistry. Alchemy was a distinctly unclassical subject, becoming a common method of analysis only after the time of Hippocrates, Plato, Aristotle, and even Galen. Moreover, its results seemed to support a view that things were made up from a variety of kinds of corpuscles, a view that William Newman has now shown was present in the twelfth-century author known as Geber.[43] Iatrochemists therefore found fundamental axioms of classical theory to be wrong or perverse, they experimented (often working their furnaces in person) in order to discover new things, and they had a vision of forces working throughout the universe that remained beyond the grasp of reason while yet believing that they could ultimately bend them to human use, holding that knowledge is power.[44] Thanks to the work of a Dane, Petrus Severinus, who had studied in Basel, Paracelsus's works were revised and given an overall coherence and proper Latin phrasing in *Idea medicinae* (1571), making it possible for people used to academic discourse to study Paracelsus's views.[45] Consequently, the medical professors of some universities beyond Basel began to offer teaching in iatrochemistry.[46]

In The Netherlands, however, one of the most vociferous enemies of iatrochemistry was Forestus, one of the first two professors appointed to the medical faculty of Leiden, so although that university may have had ambitions to give instruction in that subject, it would not do so until the mid-seventeenth century. Iatrochemistry did, however, become part of the offerings at Harderwijk (where the Latin school was made into a university in 1600). There Johannes Isacius Pontanus was appointed professor of philosophy and medicine in 1605. After studying at Franeker, he had traveled widely in Italy and Germany and then spent several years on the Danish island of Hven working with the famous astronomer and alchemist Tycho Brahe. After further travels, he took his medical doctorate at Basel in 1600, undertook further travels in Switzerland and France, and finally took up the appointment at Harderwijk, where he incorporated iatrochemistry into his teaching. Pontanus became one of the most distinguished scholars of his generation, turning down many invitations to teach elsewhere.[47]

Given their education in the ways of nature, physicians considered themselves to have the best understanding of the human body, environmental and other causes of disease, and methods of prevention and treatment; other well-educated people, such as the magistrates, also appreciated the council and advice of physicians. Many cities therefore appointed one or more town physicians, who would be paid a salary by the town council to treat the poor without charge and to advise on matters of health and disease (especially during epidemics of plague); they could carry on in private practice in the rest of their time.[48] Some of them could be quite distinguished, even famous, like Paludanus of Enkhuizen or Forestus of Delft. Another was Johan van Beverwijck, who studied medicine at Leiden from 1611 to 1614, traveled in France and Italy, where he graduated from Padua in 1616, then settled in his hometown of Dordrecht, where he was appointed city physician in 1625. Among other activities, from 1634 to 1643 he gave lessons in medicine and anatomy to the town's surgeons, midwives, and plague masters under the auspices of the surgeons' guild. He was a vocal defender of women's equality with men, carried on a large correspondence with other learned men throughout Europe, and published numerous medical works, many of which had several editions, the most famous being his work on plague, *Kort bericht om de pest te voorkomen* ("Short report on prevention of the plague," 1636), and his medical handbook, *Schat der Gesontheyt* ("Treasury of health," first printed in 1636).[49] *Treasury of Health* was soon supplemented by his *Treasury of Disease.* The two-volume *Treasury* constituted what one physician called a great "rhapsody of almost everything that had been said on that subject before that time, by doctors and especially by poets."[50] And Beverwijck held positions on the town council and on occasion represented the city in other assemblies.[51]

Booming cities like Amsterdam attracted many physicians who wished to make their reputations and fortunes. One of the most interesting was François dele Boë Sylvius. During the revolt, his grandfather had fled from the southern netherlands to Frankfurt; François was born nearby, at Hanau. As a youth, he traveled to many Protestant and iatrochemically influenced universities in the northern netherlands, Germany, and Switzerland for his education in medicine, graduating in 1637 from Basel, still a seminary of medical chemistry. After further travel, Sylvius settled briefly in Leiden, where in 1639 and 1640 he gave private lectures in anatomy and physiology in the gallery of the hortus, being among the first on the Continent to demonstrate the circulation of the blood and to examine the lymph system. At Leiden he also became acquainted with René Descartes. Then from 1641 to 1658 Sylvius took up residence in Amsterdam, where he developed a successful practice. He worked closely with such other experimental physicians as Paulus Barbette (whose posthumous *Praxis Barbettiana*, 1669, was edited by one of Sylvius's later pupils and became a standard text on surgery)[52] and Franciscus van der Schagen (who became one of Descartes's friends), as well as excellent surgeons like Job van Meekren and Hendrick van Roonhuyse. In Amsterdam he also befriended the German physician Otto Sperling and the German iatrochemist Johann Glauber, famous for his chemical salts, who spent 1640 to 1644, 1646 to 1650, and 1656 to 1670 in Amsterdam, where he is buried.[53] According to one memoir, despite a busy practice Sylvius continued to dissect at least once a week and otherwise often spent time at his distillation apparatus or at chemical furnaces he had set up in his house. In 1658 Sylvius was attracted to a professorship in Leiden at an unheard-of salary. Given that the house he purchased in Leiden on leaving Amsterdam fronted on the main canal just a few doors down from the university itself, he must have already made a small fortune in his private practice.[54]

In Leiden Sylvius pressed hard for improving its clinical teaching by connecting the latest ideas about the causes and treatments of disease with instruction on real patients. In most universities, for some centuries, it had been common enough to expect any intending physician to have followed an experienced practitioner in order to learn about diagnosis and prognosis at the bedside, but this was done beyond the professors' instruction. In the 1540s, in Padua, a medical professor named Da Monte began to take students to a hospital to show them clinical cases.[55] At Leiden, there were plans to do the same when the medical faculty was founded.[56] But the plans remained only that until the new university in Utrecht made a move. Willem van der Straten, in his inaugural address in Utrecht in March 1636, announced that he would be taking students with him to the local hospital (*Nosocomium*) to see the patients under his care. (After a few years, this ambition was quietly dropped.)[57] Not to be outdone, in Leiden Professor Otto Heurnius wrote immediately to the curators requesting facilities to do the same, and by the end of the year they were able to arrange for their professors to give practical instruction on one of the wards of a city hospital (the Caeciliagasthuis) in cooperation with the town physicians. The teaching there, in the "Collegium medico-practicum," went through periods of on-again, off-again, until Sylvius threw himself into practical instruction with energy. He revived the Socratic Method by asking the students themselves to describe and recommend treatment for the patients they saw while he questioned them closely, and he continued to prescribe his favorite chemical medicines for the patients. When the city council objected to the great expense of this latter practice, the curators subsidized it, and Sylvius continued to prescribe what he wished to the inmates under his care. He also dissected them if they died in the hospital, so that their morbid anatomy could be compared to the symptoms they showed while alive.[58] (In Amsterdam, as well, from at least 1662, under the direction of the physician Gerard Blasius, the bodies of patients who died in the municipal hospital in Amsterdam, the Binnengasthuis—or Pietersgasthuis, as the traditionalists called it—were being used for anatomical study.)[59] To study treatment further, Sylvius invited surgeons to visit him and his students to perform demonstrations of their methods.[60]

As Sylvius's example indicates, cooperation among physicians, surgeons, and apothecaries could be very good in cities like Amsterdam. For instance, one of Sylvius's associates, Barbette, a physician and son of a surgeon, worked closely with Job van Meekeren, one of the best-known surgeons of his generation. Another of Sylvius's pupils, Frederik Ruysch, became the lecturer on anatomy to the Amsterdam surgeons' guild, in 1669 became an anatomical examiner of the midwives, and in 1672 their anatomical lecturer. Ruysch took over the instruction of midwives after the death of Hendrick Roonhuyze, another noted surgeon who worked closely with physicians and published several books, including a study of childbirth. Roonhuyze's son in turn purchased the secret of obstetrical forceps from the English physician Hugh Chamberlen, Sr., in the mid-1690s, selling it on to Ruysch.[61] At the same time, some Amsterdam surgeons, like the noted Johannes Rau, were earning medical degrees in the later seventeenth century. Even Dutch medical institutions indicate the possibilities for cooperation, since many continued to assemble, rather than separate, educated physicians, apothecaries, and surgeons. In The Hague, for instance, from 1629 the so-called Collegium Pharmaceuticum and Collegium Chirurgicum were names for meetings that took place between the apothecaries or surgeons and the city physicians, while at Haarlem a joint Collegium Medico-Pharmaceuticum was set up for the physicians and apothecaries in 1692, with the surgeons joining it in 1693.[62] Dutch surgeon-physicians have consequently been identified by at least one influential medical historian as being in the forefront of a revolution in clinical knowledge that resulted from combining surgical and medical knowledge.[63]

The association between Dutch physicians and apothecaries also seems to have been relatively strong, at least in the late sixteenth and early seventeenth centuries. Many notable early physicians were in fact sons of apothecaries. In 1609, the Amsterdam city council (*vroedschap*) received petitions seeking to establish a common medical guild containing both physicians and apothecaries. The physicians would have to show their university degrees and the apothecaries would have to perform a master proof using their knowledge of medicinals. The physicians would be formally required to help the poor whenever they were requested and would be paid an honorarium of one gilder for each patient they helped who was sick with plague and ten stuivers in all other cases (in return for which the city would also excuse them serving on the night watch); they would not prescribe by inspecting the urine alone (that is, the patient would also have to be present); in order to prevent accidents, a number of dangerous medicines, such as antimony, turbinth, mineral essences (*mineralis precipitatum*), and white hellebore could be prescribed only after a consultation of at least two doctors; and to prevent the false preparation of medicinals, they would have to write out a receipt for someone else to prepare, unless it concerned a secret remedy. At the same time, the apothecaries would have to abide by various other rules, including compounding all ingredients according to the "antidotarium" (the relatively short medieval *Antidotarium Nicolai*, stipulated by an ordinance of 1550), with the preparation of most important medicines being overseen by the dean or other leaders of the group. There would be strict regulations on the sale of poisons, and various regulations would be enforced over nonguild druggists, grocers, herbalists, sugar refiners, distillers, and so forth. Yet other regulations required the obedience of the surgeons to the physicians, who were threatened with the abolition of their guild unless they obeyed. Although the doctors and apothecaries may have been happy with this proposal, the surgeons certainly were not, and opposed it, ending the plans for the moment.[64]

Near the surface of the common interests of the physicians and apothecaries was, of course, botany. As elsewhere, apothecaries often had small botanical gardens, as Dirk Outgers Cluyt had in Delft when he was recruited to manage the garden in Leiden. In Amsterdam, Thonis Jansz. owned a shop called "The Fountain" (de Fonteyn) in the middle of the city, on the Dam, while also possessing a well-known garden just outside the city walls in which he grew many of his medicines. He gave the garden to his son, Rem Anthonisz. Fonteyn, on his wedding in 1609 to enable him to continue the shop. (The garden came within the confines of the city during Amsterdam's expansion in 1612.)[65] But for apothecaries who did not have their own gardens, and for physicians, given the revolution occurring in botany and pharmacy and the widening availability of medicinal substances from exotic places—to say nothing of rapid developments in iatrochemistry—a strong need was felt to have a place where they could further educate themselves and their apprentices. Twenty-three physicians and twenty-one apothecaries therefore joined in petitioning the city council in 1618 for a civic botanical garden for the education of apprentice apothecaries and surgeons. Among the petitioners was Dr. Augerius Cluyt (son of Dirk Outgers Cluyt, of Delft and Leiden), who had since moved to Amsterdam. The request was put into the hands of three senior medical figures: an apothecary, Steven Jansz., a physician, Dr. Johannes Fonteyn (another son of the apothecary Thonis Jansz.), and Dr. Sebastiaen Egbertsz., a former burgomaster and the current anatomical lecturer to the surgeons. Not long after the petition for a botanical garden, there is evidence in a map book of 1628 of its presence (a *stadsartsenijtuin*) on the east side of the Amstel, although it was probably not a new garden but a garden in the old cloister where the municipal hospital, the Binnengasthuis (or Pietersgasthuis) now stood; in 1627 a city gardener (*stadsgardenier*) was also mentioned.[66]

Another request for support from the city came in 1629, when petitioners argued that a substantial medical garden was necessary for study by local doctors, apothecaries, and surgeons. This time, the burgomasters were favorable, but they had difficulty deciding who should be prefect. One of the chief candidates was the friend of Constantijn Huygens, Johannes Brosterhuysen, an enthusiastic botanist (as well as poet, musician, and engraver) who in 1627 had helped Huygens lay out his private garden at his house on the Lange Houtstraat in The Hague. By the early 1620s, the Prince of Orange and people around him were laying out large gardens, partly in imitation of classical, Italian, and French garden styles but partly also in response to local requirements and concerns that gave rise to a Dutch form, the canal garden, divided internally into separate rectilinear spaces by the drainage canals.[67] Huygens, as chief secretary to the prince, laid out similar gardens, first at his house in The Hague and then around 1640 at his country house, Hofwijck ("Away from the court"). He and "Father" Cats were only the most famous to write poems about their estate gardens (*hofdichten*) as places of ordered natural beauty, healthy exercise, and virtuous respite from the evils and disorder of city life. Tending one's garden brought one as close to Adam and the Virgilian *Georgics* as modern life allowed.[68]

Given that they were being asked to spend money on a more utilitarian project, however, the burgomasters finally appointed the more academically qualified Cluyt. He had traveled extensively in Germany, France, Spain, and North Africa, sending specimens back to the Leiden hortus as he went.[69] (In 1636 Cluyt was made the inspector of the Leiden Hortus and left Amsterdam.) In addition, the burgomasters looked into purchasing land on Nieuwland by the Swanenburghwal for an expanded hortus but decided that it would be too expensive. Instead, it was set up in 1630 in the former Regulieren cloisters. The plans for founding

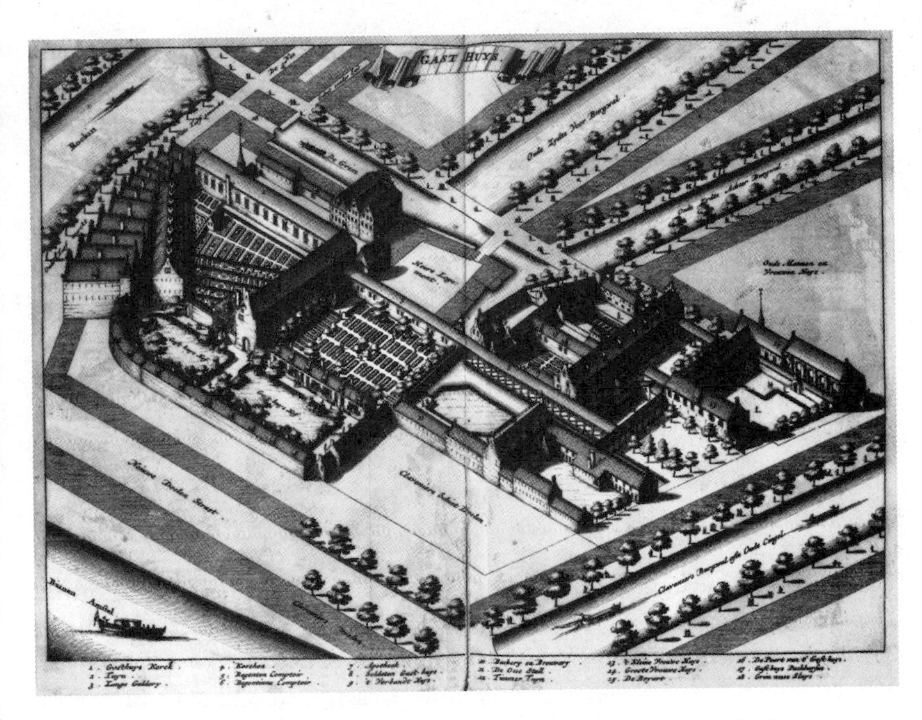

Binnengasthuis (hospital) of Amsterdam, with garden depicted within.
From Caspar Commelin et al., *Beschryvinge van Amsterdam* (1693).
Courtesy of the Wellcome Library, London

the Athenaeum gave Brosterhuysen another opportunity. In 1631 he sent Huygens a plan for a more substantial garden with a note suggesting that it should be made a part of the intended school.[70]

Given the available opportunities, the number of medical practitioners in a city with a booming economy like Amsterdam's was very great. Of master surgeons the city had perhaps about 250 from 1600 to 1670[71]—perhaps we can conservatively estimate a peak number of one hundred practicing at any one time. In 1585 there were 22 identifiable apothecaries and in 1618 about the same number, but their number trebled (equal to the increase in population generally) to 66 in 1636.[72] Until the end of the 1630s physicians had no common institution in Amsterdam, and so their numbers cannot be estimated before then, but in 1636 there were about 58, rising to perhaps 80 by the 1650s.[73] In rough terms, then, combining the approximate number of formally certified practitioners in 1636 (roughly 225) to service the medical needs of perhaps 120,000 residents of Amsterdam, one finds a ratio of about two regular practitioners to every thousand people or even somewhat more (not so different from modern Europe).[74] If this estimate is correct, there would seem to be about half as many practitioners offering their services to the public in Amsterdam as in an English city across the channel, Norwich.[75] But the estimate for Norwich is based on *all* those practicing, not just the regulated physicians, apothecaries, and surgeons. When one allows for the large (unknown) population of vrije meesters, market-stall empirici, and more covert piskijkers, cunningwomen and men, midwives, and their like, one might well imagine the medical marketplace in Amsterdam to have been at least as rich in number and variety of practitioners as other North Sea cities. This seems to have been true of other places in the provinces of Holland, Zeeland, and Utrecht, too, although in the more eastern provinces a sparser population meant that practitioners in a service industry like medicine had a harder time becoming well-off.[76]

REORDERING AMSTERDAM'S MEDICAL LANDSCAPE

Into the rich medical environment of Amsterdam Dr. Tulp brought further order in conformance with his views. His circle of friends, along with the wealth he earned as a physician, elevated him to the regenten, a rank few physicians of the period (such as Beverwijck in Dordrecht) achieved. Although his private patients seem to have appreciated his abilities, Tulp's public face was most clearly illuminated when he acted as a medical politician. Born in Amsterdam in 1593 as Claes Pietersz. (Claes, son of Pieter), he came from a line of merchants involved in the Baltic trade; his father had also been a firm supporter of the committed Reformed group in Amsterdam. The family of Claes's mother, Gherytgen Dircksdr., had been connected to the orthodox and pro-Spanish regents who ruled Amsterdam before 1578. But like so many families of the time of the Revolt, they had divided religiously, and since his mother's side joined the most zealous Calvinists (the *vromen*), when Calvinist refugees flooded that year into the city and, with the support of the populist civic guard, forced the "Alteration" on the city council and sent many leading Catholics into exile, she grew up in the dominant party.[77] Claes therefore came from a family of some substance associated with powerful religious and political networks, which in time allowed him to join the regenten himself.

Several relations on Tulp's mother's side including an uncle, Dr. Wouter Moerselaar, were physicians, which undoubtedly caused him to consider a career in medicine. The young Tulp matriculated in the medical faculty at Leiden in 1611 at the fairly late age of about eighteen, followed by his brother, who went to Leiden to study theology. Because of the death of his father in 1612, he did not travel for further education, although his family could afford to keep him in the university until he completed his studies in 1614.[78] At Leiden, he absorbed anatomy and botany from Pauw and practical, Hippocratically oriented medicine from Reinier Bontius, Otto Heurnius, and Aelius Everhardus Vorstius. Like his mentors in Leiden, Tulp took a deep interest in natural history and clinical medicine. His medical book, *Observationes medicarum* ("Medical observations," 1641), shows his outlook clearly: it collected many of his cases and observations, was richly illustrated, and went through several subsequent Latin editions because of the intrinsic interest of his descriptions. It considered tobacco to be dangerous rather than of medicinal benefit, reported the case of Jan de Dood, who had surgically extracted his own bladder stone, gave details of a woman whose body spouted liquid through her navel, accurately depicted an "orang outang" brought back to Amsterdam (although the ape was clearly what would now be called a chimpanzee), and wrote of many other curious matters.[79] A Dutch translation of his *Observationes* appeared from the shop of the Amsterdam bookseller Jacob Benjamin in 1650 without Tulp's involvement, and he was apparently unhappy enough with this pirated version to begin work on his own Dutch edition, although it was never printed.[80] His medical writing, therefore, like his anatomical demonstrating, paid close attention to careful description of the observable world rather than to theorizing. The legacy of Leiden is also to be found in his keen interest in medicinals but not iatrochemistry and in his excellent ability in anatomy: Tulp's most notable anatomical demonstration occurred during the winter of 1638–39, when he demonstrated the lacteal vessels, which had been discovered in a dog by Caspar Aselli in 1622 (published in 1627).[81]

While he was studying in Leiden, the university and the country were going through the political upheavals of a powerful conservative religious program. A theologian appointed at the university in 1603 to balance the conservatives, Jacobus Arminius, had taken the view that people had a modicum of free will with which they could chose to do good works in this life and thereby help themselves obtain salvation. But a colleague in theology who had been at Leiden for a longer time, Franz Gomarus, strongly opposed the newcomer on this, arguing that salvation depended entirely on God's will, not ours, with the consequence that personal salvation was predestined and could not be altered through good works. According to Gomarus, the Arminian position sailed too close to the Catholic view favoring free will and therefore not only led individual people astray but gave aid and comfort to the enemies of the Republic. Although in earlier years arguments about predestination had not been important to Calvinist doctrine, by 1604 the wrangling of the theologians on this point had become public and angry, bringing the question to the forefront. Within a few years, even ordinary people began to take the matter seriously and to divide sharply over it.

When the Twelve Years' Truce between the United Provinces and the Hapsburgs commenced in April 1609, even greater anger arose among those who wished to continue the war and recover the southern provinces, many of them militant Calvinists who now gave their full support to the Gomarists. The dispute entered the debating chambers of the States of Holland, with the supporters of Arminius finally putting a petition (the Remonstrance) before the States General in 1610 asking that their views be allowed, only to be countered by a counterpetition by the Gomarists (Counter-Remonstrants), who wanted them silenced. Politically speaking, if the Counter-Remonstrants won their case they would obtain state backing for a theological formulation, threatening to create a theocracy. The person at the top of the civil service, the advocate of the States of Holland and Zeeland, Johan van Oldenbarnevelt, strongly opposed such clerical precedence. By the middle of the 1610s armed supporters of the two sides had come to blows.

In 1617, Prince Maurits, stadholder of Holland and Zeeland, threw his weight behind the Counter-Remonstrants: he was intent on reconquering the southern provinces and, like Leicester before him, needed the support of the most determined anti-Catholics in the Republic. But Advocate Oldenbarnevelt felt equally strongly that the Republic needed to be protected against the union of Orangist and Counter-Remonstrant parties that threatened to transform power in the Republic, and he in turn organized his own armed forces (*waardgelders*). With the country on the verge of civil war, Maurits launched a coup by arresting Oldenbarnevelt in August 1618 and by purging local governments and intruding his supporters in their places. A special court of the States General set up by Maurits tried Oldenbarnevelt. At the same time, with Maurits's backing, a national synod of the Reformed church met in the city of Dordrecht from 13 November 1618 to 9 May 1619 (in shortened form called the Synod of Dort), which declared against the Remonstrants. Four days after the Synod of Dort disbanded, Oldenbarnevelt was publicly executed. Maurits was also intent on supporting the Reformed cause in Europe more generally. He urged his cousin, the Calvinist Palatine elector Frederick V, to accept the crown of Bohemia in 1619, which would have given the Protestant interest in the Holy Roman Empire the upper hand in deciding the next emperor. But Frederick was deposed at the Battle of White Mountain in 1620 by Emperor Ferdinand II and Duke Maximilian I of Bavaria, effectively beginning the Thirty Years' War that would devastate the German lands. After the defeat, Maurits welcomed Frederick and his wife, Elizabeth Stuart, to his court, and did all he could to help them regain the throne. The war party soon also succeeded in reviving the battle against the Hapsburgs by not extending the Twelve Years' Truce when it ended in 1621.

Domestically, the Counter-Remonstrant party demanded obedience. In Leiden committed Arminians were banned. Barlaeus, then a brilliant young philosopher, was removed from his position as professor of logic because of his open support for the Remonstrants. Others, like the learned humanist and Barlaeus's future colleague in Amsterdam, Gerhard Joannes Vossius, objected to the overly doctrinal view of religion held by the Counter-Remonstrants, considering them to have built a doctrine on a few passages of scripture taken out of context. He was forced to give up the position of rector of the theological school (the States College), although because he remained on good personal terms with Gomarus and many of that party, he held onto his teaching position. Another victim was the famous Hugo Grotius. At Leiden, Grotius had developed into one of the best Latinists, historians, and learned commentators of his generation. After taking a doctorate of laws he could have become a professor, but like others before him, Grotius preferred to make his way in the world, becoming one of the most penetrating analysts of the early seventeenth century. He became a firm supporter of Oldenbarnevelt, was imprisoned, and escaped only to suffer permanent exile at the hands of the Counter-Remonstrants. The Counter-Remonstrants attacked other schools as well. Samuel Coster had set up the Nederduysche Academie ("Dutch Academy") in Amsterdam in 1618 to instruct not in religion but in useful arts, such as comedy and tragedy, mathematics, astronomy, surveying, navigation, history, Hebrew, philosophy, and letters—although only navigation and letters were, in the end, actually taught. Yet although it was intended as an example of useful civic enterprise, by failing to teach religion Coster's academy appeared to the Counter-Remonstrants to be a bastion of libertines and Remonstrants, and they forced its closing in 1622.[82]

Claes Pietersz., the future Dr. Tulp, had completed his medical studies in 1614, just as the two parties were organizing their forces. He must have sided with the Counter-Remonstrants, for his brother went on to represent that party in his capacity as an ordained clergyman, and when Tulp returned to Amsterdam, his family connections and reputation as a staunch member of the true Reformed undoubtedly helped make him a trusted member of that community. In 1617, at the time the Prince of Orange threw his weight behind them, Claes entered into marriage with Eva van der Vroech, whose uncles had also been among the leaders of the Reformed party in Amsterdam. The young couple set up house on the Prinsengracht, the newest canal in the expanding city, lined with grand houses for the well-to-do.[83] Two years later, he could even afford to purchase land along the even more imposing Keizersgracht near the stately church built on the west side of the city, the Westerkerk. He was also caught up in the growing interest in tulips, as well, for by 1620 he was sealing documents using the sign of a tulip and hung the sign of a tulip outside his residence to make it easier for patients to find him. When he moved to the new house on the Keizersgracht in 1621—which had room for his horse and coach—he had a tulip engraved into the stone outside, thus to patients, family, friends, and even himself becoming "Dr. Tulip" or, as the Dutch has it, "Tulp."[84]

As he grew in wealth and reputation, Tulp's political activities grew in scope. He obtained patronage from an influential clan of energetic Calvinists, the Vrij family.[85] Burgomaster Frederick de Vrij urged his fellows not only to accept Counter-Remonstrant views about predestination but to lead godly lives while energetically building a godly society. In other words, Vrij was an early supporter of the so-called Later Reformation or Further Reformation (*nadere reformatie*), by which Calvinists sought to morally reform both public and private life in The Netherlands. Partly inspired by English Puritanism, and led by ministers like the prolific Willem Teellinck from Zeeland, the movement urged godliness in all things.[86] In 1622, Tulp gained election to a seat on the vroedschap and was chosen a magistrate from the list of nominees submitted to Prince Maurits.[87]

But Tulp was one of the last of the true Reformed candidates elected to the council. By 1622, with resentment at the execution of Oldenbarnevelt growing and with negative economic consequences for the trading interests of Amsterdam from the renewal of war, the city began to turn against his party. The death of Prince Maurits in 1625 gave even more courage to the "libertines." When the Counter-Remonstrant faction resorted to intimidation of their opponents by storming and plundering a house on the Monckelbaansburgwal in April 1626, it only increased the backlash against them, with the elections in February 1627 resulting in two more "libertines" being added to the vroedschap, tipping power in Amsterdam for good in the direction of the liberals. Renewed threats of violence by the Counter-Remonstrants in 1628, aimed at purging the government of Amsterdam, failed to turn the tide now fast flowing against them. The less doctrinaire party led by Andries Bicker (and called the "Bicker League" after him) consolidated its grip and for thirty years made Amsterdam a bastion of support for the so-called States Party that was ranged against the Orangists and strict Reformed, despite some powerful voices on the vroedschap still in favor of prince and cleric. When the Bicker League took power in 1627, Tulp "wobbled" in trying to mediate between them and the true Calvinists, but Tulp remained true to his colors as a conservative religious moralist when, two years after his first wife died (in 1628), he married Margaretha de Vlaming van Outshoorn, who also came from among the "devout" Reformed in the city.[88]

Throughout his life Tulp remained committed to public service. At forty, he had a notable portrait painter of the city depict him gesturing toward a candle flame while underneath stood a motto on a shield bearing a skull, "I am consumed in the service of others" (*Aliis inserviendo consumor*). Some of this was in the interest of his religious convictions. In 1642, as presiding magistrate on the city's judicial council (*magistraat*), he led his colleagues to support the Walloon (French-speaking Calvinist) church, which had excommunicated a member for heresy. The magistrates, headed by Tulp, also secretly tried the offender and found him guilty, sentencing him to eternal incarceration in solitary confinement, in a room of the workhouse (*Tuchthuis*) whose windows were nailed shut. When sometime later the matter became known, Tulp lost the offices usually held by members of the vroedschap. But when the pious Calvinists regained political influence in the mid-1650s, Tulp gained a seat as one of the four burgomasters of the city and immediately took up his old cudgels, having Joost van den Vondel's play *Lucifer*—a model for John Milton's *Paradise Lost*—banned from public performance. Perhaps his most notorious effort to impose his moral compass on the rest of the city came in getting antisumptuary legislation passed in 1655 aimed particularly at wedding celebrations. In 1659, Tulp's religious bigotry emerged again, when he was the only member of the city council to argue against allowing the Lutheran congregation to relocate and expand their church, and shortly afterward he spoke out against the "pagan" ceremonies welcoming such dignitaries as the elector of Brandenburg and Princess Mary Stuart to Amsterdam.[89] Yet he remained active in affairs of the city, eventually serving as a burgomaster four times, a city treasurer seven times, a trustee of the orphanage twice, and a curator of the Athenaeum from 1666. When the French invaded the Republic in 1672, despite having recently resigned from the vroedschap because of frail health, Tulp gave a rousing public speech about courage in the face of what seemed certain defeat.

Tulp also established new medical institutions in Amsterdam in line with his sense of correct moral order. Many physicians had long advocated tougher measures to give them legal powers to control medical practice and practitioners. One of the most public advocates for such a position had been Forestus, who believed that he had been forced to stand by helplessly and watch patients die who had taken remedies administered by people without a proper medical education. In his view empirics and iatrochemists understood little of what they were doing or—equally dangerous—did not understand things properly. He published one of the best-known tirades against empirics of the period: *De incerto, fallaci, urinarum judicio* (1589, translated into Dutch in 1626; known in English via its translation as *The Arraignment of Urines*, 1623). It began with a discussion of one of the most common methods of diagnosis, that of examining a patient's urine. Since classical theory had discussed physiology in terms of a series of digestions in the body, the waste products given off from digestion could give a good indi-

Portrait of Nicolaes Tulp. Line engraving by C. van Dalen after N. Eliasz (Pickenoy), 1634. Courtesy of the Wellcome Library, London

cation of how the body was functioning.[90] Indeed, many medieval treatises explained how to use the inspection of urine ("uroscopy") in diagnosis, carefully observing the color, texture, smell, and taste, the matter floating in suspension, and the surface scum that might appear after setting the urine flask aside for a time. Patients had become accustomed to believing that no diagnosis was complete without uroscopy and indeed that it was the only necessary element. When they were ill, they therefore sometimes sent out their servants with a flask of their urine for diagnosis without bothering to pay a visit in person, and sometimes they did this to test whether the practitioner could tell that the flask contained only white wine or the urine of an ass or that of a man instead of woman, and so forth. For learned physicians like Forestus, clear collusion existed between such patients and irresponsible practitioners who told them what they wanted to hear and then got them to take the useless or harmful medicines they sold. His treatise first warned against the common practice of diagnosis by uroscopy without also a more general consultation with and inspection of the patient, then going on to list the impostures, frauds, crafts, and cunning tricks of empirics and "wandering water-mongers," and finally telling learned physicians themselves how to avoid errors in diagnosis and prescription, emphasizing that they should never prescribe medicines in the vernacular. If they prescribed only in Latin, respectable and well-educated physicians would remain tied to a similar group of virtuous and Latinate apothecaries, separating their medical knowledge from that of the common empirics and iatrochemists.

It may have been this proposed union of physicians and apothecaries that first moved Tulp to action. In 1633 Dr. Nicolaes Fonteyn published a list of suitable drugs, a pharmacopoeia (the *Institutiones pharmaceuticæ*). He dedicated his work to the burgomasters and members of the city council—Tulp being one of them, of course—arguing for the proposition common in many other countries and cities, that the apothecaries should only dispense medicines prescribed by physicians and in accordance with an official pharmacopoeia. Fonteyn was a grandson of Thonis Jansz., who had owned the apothecary shop "The Fountain," and the son of Dr. Johannes Fonteyn, Tulp's predecessor as anatomical demonstrator for the surgeon's guild. But he was also a Catholic. Fonteyn's new work therefore derived its pharmaceutical content mainly from French works, reflecting his religious background and his education in Paris.[91] He must have moved in somewhat different Amsterdam circles than Tulp and was probably distrusted by him.

Tulp therefore invited a group of physician friends to a lavish dinner at his house on 18 April 1635, at which he proposed that they draw up a new pharmacopoeia that would regulate the practices of apothecaries. They agreed to set up a committee of seven, with Tulp as chairman, to compose a pharmacopoeia that better suited their views.[92] Shortly after Tulp proposed his initiative, the plague that had festered for a few months broke out into a severe epidemic that lasted until 1636, which is sometimes said to have prompted Tulp's initiative.[93] Even during the plague, however, which caused about twenty-five thousand deaths in Amsterdam alone during 1635–36, their work went forward. The committee of physicians also decided to co-opt three apothecaries as consultants, including one of Fonteyn's relatives, his uncle Rem Anthonisz. Fonteyn (he of the shop on the Dam and the medicinal garden). The resulting document apparently grew out of Tulp's appreciation for the Augsburg pharmacopoeia. After the committee had agreed on the contents, Tulp invited all the physicians in Amsterdam to meet in the surgeons' rooms in the Sint Anthonispoort, where he obtained their general agreement that a new official pharmacopoeia for the city was necessary and that the one he had caused to be drawn up was a good one, although they were also all invited to comment on its details before agreeing to the final version. The work was then handed to the burgomasters, and after they consulted with the city council, it became the official manual on 29 April 1636. After Tulp's publication, Fonteyn left the city.[94] (Already, in late 1633, another senior Catholic physician, Vopiscus Fortunatus Plemp, had left Amsterdam to take up a position at the university in Louvain as president of Breugel College and soon thereafter as professor of medicine.)[95]

The new pharmaceutical ordinances of Tulp's soon also provided the occasion to bring about the establishment of a college of physicians and a guild of apothecaries, institutions Tulp undoubtedly welcomed. The city issued further regulations relating to the enforcement of the pharmacopoeia in January 1637, naming two physicians and two apothecaries as inspectors of the shops.[96] Their duties proved greater than expected, however, and other physicians wished to make further amendments to the pharmacopoeia. Various proposals were therefore made to the vroedschap, resulting in statutes of 15 January 1638 that established both a Collegium Medicum and an apothecaries guild.[97] These two institutions regulated the activities of the physicians and apothecaries of Amsterdam, while the collegium also held supervisory responsibilities for the surgeons' guild and gradually came to oversee the midwives as well. (It met in the surgeons' quarters over the Kleine Vleeshal, or Sint Margarietenklooster, sitting from 11:30 A.M. to 2:00 P.M. every Tuesday.) Because in Amsterdam the guilds had no special political relationships to the city council—unlike, say, in London, where the guild companies more or less ran the City—these groups never had the authority to regulate anyone other than their members. But for the physicians, apothecaries, surgeons, and soon the midwives, the Collegium Medicum and the associated city ordinances were important, indeed, and the plan was imitated elsewhere: Collegia Medica came into being in The Hague in 1658, in Middelburg in 1668, in Haarlem (a Collegium Medico-Pharmaceuticum) in 1692, in Leeuwarden in 1700, in Utrecht in 1705.[98]

The Collegium Medicum also made further efforts on behalf of building up an Amsterdam botanical garden. After the founding of the Athenaeum, Huygens and others had campaigned to get Brosterhuysen made the prefect of a municipal hortus tied to the Athenaeum, but the burgomasters procrastinated. The inspectors of the collegium, however, continued to believe that a great need existed for a large and proper garden where the apprentices could be properly trained, and they pushed hard for one, although again the hesitations of the city council required the original design for an extensive hortus to be scaled back to a medicinal garden. A "hortus pharmaceuticum" in the former cloisters in which inspectors could give lessons was turned over to them in 1638. But inadequacies remained because there was still no head of the garden. Inspectors Drs. Joannes Hartogvelt and Franciscus de Vick requested that such a person be appointed in 1643, and a year later a similar request was made by Drs. Arnoldus Tholinx and Egbertus Bodaeus. Finally, in 1645, Johannes Snippendael took over as unofficial prefect. He had studied philosophy in Leiden, although it is not known that he ever took a medical degree. When he took over, the garden contained more than 330 plants, but within a year he had expanded it to more than 800, according to the catalog of the new and old plants he published in 1646.[99] In March 1646, the city officially appointed him to be head of the garden at a maximum salary of four hundred gilders per year, with free housing in the garden at the Reguliershof, and in one document he is titled Praefectus Horti Gymnasii Amstelodamensis ("Prefect of the Garden of the Amsterdam Athenaeum"), which clearly indicates that a formal connection with the Athenaeum existed by then. In that year Snippendael also began for the first time to offer public botanical lessons, not confined to apprentices. He continued to agitate for the city to buy a larger plot of land that would allow the garden to expand to 1,200 specimens. For unknown reasons, he was dismissed in 1656. Two years later, the garden was indeed moved, though not to a grand new space but to the garden of the main city hospital, the Binnengasthuis, where there is evidence of a medicinal garden as early as 1628. In 1660, a professor of medicine (without salary) was named at the Athenaeum, Gerard Blasius, who also looked after the garden.[100]

ANATOMICAL LESSONS

Not long after the founding of the Collegium Medicum the city opened a new anatomy theater, with Tulp as chief demonstrator. The chamber of rhetoric "The Eglantine," which had shared accommodation with the surgeons' guild, received new quarters, and in 1639 the surgeons—once more situated in an upper room of the former Saint Margaret's convent, now a meat market—were able to construct an impressive permanent theater. Large numbers of citizens could now come to the public demonstrations. One lesson, given in October 1647 over a five-day period, brought in 229 gilders and 9 stuivers (or 4,589 stuivers), which not only paid the expenses of the event but underwrote a fine torchlight procession and grand dinner upon conclusion.[101] If, say, 150 surgeons and their apprentices paid their 6 stuivers each to freely come and go (amounting to 900 stuivers), the number of other visitors who paid 4 stuivers per day for admission would have been about 922. How many paid to attend more than one day of the demonstration cannot be known, nor can the number of attending dignitaries who did not have to pay. But if the number of paid visits were spread out evenly over the five days (although undoubtedly the opening and closing lessons packed in a larger audience than the middle ones), a conservative estimate would indicate well over two hundred surgeons, apprentices, and other onlookers watching Tulp dissect and explain each day. Tulp was preaching annually from one of the great pulpits of Amsterdam.

What should be taught from the pulpit seems to have been in quiet dispute, however. When he had been a student, Tulp had studied with Professor Pauw, who made the moral lessons of anatomy suit his family's support for the strict Calvinists. His uncle Reinier Pauw not only continued the family business of trading in Baltic commodities but was one of the first Amsterdammers to organize a trading venture to the East Indies (Cornelis and Frederick de Houtman were related to Pauw),[102] and so became one of the first members of the Dutch East India Company when it was established in 1602; he also took shares in voyages to Guyana and Brazil in 1597 and remained deeply involved in long-distance trading ventures, proposing trade with the tsar of Russia in 1614, for instance, and helping to set up the West India Company in the early 1620s. He served eight times as one of the Amsterdam burgomasters between 1605 and 1620, represented the city in the States of Holland, and for two years represented the province in the council of state. Together with his business partner Gerrit Jacob Witsen, he also led the Counter-Remonstrant faction in the city from 1611 onward. In that capacity, he exercised his influence to help Maurits purge the Amsterdam vroedschap of those not fully committed to war against Spain, powerfully advocated holding the Synod of Dort, and took a seat on the judicial body that tried and executed Oldenbarnevelt, even helping to make the case against him.

The professor was certainly not politically active like his uncle. But his religious sympathies came out in the two moral messages he conveyed to the crowds at his anatomy lessons: all humans are mortal, and the glory of God could be seen in the extraordinarily complex handiwork of his creation. From long before the time of Vesalius, illustrators of anatomical books had made a point of weaving a personification of death into their pictures—usually in the form of a skeleton, as in the popular "dance of death" genre. In the famous frontispiece to Vesalius's *De fabrica* itself, a human skeleton presides over the show in which onlookers struggle for vantage points from which to see the opened corpse. A dog

Anatomical dissection by Pieter Pauw. Line engraving by Andries Stock after a drawing by Jacques de Gheyn II. Courtesy of the Wellcome Library, London

on the lower right is held back, presumably wishing to devour the scraps from the dissecting table, emphasizing that our bodies are mere flesh. In imitation of Vesalius's famous frontispiece, an early depiction of Pauw at work has the skeleton of death looming over the scene while dogs await. As one of Pauw's admirers commented, he taught the Dutch the beginning of Spartan wisdom: recognize that you are mortal, *nosce teipsum* ("know thyself").[103]

To reinforce the connection of the memento mori with anatomy, Pauw filled the anatomy theater when it was not in use with human and animal skeletons representing both the quick passage of life and various moral themes. In one image of his theater, a human skeleton toward the rear sits prominently on a horse while carrying a sword and wearing a cap from which flew a large feather, a sign of pride and vanity. Several other standing male and female skeletons held flags lettered with mottoes such as *Pulvis et umbra sumus* ("We are dust and darkness") or *Homo bullus* ("Human life is like a bubble"). Skeletons of animals signifying moral themes were also mounted and displayed throughout the theater. At the front were human skeletons, one holding a spade and one a fruit, with a tree between on which was wound a snake, reminding onlookers of Adam and Eve's original sin, which had introduced inescapable, mostly irredeemable death to humanity.[104] In the middle of all this is shown a kind of communion table, on which was set a human body rather than the body of Christ, one arm stuck out in a semi-crucifixion. The anatomized body not only provided a reminder of the consequences of Adam and Eve's original sin but suggested that the only way of overcoming their deed was to walk away from the tree of the knowledge of good and evil, allowing our bodies to add to the collective knowledge of the corporeal fabric with which God had clothed himself. The criminal redeemed his life, if not his soul, only when his body suffered further torture and mutilation from the anatomist for demonstrating universal truths to be found only in the interstices of creation itself. It may be that the public dissections and displays affected local sensibilities deeply. One of the favorite subjects for local painters in Leiden were still-life scenes containing human skulls, extinguished candles, open books, and other reminders of the shortness of life, which they sold in large numbers to students in religion and philosophy.[105]

There were alternatives to Pauw's views. William Schupbach has shown that, following André du Laurens's anatomical text at the end of the sixteenth century, the expression *nosce teipsum* might be used to indicate a more optimistic set of messages than the ones Pauw emphasized.[106] This other sense spoke to the idea that God had made humankind in his own image, that a wonderful collaboration of organs and faculties created the conditions for life, that the human body was, therefore, a microcosm of the larger cosmos. According to this line of interpretation, to know oneself was to be led to a knowledge of God the creator. As Ovid put it, *Est Deus in nobis:* "God is in us." Pauw was aware of this view of the importance of anatomy: as he later confessed, he gave more effort to his anatomical studies than his botanical ones "because I judged that God himself intended greater, and more certain, evidence of His wisdom, power, and goodness, to appear in the formation of the human body than elsewhere."[107] But Pauw had chosen to emphasize the more pessimistic interpretation of the lessons in the

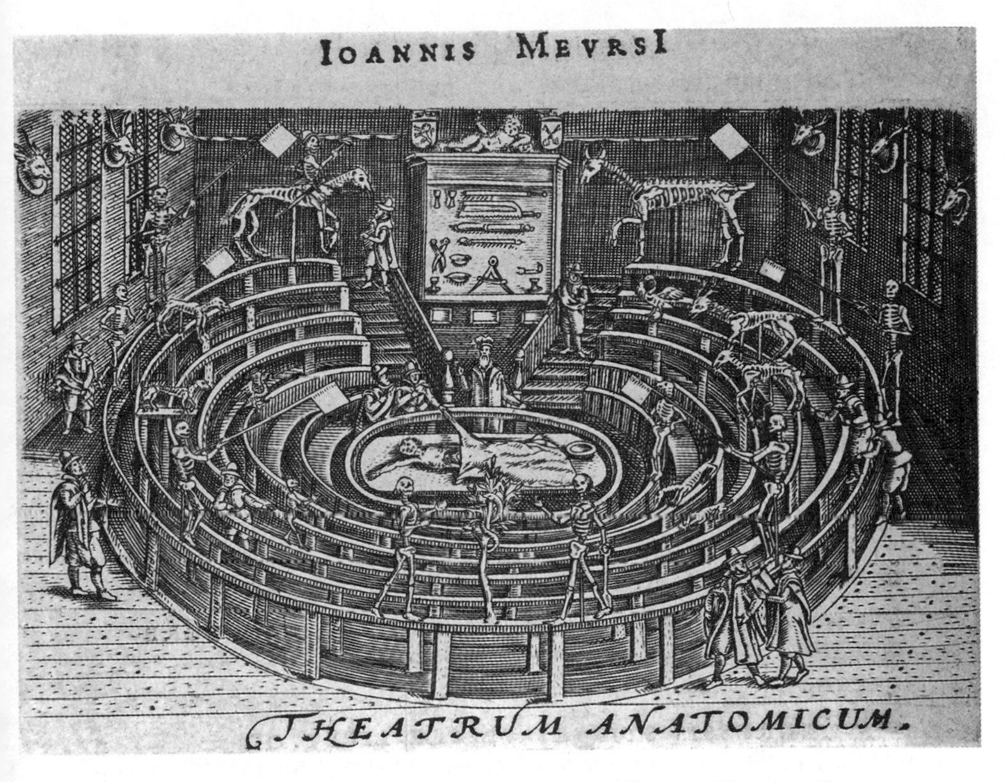

Engraving of Leiden anatomy theater, c. 1614. Johannes Meursius after C. van de Passe. Courtesy of the Wellcome Library, London

way the anatomical theater was depicted, placing the emphasis on our mortal frame. Whereas all religious groups, including Catholics, Lutherans, and Jews, accepted the fundamental message that to live rightly we must keep our mortality constantly in mind, Pauw's enthusiasm for emphasizing the shortness and vanity of human life without a hint of redemption had sympathetic resonance with the Counter-Remonstrant position that was coming to dominate the university and the country and that his Amsterdam relations were energetically supporting. It would seem, then, that Professor Pauw's anatomical lessons may have been intended to underpin a particular theological position that emphasized the largely unredeemed nature of human mortality.

When Pauw died in August 1617, on the eve of the religio-political crisis of the Synod of Dort and the execution of Oldenbarnvelt, the university's curators faced a difficult decision about how best to replace him. They considered several possible candidates for his position in anatomy. Four outsiders were invited to perform public anatomy lessons as a trial during the search for a successor.[108]

One was Adrian van Valckenburg, a former theology student who had turned to medicine to escape the religious civil war and who later gained a medical professorship in Leiden; another, Henricus Florentius, later practiced medicine in the city of Leiden and in 1636 edited and published Pauw's posthumous work on plague; two are little known: H. Rijkius and Hubertus Bijlius.[109] In the end, however, the curators chose a safer course, settling on Pauw's former colleague Otto Heurnius. In 1601, on the death of his father, Johannes Heurnius, Otto had been made a medical professor extraordinary (that is, without salary), finally gaining a regular medical professorship in 1611, two years after the death of Clusius. In 1617, he simply added the responsibility for anatomical demonstrations.

Once Heurnius took up the post, however, he expanded the moral messages to be found in the anatomy theater into a profusion of images, emblems, and signifiers that went well beyond the core lessons of Pauw. He brought into the anatomy theater a natural history collection (and a collection of anatomical books for the library) that added to the skeletons examples of the wonders of God's creation. The curators approved Heurnius's writing to one Dr. Laurentius de Croix in Aleppo requesting that he send rarities to the university. He also corresponded with a former Leiden student of philosophy, law, and Oriental languages who had gone to Aleppo, named David Le Leu de Wilhelm, also asking him to send artifacts and information from Egypt and other places. In 1620, he purchased for the university some monstrous, embalmed bodies of little children from Humphrey Bromley, an Englishman who had become a citizen of Amsterdam. He also asked the curators to use their influence in having skeletons from people of the East and West Indies sent to him, at the same time particularly urging that the regions of the Patagonian giants and of the Amazons be explored, where he still thought that people such as those with no head but faces in their chests (as described by Pliny) would be found. Heurnius collected specimens using money not only from the university but on his own account, although he later tried to get the curators to pay him twenty-nine gilders for them.[110]

The result of Heurnius's efforts are plain from an inventory he made in 1620, supplemented in 1622 and 1628. In addition to a host of human and animal skeletons, the collection now included various lists and placards; surgical and anatomical instruments; a shoulder blade, rib, and twenty-nine other bones from a whale stranded in 1600; seven stones surgically removed from the bladder of his father, Joannes Heurnius (the first medical professor of Leiden); a unicorn and other horns; a stone from the kidney of a young girl; mummies, windings of linen, and other items collected for him in Egypt by David de Wilhelm; a roll of red and a roll of white Chinese paper; Japanese utensils for serving tea; paintings of exotic fruits, nuts, woods, stones, and so forth; the liver of a young woman of seventeen anatomized by Heurnius in 1620; the vital organs of pigs; nuts, leaves, and other dried items "from India"; a little pot of Chinese beer; and other odds and ends.[111] Some of these items had earlier been collected by Pauw and placed in the covered gallery that was erected along the south side of the garden in 1599; specimens were readily shifted from the hortus's ambulacrum to the *theatrum anatomicum* and back again.[112] All these added to the overwhelming impression of the wondrous nature of creation that included finding beauty of structure and function even in the midst of anatomical gore.

But Heurnius also hung many prints and engravings in the theater, which according to the most thorough interpretation show him to be reaching out for connections to the tradition of Erasmus, Lipsius, and others who taught that the choices people made during life on earth made a difference, suggesting a less than Counter-Remonstrant outlook. In other words, Heurnius was moving anatomical teaching back into conformity with a classical view about how the choices made in life on this earth affected both the future and one's salvation. Some images certainly continued the theme of vanitas, showing the inevitability of death. For instance, he presented an engraving by Goltzius representing a child blowing bubbles while leaning on a skull, with the motto beneath making the meaning plain: *quis evadet* ("who can escape?"). Others demonstrated the possibility of eternal life through Christ in biblical scenes. There were also engravings of the Princes of Orange: William the Silent, Prince Maurits, Frederick-Henry, and other members of the house of Nassau, with depictions of other princes placed opposite them without regard to religion, all representing the political virtues that a classical moralist like Lipsius would have applauded. Heurnius also included a huge scene of the Battle of Nieuwpoort, a crucial battle won by Prince Maurits in 1600, in which the discipline of the Dutch troops finally resulted in the defeat of a veteran Spanish army; across from this he placed a grand engraving of the pharaoh's charioteers being swamped by the Red Sea as Moses and his people look on. Heurnius showed other providential events, too, such as the stranding of the whale on the beach near Katwijk in 1600, a portrait and description of Eva Vliegen, from Meurs in Germany, who was reputed to have gone without food (and only a little liquid) for twenty years, and a herring fished from the North Sea on whose two sides were marks interpreted as a message about vanquishing evil and keeping faith in God. There was a series of four engravings showing famous classical episodes of men falling from the sky to chastise them for faults; an allegorical map of the world on the theme "What shall it profit a man, if he gain the whole world, and lose his own soul?" (Mk. 8:36); depictions of the four historical ages (golden, silver, copper, and iron—the last being the modern age); an engraving by Hieronymous Cock after Pieter Brue-

Phaeton falling from the sky after driving the chariot of the sun too close to earth. From the series *The Four Disgracers*, 1588. Engraved by Hendrick Goltzius after Cornelis Cornelisz van Haarlem. By permission of the Metropolitan Museum of Art, New York

gel's *The Alchemist*, mocking the subject; depictions of the four temperaments and five senses; and a series of four allegories on how patients view the doctor: as Christ when the illness is life-threatening, as an angel when being treated, as a man during convalescence, and as a devil when recovered and payment for services rendered is requested.[113] Many of these created a context of interpretation suggesting the idea that the teaching of anatomy could move onlookers in ways that would help them live more virtuously and so help them obtain salvation. It was not a theme that the most committed predestinarians would have appreciated, so it was said quietly and by suggestion, but it was there.

If most of the modern interpretations of Rembrandt's painting of Tulp's anatomy lesson of 1632 are correct, Tulp wanted his audience to leave convinced of the wonderful ways in which God had created the human frame. Among other wonders was the construction that allowed thumb to oppose forefinger, making the human hand possible, giving rise to all the human enterprise that followed therefrom. It certainly suggested God's wisdom. The theme of Tulp's sermons went beyond the Counter-Remonstrant lesson he must have learned again and again from Professor Pauw—that we are all mortal—to some of the larger themes about the wonders of creation. But he gave no sign that the lessons of anatomy might be redemptive in making people better. There are hints that some of the Amsterdam literati, however, thought like Heurnius that the lessons of anatomy would at least make people into better citizens.

When the new anatomy theater opened in Amsterdam in 1639, Barlaeus was commissioned to commemorate it with an oration and poetry. In the years after being expelled from his professorship at Leiden because he had not supported the Counter-Remonstrants, he had continued to develop ideas about the moral foundations of life in conjunction with many other nondoctrinaire intellectuals. He became a member of a group who met to speak freely without supervision by the most orthodox at the Castle of Muiden, not far from the city (from which they earned the name of the Muider Kring, or "Muiden Circle"). Pieter Cornelisz. Hooft was the keeper of the castle and so their host. Hooft, arguably the most important poet and playwright in The Netherlands before Vondel, had also found himself disadvantaged the midst of the cultural civil war. A son of one of the most prominent of the liberal Calvinist magistrates of the city, like Dousa of Leiden Hooft had toured France and Italy and absorbed the style of the French Pléiade and Italian poets. He became a master of the love sonnet and strove to promote virtues that emerged from nature rather than being imposed from an explicitly doctrinal point of view.[114] But as the Counter-Remonstrants seized control of country and city, Hooft found his best outlet to be writing dramas based on distant historical events, a tack that allowed him to explore political and theological dangers without explicit comparison to the contemporary scene, offering the possibility of more nuance and less opposition. In the years following the Counter-Remonstrant purge of the Amsterdam vroedschap, not even this was possible, and Hooft turned to writing books of history per se, where he could speak from behind ramparts fortified with fact.

Other members of the circle included Hooft's very well read wife, Christina van Erp, the former Leiden professors Barlaeus and Vossius, Roemer Visscher, famous for his emblem book (*Sinnepoppen*) and frequent host himself to literary soirees in Amsterdam, and Visscher's famous philosopher-daughters Anna and Maria. One of them, better known by her married name of Maria Tesselschade (she was widowed in 1634), grew to become one of the most famous female intellectuals of the day. Barlaeus praised her to the skies for her writing, painting, sculpting, weaving and spinning, singing, composing of music, flower arranging, and other talents, and above all for her virtu.[115] They were sometimes joined by the very learned secretary to the Prince of Orange and enthusiast for the work of Sir Francis Bacon, Constantijn Huygens (who wrote Tesselschade some rather suggestive poems in Latin, which she understood poorly—or said she understood poorly—and which Barlaeus would not fully translate for her). Still others were the grand pensionary and author of emblem books Jacob Cats and the soon-to-be famous playwright Vondel, who later converted to Catholicism.[116]

After the Bicker League took power in the late 1620s, many members of this Muider Circle promoted a plan to establish a university in Amsterdam, although their plans were trimmed back to setting up the Athenaeum in 1632 (with Gerardus Joannes Vossius as rector and Casparus Barlaeus as his professorial colleague). Within a few years, separate colleges for the support of Remonstrant and Lutheran students were added and further professorships beyond the first two were created, but without a professor of theology until 1686. In 1637, the city even opened a permanent theater to offer plays to the public, the Schouwburg, much against the wishes of the strict Reformed. Its first run was a history play on the much earlier, Trojan-like ruin of Amsterdam, *Gijsbrecht van Amstel*, written by Vondel. The play was a plea for peace, dedicated to the brilliant Hugo Grotius, who had been sent into exile by the Counter-Remonstrants years before, never to be able to return to his beloved homeland: one can easily see Vondel pointing his finger at the men of war and dogma who still threatened to bring the city down.[117] Both the Amsterdam Athenaeum and the Schouwburg expressed an open-minded civic humanism rather than a doctrinal line.

Given that the new anatomical theater was a part of the development of public cultural institutions in Amsterdam and given that anatomical lessons could teach a variety of meanings, Barlaeus must have weighed what he would say very carefully. The main dedicatory poem, which was painted in gold letters over the demonstrating stage as a constant reminder to the audience, stated themes that both he and Tulp would have agreed on. It began by remarking on the benefit that criminals can yield to the art of healing after their death both by transmitting knowledge of their bodies from their dissection and by teaching the living "not to die for crimes." It then went on to praise the eloquence of Tulp's hands in performing the anatomies, again placing emphasis on the body part to whose marvelous structure Rembrandt had drawn attention in his painting of Tulp. Barlaeus's poem closed with the following verse: "Listener, learn [about] yourself, and while you proceed by examining one part after another, believe that God lies hidden in even the smallest fragment."[118] Another of Barlaeus's poems—although not reproduced in the theater—was written on the subject of the anatomical table. It, too, would probably have been in keeping with views Barlaeus and Tulp both shared, beginning by commenting on the pitiful state of the naked corpse, the bloody table, and the bodily parts, then speaking of how many evils attack us, whether personal failings or natural elements, each life being cut down by a different enemy. Consequently, one should "learn that it is through God that you are given health."[119]

But a third poem by Barlaeus introduces another note. Written on the subject of the anatomical theater itself, it concludes that "once the light [of life] has been extinguished, man disintegrates in putrefaction. Behold this, O citizens, and tell your Magistrates that here you can learn the ways of death and a desire to shun death."[120] The thought that citizens might be capable of becoming less sinful by learning the desire to shun death through anatomy suggests that Barlaeus was here setting out a moral religion similar to that crafted by Heurnius in Leiden. Strict Counter-Remonstrants, committed to the doctrine of double-predestination, would not have thought that people could change their moral state. Remonstrants and their fellow travelers, like Barlaeus, held a position more sympathetic to the humanist tradition in which the emulation of virtue could reform one's inner life and help on the path toward salvation. As for Tulp: pietists such as Tulp's political mentor, Vrij, also emphasized the theme of living according to God's dictates, stressing behavior and sensibility over doctrine. Vrij even published a long heroic poem titled *Anatomia* (1622), perhaps under the influence of Tulp, that urged his fellow citizens to live godly lives.[121] In other words, if Barlaeus's poems are indicative, many wished the Amsterdam anatomy theater to teach not only that the fabric of the body could thunder about death and show the wonders of God's creation but that its lessons might help support moral reform, too.

Tulp's lectures, then, were apparently intended to remind the onlookers that the magnificence of the creator was to be found not in his unknown and unknowable grand plan but in the fine material details of his creation. The fabric of the body repaid investigation, allowing insight even into how that most human of gestures—the forefinger and thumb reaching across the palm to touch each other—was accomplished by the movement of particular muscles and ligaments, a movement that allowed the dissector himself agency: it was he who probed the structure of things, making use of the very structures he examined. This was therefore not a pitch for moral reformation tout court but a plea for humility, hard work, keen attention to the particulars, and the marvelous nature of God's creative powers, something on which all the anatomists could agree.

Yet something still escaped. The anatomist might self-reflexively describe himself as a marvelous work, an embodied mind probing a corpse; but that corpse is not just like himself, for the pale, lifeless body in front of him obviously betrays the absence of life. This magnificence of the details of bodily form coupled with the absence of some additional spirit created a vast field for knowledge and interpretation, a field that could simply not remain under the control of any single party. But on one thing both Tulp and Barlaeus could agree: as Barlaeus said, "Believe that God lies hidden in even the smallest fragment," and as Tulp advised his son (who died before launching a medical practice of his own), "In medicine, there are no tenable pronouncements which are not founded in experience, which is worth more than the teaching of all masters. Do not float too freely on your wings, but limit your studies to the circle of reason and experience."[122] Descriptive naturalism remained the basis on which many people could come together despite their various interpretations of what the creator intended.

Even with differences in religious outlook, then, Dutch medical practitioners and their fellows among the urban regenten and burgers could agree on the importance of exact clinical and anatomical description. This method seemed to hold out the possibility of new therapies as well as new knowledge, bringing together patients and practitioners. It also created the possibility for building common bonds among different practitioners, whether apothecaries, surgeons, physicians, or chemists. A person like Sylvius, who became a wealthy private practitioner in Amsterdam and then a professor at Leiden, who indulged his passions for both chemistry and anatomy, and who listened carefully to other practitioners no matter their background or affiliation, is almost unimaginable in a city like London or Paris. So is someone like Beverwijck or Tulp, publishing medical books and leading public anatomical lessons while also debating the political issues of the day as senior figures on their city councils. In London or Paris, the aristocratic framework for decision-making and personal display, the scholastic disputations that dominated their university medical educations, and their medical institutions that clearly demarcated physician from surgeon from apothecary, and all from chemists and empirics, created bitter jealousies and rivalries among practitioners.[123] The Dutch world was no paradise, and there were many medical disputes, but they tended to be focused on particular problems and issues or on personal rivalries rather than being generalized into institutional parties. For even the most fundamental philosophical positions could be subsumed under the general rubric of objectivity—of uncovering more information about the things of nature—in the hope and expectation that they would be of use.

5

Truths and Untruths from the Indies

In answer to a question about whether the living or the dead were more numerous, he answered: in which number do you reckon those who travel on the seas?

—ANACHARSIDIS, *Apophthegmata*

It was from his own sickbed when, at about eight in the evening of 19 September 1629, he was called to attend the governor-general, who had collapsed. The doctor had been worried about Coen's health and had advised him to cut back on his duties, but Coen was a single-minded man who insisted on continuing his rounds, inspecting the defenses despite prolonged and severe diarrhea, remarking that he did not have the time to keep to his bed, "as if he had said in a kind of prediction, that a general ought to die in the discharge of his duty." When he arrived on the scene, Dr. Jacobus Bontius found Coen prone, lying amid some of his counselors, laboring for breath and coughing dryly. On examination, he found Coen's extremities to be cold, and he was covered by "a cold and clammy sweat. His stools, which had been preceded by billious vomiting, were very copious, flatulent, watery, and full of bubbles," while at his wrist could barely be felt a languid "creeping." These symptoms fit with what Bontius later identified as the "cholera morbus," which "is likewise extremely frequent," often causing those attacked by it to die quickly, sometimes within twenty-four hours, as with Cornelis van Royen, steward of the hospital: the disease "is attended with a weak pulse, difficult respiration, and coldness of the extreme parts; to which are joined, great internal heat, insatiable thirst, perpetual watching, and restless and incessant tossing of the body. If together with these symptoms, a cold and fetid sweat should break forth"—as it did in Coen's case—"it is certain that death is at hand."[1] Bontius told the others present that the governor was dying. When the counselors protested that Coen had merely fainted after a bout of vomiting, the doctor replied that he hoped he was wrong but had never discredited his profession "by passing any rash and unwarrantable judgment." About one o'clock in the morning Bontius's prognosis proved right when the governor experienced convulsions and died of a "suffocating catarrh."[2]

The knowledge and wealth of the East Indies was acquired by the Dutch at a high cost, one borne most heavily by islanders but shared even by the VOC's chief administrators. Yet in the midst of human tragedies that underpinned the drug trade, the pursuit of natural knowledge continued, and that written down by Bontius was truly remarkable. Despite a punishing schedule of administrative duties and having to endure two sieges in which his health suffered terribly, Bontius gathered a great deal of material about the medicine and natural history of the region from local informants, from which he composed four major studies. He worked on Java during years when the VOC was virtually exterminating the population on some of the Spice Islands. The Dutch East India Company was certainly no scientific society. Its servants worked hard to establish complete control over the trade in fine spices, in the process bringing devastation to many, from their own sailors to whole groups of islanders. Bontius's methods, moreover, reflected the VOC's dearly held values. He turned the strange and exotic into parcels of useful and accurate information, information that could be transported easily from Java to literal-minded readers. And yet Bontius grew to think so well of many of his local acquaintances that he considered them superior in knowledge and skill to any European medical practitioner. The evidence from Asia continued to be full of fascinating wonders and, even better, useful skills and products.

But a question about the evidence quickly surfaces: How was the credibility of information about exotic objects to be established? "Credibility" may have its roots in the Latin *credo* ("I believe"), but is belief sufficient? In subjects such as anatomy, many expert eyewitnesses could gather around the dissecting table and agree that they witnessed something in common. Indeed, many recent historians have focused their studies on the production of knowledge at particular sites, making all knowledge appear to be local knowledge—that is, the knowledge of the social group personally involved in eliciting a particular event and the agreed interpretation of it.[3] But of course, others might later see the same thing somewhere else. Bruno Latour and other historians have therefore explored how local sites are connected—that is, how knowledge networks get built that can transport information from place to place. Local communities are embedded in larger webs of relationships, which all together affect the development of provisional universal knowledge.[4] Others have tried to solve the question of why a claim to know something is accepted or not by invoking social status. Steven Shapin, for instance, has argued that although the establishment of matters of fact depended on credibility, the essence of credibility in turn is "trust," which depended on social authority. In his study of Robert Boyle and seventeenth-century England, Shapin finds that the only people with enough social authority to arbitrate early modern natural knowledge claims were gentlemen.[5] But as we shall see, even when people of high social standing stood behind claims to matters of fact, they could not make something false true. Bontius struggled to provide information that was accurate, and he criticized and corrected his predecessors when he found that they erred. Yet even he made mistakes, and his learned editor introduced further errors, some recognized at the time and others found to be wrong only later. Matters of fact, especially when coming from afar, from only one or a few eyewitnesses, were always subject to revision.

HEALTH AND ILLNESS IN THE VOC

The collection and transportation of the natural wealth of the Indies came at a terrible cost in human life and suffering. It began with the notoriously tough lives of the VOC's employees. At the bottom of the heap were the common sailors and soldiers, whose contracts committed them to five years' service before returning home. About 60 percent of the people aboard VOC ships were seafarers (from captain to able-bodied seaman), and 30 percent were soldiers. Although they never outnumbered the Dutch in the seventeenth century, men from Germany and Scandinavia were numerous, especially in periods of prosperity when the Dutch had more alternatives for employment.[6] The VOC's constantly high manpower needs were met by local chambers using private recruiters (*volkhouders*), who sought out men looking for employment, lodged them until they boarded ship, and kitted them out on leaving, all for a bond committing the sailor or soldier to repay them about 150 gilders—the VOC would usually advance two months' wages (about 20 gilders), which was used to meet the first installment on the debt. The volkhouders earned a bad reputation, often being called "soul-sellers" (*zielverkopers*), but by resorting to this commercial arrangement, the VOC never had to rely on the press-gang or other forms of coercion to meet its manpower needs. Clothing was the sailor's responsibility and was often inadequate; the space available below decks for members of a watch to lie down or hang their hammocks was cramped and usually fetid; and unlike the officers and merchants, ordinary seamen could seldom afford to supplement the rations aboard from their own stocks, so their food was generally stinking and monotonous within a few weeks of sailing, bad even by contemporary standards.

Most crews were likely in ill health for much of their time on ship. Scurvy was common, and typhus and other infectious diseases might all too quickly lay low much of a crew. Malaria, beriberi, the bloody flux, and many other diseases struck frequently once Asia was reached. Since pay was higher in other parts of the merchant fleet, and even in the navy and army, VOC sailors and soldiers tended to be even more motley than usual, most down on their luck, a few seeking wealth or adventure, many living on the edge and quick to turn violent. Aboard ship, even small infractions could result in severe lashings. For more serious offenses, the punishments could be frightful: fighting with a knife, for instance, was dealt with by having the ship's surgeon pound a knife firmly through the offender's hand into the mast, positioning the knife in the hand according to the seriousness of the offense. (The victim of this punishment was left to work his hand free by pulling it off the knife.)[7] Keelhauling and capital punishment could be meted out for more serious trouble. Even despite such corporal powers, VOC officers sometimes faced mutiny.[8]

Mortality was very high indeed. Although each ship's history was different according to circumstance, it seems that on average 10 to 15 percent of all shipboard residents died of accident or sickness before reaching the Indies.[9] Once VOC employees arrived in the East, the dangers to health during the ordinary five-year service remained steep. Roughly two of three never made it home—the highest death rate by far among all the European trading companies. Survival rates for officers and merchants were higher, but they often suffered ill-health, too. Of those who survived, almost all must have endured a serious illness at least once. By the late seventeenth century, the trade with Asia was costing about six or seven thousand European lives per year.[10]

The VOC chambers, however, did not overlook the provision of medical care for their employees. Ships had one or more surgeons and medicine chests aboard according to their size. The chamber of Middelburg, in Zeeland (the second largest of the chambers), for instance, appointed 114 surgeons and three physicians for service aboard ship or in the Indies from 1602 until 1632. At first, doctors and master surgeons of the city recommended candidates for the posts, but from about 1610, as more candidates from beyond Zeeland applied, the chamber itself saw to their examination; applicants were turned down in some cases, while in others they were found to be capable only of the post of assistant-surgeon. The chamber employed two physicians and two surgeons to help with these medical matters. It also provisioned the medical chests sent with each ship, an important task. Another duty was assessing whether people in the company's service had been disabled at work, in which case the chamber paid compensation. (In August 1618, the Heren XVII unified the compensation payments among the chambers and in the Indies, paying 800 gilders for the loss of a right arm and 500 for a left, 450 for the loss of a leg, 800 for the loss of both legs, 600 for the loss of a right hand and 400 for the left, and 1,000 gilders for the loss of both hands. When they later reviewed the system, they added payments for the loss of one or both eyes.) The chamber also assessed claims for payment from outside surgeons who had treated injured VOC members, deciding whether the treatment had been carried out properly and whether the charges were sensible. Last, the chamber reimbursed the municipal hospital for caring for sick and wounded sailors in the VOC's employ.[11] Out in the Indies, the VOC provided surgeons—and occasionally physicians—at trading stations and at the hospitals built for major garrisons, and in the mid-1660s a "medical shop" came to be established at the main Dutch port of call in Batavia (now Jakarta) in order to see to the medical fitting out of ships before the return voyage.

Although they earned a poor reputation posthumously, the VOC surgeons appear to have been generally well trained and capable.[12] Many spent much of their life in service to the Company. A few became noted. Nicolaas de Graaff is probably the best known. He made sixteen voyages, five to the Far East. An account of his experiences, based on a record he kept, was published posthumously in 1701 as *Reisen van Nicolaus de Graaff, na de vier gedeeltens des werelds* ("Voyages of Nicolaus de Graaff, to the four corners of the world"). Born in Alkmaar in 1619, he apprenticed with a local surgeon, IJsbrand Coppier, and then took service with the VOC in 1639 after examination in the chamber at Hoorn. His first voyage, as assistant surgeon aboard the *Nassau*, lasted four years and earned him a broken skull and a large head wound at the siege of Malacca. He quickly signed up again, departing in 1644 on the *West Friesland*, now as a full surgeon, returning in 1646. He made subsequent trips on a whaling vessel to Greenland, a Mediterranean trading voyage, and on Dutch navy ships (seven times during the various wars of the 1650s and 1660s). He took service with the VOC again in the late 1660s, traveling to Ceylon, where he stayed two years. He then traveled up the Ganges to Chiopra (on his way spending seven weeks in prison in Mongeer under suspicion of espionage), where he practiced medicine on the Muslim governors as well as on his compatriots, which allowed him to travel freely by horse and boat and even hunt tigers. It was a period of terrible famine, however—so bad that he heard of mothers eating their children and people selling themselves into slavery quite cheaply to stay alive. After about two years in Bengal—an important source of slaves for the VOC[13]—De Graaff returned to Ceylon in November 1671, apparently very wealthy from private trading, since he sent many goods (including saltpeter, opium, nutmeg, bales of silk and cotton clothing, and fifty-seven slaves) as a "present" to the chief Dutch port of Batavia aboard a ship that went down at sea. (Shortly thereafter he received another shipment of eleven slaves, which this time he sent along in different vessels.) He sailed home in 1672. Finally he seemed settled, with his wealth and experience securing him the office of alderman (*schepen*) of his town of Egmond aan Zee. But at the end of 1675, now aged fifty-six, he took service with the VOC yet again, returning to The Netherlands in 1679. He and his second wife held various local offices in Egmond aan Zee, but he left for the East again, for the last time, at the age of sixty-four, in May 1683, taking service as a lowly chief barber-surgeon (*opperbarbier*). The ship went as far as Macao, in China, in the summer of 1684 before returning to Batavia in early 1685. De Graaff accompanied a delegation to the emperor of China, visited Bengal again, returned to Batavia in November 1685 via Molucca and Bantam, and finally returned to the Netherlands in August 1687. It was then that he probably composed his travelogues. He died in the autumn of 1688, a year short of seventy. In 1701, his *Reisen* ("Travels") and *Oost-Indise spiegel* ("Mirror of the East Indies") appeared from a printer of his native city, Hoorn.[14]

Clearly, not all surgeons had as much stomach for adventure as De Graaff, nor did all survive the rigors of such service for as long. But it was a big world, in which plenty of adventure could be found and riches obtained for those willing to take risks and gamble on good luck. De Graaff and his like were clearly tough in body and spirit, hardened by experience, but also capable of helping others recover from their own assaults of fortune when the inclination or duty called. Because almost all of the Dutch medical services in the East were staffed by such men, what Asians learned about European medicine was acquired mainly through surgeons such as De Graaff. For instance, VOC trade negotiations with Chinese officials in the early 1660s were aided by the skills of the delegation's surgeon, who successfully treated several of the Qing officials. But surgeons might also be used for purposes other than healing: a year earlier, during the siege at Casteel Zeelandia on Taiwan, a Dutch surgeon vivisected a Chinese prisoner before a large crowd, no doubt to instill terror (perhaps in imitation of the Chinese method of torture-execution "by a thousand cuts").[15] More generally, as the words about De Graaff possessing and shipping slaves also indicates, the VOC's trading arrangements themselves inflicted a huge toll of misery and death not only on their employees but on many of the peoples of Asia.

TERRORS ON THE BANDA ISLANDS

VOC activities in the Spice Islands, especially, led to terrible destruction. This was especially so on the five small Banda Islands where the nutmeg trees grew. The inhabitants of Banda had a relatively egalitarian form of social and political life, dividing up the nutmeg groves and having rights to the nutmeg they each collected from their trees. Like other peoples of the Moluccas, they traded their spices for staples and luxuries. Many traders—from China, Southeast Asia, South Asia, and elsewhere—came to their archipelago to buy small lots of their spices, while the Bandanese themselves sometimes sailed to other islands in the Moluccas, to the city of Makassar (on the southeastern tip of the Celebes—now Sulawesi), and even as far as Java in order to exchange goods. The Dutch presumably followed regional custom for dealing in nutmeg and mace: on collection, the spices were placed in vinegar and salt in large earthen pots and then transported to fortified compounds ("factories" in the language of the day) where the contents were macerated for a day or two, gently boiled, and preserved in sugar before being shipped onward.[16] They also shipped whole nutmegs preserved with a dusting of lime.

The VOC had concluded treaties with the Bandanese granting the Dutch a monopoly on the purchase of their nutmeg and mace. To do so, they had dealt with the numerous men among the Bandanese with nominal authority (called *orang kaya*), whom the officers of the VOC considered to be lords or princes. In 1602, officers of the newly formed Company signed a treaty with the orang kaya that gave the VOC sole rights to the islands' nutmeg and mace at a common price negotiated for all, in return for Dutch-supplied rice, cloth, and other goods, as well as protection against the Portuguese and English. The treaties not only spelled out what would be exchanged for the nutmeg, and for how much, but set out how the exchange would be arranged, so that the VOC merchants could deal with one group who supplied them with large quantities rather than having to make countless bargains with individuals to accumulate a shipload. This accorded with instructions from the Heren XVII.[17] The officers of the VOC conceived of their work in terms of legalities, the most important of which was contract—as we have seen, an almost sacred concept in the Netherlands. But Dutch views about contract and property did not accord with local sensibilities. People on the Bandas and elsewhere acted as though they owned property by right of custom and labor, which was irrevocable; the VOC, by contrast, assumed that if the same people had committed themselves to handing over parts of their property for a fixed return but then failed to abide by their contract, severe penalties might follow, including confiscation of all that was theirs. During the later 1630s and after, sagely advised by Dutch missionaries, VOC officials on Taiwan employed methods of ruling by spectacle to manage the local people, supplementing the "legal-rational rule" they used to deal with the Chinese and their own employees.[18] But in the 1610s and 1620s, in the Banda Islands, before any missionary activities, Dutch officers expected the local "princes" to understand what they were doing by entering into contracts and to live up to the consequences: they were not treated as children nor as inferiors. As it happened, however, there were clearly incommensurable views of political authority at work. The orang kaya had few powers to force other Bandanese to obey the agreements they struck with the Dutch, who still found themselves negotiating with many individuals over relatively small quantities of spices. They also found few takers for the terms that had been settled by treaty. The VOC's purchase price for nutmeg and mace was lower than the Bandanese could get from others, the rice offered was not of the desired kind, much of the cloth traded to them was of European manufacture rather than Indian weave—and woolen rather than cotton—and the sale price of the unwanted rice and cloth was higher than market rates.

From the point of view of the VOC's officers, a contract was a contract, and they needed to make profits from the spices over and above the going local rate in order to be able to spend money fighting off their rivals. But ordinary Bandanese refused to go along with the new rules because they were not in their interest. They argued with and even attacked orang kaya who tried to enforce the treaties or more often simply went about their business as before, selling to non-VOC merchants by sailing elsewhere, which continued to put some of the nutmeg trade into other hands. Moreover, the English soon established a small settlement serving the two westernmost Banda Islands (Ai and Run), from where they traded with anyone who came. But the contractual principle in which the Dutch believed meant that the Bandanese sale of nutmeg and mace to others constituted a breach of agreement ("smuggling"), and they were determined to end it. In April 1609 a large VOC naval and military force led by Admiral Pieter Verhoeff arrived in Banda, ordered the English out, and began negotiations with the local people to build a fortress. When negotiations broke down, Verhoeff began building the fort anyway, which seriously threatened the Bandanese. They asked to restart negotiations and invited Verhoeff and his counselors to a meeting place in a grove of trees, where they attacked and killed the admiral and almost thirty other Dutchmen—many of them high-ranking officers—before retreating into the hills. This caused the remaining Dutch to speed up construction of their fort while undertaking punitive expeditions. Once the fortress was complete and manned, the Dutch set about blockading the islands with their gunboats in order to stop trading outside the contractual framework, which finally forced the Bandanese to agree to a new treaty giving the VOC the right to inspect all waterborne craft sailing to or among the islands.

One of those who had escaped the 1609 ambush on Banda was Jan Pietersz. Coen. He was undoubtedly the most influential figure in the early history of the VOC, securing the Company's place in the Asia trade, though at a terrible cost that was, ironically, also to the long-term detriment of the VOC. Coen had learned the mysteries of business working for seven years as an apprentice to a Flemish merchant in Rome, then struck out on his own, joining the VOC in 1607 when twenty years old as an under-merchant. He dedicated long years to rough service in Asia on behalf of the Company, and he showed little patience for directives from the governors on the other side of the world who had no experience of realities on the ground. A committed Calvinist, perhaps even harder on himself then he was on others, he quickly rose in the esteem of the first governor-general, Pieter Both, who called him "a delightful and God-fearing young gentleman, very modest in his living, chaste and of good heart, no drunkard, not self-centered [*hoovaerdich*], very capable in council, and very able in business."[19] In the VOC's service Coen participated in the fighting that sometimes accompanied efforts to secure supplies of spices, including events on Banda that caused large losses among the many Dutch officers and merchants gathered there. He returned to the Netherlands in 1610, where he submitted a report on the mismanagement and corruption he had encountered in the East that got the attention of the Heren XVII.

By 1612, Coen and others had forced the Heren XVII to recognize that in pursuing large profits negotiated by treaties that were imbalanced to their advantage they would have to enforce the agreements with military power, despite the financial and other costs. He firmly believed that to be financially secure the VOC needed to acquire a monopoly over the spice trade and to enforce it, which would require coercion against rival merchants and sometimes against the suppliers. As one historian has put it, whereas the Heren XVII back in the Netherlands thought of the VOC as a trading company, their officers in Asia emphasized its sovereign powers; whereas the Heren XVII had the idea that their merchants would buy and sell in Asia according to local custom and law, so that the Company would merely need to supply ships and a few fortified bases for protection of their personnel and storage of their goods, events in the region moved the VOC toward a policy of establishing a monopoly over the supply of spices that, because it would not be welcomed by local traders, would require the active use of violence to institute and police. Coen pushed the Heren XVII toward the monopoly policy not only by sending a stream of letters and position papers to the Netherlands but by his actions on land and sea.[20] He was returned to Asia in command of two VOC ships in 1613. Not long after, in 1614, he was appointed director general, the second in command to the governor-general, a position that placed him in charge of VOC trade. In advocating that the VOC establish monopolies over supply, Coen sharply criticized and undermined the activities of his superior, Governor-General Laurens Reael (a doctor of law), and his supporter Steven van der Hagen, who operated in Asia by negotiation and contract whenever possible.[21] When both of them left the Indies in 1619 in the wake of the Synod of Dort and the execution of Oldenbarnevelt, Coen himself obtained the position of governor-general.

Coen had no compunction about using violence to secure his ends. The Dutch borrowed many of their war-making methods from local custom and made use of local allies, but they also had a superior technology in firearms and cannon, the ability of an outsider to use divide and conquer strategies that made good use of local resentments among the locally warring states, and, perhaps above all, a commitment to "total war": the inflicting of crushing casualties intended to eliminate opposition and demoralize the survivors. Such methods had worked for the Spanish, Portuguese, English, and French monarchs in dominating local people in Europe and elsewhere, and the VOC also employed them purposively, while the company's ability to coordinate their campaigns at sea more flexibly than their European rivals often enabled them to overcome them, as well.[22] Since the English presence on the western Banda Islands continued to represent an alternative to VOC rule—in 1615 some of the Bandanese even wrote to King James I seeking English protection while English ships were sailing to the Moluccas to foment rebellion against the Dutch—the VOC again resorted to force, conquering one of the Banda Islands occupied by the English (Pulau Ai) and taking hundreds of local hostages to enforce obedience. By the mid-1610s, then, Coen and his officers clearly were comfortable getting what they wanted by treaty and contract, backed by coercion.

The Heren XVII meanwhile increasingly desired to have a place in the area that would serve them as Goa served the Portuguese: a first port of call for ships arriving in the region from the Netherlands, the place where all homebound ships could take on final cargoes, and the location for the main warehouses, arsenal, and administration for the VOC in Asia. They therefore sent out Governor-General Pieter Both to establish a base of operations. In 1610 he set up at Bantam, a principality on the north coast of Java overlooking the Sunda Strait, where the sultan resisted the claims of the central Javanese king of Mataram to suzerainty over the whole island, using foreign merchants to bolster his position. Both and his successors never succeeded in making Bantam into a general rendezvous, however, because of the insecurities of the VOC's position there. The town of Bantam also included English and Chinese factories, the latter of which controlled huge quantities of the pepper trade. Wanting to drive out the competition, the Dutch requested various privileges, but the regent, acting on behalf of the underage sultan, refused. Having tried to act against their rivals but failing, the Dutch now felt quite concerned about their security in the face of those they had made enemies, who might strike back with force.

As soon as he was able, Coen took matters into his own hands. About fifty miles to the east and nominally under the control of the sultan of Bantam was a coastal town called Jakatra. Not getting what he wanted from the sultan himself, in 1617 Coen went behind his back and negotiated privileges from the local prince (*Pangeran*) of Jakatra, who allowed him to build a warehouse in the Chinese quarter. Coen developed this into a fortified stone compound—despite the Pangeran's wishes—and started construction of a shipyard and hospital on the small island of Onrust in the harbor; he then moved the main VOC operations from Bantam to this new post. At the end of the year, a large English fleet put into Jakatra, arriving from the Banda Islands, where they had unsuccessfully tried to reinforce their foothold in the face of Dutch opposition. Seeing the Dutch in a new and fortified position, the English decided to build a fort across the river. On 14 December they captured a VOC ship, the *De Swarte Leeuw*, richly laden with spices. When Coen demanded its return and was rebuffed, he opened fire but, being outmanned two to one, was forced to sail for distant Ambon to assemble a force large enough to fight back. The VOC employees who remained in the compound were besieged by the English, Chinese, and Jakatrians, but thanks to divisions in the enemy ranks, they held out for almost sixteen months. Coen meanwhile gathered his forces at sea, captured the smaller English fleet designed to relieve the remaining forces on Run in the Bandas in 1618, began snapping up all the English shipping he could, and finally at the end of May 1619 returned to Jakatra and took the town by force, despite contrary orders from the sultan of Bantam.[23] Coen had landed a fatal blow to English attempts to get a piece of the trade in cloves and nutmeg. After fortifying the Dutch compound, Coen laid claim to the lands around the town. This was "Batavia," the first Dutch territory in Asia. The VOC finally had its eastern headquarters.

Not long after his triumphal conquest of Jakatra, however, Coen received word of an agreement concluded between the States General and King James a year before, in 1617. Back in the Netherlands, the Heren XVII had been worrying about a larger strategy: the Twelve Years' Truce with Spain would soon be ending, and an Anglo-Dutch alliance would need to be restored for the impending hostilities—or at least the Dutch could not afford to push James I into the Spanish camp, which some at the English court were urging. Earlier conferences at London and The Hague had failed to prevent Anglo-Dutch competition in Asia, which had led to the outbreak of open conflict there. The treaty therefore agreed that the VOC and the East Indies Company (EIC) would collectively purchase as much pepper as they could and each would dispose of half of it, and the EIC would also get 30 percent of the cloves and nutmeg to sell in return for subsidizing 30 percent of VOC military activities. It was the kind of agreement made by lawyers and diplomats not concerned with the day-to-day business of making a profit. From Coen's point of view, he had finally placed the VOC in a position to enforce a monopoly over the spice trade only to have the defeated English cut in on the trade for reasons of state. Of course Coen objected strenuously, dressing down his superiors by letter.

He also continued to act according to what he considered the best interests of the Company, allowing the English only the letter of the law. Coen managed to outlast his superiors and gain the post of governor-general himself in 1619, when he quickly moved to get a better position in the pepper trade (which had its source in both the Indian subcontinent and Sumatra). He aimed his first attack at the Chinese, major competitors in purchasing the spice. He seized pepper from Chinese junks trading at Bantam on the grounds that the pepper was owed to the VOC in return for monetary advances the Company had made; encountering resistance from the regent of Bantam, he blockaded the port, forcing the Chinese to leave for other places, especially Jambi on the east coast of Sumatra. Coen then attacked the Chinese trade there, forcing it to close. In 1620, fresh from his success at Batavia, he sacked Japara, on the central Javanese coast, executing all the Gujaratis trading there, destroying the English factory, and forcibly carrying back large numbers of Chinese to Batavia.[24] The only place left to the Chinese to trade in pepper outside the control of the VOC was the relatively insignificant port of Banjarmasin in Borneo.[25] They were therefore forced to come to Batavia under license from the VOC to exchange cloth, porcelain, tea, and other products of China for the spices gathered up by the Company, coming to constitute a major part of the inhabitants of the city, but under Dutch suzerainty.[26]

With much of the pepper trade now in VOC hands—although because it grew in too many places he could never quite monopolize it—Coen reinforced the Company's monopoly over the trade in fine spices. At the first opportunity he moved to sort out affairs on his terms in the Bandas. The remaining English garrison on Run had finally surrendered in late 1620, but signs of resistance to the Dutch by the Bandanese remained, which Coen considered further evidence of bad faith. In 1621 Coen therefore sent in additional forces and all but exterminated the population. Men, women, and children were put to the sword or worse, villages were razed, crops were dug up and set alight. Of the few hundred survivors out of a population of perhaps fifteen thousand, most were transported back to Batavia in slavery, while a few were kept on the islands under strict watch to teach the Dutch how to grow nutmeg. The nutmeg groves were divided into sixty-eight parcels (*perken*), which were given in loan to former employees of the VOC; they could in turn buy slaves at a fixed price from the VOC (mostly provided from the Celebes and Ceram), the VOC promising to supply enough slaves to cultivate the trees. Given the high mortality rates and occasional escapes among slaves, a continual importation of them was found to be necessary. The *perkeniers* also purchased rice from the VOC to feed the slaves. Of course they sold their nutmeg and mace only to the VOC—at fixed prices. After horrific violence, then, Coen had in effect committed the VOC to ownership of the nutmeg groves and a system of plantation slavery.[27]

After completing his campaign of extermination on Banda, Coen headed for Ambon, the key to the central and northern Moluccas, where the cloves grew. While here, as elsewhere, the Dutch had originally been welcomed for their help against the Portuguese, they soon found themselves unwelcome. Members of the Dutch garrison started preying on local girls and women, and at first they brought no religious teachers of their own to replace the Portuguese Jesuits, resulting in even more conversions to Islam, which the Dutch considered to be ungodly acts of subversion. They also imposed a system of forced labor for construction of works. Moreover, the VOC's monopoly meant, here as elsewhere, that the Dutch bought cloves very cheaply, dealt with private trade brutally, and in return sold inferior rice and textiles to the Ambonese at high rates. By 1616, these and other grievances led to local resistance to the VOC, which was put down by an expeditionary force. By the end of the 1610s, with the Dutch having obtained the upper hand vis-à-vis the English in the region, they had become widely and deeply hated on the island. In 1621, then, after virtually wiping out the Bandanese, Coen turned his attention to Ambon and threatened to do the same to them. That he did not is only because it was too late in the season and his fleet had to catch the trade winds back to Batavia.[28]

Coen's threats quieted the chance of rebellion for a while, although the Heren XVII recalled him in 1622 for overstepping his authority. His successors on Ambon, however, were increasingly nervous and lashed out in a panic. A small group of English servants of the English East India Company resided on Ambon, overseeing the acquisition of cloves as guaranteed them by the 1617 Anglo-Dutch treaty. Social relations between the English and Dutch seemed good on the surface, but underneath distrust and fierce rivalry continued to brew: the English felt the grievances of a dependent, while the Dutch both hated the truce that gave the English rights to much of their spice trade and suspected that they plotted with the locals against them, the despised overlords. The fear and suspicion must have been palpable. On 23 February 1623, a Japanese soldier in English service was found investigating a forbidden place along the fortifications while also pumping a Dutch sentry for information. Others were alleged to have been inquiring on the previous day about the strength of the guard. The soldier in question was arrested and tortured, finally confessing to the charge of his inquisitors that he had been spying for the English and that an uprising was planned after the arrival of an English ship in the harbor.

The rest of the Japanese and the English were arrested, and over a period of a week each man was individually subjected to the water torture, in which a cone was placed around the victim's neck above his mouth and nose; water was then poured in, forcing him to gulp water to avoid drowning, which not only choked him but caused his tissues to swell out of all proportion from excess water in the body, causing severe agony; this torture was supplemented in some cases by burning the victim's armpits, feet, and hands with a candle or pulling out his fingernails. The investigators thus obtained signed confessions of participation in a plot from almost everyone. On 8 March, the Dutch governor of Ambon, Herman van Speult, called a meeting of the local VOC Council and asked the *fiscaal* (their legal council, who had supervised most of the interrogations) to draw up a document recommending that the guilty be executed for lèse-majesté. The next day, ten Englishmen and all ten Japanese, as well as the Portuguese overseer of the VOC's slaves on Ambon, were executed by sword, with the chief of the English station being subjected to the horrors of drawing and quartering. Yet the VOC officials took great care about the contractual arrangements they had agreed upon: two of the Englishmen were granted temporary pardons to see that the EIC goods were transported to Batavia. There, one escaped on the entry of the ship into harbor; the other was set free by the government at Batavia, which considered the judicial proceedings to have been mishandled.

When news of the "Ambona massacre" arrived in England, it created a tremendous stir, and only great diplomatic exertions and temporizing on the part of the VOC and the States General prevented war between the two nations.[29] For some it showed members of the Counter-Remonstrant and Orangist party, who had eagerly gathered strength for renewed war against the Spanish at the end of the truce, as singularly self-interested and self-righteous deceivers. The incident was also used to excite English feelings against the Dutch many times later in the century. It did, however, force an end to the contract that had been arranged between the VOC and the EIC and allowed the VOC to expel the English from many of their settlements in Asia, while the example made of the English is said to have deepened the "respect" of the Ambonese for the Dutch.[30] Events on Ambon also convinced the Heren XVII of the need for a strong governor-general, and so, after a decent interval following the political uproar surrounding the Ambona massacre, they secretly returned Coen to Batavia in 1627 aboard a powerful fleet.

The VOC brought further suffering to Ambon when they found that smuggling was continuing. In 1625 a fleet arrived via the Pacific and carried out expeditions against the island of Ceram to destroy all the boats and cloves, as well as the food crops of coconut and sago palm, which led to terrible famine. (Ceram was an important destination for many of the illegally exported cloves.) From 1630 onward, annual attacks on Ceram and rebellious villages were launched to destroy all rival clove and fruit trees as well as to burn all villages and rob people of their means of living. When the VOC captured all the local leaders on Ambon via treachery in 1634 in order to enforce obedience, there was a general rising, put down not only by Dutch ships and troops but by headhunters imported and paid by the VOC, which further shocked the Ambonese. Another and more widespread rebellion in early 1636 forced the Dutch to remain in their forts, although a VOC expeditionary force arrived late in the year to relieve their compatriots and spread death and destruction among their enemies. Yet another rebellion in 1638 was quelled by force. Finally, in the summer of 1646, the Dutch forced all Ambonese to settle along the seashore, keeping them from the mountains where they had previously sought refuge, while also forcibly mixing families among the settlements to weaken kinship ties. All firearms were confiscated, and for half of each year one-half of the local leaders had to live as hostages in the Dutch bastion of Fort Victoria under pretense of advising the VOC council, the remaining leaders taking their place for the other six months. People raising cloves outside areas controlled by the VOC were exterminated during the late 1640s and the 1650s. No further resistance surfaced until 1817. "This brought an end to a stirring period in the history of Ambon" was the unfortunate formula of one historian fifty years ago.[31]

The VOC's enterprises in Asia destroyed countless lives, which no amount of medical assistance could alter. Moreover, in a terrible irony, the efforts of Coen and others to gain a monopoly for the VOC on the export of fine spices to Europe may not have secured the Company's future. They did, of course, minimize the purchase price while giving the VOC the power to hold goods back from the market if anticipated sale prices were too low. Considered in terms of income, the VOC's control of the purchase price on many spices enabled it to achieve gross profits sometimes reaching as high as 1,000 percent.[32] Yet the financial costs imposed by policing the monopoly—building fortifications, garrisoning them, launching punitive raids, and patrolling the waters—were substantial.[33] At the same time, conditions of forced labor no doubt had a real effect on causing the net output of clove production to fall after the VOC established its monopoly. Thus, following the effective monopolization of nutmeg, mace, and cloves by the VOC in the 1650s, "there followed a prolonged slump in export values."[34] As the

VOC took control of the spice trade by midcentury, merchants of Yemen and Cairo found a substitute in a new product, coffee, which the VOC again came to dominate by the century's end.[35] In other instances, as in the case of pepper—the one spice the Company could never monopolize because of its widespread cultivation—falling prices increased demand. Pepper remained the most important commodity returned in VOC bottoms to Europe, with the quantities imported even rising through the century while prices fell. Indeed, when the price of pepper threatened to become unprofitable for the European companies in Asia, this condition "gave rise to a more or less voluntary cessation of hostilities" related to the pursuit of supplies.[36] In the 1680s, the Heren XVII "reckoned total European demand for pepper at 8.6 million pond (4.3 million kg), only 25 percent more than the market of 1620, or an average annual growth rate of 0.32 percent. They had long before come to the conclusion that European demand was inelastic with respect to price and struggled to bring their English competitors to heel so that supplies could be controlled and price discipline established, as they had done with the fine-spice markets, where demand also appeared inelastic but where the VOC held an effective monopoly." The roughly one-quarter of all VOC revenues raised from selling spices had to be supplemented by a range of other products, from dyestuffs and drugs to porcelain, cotton textiles, saltpeter, and (at the end of the century) tea and coffee.[37] The VOC monopoly positions may therefore not only have led to more and more coercion of the producers but may also have caused consumers to shift to other comestibles, such as sugar. With cheap cultivable land and slave labor, West Indies sugar flooded the European market while prices dropped, making it available to almost everyone; and although a sugar industry developed rapidly on Java after 1640 and provided much of the sugar on the Asian market, it could not compete on the European market.[38]

Coen was clear-sighted enough, however, to see that in the short to medium term, a monopoly position would guarantee the VOC a strong profit margin, high enough to finance the military ventures and settlement policies that drove the Spanish and Portuguese out of many places in Asia and kept the English in check. As one economic historian has put it, "One could theorize that without monopolies, the market in Europe would have grown still faster. . . . But it is not at all clear how settlement and development would have been financed had free trade prevailed from the beginning."[39] That is, the added income from the monopoly trades both required and supported the presence of large Company establishments in Asia. Economically speaking, the Company's monopoly in some spices might have been viewed internally as something to support stable rather than "excessive" prices.[40] But the acts of violence Coen perpetrated were not only horrifying; in the long run they were likely to have been economically counterproductive.

A PHYSICIAN IN THE DUTCH EAST INDIES

Dr. Jacobus Bontius had taken ship for the East Indies with his wife and two sons on 19 March 1627 aboard the powerful fleet that was secretly returning Governor-General Jan Pietersz. Coen to the Indies.[41] Like so many others, Bontius had joined the VOC because of frustrated ambitions at home. A son of Geraerdt Bontius, the first professor of medicine at the university of Leiden, he was only seven years old when his father died. His eldest brother, Reinier, later obtained a medical professorship at Leiden and assumed a post as physician-in-ordinary to Prince Maurits. The next eldest, Jan, also studied medicine before settling in Rotterdam, where he additionally served as a tax collector. The third brother, Willem, with whom Jacobus was on good terms, sat on the university senate as a professor of law while serving as sheriff for Leiden in 1619, during which time he coercively suppressed Remonstrant activities. (The noted poet and playwright Joost van den Vondel, who hated the Counter-Remonstrants, made a public mockery of the grand ceremony Willem Bontius organized to bury his beloved dog Tyter.)[42] Jacobus also had at least three sisters, about whom nothing is yet known. He followed the main family tradition, matriculating in philosophy at Leiden at age twelve, also earning a medical doctorate there at about twenty-two years of age (in 1614). He then tried to establish himself in practice in the city of Leiden. During the terrible plague that struck Leiden and other Dutch cities in 1624–25, Bontius invented a remedy: because of the scarcity of bezoar stones (concretions expelled from the stomachs of goats from a region of Persia and islands off the Coromandel coast, considered an antidote to many poisonous diseases), he used human bladder stones mixed with theriac or mithridate and a few drops of oil of amber or juniper, which he thought worked excellently.[43] But it did not, apparently, make his fortune, for in later letters to his brother Willem, Jacobus complains that he had not been able to make much of a living practicing medicine in the competitive university town.[44] He pursued botany with enthusiasm, but late in 1624, when the head of the botanical garden and professor of medicine died (Aelius Everhardus Vorstius, who had succeeded his father), the curators replaced Vorstius not with Jacobus Bontius but with his son, Adolphus Vorstius, perhaps because they already had a Bontius on the faculty.

On Monday, 24 August 1626, therefore, Bontius, his wife, and his children appeared before the Heren XVII to be granted a new position with the VOC: physician, apothecary, and overseer of surgeons in the VOC territories.[45] He was being given a general remit by the Company at large rather than by a single chamber to oversee all of the VOC's medical affairs in Asia: to run the hospital at Batavia, to supervise the supplying of ships calling there with proper surgical and medical chests, to inspect the medical personnel of the Company, and to see to the medical needs of the higher-ranking employees of the VOC.[46] From the point of view of the VOC governors, Bontius's knowledge of botany must have made him especially attractive for the new position. There was widespread interest in the natural history of the Indies among the Heren XVII, and in previous decades they had helped Clusius, Pauw, Heurnius, and others to obtain information and specimens from Asia. Clusius and other botanists at Leiden, private gardeners, collectors of specimens of exotica such as Paludanus, and the merchant investors themselves had rapidly assembled new bits of information about the natural world of Asia, and they wanted more from an expert on the ground. From later remarks by Bontius it is clear that the governors of the VOC expected him to compile a natural history of their possessions in Asia. As he later wrote his brother, Bontius in turn hoped that his travels, writings, collection of books, and exotic botanicals would make him the natural candidate for a position on the Leiden faculty.[47] Perhaps for the same reason, Justus Heurnius—son of Johannes and brother of Otto, both professors of medicine at Leiden—had set out for the Dutch East Indies in 1624, where he collected plants for his brother back in Leiden, although in the end he spent most of his time preaching the gospel there.[48]

On the voyage out, Bontius was consulted often by surgeons of the various ships in the fleet because of the constant illnesses that attended such voyages. He himself lost his wife, Agenita van Bergen, even before reaching the Cape of Good Hope (he later wrote that his travels to the Indies had freed him from the labyrinth of a most inauspicious earlier marriage).[49] At the Cape—not yet settled by Europeans—they stopped to take aboard fresh water and food, allowing many of the victims of scurvy to recover, then sailing on to arrive in Batavia on 13 September after a six-month journey, having lost about forty-four lives at sea (about 4.5 percent of those who embarked, not bad for the time).[50] Roughly three months after his arrival in Batavia, on 16 December 1627, Bontius married Sara Geraerts., widow of Hendrik Pauwels.[51] Yet tragedy continued in his personal life: he lost his second wife on 8 June 1630 to "cholera"—an acute intestinal disease. He married again on 14 September 1630, joining with Maria Adams, widow of the Rev. Johannes Cavalier. And in early 1631 his eldest son died of "kinderpoxkens" (perhaps measles). He was also charged with many responsibilities. At the end of 1628, for example, he was made a member of the Court of Justice, the highest judicial body in the Dutch East. During his service on the court he heard the case of a couple of Dutch teenagers who had had sexual relations (in the governor-general's quarters), incensing Coen, who ordered his chief law officer (*advocaat fiscaal*) to prosecute them without mercy. The court had the fifteen-year-old boy beheaded, although they saved the thirteen-year-old girl from drowning by sentencing her to a severe public whipping. Perhaps Bontius was one of the members of the court who saved the girl's life, for she was the daughter of Coen's successor, Jacobus Specx, who on 1 May 1630 chose Bontius to be his advocaat fiscaal. Bontius also served Specx as bailiff of Batavia from 15 October 1630 to 18 January 1631. Later that year he succumbed to the diseases and rigors of service in the East: when he made out his will on 16 November 1631, he was bedridden but "healthy in mind, memory, senses and language"; he died on the thirtieth.[52] His only living son, Cornelis, received a legacy (his third and surviving wife inherited his chattels and real estate) and was sent back to the Netherlands to be raised by his uncle Willem.[53]

By the time of his death, however, Bontius had not only accomplished a great deal but survived two sieges of Batavia in 1628 and 1629 despite falling seriously ill during both campaigns.[54] The sieges were the consequence of attacks by the exasperated king of Mataram in an attempt to rid himself of the Dutch. The initial attack at the end of August 1628 was repulsed, and the king found himself forced to settle down to a siege, which lasted until the beginning of December.[55] Bontius's brief account of the siege mentioned being shut up in their stone compound by "34,000 troops" of the Javanese commanded by a brave and experienced commander, "Tommagom Bauraxa" (the "tumenggung" Bau-Reksa of Kendal), with whom there were daily skirmishes. The bodies of people killed in these fights often ended up in the river that flowed by the Dutch compound, polluting the water and breeding swarms of worms, while the stench of unburied human and animal corpses pervaded the humid air. At the same time, since the Javanese had not prepared for a long siege and were running out of rice, they resorted to eating the root of serpentaria, which they pounded and soaked in the river to make it edible. This left a glutinous substance in the water that the Dutch believed to be poisonous. Moreover, since the Dutch compound lay at the harbor's edge, the river water was subject to tide and wind carrying salt water from the sea and was often brackish. Topping everything off, it was the rainy season, accompanied by great heat.[56]

An epidemic of severe dysentery sowed death among the Dutch garrison, and Bontius mentioned several of the fatalities by name: the "honorable Jeremiah de Meester, a member of the Indian council," Jacob a Dooreslaar, secretary to the governor, Willem Wyntgis, the young lawyer for the treasury and Bontius's very close friend, the most learned Reverend Joannes Cavalier (whose widow Bontius later married), and even the lovely children of Adrian Blocq. These were, however, just a few examples of more than six hundred sufferers. It is not surprising that a year later Bontius described dysentery as a "horrible" and destructive disease, "killing more people" in the Indies "than any other affliction"; the symptoms were "an ulceration of the intestines" with constant stools, "at first mucous, afterwards bloody, and lastly, purulent, intermixed with filaments and the very substance of the bowels, with intolerable pain and griping of the belly." He knew the agonies of the siege firsthand, having been bedridden for four months under so severe an illness that he himself nearly died, being seized first with an ardent fever, then dysentery, and finally beriberi, which his family was still suffering from when he wrote about the experience. In a more detailed account of beriberi, Bontius wrote that during the worst of its onslaught, "for a whole month" his voice was "so weak, that people who sat close to me, could with difficulty understand what I spoke." The Javanese attacked a second time at the end of August 1629 before breaking off for the last time in November. Again, the doctor commented, "during the time when we were besieged by the Javans" the epidemic disease of dysentery flourished among them—and finally ended Coen's life.[57] Bontius was himself again laid low, this time of a "tenesmus," which was even more dreaded than the dysentery because of the great discomfort accompanying it: "an ulceration of the intestinum rectum, with constant pain and a desire of going to stool, when first a little mucus, mixt with some drops of blood, and afterwards purulent matter is discharged." It had confined him for the four months before he signed the dedication to one of his books on 19 November 1629.[58]

MEDICINE AND NATURAL HISTORY OF THE INDIES

Amid such events and illnesses, what Bontius accomplished by way of gathering information on local medicine and natural history during the four years and two months he lived in the Indies is remarkable. Once he had become fit enough after the first siege and his first bout of dysentery and beriberi, Bontius put the fine knowledge of anatomy he had learned at Leiden to work, and backed by an order from Coen, he and two resident surgeons—Andrew Durie, surgeon of the hospital, and one Adam, surgeon of the Dutch garrison—began to conduct postmortem dissections to see if they could understand better the course of dysentery and other tropical diseases in the body. In one case, for example, the body of a soldier was examined, showing that the intestines "were greatly inflated, and deprived of their inner coat," while the gall bladder was distended because of being overfull with a starchy white substance.[59] Although we cannot be certain how many autopsies they performed, Bontius recorded nine between February and November 1629 (along with three other medical cases) in a manuscript published

Frontispiece of *Oost en West Indische warende* (1693; Dutch translation of W. Piso's *De Indiae utriusque*, 1658), showing Bontius and Durie conversing outside the main hospital in Batavia with "Indian" and Chinese figures. Probably based on a lost painting. Courtesy of the Wellcome Library, London

many years later as *Observationes aliquot selectae* ("Some select observations"). Together with four other "Observations" on the medical consequences of the siege of 1628 (*Observationes: Aliquot ex plurimis selectae*), these constituted a sort of appendix to a manuscript treatise eventually published as *Methodus medendi qua in Indiis Orientalibus oportet* ("On the proper treatment of diseases of the East Indies") which, according to the date of its dedication (19 November 1629), Bontius must have completed immediately after the lifting of the second siege.

The *Methodus medendi* described nineteen major diseases of the belly, chest, and skin found in the East Indies but unknown in the Netherlands. Bontius claimed that his diagnoses were based on the best method: all the treatments had been affirmed by experience, so that he proposed nothing he had not seen with his own eyes and judged to be true.[60] But at the same time, the work celebrated how local plants could be used medicinally to treat local diseases, a common medical point of view that emphasized the beneficence of nature. "Where the diseases . . . are endemic, there the bountiful hand of Nature has profusely planted herbs whose virtues are adapted to counteract them," he commented.[61] For instance, the "palsy" locally called beriberi is described according to its symptoms, humoral pathology, causes, and cure. Bontius recommended moderate exercise when possible and strong rubdowns—a method practiced by the Bengali slaves and Malayan women although, like bathing, not by the Dutch themselves—which should be accompanied by fomentations and baths made from the "noble" herb called "lagondi" (or eastern privet), while the feet and hands are anointed with oil of cloves and mace mixed with oil of roses.[62] He later also commented that in his own case of the disease, before he began to take China root—"a medicine much used by all the Indians, and particularly the Chinese"—he "could not move my legs or arms but with difficulty."[63] China root was also generally useful in venereal and all chronic diseases, dropsy, and the severe skin disease common on Ambon that the Dutch called "Amboynse pocken." Lagondi was an even better herb, seeming to be a divine remedy to the people of the Indies, who used it in the form of fomentations, baths, and poultices to treat almost all diseases, promote menstruation, ease childbirth, cure all ills of the uterus, promote urination, relieve complaints of the bladder and kidneys, and ease the pains of colic. "In a word, the panacea of the ancients was nothing to this of ours."[64]

Within a couple of months Bontius had also finished his "Dialogue," later published as *De conservanda valetudine: seu de diaeta sanorum in Indiis hisce observanda dialogi* ("On the preservation of health: or observations on a sound way of life in the Indies in the form of a dialogue"). It was modeled on the famous work of Garcia da Orta, although more stilted in style. It took the form of a conversation between Bontius and his surgical colleague Durie ("Duraeus"), during a morning break after an early round of visits to patients in the hospital. Bontius gave Durie the role of questioner, to which Bontius replies as the expert, writing as if everything he knew had been learned from experience. Much of what he wrote must have been learned from Durie himself, however. Durie, son of the famous Protestant ecumenical advocate Robert Durie, had been born in Scotland four years before Bontius and had enrolled in the faculty of medicine at Leiden on 18 October 1612, where his period of study overlapped with Bontius.[65]

Before completing his degree, however, Durie took a position as surgeon with the VOC, leaving for the Indies on 26 December 1619. He became chief surgeon to the hospital in Batavia and chief surgeon of the fort (the "Castle"), and he assisted Bontius with many of the postmortems conducted in 1629. As a well-known person of sound character, he also came to hold positions such as elder of the consistory, sheriff of the city of Batavia, head of the orphanage, regent of the female house of correction (*Vrouwen-Tugthuys*), and surgeon to the poor (*chirurgijn van de Diaconie van de stad*). From several of these positions he collected a good salary, which supplemented what he received from the VOC: the free room and board of an upper-merchant, plus the sixty-five gilders per month he earned as surgeon of the hospital. Durie was, then, both a well-educated and experienced surgeon and a man of influence and modest wealth near the center of power in Batavia.[66] Bontius wrote of him with respect and friendship, and Durie appears to have been present at Bontius's deathbed.[67] He long outlived Bontius and at least two of his three wives, surviving until about 1655. As a relatively old hand at the medicine of the Dutch Indies, Durie must have been one of Bontius's chief informants. But Bontius reverses their roles in the text, giving himself the part of instructing Durie.

Like Orta's work, the discussion in the "Dialogue" is organized around the common medical theme of the six non-naturals: the six topics according to which learned medicine had understood the effects of environment and behavior on personal health. As usual, they begin with the subject of air and move on to food and drink, evacuations and retentions, motion and rest, sleep and waking, and passions of the mind. The topic "air" allows for a consideration of the climate, seasons, and periods of the day. The air of the country generally is both hot and "exceeding moist" and around Batavia in particular is "not very wholesome," since the land near the city "abounds with stagnant water and marshes," while the winds off the mountains bring "gross and fetid vapors, not to say poisonous, on account of the multitude of insects"; "besides, the penetrating nature of this air produces that miserable species of palsy called the beri-beri." Fortunately, the sea breezes were far more healthful. Only two seasons occurred: the rainy one from November to early May and the (comparatively) dry one from late May to the end of October; Bontius also thought that because Java was just below the equator, the rainy season constituted "summer." Nevertheless, from dawn until about nine in the morning, and from four in the afternoon until nightfall, the air was "temperate" and breezy, allowing healthy exercise (or even labor); people had to be careful in their movements during the forenoon (the most dangerous time) and the early afternoon.[68]

Bontius often found fault with Orta in this manuscript, especially in discussing local food and drink. Orta had apparently never traveled east of Ceylon or visited the Persian Gulf region, having relied on other oral or written testimony for information about those places. Bontius is therefore almost as critical of Orta as Orta had been of the ancients. For instance, he declares that in saying that the Javans and Indians think of pepper as containing a cold quality, Orta was "again ridiculous." Only the ancient sophists could hold such a view, and they had been refuted by Aristotle's comment that in such cases experience had to be consulted. In the same dialogue, Bontius says that Orta got the uses of *Calamus aromaticus* all wrong, for although he "acknowledges no other use, either of it, or the sweet smelling reed in India, than for bedding horses . . . had he been as diligent in investigating the qualities of aromatics, as [he was] discerning in reading Arabian physicians, he would not have been ignorant of the uses of that plant: for throughout India both fish and meat are cooked with a bit of calamus aromaticus or the sweet smelling reed, both to improve their flavor and to invigorate the stomach."[69] Such comments suggest that, as the VOC struggled to displace the Portuguese as the chief long-distance merchants of Asia, Bontius was claiming the superiority of his knowledge over Orta's and, more generally, of Dutch knowledge over that of the Portuguese.

Bontius went on to prove at greater length Orta's failings by reviewing and commenting on Orta's book, a manuscript he finished quickly, by February 1631. It was later published as *Notae in Garciam ab Orta* ("Notes on Garcia da Orta").[70] He seems to have worked from a copy of the 1605 folio of Clusius's edition of Orta's *Colloquies*.[71] He repeated his view that Orta had been "ridiculous" when considering the qualities of pepper and should have considered Aristotle's objections to such things more carefully; and in commenting on Orta's description of cardamom, he remarks that he "committed a great mistake" in describing the seedpods as hanging down like peas, "For I, who have seen the cardamom grow in great quantity a thousand times, can affirm that it resembles reeds."[72] Yet when he considered Orta's work as a whole, Bontius was much less critical of Orta than he had been when he wrote the "Dialogue." Mostly he simply, rather temperately, corrected or supplemented Orta. For instance, where Orta remarks that those who use opium appear drowsy, Bontius wanted to remove the implied criticism of the drug, for "if we did not have this opium and opiates the prospect in this very hot region of making medicines to treat dysentery, cholera, ardent fevers, or other bilious diseases that swell the organs would be frustrated." Similarly for the tree that produced gum benzoin: Orta says it is large and tall, but Bontius had seen it in Java for himself, where it is composed by "the union of several suckers, like the smilax aspera, or sarsaparilla," although the suckers can each exceed the thickness of an arm. In addition to such corrections, he added considerable information. For example, Orta confessed that he had not seen asafetida, "called 'Hin' by the Javans and Malaians," so Bontius described it. Likewise, when discussing ivory, Orta confessed that he had never seen a rhinoceros, whereas Bontius had "not only seen them a hundred times hiding in their lairs, but also wandering in the woods," which gives him an opportunity to tell of a dangerous encounter he had witnessed. He also repeated some of the additions mentioned in the "Dialogue," but more moderately: in the case of *Calamus aromaticus*, or sweet flag, for instance, he refers the reader to Orta's remarks on its medicinal usefulness and simply adds that Malayan women use it in the kitchen by adding it to fish and flesh.[73]

Bontius sent all three treatises he had written—the *Methodus medendi* and the two sets of observations, together with the "Dialogue" and "Notes on Garcia da Orta"—to his brother Willem along with a cover letter on 18 February 1631. He hoped that they would "be judged worthy of being committed to the press," in which case "let them see the light: but if they appear not to be sufficiently polished, keep them at home with yourself, as a token of my sincere affection."[74] Unfortunately, his death in November meant that he could not lobby for them further, and they disappeared from view for more than a decade.

According to that letter to his brother, however, Bontius was at work on yet another book, this one on the natural history of the region, which he expected to finish soon: "expect next year, if the power of life remains, a full description of plants, shrubs and trees, with a delineation of each drawn from life."[75] He seems to have considered it to be his major task on behalf of the Heren XVII. The letter addressed to them in November 1629 (later printed as the dedication to the *Methodus medendi*) expressed his devotion to their service, which would be even more evident when he had finished his "commentaries on the shrubs, trees, and herbs which grow in Java." That he had been working on it even before completing his first manuscript is clear from a comment he makes on the disease tenesmus in the *Methodus medendi*, adding: "And would at this moment this disease, which has laid me low for about four months, as well as the Javanese who surround us on all sides, permit me to travel around the countryside to freely explore the delightful woods of Java and gain an exact knowledge of the many most noble herbs that are here!" He promised that a future volume would give the names of a great number of tress, shrubs, and herbs. "I shall, likewise, give you an account of birds and fishes (a subject I was always fond of) which are caught here; explain to you their nature, and show what are their peculiarities, or in what they differ or agree with those of our own country." Additionally, in the entry on Indian sage in the same book he says that he will give more information on the medical uses of the plant in his forthcoming study. He mentioned his projected work in other manuscripts, too: in the "Notes on Da Orta," for instance, he noted, "You may see the figure and description of it [that is, a fruit locally called *focqui*] among my *Exotic plants*, which I shall endeavor to let you have next year."[76] In subsequent writing, moreover, he sometimes repeats observations, as if drawing from a common set of expanding notes.[77] Perhaps even from the start, but certainly no later than his recovery from the effects of the second siege, Bontius was keeping a record of his observations on natural history, in both words and pictures. It may even be that Bontius's last work, unfinished on his death, was the one he began first.

Bontius's descriptions of the natural history of Java often contain interesting anecdotes as well as natural details. He comments on the internal organs of a few animals, suggesting dissection. He kept at home the skin of a thirty-six-foot snake that he killed in the woods. Others he kept alive: his observations of the chameleon were based on one living "in a case [*cavae*] at home," and he had at his house a flying lizard "which measured three quarters of an ell" (probably meaning a Flemish ell, which would make the lizard about twenty inches long). "It can fly, but does not persist in flight for long, . . . reaching forty paces, or in turn thirty, just like flying fishes." He kept various birds in his back garden. He also thought it a "great pleasure" to see the speed of house lizards when they chased flies and ants.[78] Other descriptions were based on observations in the woods near Batavia. He vividly described an encounter with a tiger and another with a rhinoceros (trying to protect her young). As for plants, he concentrated almost entirely on medicinals and a few cooking herbs. A few lesser-known plants were also described, including a thorny shrub that had no known name, "not growing very high above the earth, which has yellow flowers like the night-shade." When the fruits were "rubbed between the hands they emit a strong fetid odor, surpassing even assa foetida," for which reason "Indian women apply it to the nostrils of the hysteric."[79]

SOURCES OF KNOWLEDGE

In acquiring his information, Bontius had the advantage of Orta's and other texts as a starting point for his investigations. A list of citations taken from all his works (including that of the natural history discussed below) includes not only Orta, but Pliny's *Natural History*; Pierre Belon's *Les observations de plusieurs singularités* (first published in 1553) on the natural history of the Near East; Christoval Acosta's *Trata de las drogas y medicinas de las Indias Orientales* (1578), which is almost entirely derivative of Orta; Prosper Alpino's *De medicina Aegyptiorum* on Egyptian medicine and botany (first appearing in 1591); and of course Linschoten's *Itinerario* (1596) on Portuguese Asia. He also mentioned Pedro Tei-

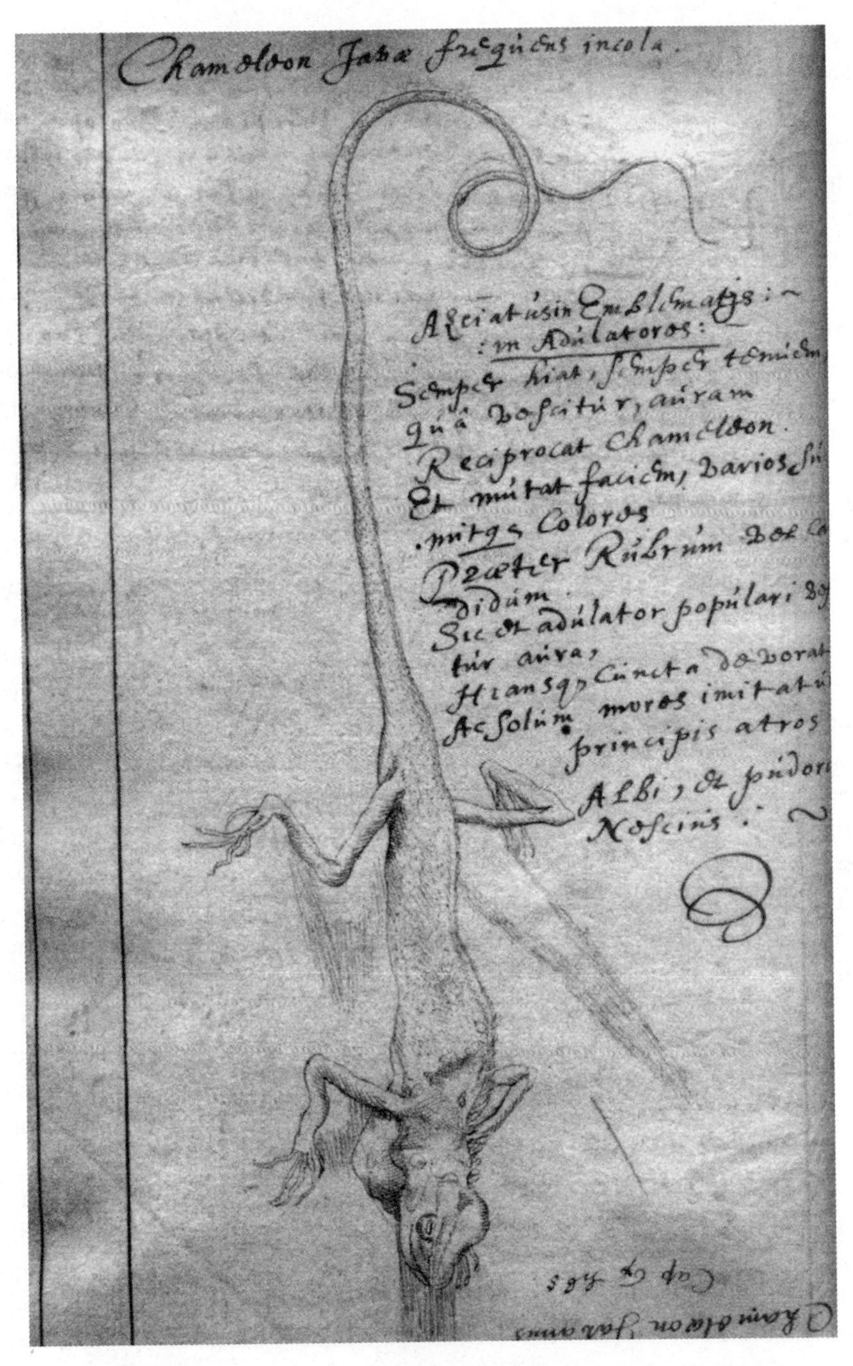

Drawing of a chameleon. (MS Sherard 186, fol. 44v). By permission of the Plant Sciences Library, Oxford University Library Services

xeira's *Relaciones* (a detailed account of the land route from Basra to Aleppo), and quoted from Horace, Juvenal, Martial, Virgil, and—several times—the Roman playwright Plautus. He clearly had a small library with him; perhaps he had even brought along the whole of his very sizable collection of more than two thousand books.[80] But he also repeatedly noted how he had been an eyewitness to many things Orta and other European authors had never seen. After all, he said, "what is shown to one eye witness is truer than ten things heard."[81]

In keeping with his emphasis on eyesight, most descriptions of animals and plants in his natural history were accompanied by illustrations. In a few places Bontius refers to making drawings himself, but those intended for publication were done with a better hand, that of a young relative who had come to Batavia. In a letter to his brother, Bontius mentioned that he used Adriaen Minten as his draftsman for the drawings, with the permission of the governor-general (as would have been necessary for getting a VOC employee transferred to another job). Minten was the son of their cousin Anneken Screvels, who was the sister of Ewaldus Schrevelius, another professor of medicine at Leiden. Presumably some of the drawings, at least, were done by placing a specimen on a piece of paper to trace the outline, after which the details were filled in, since Bontius describes this method going awry in the case of the gecko (because of the lizard's sticky feet). Bontius explains that he wanted to help Minten get ahead in the Indies, but he neglected the work he was doing for Bontius ("dan hij heeft hem verlopen"), so Bontius arranged to have him sent back home. The extant drawings seem to be by the same hand with few exceptions, one being a tiger drawn by Governor-General Specx himself, in whose presence a "tremendous" tiger was caught and killed just outside the city wall in May 1630.[82]

Clearly, however, most of Bontius's information came from things heard rather than observed.[83] Bontius, like Orta, had many opportunities to meet a variety of people of different backgrounds, places, and experiences, not only learned Hindus and Muslims but merchants, sailors, and ordinary people, even slaves. It is well to see for oneself, in order to make careful observations, but information about how things are used can only come from noting the behavior of other people or, better, discussing things with them directly.

One pool of useful knowledge immediately available to Bontius was the oral tradition of European medical people sent to the East. One who undoubtedly knew much and helped Bontius was his friend Durie. Also, though he makes no mention of her, Bontius might have drawn on the knowledge of the official midwife appointed by the council in 1625 and paid to serve the needs of poor Dutch women.[84] At other times, Bontius refers to merchants as sources of knowledge. For instance, when he corrects Orta on the subject of asafetida, he says that he had in his custody some of the second of its two varieties, "given to me by an Armenian merchant friend, who brought them out of Persia." When writing of bezoar stones, he relates their origin as told to him by Persian and Armenian merchants, who gave him information "completely faithfully," and buttressed it with the written testimony of "P[edro] Texeira, a Portuguese," in his Spanish account of the kings of Persia, as confirmed by Dutch and English merchants. In writing of a medical substance called "tutty" (usually now considered to be zinc oxide), Orta admitted he had it on hearsay that it was produced from the ashes of a tree, while Bontius knew it to be made by calcining a "glutinous earth," which "Persian and Armenian merchants who come here to negotiate business" assured him was found in great quantities in Persia.[85]

Often, however, Bontius refers to how the "Indians" or more particularly the "Javans" or "Malays" did things, which he clearly admired, or came to admire.[86] He made the following remarkable comment when writing of Indian loosestrife: "Nor am I ashamed to say I have learned this from those called barbarians and unlearned, for these people, especially the Gujurates and those from the Coromandel coast, sagaciously distinguish between medicinal, edible, and poisonous herbs better than the most expert botanist in our country."[87] When treating *Calamus aromaticus*, or sweet flag, he mentioned that Malayan women use it in the kitchen, going on to chastise his European brethren for their self-appointed superiority: "And here, by the way, I note that these nations, though many among us call them barbarian, are superior to the Poles and Germans in pickling fish, who nevertheless are awarded these plaudits without blushing." In another case he thought that the people who come from Surat and the Coromandel coast must be followers of Pythagoras, since they were vegetarians who abstained even from red beans and herbs because of the color. "Thus it is that those who in other things are illiterate have an exact knowledge of herbs and shrubs, such that if the most learned [Pieter] Pauw, prince among the botanists of our age, came back from the dead and travelled here, he would be surprised that these barbarian peoples could instruct him." At another point, he burst out: "Besides, every Malayan woman practices medicine and midwifery with facility; so (I confess that it is the case) I would prefer to submit myself to such hands than to a half-taught doctor or arrogant surgeon, whose shadow of education was acquired in schools, being inflated with presumption while having no real experience."[88]

Interestingly, much of the information he acquired and admired concerned women's medicine and cooking methods. To give just two instances: the "hog stone" or Malacca stone (a soft and fat stone "that feels like Spanish soap"), which came from the gall of hogs and the stomachs of porcupines, was infused with wine to treat "mordexi." But it was dangerous to pregnant women and caused abortions. "I have been told by Malayan women that it is certain to provoke an abortion, and if their menstrual purgations do not come at the right time, if they only hold this stone in their hand they are rejuvenated." Similarly for Indian saffron, or turmeric, locally called "borbory": "throughout the Indies no plant is more frequently used." It was taken internally and applied topically for obstructions of the bowels and the mysentary and for urinary complaints. Moreover, "in diseases of women nothing is so much celebrated by the Malayan women than this borbory. It has a divine effect in easing childbirth, in cases of difficult urination, and in kidney problems. For problems of the uterus it is a specific. And to make sure of this truth, among all medicines I have myself found nothing better in these afflictions than this remedy."[89]

How he obtained such information about women's use of local substances is less clear. Bontius may of course have been listening to patients, some of whom were undoubtedly women. In Bontius's day, about one hundred European women were present as the wives of the upper-merchants or administrators, and some may have been on good terms with Asian women.[90] Large numbers of VOC employees took local female partners, a few entering into long-term relationships, even buying out their contracts with the Company to stay with their partners in Asia, where they learned much about regional customs and languages. A growing population of people with mixed heritage and multilingual abilities was beginning to take shape in Batavia and other VOC settlements, many of whom became crucial brokers of information. Bontius may therefore have obtained information about the practices of local women through his wives and their female networks, as well as his male patients and their networks.[91]

He had other sources, too. At least one piece of information came from a local professional counterpart: when describing the uses of the shrub *sedo*, Bontius commented, "While I am writing this an old Malayan, who practiced medicine among his people not unsuccessfully, tells me that nothing is better against poisoned wounds inflicted by Javanese daggers and lances."[92] Bontius may have been referring to one of the physicians of Java who adapted the learned medical traditions: texts deriving from Sanskrit were translated and rewritten in Java, Bali, and other places in Southeast Asia, although they were much rarer in Malay than in Javanese or Balinese languages. Most local practitioners, however, relied not on learned discourses but on a wealth of experience about the use of botanicals in recipes handed down both orally and in writing.[93] Whether Bontius obtained information from indigenous medical practitioners by familiarity and friendship or by payment, it is likely that they were the source of much of his information. Even slaves could contribute. Like many well-to-do residents of Batavia, Bontius owned slaves, referring at one point to "my Moorish slaves," confirming that the "chattel" (*goederen wezende*) mentioned in his will were enslaved persons. They certainly told him about tigers, so it is likely that his servants and slaves were also important informants about the uses of herbs in the kitchen.[94] He may also have obtained information from women selling vegetables in the markets of Batavia, since in traditional societies even today, women tend to be the main retailers of local produce, not only bringing things to market that university-educated ethnobotanists and others have not seen before but providing information about their uses.

There were sometimes suspicions to overcome. The one instance in which he gives us some detail makes it clear that although he was willing to pay for information, or at least specimens, fear on the part of his informants sometimes created problems that money could not overcome: he had to create a sense of trust. One of his neighbors happened to be an elderly Javanese woman—and a slave of a Chinese gardener—who possessed a "woodpecker" that could speak even more like a human than a parrot could (undoubtedly a mynah). Bontius several times tried to purchase it from her in order to draw it. He then tried to get her to lend it to him. After many conversations, she finally agreed, but with the stipulation that he would not feed it any pork. When Bontius and the artist got it home, the bird began saying, "Orang nasarani catjor macan babi," or "Dog of a Christian, eater of pork." The woman clearly feared that Bontius might kill the bird or feed it pork on hearing this. He implies that he did neither.[95]

He, too, had suspicions. When he was working in Batavia, the Company recorded about 2,400 Chinese and 35 of their slaves resident in the city, 1,900 European-born employees of the VOC and their 180 slaves, 630 free citizens and their 730 slaves, 650 "Mardijkers" (indigenous people who had taken up Christianity) and their 150 slaves, and 80 Japanese and their 25 slaves, for a total population of around 8,000 (of whom about one-third were enslaved).[96] The ethnic Chinese were so numerous in Batavia that in 1619 Governor Coen appointed one of them the "headman" or "captain" to deal with their civil affairs and to represent them on the city council and elsewhere, while after 1635 a (Dutch-born) medical practitioner was appointed by the VOC to look after the many Chinese in VOC employ; after 1640, a hospital for the Chinese was also built.[97] Bontius's one comment slurring an entire group, however, was about the Chinese: in writing of a locally concocted alcoholic drink the Dutch called "quallen," which he considered to be so unhealthful that it "is to be avoided as death itself," he remarked that it was made by the Chinese, "the most avaricious and crafty wretches on the face of the earth."[98] Equally, the only Chinese medicine he praised, China root, was in use throughout the region, so he could well have learned of its uses for beriberi and other conditions through Javanese rather than Chinese sources.

Coen had been trying to drive the Chinese out of the pepper trade, and the VOC was having trouble making useful trading contacts with China proper, since the Spanish and Portuguese managed to keep suspicions about the Dutch high among Ming officials. Bontius's generalization may suggest, then, the kind of lumping remark he might have made about the Portuguese or other potential adversaries who had a national and linguistic identity: although it was an ethnic slur, it may not have been exactly "racist," which is a matter more for the nineteenth century than the seventeenth.[99]

When it came to the "Javanese," however, even personal experience with suffering during the prolonged sieges they mounted against Batavia could not dissuade Bontius from praising them to the skies. He added a comment to Clusius's annotation on Orta about their method of writing on palm leaves, noting that "they write so elegantly as to excel us by a long way; and when they draw the characters they delineate on these leaves (which are Arabic) then my indignation rises against those of our Europeans, and especially our compatriots, who admire nothing unless it is their own, even calling these peoples barbarian who, of a more laconic mind, can express more of their meaning in only a few significant characters than ours can with long phrases and useless multiplicity of words." This suggests that he had been imbued with Lipsius's admiration for Tacitus, who cultivated brevity of expression. He also echoed Lipsius's political views in praising the Mataramese: "Although it appears that the kingdom of Java is despotic [*Tyrannicum imperium*], they exercise their authority in light of the condition of the people so that everyone, unless blinded or completely thick, will quickly satisfy their understanding to the effect that the political life here is supportive, the government ruling well and the people obeying even better." "I often marvel at the carelessness of our people, who without respect call these people barbarians," he concluded elsewhere, who "not only in their knowledge of herbs but in all aspects of their economic arrangements [*oeconomica administratione*] leave our own far behind."[100]

One can speculate that Bontius was developing a sense of owing the Javanese and others respect for the knowledge they shared with him. As his respect for Orta's text grew according to the time he spent with it, so, too, his respect for the knowledge of the Javans seems to have grown with time. One privilege of being a well-placed physician was the ability to get to know all kinds of people if he desired. At the same time, he came under obligations to them when they gave (or lent) him objects and information. The exchanges between Bontius and his informants might therefore be said to fall into the category of what Marcel Mauss termed "gift exchanges."[101] The simplest part of Mauss's message is that gifts are special and personal, imbuing in their recipients a feeling of obligation toward the giver and weaving together the lives of all participants in the exchanges. Moreover, "it goes without saying," comments one of Mauss's interpreters, that the gifts given "are not necessarily 'things' in the sense of material objects having a cultural significance. The 'thing' may also be a dance, a spell, a human being, support in a dispute or a war, and so forth" or, we might add, even a bit of knowledge.[102] Similarly, Bontius considered the Heren XVII to be more than mere employers. In the preface to his *Methodus medendi*, he wrote of how he had devoted his medical labors to their excellencies for almost three years, "which on my arrival in your Indies, I did not permit to remain sterile, but began to exert with activity, to demonstrate my endeavors, that the emoluments you have conferred upon me should be productive of public advantage." His work was therefore meant "as a small return for the many obligations I lie under to your excellencies, and which I can never fully repay." The "small paper present" that was his manuscript was "all I can afford" but was meant as a gift he hoped they would deign to accept.[103] In other words, Bontius considered his position in the Company to be a gift from the Heren XVII, which he repaid through his efforts to put on paper a work on the medical world of their Eastern "possessions." We are entitled to extend his own categories, then, in thinking that he in turn recognized the gifts of knowledge he received from other people. He repaid them partly by defending them against defamatory remarks such as being called "barbarians."

CONVEYING MATTERS OF FACT

Yet if the personal relations Bontius developed in his work brought the obligations of the gift, he also wished to convey what he learned to others in Europe, and in so doing some important things went missing. He loudly praised the God-given medicinal effects of local plants and local people's uses of them, but at the same time he dismissed out of hand the local meanings attached to the plants and practices. That is, his concern for descriptive information—empirical matters of fact—gave him a blind spot when it came to what we would call the intellectual system of, or cultural assumptions behind, local medical practices. He says nothing about any learned medical traditions he might have encountered, nor does he reveal much about the "magical" practices that were deeply embedded in local medical ritual.[104] Mainly he simply remained silent about such matters. But he may have come across a fair amount of such information that he simply discarded as superstition. He gives an indication of this outlook in one instance, in commenting on Indian verbena. "This herb is considered sacred among the old Indian women (which they have in common with our own old women)," he wrote. But then he immediately apologized for saying even this much. He had mentioned it only to "demonstrate the foolish habit" of mind that considers such things to be true, but "I am not one of those who has a propensity to superstitious belief about [what are] the natural powers of medicines."[105] This went for superstitious old Dutch women as much as his local informants.

In other words, Bontius did not simply encounter and speak with various people, exchanging gifts of respect and assistance with them; he did so in light of strong assumptions about valuing "matters of fact" over belief. He converted local words and things into parcels of information that could be packaged in Dutch or Latin words and syntax. At the same time he discarded the contexts in which he found the information, either not understanding them or self-consciously stripping them away as religious "idolatry." He privileged certain kinds of knowledge: information about things and the material uses of those things. Foreign nouns, adjectives, and verbs that were concrete—the simple things that came from the five senses rather than the mind's eye—were readily transferable; abstract concepts he ignored, misunderstood, or dismissed. Many ideas about medicine held by indigenous people remained incommensurable with Bontius's concepts in large part because he had no interest in their speculative ideas. He therefore transformed the knowledge he acquired into descriptive statements about the things and practices themselves. As he acquired this kind of information, either from his own sensations or from translation, he set it down on the page in the form of bundles of information. Bontius's works therefore clearly represent the major form of natural historical investigation in the period: a palimpsest. His manuscripts demonstrably built on the knowledge of his predecessors, both European and "Indian," who were often obscured under his layers of information, despite his own sense of owing them a great deal.

Proceeding as he did, however, he expected his information to be of use to almost anyone else. Bontius was boiling things down to their lowest common denominator, information units that could be circulated in just about any context. He (re)produced knowledge, accumulated it, and exchanged it, making information—if not theories—commensurable. For Bontius, matters of fact were the coin of his realm. In transforming local knowledge into transferable matters of fact, he was therefore doing something that had similarities to taking the production of nutmegs away from the Bandanese. One might even say that nutmegs, properly cultivated, are as renewable a resource as information about them: there are countless potential nutmegs, just as there are countless potential units of information about them. The question is not so much whether Bontius was shipping all the potential information about the Indies to the Netherlands. Clearly he was not, any more than Coen exported the last nutmegs. But both were putting the means of accumulation and exchange in the hands of members of the VOC, appropriating the benefit of the exchange to themselves. Bontius honored his informants and spoke of (most of) them well, but he alone could put the information he acquired from them onto the page in a language that would be understood well beyond the geographical limits of their culture.

To be successful in this information economy, however, Bontius had not only to write down and pass on the information he accumulated; he had to put it into circulation. That took longer than he must have expected. The medical manuscripts he had sent to his brother in early 1631 appeared together in the form of a book only in 1642. Franciscus Hackius, one of the major Leiden publishers, and one who dealt with medical books, published them under the title *De medicina Indorum* ("Medicine of the Indies").[106] Although printed in a small format (12mo) volume with a small, inexpensive typeface, the book did have a little, simple engraved frontispiece done by Cornelis van Dalen. The frontispiece confused images of the West and East Indies but clearly portrays Bontius learning from the local people: an "Indian" wearing a feathered headdress presents the book to a man in academic gown and fur hat holding a staff with an entwined serpent in his left hand (the caduceus, sign of medicine) and a pinecone-like fruit from the Indies in his right hand, while a bird appears above his right shoulder and head. The several treatises were printed in an order reversed from the date of their completion: the cover letter Bontius sent to his brother in 1631 became the preface; his commentaries on Orta, composed last, constituted the first book; the dialogues on the preservation of health made up the second part; the *Methodus medendi*, his first essay, became the third book (with its own dedication to the Heren XVII dated 19 November 1629); and the two sets of accompanying observations on disease and dissection became the fourth. With their subject matter of medicines as well as diseases of the Indies, Bontius's works attracted wide interest among the medical community, and in 1645 the *De medicina Indorum* was reset (with the contents in the same order) and printed again, this time in Paris in a much larger format and finer typeface, paired with the work of the Italian traveler and botanist Prosper Alpino's *De medicina Aegyptiorum* ("On Egyptian medicine," 1591). This edition of Alpino and Bontius was reprinted yet again a year later. The Latin version of *De medicina Indorum* continued to be printed additional times over the centuries, was translated into Dutch in 1694, and even appeared in English in 1769 as a "great work of public utility" to accompany the British expansion into India.[107] His legacy was important enough that three centuries after his death Bontius remained heralded as one of the fathers of "tropical medicine."[108] His work on natural history, however, would not appear for over a quarter of a century, and then because of propaganda for Dutch activities in the West Indies.

NEWS FROM THE WEST INDIES

Bontius's natural history of Java was first published as part of a large and lavish book, *De Indiae utriusque re naturali et medica libri quatuordecim* ("On the natural history and medicine of both Indies in fourteen books," 1658). It was edited—and altered—by an Amsterdam physician, Willem Piso, who with the assistance of a number of other physicians, liefhebbers, and the learned Caspar Barlaeus, at the urging of Maurits of Nassau, used the investigations into the medicine and natural history of both the East Indies and Dutch Brazil to promote public interest in the Dutch West Indies. But Piso played fast and loose with the formerly stringent editorial standards in natural history, taking credit for things accomplished by another physician in Brazil, Georg Marcgraf, and adding other bits of information to Bontius's work without attribution. When this confusion about matters of fact became known, it caused Piso to become the butt of humor. Even when used in a work publicizing the achievements of the Dutch West India Company (WIC), then, being exact about matters of fact was of prime importance, but not even the reputation of an eminent physician working for a prince could guarantee its complete accuracy.

Piso had served in Brazil during the period when much of the sugar colony seized from the Portuguese was governed by Maurits of Nassau, a nephew of the former Stadholder of the same name. Maurits has been put in place by the West India Company, which had been founded shortly after the United Provinces resumed war against Spain in 1621. The WIC was meant to be the equivalent for the Dutch Atlantic trade of the VOC for Asian trade, a joint-stock company uniting previous ventures, organized according to Chambers and having a general governing board, the Heren XIX ("Gentlemen Nineteen").[109] But the Atlantic world differed from the worlds of the Indian and Pacific Oceans. For one thing, it was easier to get to, and because the journeys were shorter, initial investment costs for outfitting ships could be lower, promising greater returns and attracting more investors, making it harder to monopolize the trade. Indeed, although there had been talk of getting the States General to charter a West Indies company almost from the time the VOC was established, almost two decades passed before anything was done, partly because so many diverse business interests did not want to submit to a monopoly. For another thing, the riches of the Atlantic economy were much more varied than the spice trade: fortunes could be made in trading ivory and gold on the Guinea Coast of West Africa; across the sea, in northeastern South America and into the Caribbean, were productive sugar plantations whose labor force could be supplied from Africa through the horrific commodification of people in the slave trade; on the northern coast of South America were salt pans critical for Dutch fisheries during times when Portuguese salt was prohibited to the Dutch; from Mexico and Peru came huge amounts of silver, as well as gold, cochineal (a dyestuff), cacao, and other goods; and from the northern parts of the West Indies (what we call North America) came large quantities of furs and a growing amount of tobacco and other botanicals. Trying to bring all these various exchanges under one umbrella would create a possibly conflicted business strategy. In addition, although many Jews and former New Christians who fled from Portugal after 1600 because of increased prosecution by the Inquisition had business connections with friends and relatives in the New World, especially Brazil, they sometimes found the Counter-Remonstrant supporters of the WIC to be anti-Semitic.[110] Finally, the Spaniards and Portuguese (united under the Spanish crown since 1580) claimed a monopoly on trade with these lands, were present in force, and could quickly be reinforced from Iberia, which led to heavy losses among the Dutch interlopers in the Atlantic.

With the renewal of the fight against Spain, however, the States General chartered the WIC in June 1621 as an instrument of war, hoping that the company could attack and displace one of the most important sources of Spanish financial power while paying for the campaigns from the profit of the ventures. Many of the company's early investors (although by no means all) were Counter-Remonstrants and others who had an ideological commitment to winning back the world from the Catholics, and especially the Spaniards. They believed, for instance, that the West Indians (or just "Indians" for short) were oppressed by the Spaniards and that they would eagerly rise up against their overlords if given half a chance. According to this hope, small Dutch military forces could have huge effects when coupled with the ambitions of their natural allies. Over the years, however, bitter experience would create disillusionment among those who had believed these ideological dreams.[111] The WIC did gradually attract investment from others who simply hoped to make money. But the venture was never a great financial success; the old WIC was disbanded in 1674 and a new one was formed to concentrate on developing the sugar and slave trade of Africa and the Caribbean.[112]

Things did not start well for the company. Assembling a large fleet intended to strike awe into the Spaniards took two years, and after the fleet sailed problems slowed it down and allowed its enemies to prepare. In 1623 the fleet set sail to circle into the Pacific through the Straits of Magellan and seize the silver fleet that annually sailed up the coast from Chile and Peru to Mexico and then over to the Philippines (to fuel the trade with China and Japan), but the slow start and missed opportunities led to failure. Another large fleet set out in 1624 to capture Brazil, taking Bahia (or Salvadore), the capital of Portuguese Brazil, and seizing a good deal of silver, tobacco, and sugar. But an even larger fleet from the Iberian lands soon recaptured the town. There were better successes at taking and holding forts along the Angolan coast. When in 1628 Piet Heyn captured the Mexican silver fleet that annually sailed to Spain (although the Atlantic Peru fleet escaped capture), the amount of booty flowing into WIC coffers as a result of grabbing these ships loaded with silver, gold, pearls, silk, hides, cochineal, indigo, and dyewood was valued at 11.5 million gilders, causing the price of WIC stock to double and a large dividend to be paid to investors (the only one in its history), the one real financial success of the company. These funds helped to underwrite yet another attempt on Brazil, which in 1630 captured the northern cities of Recife and Olinda, the doors to the sugar plantations of Pernambuco.[113] The Portuguese and their local allies, however, retreated into the hinterland to wage a guerrilla war, which lasted for years, draining WIC finances, wearing down the Dutch, and causing infighting among the governing council in Brazil.

In 1636, the Heren XIX therefore decided to send out Johan Maurits as governor. Within a few years he had successfully restored order and got the sugar plantations working again, although he returned in 1644 because of frictions with the governors of the WIC. Ostensibly they recalled him because Portugal had freed itself from Spanish rule in 1640 and negotiated a truce with the Netherlands, but they probably also blamed Maurits's extravagant expenditures for some of the WIC's continuing financial difficulties. Despite a flow of new settlers to Brazil from the Netherlands to labor in the sugar plantations, the WIC revitalized the importation of slaves from Africa, transporting more than a thousand every year, for which they took over Luanda and other places south of the Gold Coast.[114] After Maurits's departure, however, conflict between the local Portuguese plantation owners and the Dutch settlers led to rebellion, and by 1648 the Dutch were confined to their fortified areas and suffered defeat whenever they tried to break out. The Peace of Münster that ended the war with Spain in May 1648 made the struggle seem even less politically important, and since neither the WIC nor the VOC was ever subsidized directly by the Dutch state, the company looked a worse financial investment than ever. In January 1654, the Dutch were forced to capitulate and leave Brazil, weakening their strategic position in the Atlantic.

During a period of fighting in 1637, Maurits's physician-in-ordinary, Willem Milanen, died. Maurtis immediately requested a replacement from the Heren XIX, and Willem Piso got the job. Born in Leiden in 1611, Piso had studied in the local university from an early age before acquiring his medical doctorate in 1633 from the university in Caen (a favorite place for educated Dutch professionals to acquire a degree cheaply),[115] after which he tried to develop a medical practice in Amsterdam. There he obtained introductions to the Muider Kring, including

Barlaeus and Vondel, and probably got to know another of its members, one of the most learned governors of the WIC, Johannes de Laet, a man of enormous interest in natural history and geography. When Piso left for Brazil, Vondel wrote a six-stanza poem as a farewell. The fleet left on 1 January 1638, landing in Brazil a few weeks later. On arriving, Piso worked hard to preserve the life and health not only of Maurits but of many others, became the head of the local hospital (with two surgical assistants) near the town of Maurits, ran the hospital erected by the governor for the treatment of African slaves, grew many native plants in the large garden Maurits had established at his Vrijburg Palace, and sat on the council of justice. Although Piso took an interest in local medicines and medical practices, there is no evidence that he traveled beyond the main Dutch settlements.[116]

Also sailing with Piso in the fleet to Brazil was the artist Frans Post, who would most impressively paint some of the landscapes of Brazil, and the physician Georg Marcgraf. A year older than Piso, Marcgraf came from a well-educated clerical family in Liebstadt, Saxony, bordering today's Czech Republic.[117] After a fine early education in Greek and Latin, music, and drawing, he left home at seventeen on a *peregrinatio academica* to many German and Baltic schools, matriculating in the medical faculty of Leiden in 1636. A traveler by nature and eager to go to Brazil to conduct observations of the southern skies as well as investigations into the natural history of the region, Marcgraf wrangled an appointment as the astronomer of the West India Company. According to what little information is available about him, not long after his arrival Marcgraf was present at one of Maurits's sieges against the Portuguese but managed to survive not only the grievous diarrhea then raging in the army but a cannonball that just missed his head. Putting his mathematical skills at the governor-general's command, he helped Maurits with the military engineering during the siege and became one of Maurits's favorites. He was given a guard to accompany him on his movements, was constantly in attendance at the prince's residence, constructed an observatory at Vrijburg Palace, and was placed in charge of the medical shop in the new city of Mauritstad. He took advantage of these favors to chart the southern skies and to conduct at least three journeys into the hinterland, exploring the fauna and flora and speaking to the people (whether of European, Africa, or indigenous ancestry), like Bontius often obtaining information about their medical practices. All this he wrote down in notebooks in a cipher of his own devising so that no one else would steal his work should he die.

Marcgraf appears to have collected assiduously, too, and some of his specimens made it back to the Netherlands. His brother later reported being told by Samuel Kechelius (a former roommate of Marcgraf's and curator of the cabinet of curiosities in Leiden) that he once saw "a book of Dryed Insects of Brasile sold att Harlem for 4000 Florins, ye names to all which were written in Marcgarves hand,"[118] while some specimens from his herbarium remain today in Copenhagen and Oxford. Maurits, too, collected and later gave away many specimens and objects as presents, including a large number that went into the cabinet of the university in Leiden.[119] According to his brother, Marcgraf was packed up and preparing to leave with the prince when Maurits pulled out of Brazil in 1644 but was unexpectedly sent to Angola, where he died. The brother inferred that Marcgraf was more or less killed in this way by one or more jealous persons, even implying it was Piso, although the evidence adduced by a modern investigator would seem to show that Marcgraf died a year earlier on a mission to Angola.[120]

Nothing like Maurits's sponsorship of artists and naturalists occurred elsewhere in the WIC territories. In the north, in 1609 Henry Hudson had accidentally discovered a large river while seeking a northwest passage to Asia for the VOC, and the region seemed a good place from which to get furs. Various partnerships formed to exploit the new trade and merged in 1614 into the New Netherland Company, which was merged with the WIC when it was created in 1621. By 1625 a fortress had been started on the island of Manhattan, and Dutch settlements soon spread up the Hudson and down toward the Delaware River.[121] In this early period, Nieuwe Amsterdam gradually attracted more and more interest and cooperation from the new English settlers to the east, who were also there to flank the Spaniards and who found the Dutch settlement a fine place to vend their goods outside the constraints of the London trade.[122] As the settlements grew, medical people appeared, from midwives to surgeons and physicians.[123] Yet although at least some of them must have learned something about local medicines and practices of the indigenous inhabitants, none wrote up anything substantial about the region's medicine and natural history. The same can be said of the inhabitants of the sparse Dutch settlements along the Guiana coast and Suriname (although we will later note the remarkable late-seventeenth-century work of Maria Sibylla Merian in the region). Nor were there any significant studies of Angola, although what Marcgraf might have accomplished had he lived is a question to ponder. In spite of the interest of some of the governors of the WIC in natural history, its servants were too preoccupied to spend time on studies of naturalia, with the brilliant exception of Prince Maurits and some of his people.

Following Maurits's departure in 1644, a remarkable group of publications appeared, celebrating Dutch Brazil during Maurits's era, no doubt as part of the campaign to remind everyone of the importance of Brazil in the years leading up to the Peace of Münster. A huge neo-Latin epic poem titled the *Mauritias*, written by Maurits's chaplain, Franciscus Plante, appeared in 1647, as did Barlaeus's monumental *Rerum per octennium in Brasilia* ("What happened during

Landscape by F. Post. In Caspar Barleus, *Rerum per octennium Brasilia* (1647). By permission of the British Library

eight years in Brazil"). Barlaeus's book included engravings based on Post's landscapes, many descriptions of places and people taken from manuscripts written by some of the prince's people, including Piso, and four pull-out maps that could be put together to create a large wall-map of the territory (the original large sheet, based on a map prepared by Marcgraf, was published in 1646).[124] The most lavish book, however, was on the medicine and natural history of Brazil, the *Historia naturalis Brasileae* (1648), the costs of which were underwritten by Maurits himself.[125] (The publisher was Franciscus Hackius, who had printed Bontius's works on the medicine of the East Indies six years earlier.)

Preparation of the publication began not long after the Maurits's return to the Netherlands. After his death Marcgraf's manuscripts were acquired by the governor of the WIC most interested in geography and natural history, Johannes de Laet. After hard work, he managed to crack Marcgraf's cipher. By October 1646, De Laet had established a text from Marcgraf's manuscripts, but the preparation of the illustrations for the huge volume took longer, delaying publication for over a year—so rather than being the first, it was the last of the great scholarly tributes to Maurits's rule.[126] The illustrations were based on drawings, sketches, and paintings by Frans Post, most famous for his landscapes, and by Albert Eckhout, whose paintings of the people and naturalia of Brazil have become deservedly famous, many being copied and even turned into tapestries. (Most of the originals for the book's illustrations—many now in Kraków—have been traced and carefully described by P. J. P. Whitehead and Martin Boeseman.)[127] The *Historia* has since served as a major resource for the study of tropical natural history and medicine and was not superseded for more than a century and a half, when a new generation of naturalists led by Alexander von Humboldt began a series of expeditions into the South American tropics. Both Piso and Marcgraf (as edited by De Laet) contributed the text. Marcgraf's work was by far the most substantial, at over double the number of pages, in the form of eight books: three books on plants, one on fish, one on birds, one on quadrupeds and serpents, one on insects, and one on the meteorology of the region.[128]

Piso offered four books on the medicine of Brazil: on the local airs, water, and places (including recommendations for a healthy regimen); on endemic diseases; on poisonous insects and animals and appropriate treatment; and on more than a hundred medicinal simples. In the medical parts of the book, Piso praises opiates, as Bontius had done. He also touts the benefits of sugar, which had been used as a medicine since antiquity. (Barlaeus thought the older kind of sugar remained the best for medicinal purposes—particularly complaints of the liver and intestines—while the more refined product of Brazil was better for use in cooking; the well-known physician from Dordrecht, Beverwijck, worried that the popularity of West Indian sugar was undermining the local honey industry.[129] Others, like the alchemist and projector Johann Joachim Becher, writing about three decades later, considered refined sugar to consist almost entirely of the distilled universal spirit that gave rise to growth and generation: "Sugar is the noblest and sweetest juice of the earth, digested and cooked through by the heat of sunbeams, and thus a noble balsamic substance that is most closely related to the human blood, the inspirited blood burning away in the fire just like sugar." Sugared wine therefore had powerful healing and nourishing properties.[130])

Even more important than sugar, however, was a new remedy against diarrhea and dysentery that Piso had learned from the local people: root of ipecacuanha, which he wrote about in two chapters on treating local disease and further described in his chapter on medicinal simples. He also learned of another treatment for an endemic disease, something he called "Indian Lues" (probably *Framboesia tropica*) to distinguish it from syphilis, with which it shared some symptoms. For

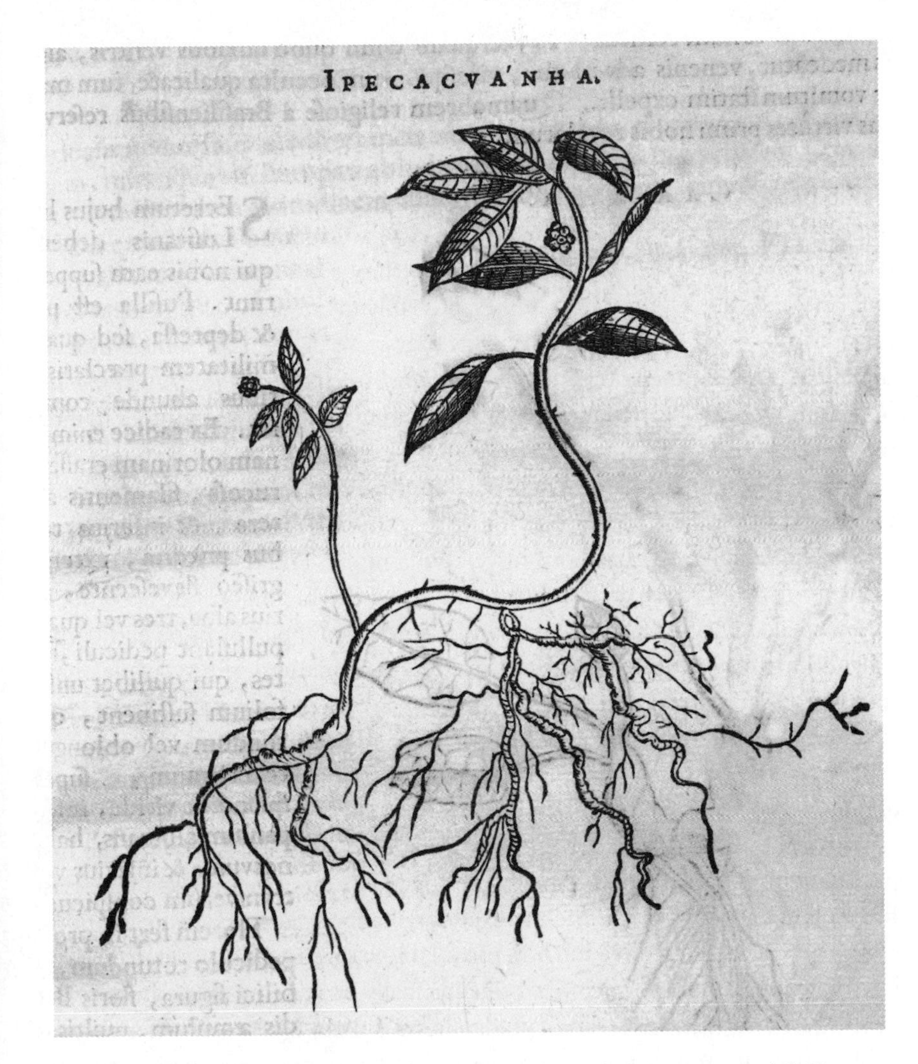

Woodcut of ipecacuanha root, from W. Piso, *De Indiae utriusque* (1658). Courtesy of the Wellcome Library, London

this lues he recommended treatment according to the local methods, which included the use of an herb called "caaróba" as well as sarsaparilla.[131] In these and other illnesses, Piso was as keen as Bontius to learn from the local people, although his chief source was probably the nurses in the hospitals he supervised in Recife.[132] Piso and Marcgraf made fewer references to their Spanish predecessors who had written on the West Indies. But they cited one important work of natural history that had not yet appeared fully in print: Francisco Hernández's great work, which had been written and painted for Philip II. They apparently obtained their information about this source from De Laet, who had acquired a copy of the Spanish edition of Francisco Ximénes (printed in Mexico City) on the natural history of New Spain, which quoted freely from Hernández.[133]

Ten years later, Piso brought out another expensive book based on the *Historia* and on the work of Bontius. Given the WIC's withdrawal from Brazil in 1654, it was pitched as a patriotic summary of the best Dutch studies on the medicine and natural history of both Indies, probably to encourage his readers to remain as interested in the West Indies as they were in the East Indies. Indeed, the title page announced the change: *De Indiae utriusque re naturali et medica libri quatuordecim* ("On the natural history and medicine of both Indies in fourteen books," 1658).[134] For this edition, Piso took credit for the first six books. The first two repeated what he had published in the 1648 *Historia* on the local airs, water, and places and the nature and cure of diseases found in Brazil. He expanded his chapter on poisonous insects and animals and the appropriate treatments to include remarks on other animals and the metamorphosis of insects, now placed at chapter 5. Chapters 3 and 4 combined much of Piso's earlier material on medical simples with much of Marcgraf's natural history, while chapter 6 was on spices and aromatics. A short second section contains two books credited to Marcgraf, one his work on topography, meteorology, and observational astronomy, the other a commentary on native Brazilian and Chilean languages and other aspects of culture. The final section, attributed to Bontius, faithfully reproduces the four medical works first printed in 1642, together with the first publication of his books and drawings on the plants and the animals of the East Indies, for which Piso admits he made editorial annotations and even additions.

Piso clearly introduced much new material into this 1658 work, but in doing so he also created confusions. Unlike humanistically educated authors of an earlier generation, such as Clusius and Paludanus, he did not demarcate where he was quoting from his source and where he was adding his own information or commentary. A comparison between the 1648 and 1658 volumes conducted by M. H. K. Lichtenstein in the 1810s and 1820s, and translated into Portuguese in 1961, "found that Piso had tried to avoid the appearance of literal copy, but in changing the words often produced a contradiction with Marcgraf's description, or even his figure."[135] Marcgraf's younger brother later wrote of Piso's "forgettfullness, boldness, & vanity" and of how the "Surgeons & Apothecarys of Amsterdam" joked about how Piso had "improved" the second edition "of the Brasile History" by coming to them for the names of Brazilian fish and birds, since he did not always know what Marcgraf was writing about. A few decades after, the famous naturalist Linnaeus took up Marcgraf's cause by naming a genus of very spiny plants after Piso (*Pisonia*), remarking when he did so that their spines were as nasty as Piso's reputation.[136] Moreover, although in his prefatory material Piso acknowledged Marcgraf as his most important source, he did not put Marcgraf's name on the title page as author or coauthor, causing Piso to be charged with plagiarism. Some modern commentators on the "Piso problem" have let him off by accepting the explanation he offered in his preface, that he had employed Marcgraf—that Marcgraf was Piso's servant or assistant (*meo domestico*), brought to Brazil at his expense—which is expanded at the beginning of chapter 4 when Piso writes that it does not matter whether the information came from him or Marcgraf, since what counts is its truth.[137] Marcgraf's brother was furious at the claim that Piso had made his well-educated brother, employed by the WIC as their astronomer, into his servant, and he, Linnaeus, and others also criticized Piso for botching some of the work through his ignorance.[138]

If one examines the material Piso printed from Bontius, one can only conclude that Piso's critics had a point. Because Piso did not signal the difference between Bontius's original and his additions, and because many of his additions are based on the reports of others, he sometimes introduced errors that have since been laid at the feet of Bontius. The natural history material attributed to Bontius in Piso's edition contained information on thirty-three animals and sixty-two plants of the East Indies, with an illustration for almost every one. An original manuscript of Bontius's on natural history has recently resurfaced, containing information on, and illustrations for almost all of, sixteen animals, birds, and fish, and forty-two plants, in no particular order.[139] It is possible that another manuscript volume, still missing, contained similar illustrations and texts on some of the additional seventeen animals and twenty plants. Clearly, however, Piso obtained much of the supplementary information from sources other than Bontius.

A comparison of the printed version and the extant manuscript makes it evident that Piso reordered Bontius's material according to his own views, did some light editing, added poems and occasionally other information, and sometimes even introduced entirely new entries. Most of the changes are editorial, if unmarked. For instance, the bird of paradise entry in the book includes a poem not written in the manuscript—a learned flourish typical of Piso but alien to Bontius—while the description of the chameleon is fleshed out by Piso with a discussion of the emblem in which it appears in Andrea Alciati's emblem book. Bontius's manuscript does not include a drawing of a rhinoceros, so for his edition Piso obtained one from Johannes Uyttenbogaert, an important Remonstrant minister and politician from Amsterdam, who must have made the drawing from a specimen brought back from the Indies (or who was the intermediary who obtained a drawing from someone who had seen a rhino).[140] There is an entry on the tiger

Governor-General Specx's drawing of a tiger. (MS Sherard 186, fol. 62v). By permission of the Plant Sciences Library, Oxford University Library Services

in Bontius's manuscript, which is reproduced faithfully up until the point where the porcupine is mentioned (which concludes the entry in the manuscript), but in Piso's book additional information is included, from a source or sources unnamed. On the other hand, the drawing of a tiger done by Specx in Bontius's notebook is clear, while the artist in the Netherlands, trying to turn the drawing into a woodcut for Piso, not only left out the background and shading but, never having seen a tiger, took Specx's rendering of the animal's stripes to be shadings of muscles, making it a strange-looking beast.

The longest of Bontius's entries is on the tea plant, which has no accompanying illustration in the manuscript. The reason for this was simple: "I could never manage to see the green leaves here," Bontius wrote, since it was grown elsewhere and imported to Batavia by the Chinese as dried crumbles. Bontius tried soaking the leaves in water and fitting them together to obtain an impression of the form of the leaf, but to no avail: the leaves were not only dried but broken in too many small bits. Chinese informants gave him contradictory accounts about whether the leaves used in the infusion they drank came from an herb or a shrub, but Governor-General Specx, who had opened trade with Japan for the VOC and had seen it growing there, was able to confirm that it was a shrub. "It is certainly true that it undoubtedly encourages good health and as a medicine acts not unhelpfully to rid the chest of thick phlegm." With its "excellent diuretic properties," tea also acted as "a fine remedy against bladder and kidney stones."[141] By

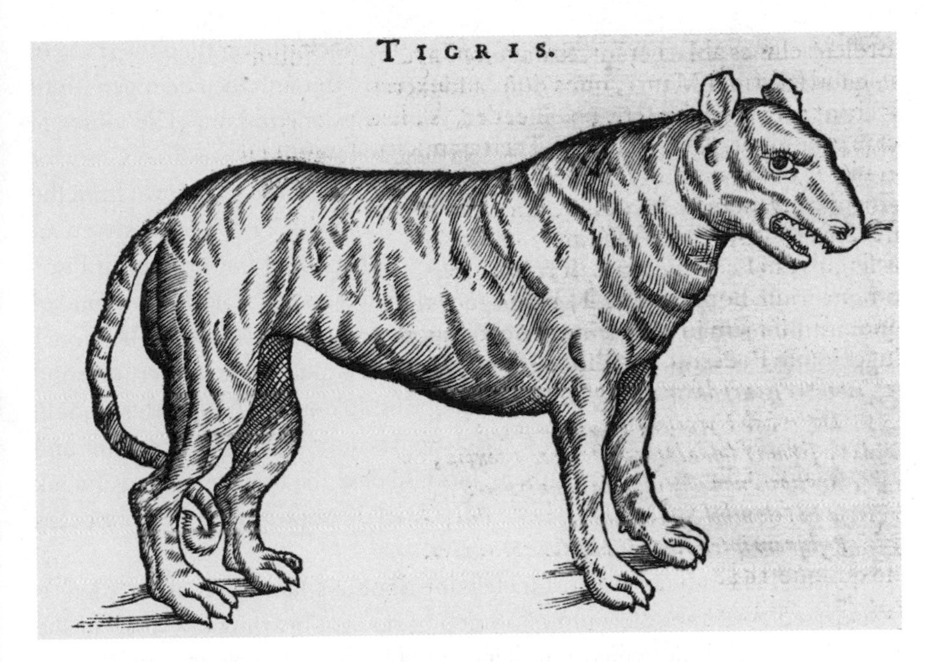

Engraving of a tiger, making the stripes into shadows of musculature, from W. Piso, *De Indiae utriusque* (1658). Courtesy of the Wellcome Library, London

the time Bontius's natural history was published, however, Piso had access to a drawing of the shrub provided by François Caron, who had been a commander of the Dutch station in Japan following Specx. And because "this beverage is getting more and more known in India [that is, Asia] and also among distinguished Europeans," Piso added considerable information that he had collected on tea and its properties. (Dr. Tulp included this information on tea from Piso's book in the revisions he was making to his own book of medical observations.)[142]

But in addition to these silent editorial additions, Piso also added entirely new entries to Bontius's book on the testimony of other witnesses now living in Amsterdam. There is an entry on the "Baby-Roussa" (*babiroesa*), an animal found on Sulawesi and the nearby island of Buru, based on "a very accurate drawing" of a skull collected by the elder Swammerdam.[143] The "Dronte, called by others Dodaers" of Mauritius—the creature better known to moderns as the dodo, which would be extinct by 1680—gets one of its earliest accounts and an often-reproduced drawing, but since there is no evidence that Bontius was ever on Mauritius, the illustration would appear to be by someone else. Piso also added a truly "wonderful" story from Japan: "In cases of chronic head-ache, or obstruction of the liver and spleen, or in cases of pleurisy, they pierce the organs mentioned with a silver or bronze pin [*stylo*] not much thicker than the string of a lute [*cythrarum*], slowly and gently being pushed through the said organs until the pin comes out the other side, which I also saw done in Java." The source of this early account of acupuncture is not clear, but perhaps it was from Caron or Specx, who had both spent time in Japan as well as Java and are known from the comments about tea to have been in conversation with Piso. There is even an odd entry among the herbs. Bontius wrote nothing about flowers, but in Piso's book there is a collective entry with little detail, in which it is noted how fond of fragrant flowers and perfumes are the Javanese and all "Muslims." "Therefore I would advise the Dutch, who cultivate flowers of widely different perfume and color and spend their entire fortune doing this, to come hither, if they should desire still more flowers. For Java has a tremendous quantity of fragrant and many-colored flowers, so many, that if I should describe them all, paper and ink should run out before I had finished." A few brief descriptions are then presented of the most strongly scented ones.[144]

Most significantly and unfortunately for Bontius's later reputation, there is a supposed eyewitness account of a troop of orangutans (literally "men of the woods"), which are described as walking upright. The accompanying illustration is clearly fantastical (although Linnaeus later used it to show how humans resembled apes), making them look simply like naked persons with thick halos of hair around their heads and a few patches elsewhere on the body. Most remarkably, the written account claimed that the orangutans clearly exhibited the range of human passions (even sadness and modesty), being, therefore, human in almost all ways aside from the inability to speak. But this account is not present in Bontius's manuscript, nor is it likely that orangutans ever lived on Java, which means that this account must have originated from Sumatra or Borneo (where they live) or, more likely, was constructed entirely from conjecture, probably from reading European myths of satyrs, pygmies, and "wild men" into stories about the forests of Java coupled with rumors about the apes. The account tested readers' credibility, so that Bontius's reputation suffered ever after for something Piso had introduced.[145] In other words, the more credulous Piso treated Bontius's manuscript as his own, creating confusion about what Bontius, he, and others reported, marring Bontius's careful work.

At the same time, however, it is also clear that Bontius himself did not always clearly understand the information he acquired for his natural history. For instance, there is a muddled entry in his manuscript on "De Jaaca, durionibus jac fructu Champidacca dicto" ("The jaaca, or jack fruit, called champidaca, of the durions"), which appeared in Piso's edition as "De Durionibus, Iaaca, & Champidaca dictis" ("Of the durions, called jaaca and champidaca"). Bontius was lump-

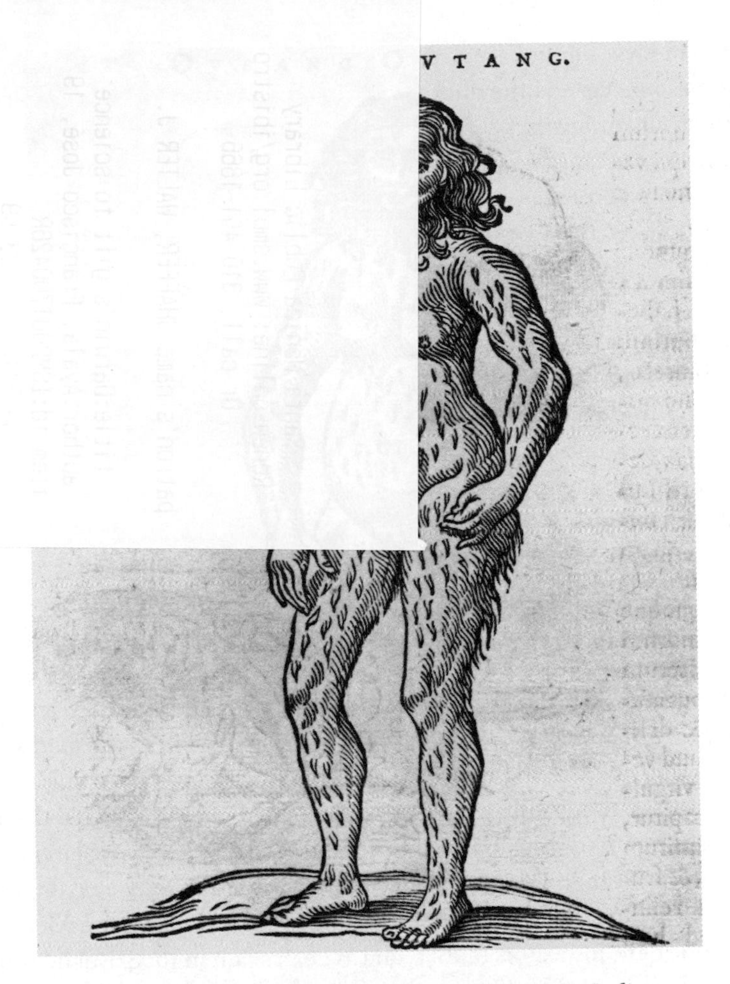

Engraving of an "orang outang," from W. Piso, *De Indiae utriusque* (1658). Courtesy of the Wellcome Library, London

ing at least two fruits (jackfruit and cempedak) into a family he called durion, which is in turn a distinct fruit in its own right. He also provided two illustrations, which in Piso's edition are reproduced upside down and labeled "the larger Durion, or Jaaca" and "the lesser Durion, or Champidaca." The "lesser Durion" appears from his manuscript drawing to have the leaves and placement on the branches typical of breadfruit (*Artocarpus altilis*), although the fruits are drawn with spiky surfaces rather than the bumpy ones of breadfruit; the illustration of the spiky "larger Durion," which Bontius called jaaca (jackfruit), resembles the durion (*Durio zibethinus*), although that fruit is thought to be native to Sumatra rather than Java; Bontius's text also emphasizes the rotten and nauseating smell but wholesome taste of the durion. Neither illustration resembles the large, oblong jackfruit (*Artocarpus heterophyllus*) or the cempedak (*Artocarpus integer*). Given the importance of all these plants in the local diet, his informants would have been unlikely to have mixed them up. The initial confusions were therefore those of Bontius. Because the illustration of the "Jaaca" is the first in the manuscript, it is likely that these confusions were a product of Bontius's initial conversations rather than his more mature reflections. At first he identified the durion but mixed its name with that of the jaaca, but then lumped with it other large, round fruits with bumpy surfaces (probably breadfruit) and further mingled the names, confusions compounded by Piso's ham-fisted editorial interventions.[146]

The episode has its lessons. One is that even sincere and careful observers may fail to see important distinctions among specimens at first sight: the urge to oversimplify is powerful, especially in the initial stages of investigation. A second is that by the middle of the seventeenth century editorial practices were sometimes falling foul of publishers and authors who wanted the potential buyers of books to like the look of them as much as the content. Using different typefaces and notations to distinguish one person's claims from another's made for expensive typesetting and less attractive books. This means that the audience for works of medicine and natural history had grown well beyond a circle of expert scholars, with consequences for the watering-down of knowledge. Publishing a popular book, moreover, could enhance one's professional reputation, even if that meant silently adding interesting material to another person's text or making it appear that the work of others was one's own. Piso had become a respected physician in Amsterdam by 1658, serving as the senior officer of the Collegium Medicum in 1648, as one of its inspectors in 1655, and as dean again in 1678. But he could use the new edition to display his connections among a number of the weightiest literati of Amsterdam, partly by obtaining laudatory letters from them to introduce the book. (Those he tapped to write words of praise included the eminent Constantijn Huygens, Dirk Graswinckel, a noted jurist, Nicolaas Heinsius, a man of learning who had worked for Queen Christina of Sweden, and several learned physicians.) This also suggests that Piso was trying to secure an even better reputation for himself, a mark of the competitive environment of a city like Amsterdam. This helps to account for why he took more credit for Marcgraf's work than he should have and why he so carelessly "improved" Marcgraf and Bontius. It is disappointing that such an otherwise glorious book, which for all its failures brought much new material to light, is so flawed at the level of keeping the authorship straight. But this, too, makes a point: though he was an eminent physician, Piso could not make his errors right simply by referring to his reputation or social rank. Just because an authority like Piso wrote something did not make it so.

It is also clear, finally, that not only the work of Bontius, Marcgraf, and those like them in the field appropriated and reinscribed information previously developed by other people in other cultures, past and contemporary: so, too, did people in the metropolis. In doing so they all universalized and objectified descriptive information, making it easy to exchange, even enabling it to pass to cultures that had not a clue about the original rationale for the coming-into-being of a practice. Bontius and Marcgraf might be said to be like the merchants who worked for the VOC in the East, acquiring and shipping information back to the metropole, while Piso might be more like the directors who governed the Company based on information received, allowing its alteration.

It is therefore too simplistic to see the VOC and the WIC as trading corporations alone. They were also arms of the Dutch state, contributing finance to a massive military build-up that destroyed Portuguese power in Asia and severely crippled it in the West Indies, and allowed the Dutch to rule the waves for several decades. This "military fiscalism" took a horrifying toll in lives and well-being, but it also created the "preconditions" for the flow of resources that placed the Dutch in a position to dominate the world economy for a time.[147] In turn, descriptive information about strange places and natural things was critical not only to decision-making but to creating audiences for the new tastes of health and pleasure being brought to The Netherlands from the Indies. The northern Dutch world had become an intellectual entrepôt as much as an entrepôt for goods and finance. In the home metropolis collections were accumulated, housed, and preserved, inventories were taken and sometimes published, and redistribution of the value-added information and objects was initiated. Material progress and utility became the watchwords of contemporary naturalists even when they reveled in curiosities. From careful investigation and reporting they wished to create enduring knowledge that could be handed down to others. The political economy of early modern commerce depended on accumulation, sorting and accounting, exchange, and a credibility that lay not only in trust but in checking and double-checking before committing to a proposition; so, too, did the natural knowledge of the day, even if it did not expunge all error.

6

Medicine and Materialism

Descartes in the Republic

It is an absolute perfection and virtually divine to know how to enjoy our being rightfully. We seek other conditions because we do not understand the use of our own, and go outside of ourselves because we do not know what it is like inside.

—MONTAIGNE, "Of Experience," *Essays*

Many netherlanders were aware of both the enormous wealth flowing in as the consequence of violence and trade and the lesson of public anatomies stressing how human abilities were linked to our material form. For some, the two were more than simply juxtaposed: they seemed to arise from our natures. Perhaps we humans are mere momentary concretions of natural elements, governed entirely by the laws of nature rather than by any divine purpose or even by our rational thoughts; perhaps powers in our bodies beyond our control—nature itself—cause us to act and even think as we do. Dutch moral philosophers were forced to contemplate a world governed not by reason but by the passions. A longtime resident in The Netherlands, René Descartes, found his attention increasingly being drawn toward these debates, with the result that he developed an explanation of how the passions arose from the material world itself and became the main source of human behavior and thought. Some Dutch physicians took his position a step farther, resulting in a full-fledged materialist view: we are governed by matter in motion, and nothing else. Nature was the source of great benefits, from health and wealth to happiness and love. This position, however, negated the view that we should be governed by rational virtue and led to an enormous backlash from the religious against the new philosophers. For a time, though, at the height of the Dutch Republic, the leading regenten, at least, governed according to a sense that their wealth and political power were built on how nature really made humans rather than according to the hypocritical preaching of virtue.

DESCARTES IN THE NETHERLANDS

After a decade of traveling around Europe, the French aristocrat returned to The Netherlands in 1628 and, except for three short visits to his homeland, remained there for more than two decades. The Dutch world changed him as much as he changed it. In The Netherlands Descartes would give voice to a view of human and animal physiology that was thoroughly materialistic, and there he would write his famous philosophical works, books deeply affected by local information and conversations: *Discourse on Method, Optics, Meteorology, Geometry, Meditations on First Philosophy, Principles of Philosophy, The Passions of the Soul*, and the posthumous *Description of the Human Body*. He has been claimed by his native country as its most famous intellect.[1] But if one were passing out awards based not on place of birth but on place of work, it might be said that he became one of the most eminent Dutch philosophers of the age. In the Republic he moved away from his early concerns for knowledge via contemplation to become absorbed in the enthusiasm for medicine and the investigation of the precise details of nature, especially animal bodies, including the human body. Many of his Dutch interpreters, too—a large proportion of whom were physicians—considered him to have provided them with a set of views that bolstered their interests in descriptive research: to them he provided philosophical arguments for the fundamental importance of empirical investigation. By the end of his long stay in The Netherlands, in conversation with Princess Elizabeth, Descartes had also come to surprising conclusions about how we are ruled largely by our passions, which are good and can teach us how to live long and happy lives. To consider him a thinker preoccupied with natural philosophy alone, and that to be a philosophy of rational "deduction" rather than one of sensory "induction"—as is often done—not only oversimplifies his work but gives a false impression of it. To picture Descartes in light of his preoccupations with the body and passions, health and disease, is to see his work through the eyes of his Dutch colleagues.

René had been born in 1596 at La Haye, between Tours and Poitiers in central France. In 1607 he followed his elder brother Pierre to the large Jesuit college at La Flèche—enrollment was between twelve hundred and fourteen hundred students—then leaving for the study of law at the university in Poitiers in 1615 and 1616, where he graduated with a *licence*, which made him eligible to hold a government position as a judge or councilor.[2] Instead, after turning twenty-one, when he could decide on his own life course, he joined the army as a gentleman volunteer. He may have aspired to move beyond the corps of French bureaucratic gentlemen (*noblesse de la robe*) for which his father intended him into the more highly regarded military aristocracy (*noblesse de l'epée*). In these days before the professionalization of arms, this life also allowed far more daily personal liberty than working as a state official bound by office routine and paperwork. For reasons that remain obscure, the army Descartes joined in 1618 to learn about war was that of the Prince of Orange, camped at Breda, near the border with the Spanish Netherlands. Perhaps he was attached to one of the two French regiments serving the States General following Richelieu's efforts to support anti-Hapsburg partners.

There in Breda, on 10 November 1618, the young cavalier met Isaac Beeckman, about ten years his elder, probably at the local chamber of rhetoric, Het Vreuchdendal, which Beeckman was known to attend and which Descartes may have visited because of his keen interest in literature.[3] The enormously ingenious and technologically adept Beeckman had stopped in Breda on his way back to his hometown of Middelburg. He and the Frenchman quickly found that they shared many interests, including music.[4] In a later letter to his friend, Descartes wrote, "It is you alone who have roused me from my state of indolence, and reawakened the learning which by then had almost disappeared from my memory."[5] Working together, Beeckman and Descartes found that they could solve certain kinds of physical problems using mathematics, working out a new law describing free fall: a falling body acquired additional motion each time a corpuscle hit it, resulting in acceleration, with a velocity proportional to the square of the time expended in the fall. The physical problem and the method of handling it (which Beeckman would call "mathematico-physics") was Beeckman's, whereas the mathematical solution was probably Descartes's.[6] It was one of the earliest demonstrations of how physical phenomena could be modeled mathematically to solve a problem in natural philosophy, contemporary with the as-yet unpublished work of Galileo. As Margaret Jacob has written, "Isaac Beeckman must be recognized as the first mechanical philosopher of the Scientific Revolution. There were other mechanists before and contemporary with him—not least of them Galileo—but none of these developed a systematic philosophical approach to mechanical problems, one that speculated as to the atomic construction of matter and designated this mechanical philosophy of contact between bodies as the key to all natural forces, to every aspect of reality from watermills to musical sound."[7]

Beeckman also urged his young friend to write down his own ideas and presented him with a parchment notebook in which to record his thoughts.[8] Descartes had earlier given Beeckman a manuscript, *Compendium musicae* ("Compendium of music"), which discussed how music moves the passions—a critical set of problems to which Descartes would return frequently in coming years. In mid-February, Descartes traveled to Middelburg to visit his friend again, and in March 1619, still in the Republic, he was writing to Beeckman about his hopes for developing new geometrical methods that could deal with discontinuous as well as continuous quantities. But perhaps because of Prince Maurits's increasingly militant moves in support of the Calvinist international, Descartes took to the road again in early 1619. Ten years later he returned and renewed the friendship.

In the years between, Descartes's life is mainly unknown to us, although he traveled widely. He returned to France sometime between later 1620 and 1622.[9] Around autumn 1623 he headed for Italy, apparently visiting Venice, Rome, and Florence at least, before returning to France in about May 1625. He was in Poitiers in late June and in Paris in July, wintering in the countryside in 1627–28, and traveled to The Netherlands in the autumn of 1628. At the time of Descartes's return to the north, there seems to have been general interest in France about the Republic, and others decided to make the trip. Pierre Gassendi was one, setting out only a few months after Descartes. His friend and patron, Nicolas-Claude Fabri de Peiresc, had traveled in the low countries—visiting Paludanus's cabinet, among others—and had come away impressed.[10] By the time Gassendi had settled back in Paris, his interest in Epicureanism, which would make him renowned and which was prompted by his conversations with Beeckman, had been redoubled.[11] Gassendi's excitement about his visit north must have been palpable, for within two weeks his friend Marin Mersenne himself started planning a trip to the low countries, which took place from April to October 1630, following the same route as Gassendi. Mersenne even sacrificed his clerical robes before crossing into the Republic in order to make discourse easier. The trip left Mersenne with a new sense of the serious common purpose of learned men of other confessions, changing his philosophical outlook in relation to the Church. It also imbued him with an awareness of the importance of empirical investigation for the pursuit of natural philosophy.[12]

On arriving in The Netherlands himself, Descartes visited Beeckman almost at once, on 8 October 1628, in Dordrecht, where his old friend was rector of the Latin school. A few months later, on 16 April 1629, he enrolled at the university in Franeker, where he would reside until the autumn. Perhaps Descartes matriculated at Franeker to have access to the university library or because it was a quiet place where he could safely attend Mass.[13] More likely it was because of the philosophical debates under way there: the generally pro-Ramist and anti-metaphysical (and anti-Aristotelian) character of its professors was being challenged by a new colleague, Johannes Maccovius, who revived the teaching of metaphysics in an attempt to fend off theological heterodoxy.[14] In his case, he valued the new Aristotelian teachings of Francisco Suárez and others of the famous school of Coimbra as essential tools in building up Counter-Remonstrant theology.[15] Descartes's intermediary, André Rivet (a Counter-Remonstrant professor of theology and minister to Prince Frederick Hendrick), later supported the arguments of Maccovius's colleague and ally, Nicolaas Vedelius, so it is likely that it was Rivet who suggested that Descartes follow the teachings at Franeker.

Although Descartes was no Aristotelian, the revived debate about metaphysics would have interested him deeply. Between his first meeting with Beeckman and his return to the Republic he had famously acquired the ambition to place all of natural knowledge on a new and sound footing using first principles. After his initial work with Beeckman in applying mathematics to physical problems, Descartes developed his thoughts on mathematics and how similar methods of reasoning could be used to explain everything known to the human mind. He began by writing down his dreams and notions about this in the notebook Beeckman had given him (which has been missing since the late seventeenth century but on which G. W. Leibniz and Descartes's first biographer, Adrien Baillet, took extensive notes). The system Descartes eventually imagined was likened to a tree, with roots consisting of metaphysics, its trunk of mathematical physics, and the branches of ethics, politics, medicine, and other subjects for which philosophy could explain and guide human behavior. Many other classical philosophical systems had shared the same basic ambitions, but as is well known, the variety of philosophical opinions circulating by the late sixteenth century led to a proliferation of views about how to obtain truth, among which were varieties of skepticism, which to a greater or lesser extent placed in doubt the possibility of finding certainty in anything.[16] Descartes believed that his method could lead to certainty rather than doubt because it founded all natural knowledge on a few clear and certain axioms known to be true by intuition, not unlike proofs in geometry; in a similar way, he and Beeckman had been able to resolve problems in natural philosophy by applying geometrical and mathematical principles. With his famous expression *cogito ergo sum* ("I think, therefore I am") Descartes envisioned a method of establishing that his intellect, at least, existed, and that from this finding it was possible to proceed by logical steps to proof of the existence of God, basic aspects of the rules God had established for the world, the existence of material nature, and so on. He believed that his method would allow him to solve real-world problems by the application of a few simple principles.

But these had been dreams of his youth, which his second residence in the Republic would transform. The problem, of course, was that a complete new system of philosophy along the lines he imagined was a dream. Working it out thoroughly, by contrast, was not only extremely difficult but unnecessary for the practical ends he had in mind. He had drafted his *Regulae ad directionem ingenii* ("Rules for the direction of the mind") just before leaving for The Netherlands, but it remained incomplete, with the third part, on how to apply his rules to the real world, only projected.[17] He quickly laid it aside. (This partially composed work of his youth would be published only long after his death for his aficionados, appearing first in Dutch in 1684 and in Latin at Amsterdam in 1701.) Perhaps Beeckman threw cold water on it, too. The two had gotten off to a warm start, but by the end of the summer of 1629 they were no longer on speaking terms, the apparent reasons for which were some inadvertent blunders on the part of Gassendi, Mersenne, and others: Beeckman came to believe that Descartes was giving no credit to his old friend for helping launch him on the road of mathematico-physics, while Descartes in turn held that Beeckman was giving no credit in his music theory to himself, despite having shared his early manuscript.[18] By then, however, Rivet had introduced Descartes—as he would Gassendi and Mersenne—to Henri Reneri and other philosophers, and so, despite falling out with his old friend, Descartes decided to remain in The Netherlands, frequently shifting his place of residence, slowly furthering his philosophical work in conversation with eminent intellects, several of whom became trusted companions. (He had also made much progress in learning Dutch during his first stay and would have been increasingly comfortable in it.)

By the autumn 1629, he had moved on to Amsterdam, with plans for an even more ambitious project related to the discarded third part of the *Regulae*. As he explained to Mersenne in a letter of 13 November, "I have resolved to explain all the phenomena of nature."[19] It was to be a revolutionary general guidebook to knowledge later referred to as *De mundo* or *Le monde* ("The world"), which would explain how one can know certain things about the intellect and God, and from them how it is possible to describe the creation of all things in the world down to the human body, reproduction, and all the things necessary for life. He almost abandoned this work many times, too, only to have the encouragement of a friend keep him at it.

His new friend Reneri (also a friend of Gassendi), a firm supporter of descriptive natural history and well known to members of the undogmatic Muider Kring, was practicing medicine in Amsterdam and teaching philosophy when Descartes met him in the late 1620s. It may even have been the attraction of working with Reneri that caused Descartes to move from Franeker to Amsterdam, for when in 1631 Reneri took up a professorship at the new Illustrious School in Deventer, Descartes followed him there (where his illegitimate daughter, Francine, would be baptized on 7 August 1635). When in 1634 Reneri moved to the Illustrious School in Utrecht (which soon became a university), Descartes again followed him. Later on, when in the summer of 1638 Reneri acquired a new colleague in the medical faculty, Henricus Regius (Le Roy), he showed Regius Descartes's manuscripts, with the result that it was in the medical faculty that the first major defenses of Cartesian philosophical principles were offered; Reneri also introduced Descartes and his ideas to people like Constantijn Huygens and Pieter Corneliz. Hooft. (By 1637 Descartes had also become well acquainted with the physician Cornelis van Hogelande, author of a work on physiology, a Catholic and a reputed Rosicrucian, son of the Johan van Hogelande who had worked so hard to get Clusius to come to Leiden. Hogelande soon became one of Descartes's most trusted friends: Descartes used him as an intermediary with his illegitimate daughter and as a recipient for private letters, stayed at his house on many occasions, and left a trunk of his papers with him when he went to Sweden in 1649.)[20] Although Reneri was excited by Descartes's work because he, too, was unsatisfied with current philosophy, he did not fully accept his friend's views. Nevertheless, when Reneri died in March 1639, his colleague Antonius Aemilius used the occasion of his funeral oration to praise Descartes's philosophy.[21] It was Reneri, then, following Beeckman, who most immediately furthered the work of Descartes and provided the occasions for it to become famous.

The philosophy Descartes wrote in The Netherlands quickly departed from his earliest visions of a new system, however. As important to Descartes as Reneri's moral support was the general Dutch orientation toward investigating phenomena using physical methods as the foundation for natural philosophy—something that had also left its mark on Gassendi and Mersenne. Beeckman's approach, for instance, had been to isolate individual problems and then work toward their solution using the best methods available for each rather than to apply a universal set of rules. This hardly compared to the thorough system Descartes dreamed was possible, but it worked. Descartes had seen that everything was interconnected, but he found that starting from first principles was a long way from solving practical problems. Daniel Garber has recently concluded that "it is indisputable that as his system grew, perhaps from the first metaphysics of 1629–30 onward, method became, first in practice, and then after 1637 in theory, less and less important to Descartes."[22] In The Netherlands the predominant method, if that is the right word, was *circumspice*, "look around you." This underpinned, for instance, the eclectic approach of his friend Constantijn Huygens, secretary to the Prince of Orange and enthusiast for Baconianism, who had been introduced to Descartes in 1632 by Jacobus Golius (the Leiden professor of mathematics and of Oriental languages); Huygens often tried to turn

Descartes's attention to practical ends, thinking that his grandest philosophical projects were overly ambitious.[23]

As he absorbed more and more of this practical and empirical outlook, Descartes did not directly confront his changing views so much as move beyond them, leaving behind the dream of his youth while turning to the study of difficult questions about how things are, like so many of the naturalists around him. Or as it was put by other authors commenting on Descartes's published views in 1637, "he says the further we advance in our physical knowledge the more important . . . experiments [*expériences*] become."[24]

One of the enthusiasms Descartes developed regarding expériences was the exploration of the material structure of animal bodies. These were years of intense public interest in the anatomy lessons performed at Leiden, Amsterdam, and Delft, with Rembrandt's portrait of Dr. Tulp painted in 1632. Descartes, too, was captivated. Almost as soon as he embarked on his new philosophical project, at the end of 1629, he wrote to Mersenne that he did not wish to be distracted with further philosophical inquiries, for "I want to begin to study anatomy." A few months later he wrote, "I am now studying chemistry and anatomy simultaneously; every day I learn something that I cannot find in any book." As he put it in a later summary of his projected *De mundo*, he recognized that "I did not yet have sufficient knowledge to speak of [animal bodies] in the same manner as I did of the other things—that is, by demonstrating effects from causes and showing from what seeds and in what manner nature must produce them." He had come to see that although his first principles could be used to construct the building blocks of creation, when it came to animals and humankind, especially, they could only explain the effects discovered, not imagine their complex design in the first place. One reason for not publishing his early *De mundo*, he told the world, was that "every day I am becoming more and more aware of the delay which my project of self-instruction is suffering because of the need for innumerable observations [*expériences*] which I cannot possibly make without the help of others."[25]

Unfortunately, Descartes is silent about those others, including who taught him anatomy.[26] It is likely that at the university of Franeker he encountered the medical professor Menelaus Winsemius, who had been trained by Pauw at Leiden and actively dissected: in fact, in 1629, the year Descartes matriculated there, Winsemius sought permission to dissect a second cadaver.[27] When he moved to Amsterdam, Descartes became well acquainted not only with Reneri, who was then practicing medicine, but with one of the foremost physicians there, Vopiscus Fortunatus Plemp. Like Descartes, Plemp was Catholic and educated by the Jesuits, but he had also been well trained in medicine at Leiden, Padua, and Bologna and was known as an expert anatomist. As one of Descartes's closest friends at the time, it was probably Plemp who showed the gentleman how to dissect for himself.[28] Accordingly, the section on human physiology prepared for *De mundo* and published posthumously as *De homine*, or *L'homme* (the "Treatise on man," also known as "Description of the human body"), instructs the reader about human anatomical parts: "I assume that if you do not already have sufficient first-hand knowledge of them, you can get a learned anatomist to show them to you—at any rate, those which are large enough to be seen with the naked eye." Once he had mastered the technique, Descartes dissected diligently on his own from animal material he obtained from butchers. The diagram of the brain given in the *Optics* (probably composed in the early 1630s) came from his study of sheep brains, because he could not himself make a study of human organs since "I am not a doctor by profession."[29] In notes about dissections he conducted that Leibniz later copied (and which take up nearly a hundred pages in the printed edition), Descartes wrote down precise observations about a large number of studies, often on the fetuses or newborns of cows: he was clearly interested in the physiology of development, among other matters, for he was still dissecting calves in the late 1640s.[30]

One of the sections from the complete system of natural knowledge Descartes abandoned after hearing of Galileo's condemnation in 1633 (but published posthumously) indicates where his anatomical and physiological studies were taking him in the early 1630s. *De homine* begins by giving a hypothetical account of how Descartes could explain the body and its functions on the assumption that it is "a statue or machine made of earth, which God forms with the explicit intention of making it as much as possible like us."[31] He then gave an anatomical and physiological account of all the bones, nerves, muscles, veins, and other parts necessary, followed by a discussion of the circulation of the blood. William Harvey's book announcing this last discovery, *De motu cordis et sanguinis*, had appeared in 1628, and soon thereafter Gassendi had read it and consulted a surgeon about it. The surgeon unfortunately led Gassendi to believe that the septum of the heart was porous, but Gassendi told Mersenne of his investigations, and Mersenne brought it up with Descartes, who encountered the book during 1632 at the latest. By June 1633, Descartes's old friend Beeckman was also thinking about the new theory from England, which he seems to have learned about not from William Harvey's *De motu cordis* itself but from the counterblast by James Primrose of 1630. Beeckman found Primrose's arguments against Harvey weaker than the passages of Harvey he was trying to refute. Perhaps Beeckman's positive view of Harvey came from Harvey's young friend George Ent, whose family had fled to England to escape the wars in the low countries but who had sent him to be schooled under the Beeckman brothers in Rotterdam and who kept in touch with his former teachers afterward.[32]

Descartes strongly supported Harvey's work, although he differed fundamentally about the cause of the motion of the blood, which he attributed not to the squeezing of the heart but to the expansion of the blood when it encountered the heart's innate heat.[33] He described this heat in a way consistent with much of contemporary iatrochemistry, being "heat without light"—as caused in fermentation—a mechanism he also invoked for digestion. (In some of the notes later described by Leibniz, Descartes mentions chemical remedies such as mercury, antimony, cream of tartar, and sulfur, making it clear that he was indeed studying chemistry along with anatomy.)[34] Descartes would try to convince others about the circulation by detailed physiological and anatomical arguments. His friend Plemp, for instance, published extracts of some of Descartes's letters on the subject only to refute them in his textbook of 1638, although by 1644 he had come to agree with the Harveian (not Cartesian) theory of the circulation. Also in 1638, Descartes attended the public vivisections on dogs demonstrating the circulation of the blood given by the young Sylvius at Leiden. These converted Johannes Walaeus, a professor of medicine, from being a critic into a supporter of the circulation, and he had one of his English students, Roger Drake, defend the principle in a public dispute of February 1640.[35] Moreover, in *De homine*, Descartes also famously discussed the pineal gland, a small organ on the midline of the brain, as the location for the interaction between the incorporeal intellect and the corporeal body.[36] Although the gland had been noted by Vesalius a century earlier, it is unlikely that Descartes rediscovered it in his dissecting himself. Where in his reading or conversation he hit on the importance of this organ is unclear. But when he attended a public anatomy demonstration at Leiden, probably in 1637, conducted by Professor Adriaan van Valkenburg, he asked to be shown the pineal gland, although Valkenburg could not find it and had to confess he had never seen it in a human brain.[37]

By the time he decided to publish some fragments of his early philosophical work, Descartes's medical pursuits had clearly become predominant. The first pieces of his youthful system were published anonymously at Leiden as *Discours de la méthod . . . plus . . . des essays . . .* ("Discourse on the method and essays on optics, meteorology, and geometry," 1637). The collected volume contained three substantial works extracted from his projected *De mundo* that applied his method to solving problems, along with a preface that gave a brief general explanation of his early ambitions, the "Discourse." As he said in a letter to Mersenne about the work, "I do not intend to teach the method but only to discuss it. As can be seen from what I say, [the book] consists much more in practice than in theory."[38] He was pointing Mersenne to the importance of three *Essays* of the book, in which he demonstrated the practical consequences of his work. In the prefatory "Discourse" itself, he gave only a cleaned-up, semiautobiographical explanation of how he came to the method and a few brief hints about it, also explaining how his ideas had moved on. He discussed his nine years of travel and sketched in his youthful dreams, describing how he had settled in The Netherlands to pursue his ideas more thoroughly, but what he wrote about the method is somewhat cryptic, something that Daniel Garber has recently noted as puzzling to those who consider theory to be of greatest importance to Descartes.[39] Descartes made it clear, however, that he now considered that his earlier work on the foundations of the method had simply been undertaken to enable a practical philosophy. The main purpose of his work now was

> the maintenance of health, which is undoubtedly the chief good and the foundation of all other goods in this life. For even the mind depends so much on the temperament and disposition of the bodily organs that if it is possible to find some means of making men in general wiser and more skilful than they have been up to now, I believe we must look for it in medicine. . . . Intending as I did to devote my life to the pursuit of such indispensable knowledge, I discovered a path which would, I thought, inevitably lead one to it.[40]

To accomplish such an ambitious goal, however, he acknowledged the necessity of constant labor: "I also noticed, regarding observations [*expériences*], that the further we advance in our knowledge, the more necessary they become." The simple principles he had discovered could be used well enough to explain everything post hoc, but only descriptions of how things are would allow them to be invoked. Yet "I see also that [such *expériences*] are of such a kind and so numerous that neither my dexterity nor my income (were it even a thousand times greater than it is) could suffice for all of them."[41]

That the work of "experience" or "experiment" he was doing was mainly medical is confirmed from the fullest argument in the "Discourse," which is about animal physiology. Here Descartes began with an account of the circulation of the blood—one of the earliest published works to publicly support Harvey's discovery—but his explanation is hardly theoretical: "first, so there may be less difficulty in understanding what I shall say, I should like anyone unversed in anatomy to take the trouble, before reading this, to have the heart of some large animal with lungs dissected before him." He finished this condensed analysis on animal physiology with the claim that he was now able to describe all the activities of the body and mind according to the laws of mechanics, as if it were a remarkable machine constructed by God. He concluded: "I will say only that I have resolved to devote the rest of my life to nothing other than trying to acquire some knowledge of nature from which we may derive rules in medicine which are more reliable than those we have up to now."[42]

FINDING TRUTH IN NATURE

Following publication of the *Discours*, however, acquaintances from France, especially, wrote asking for a fuller explanation of the method Descartes had sketched. Sometime between 1637 and 1640, while carrying on a voluminous technical correspondence about his views, living not far from Haarlem with his beloved little daughter (who died in September 1640, leaving her father broken-hearted), and still continuing his medical investigations, he decided to publish a much more expansive account of it, again as a way of getting beyond his early work by making his conclusions explicit.[43] I think it is clear that he went back to another part of the earlier manuscript of *De mundo* and pulled out the section laying out the first principles of his philosophy, which he heavily revised for publication. The result was the Latin *Meditationes de prima philosophiae* ("Meditations on first philosophy"), a long argument beautifully set out in six steps, one contemplation per day, inviting the reader to imagine a six-day path to understanding the elements of God's creation. It famously elucidates his early method of doubting and his proof of God, and his notion of how the soul is distinct from the body. In the preface he explained that he had been concerned that his views might be interpreted wrongly by ordinary people if he published it in French, which is why he had not fully explained his views in the *Discours*: that is, he admitted that this work was the product of his earlier years and might cause problems if incorrectly interpreted. He and Mersenne therefore circulated the manuscript to various scholars for comment, so that he could respond in advance to their objections. Published in Paris in 1641, the book was largely devoted to the six sets of objections and his replies to them; a second edition, from Elzevir in Amsterdam, contained a seventh set of objections with answers.

But given the path his concerns had taken, Descartes's preface also drew attention to the final (sixth) meditation, undoubtedly written last—that is, around 1640. There he introduced the term *imagination* as the property of intellect that interacts with the sensory world. Thus, although he never took back his earlier proofs for the mind being distinct from body, the mind "is shown, notwithstanding, to be so closely joined to it that the mind and the body make up a kind of unit." He also told the reader that he was especially pleased, not with the proofs for the existence of the intellect and of God, but with those for the existence of the material world. They showed (versus Montaigne and other skeptics) how one could have confidence in the knowledge about the world that came to the mind through the senses. He therefore shifted the apparent reason for publishing this work: whereas most people interpreted it (and still do) as casting doubt on our knowledge of the world via the senses, so that only pure intellect and God can be known with full clarity, Descartes argued that he wanted to show that knowledge of the intellect and God was more certain even than knowledge of the world, which should not be doubted:

> The great benefit of these arguments is not, in my view, that they prove what they establish—namely that there really is a world, and that human beings have bodies and so on—since no sane person has ever seriously doubted these things. The point is that in considering these arguments we come to realize that they are not as solid or as transparent as the arguments which lead us to knowledge of our own minds and of God, so that the latter are the most certain and evident of all possible objects of knowledge for the human intellect. Indeed, this is the one thing I set myself to prove in these Meditations.[44]

In other words, Descartes's goal in publishing the *Meditations* was to establish the certainty of our knowledge of God and the intellect, not to sow doubts about whether we have bodies intertwined with mind, that we can know about the material world, and so on, matters that he thought were self-evident in his empirical studies.

Turning to the text of the crucial sixth meditation itself, we therefore find him emphasizing how expériences are required to know the world. "Now, when I am beginning to achieve a better knowledge of myself and the author of my being, although I do not think I should heedlessly accept everything I seem to have acquired from the senses, neither do I think that everything should be called into doubt." He followed this with arguments for having confidence in most of what one knows via the body. God is not a deceiver, and he endows us with a variety of faculties by which one can check and correct knowledge that comes via the senses, which "offers me a sure hope that I can attain the truth even in these matters." "Indeed, there is no doubt that everything that I am taught by nature contains some truth." And what does nature teach us in general? "There is nothing that my own nature teaches me more vividly than that I have a body. . . . Nature also teaches me, by these sensations of pain, hunger, thirst and so on, that I am not merely present in my body as a sailor is present in a ship, but that I am very closely joined and, as it were, intermingled with it, so that I and the body form a unit."[45]

Descartes went on from these arguments about the body and sensory experience to say something even more surprising in light of his youthful meditations:

the original source of his doubts—how could he tell if he were dreaming or awake or being deceived by a demon?—had no foundation. "I should not have any further fears about the falsity of what my senses tell me every day; on the contrary, the exaggerated doubts of the last few [meditations] should be dismissed as laughable. This applies especially to the principal reason for doubt, namely my inability to distinguish being asleep and being awake. For now I notice that there is a vast difference between the two."[46] He had achieved his ambition, to lay to rest the ghost of skepticism, showing not only that the clearest and most distinct ideas one could have were about God and the intellect but also that mind was intermingled with body and that knowledge about the world was dependable.

Descartes's sixth and last meditation was carefully noted by the physicians (to whom we shall return shortly), but not by everyone. Even contemporaries asked to respond critically to the manuscript had almost nothing to say about the sixth meditation, except for Gassendi, who praised Descartes for managing to prove to himself something Gassendi never doubted: that the material world exists and that God does not deceive us about it.[47] But Gassendi spends only a few words noting his agreement here while raising numerous objections to other points. Indeed, ever since Descartes published his *Discours* and *Meditations*, commentators have been keen to consider, criticize, and elaborate the most unusual of his claims, his proof of God and definition of the mind as distinct from body. In light of these provocative problems, which even now prompt long classroom discussion, Descartes's own goals for his method, which seem almost to destroy what has come to be called the "Cartesian method" of doubting, draw little attention. In the context of his life, however, they are significant, for they show how deeply he was moved by the kind of natural knowledge via the senses he was pursuing in the Republic, especially in medicine. He tells us, in short, that we should not doubt that the pursuit of practical knowledge about the material world could yield real knowledge, whatever the skeptics might say.

In the process, however, Descartes rejected the classical category of "right reason" and so separated the knowledge of nature from moral philosophy. The technical term *right reason* spoke to the mind's ability to reach out to the *logos*, the source of all order and meaning. Since all intelligible knowledge emanates from the logos, those who spoke of right reason understood that it could distinguish not only between true and false but also between right and wrong. Most classical philosophical systems similarly treated moral and natural philosophy as two paths toward the same truths. In Plato's dialogues, for instance, Socrates argues for the identity of the true and the good. Because of the intertwining of the true and the good, moral behavior was a matter not of making the correct abstract decisions (or of applying the correct rule to a particular set of circumstances) but of acting and responding to circumstance by the exercise of good judgment. According to this view, judgment involves making discriminations that are as much moral as rational, and it is therefore an aspect of character as much as of reason. For someone like Aristotle, then, even understanding what the circumstances are—the "composition of the scene" itself—necessarily requires complex moral judgments. "Pursuing the ends of virtue does not begin with making choices, but with recognizing the circumstances relevant to specific ends."[48] To actualize our potential, we must gain knowledge of the good and the true and act according to it. Consequently, knowledge transformed: the wise person and the good person were the same; the wiser the better, and the better the wiser; and the measure of what is "really and truly human" is "the good man's pleasures."[49] The Stoics, too, argued for the identity of the moral and the rational. For them, "a wise person is a sign, a symbol that elucidates the deepest roots of the universe and its history."[50] Knowing and acting on the single source of the good and the true was the sole path to virtue, the only path that could fully actualize our human potential.

Mainstream Christian discussions also mingled the true and the good. To put it in biblical terms, when Adam and Eve ate of the fruit of the tree of knowledge planted by God in the garden they gained the knowledge of good and evil. This was the knowledge that distinguished humans from other mortal creatures: the ability to tell right from wrong. Right reason therefore remained fundamental to many Christian practices such as casuistry (giving moral guidance according to the circumstances). Similarly, many aspects of mind, such as conscience, found their source in right reason. Thomist and Aristotelian positions within the Catholic Church continued to hold to the doctrine of right reason, as did many Lutheran, Anglican, and other Protestant views.[51] Indeed, the classic historical work of Lucien Febvre in the mid-twentieth century argued that because "reason" implied knowledge of the good, whose source was in the Godhead, true unbelief was impossible even for the unreligious in the sixteenth century: even for unbelievers, reason would lead back to God.[52]

Descartes, however, consistently argued that we cannot know the purposes for which God has created the things of the universe, with humans exhibiting misleading pride when they think that he created the universe for our ends. Descartes also set aside any discussion of moral questions, which depend on knowing the right, which is another way of speaking of purposes, of ends. Although he famously argued for the presence of an immortal and rational soul in humans, he wrote of knowing about being—what existed—rather than about how the intellect could form clear and distinct concepts of the good. Indeed, his silence on this point is deafening because of the company it placed him in. For Niccolò Machiavelli, a good ruler can calculate how to act to further the prosperity of the state but not know if such actions are in keeping with, say, God's will. He therefore explicitly and famously rejected the idea that virtue could be acquired by living according to the dictates of reason, preferring "the view that humans are part of material nature like other brutes" and discussing not moral "virtue" but *virtú*, an attribute of the bodily spirits and humors.[53] According to Frans Burman, a medical student at Leiden who later took notes on a conversation between them, Descartes "does not like writing on ethics, but he was compelled to include rules [about a provisional moral code] because of people like the Schoolmen; otherwise, they would have said that he was a man without religion or faith and that he intended to use his method to subvert them."[54]

In refusing to comment on moral philosophy as much as he dared, then, Descartes was self-consciously embarking on a dangerous voyage. It has been argued that the provisional ethics mentioned in the conversation with Burman may have arisen from his reading of Pierre Charron's *Of Wisdom* (1601) during the crucial winter of 1619–20, when Descartes was first seeing how he could set out a universal method of knowledge.[55] Charron explicitly developed many of the elements of Montaigne's thought, arguing, for example, that our intellect brings us our greatest woes. Reason is limited and cannot clearly know the right. What we know of God depends on his leading us to him, not on us and our puny intellects. In separating religion from moral philosophy, then, Charron argued that religious dogma does not help in finding the right: many non-Christians of the past and present seemed as moral, or even more moral, than most Christians. He even went so far as to say that "nature is God" and that we can only know about nature—he is therefore associated with early Deism.[56] His book was placed on the index of prohibited books in 1605. Descartes's contemporary Thomaso Campanella held something similar and also ran into difficulties with the Church for it. Examples can be multiplied.[57] For such people, the best we can do in life is to learn from the natural world and to strive for material betterment. In the eighteenth century David Hume and Immanuel Kant would codify this separation as the difference between the "is" and the "ought," with the clear injunction that one could not learn about the one from the other: in other words, one could not obtain moral guidance from the study of what is. The division between the natural and the moral has been often lamented as a fundamental source of misery for moderns and is often attributed to "science," even to Descartes himself.[58]

The young Descartes was certainly adamant that the intellect had access to innate ideas. But did the rational soul tell us about the right as well as the true? As a young man studying first with the Jesuits and then with the lawyers, he would have been familiar with contemporary arguments about the abilities of reason narrowly defined and the ambiguities of right reason. His method relies on intuitions as its starting place, and in the *Meditations* he writes of wishing to be judged only by "those who combine good sense with application."[59] Then, too, in the fourth section of the *Meditations*, on "Truth and Falsity," he writes much about judgment and will. But the whole thrust of his method is to gain intellectual clarity. The lesson he draws from his meditations is that "if, whenever I have to make a judgment, I restrain my will so that it extends to what the intellect clearly and distinctly reveals, and no further, then it is quite impossible for me to go wrong." In other words, "I will unquestionably reach the truth, if only I give sufficient attention to all the things which I perfectly understand, and separate these from all the other cases where my apprehension is more confused and obscure."[60] What the intellect clearly and distinctly reveals are things that have being; by omission, then, moral choices implicitly fall into the category of things that are more confused and obscure, with which he will not deal. His contemporaries therefore placed him near those who cut off reason from moral intuitions. We might situate him with Grotius, who discusses natural law without invoking right reason but without arguing against it. They were trying to reform accepted views not by contradicting them so much as by changing the subject.

Such philosophical problems might seem a diversion from the medical studies Descartes was then pursuing, but they were not. Considerations of mind and body had always been critical to an understanding of the causes of health and disease, and Descartes considered his approach to open up possibilities for reforming medical practice and achieving long life. In the sixth meditation (again) he drew explicit attention to some of the implications for medicine of his method, when he discussed why God is not being deceitful when our sensations cause us to act against our own interests and so cause disease. For instance, when someone who is dropsical—having a great excess of fluid in the body—gets thirsty and drinks, worsening the condition, the sensations are not in error, he argues. In such circumstances, we truly feel thirsty. But on occasions like this the sensations can be caused by bodily events that are not their ordinary causes, which thus creates an error in our minds when we judge the cause incorrectly. He remained confident that "in matters regarding the well-being of the body, all my senses report the truth much more frequently than not."[61] The problem, then, is not caused by a deception of nature (or a failure of God's goodness) but by an error in our judgment. In the last sentence of the *Meditations*, he returns to the problem. Illness arises from the pressures of life and our inherently weak natures, which result in our failing to consider everything carefully; if we investigated the sources of all our sensations, we would not err and would therefore not become ill. Disease is, therefore, caused by errors in our judgment that lead us to act in ways that are not beneficial. Even two years before he published the *Meditations*

Descartes had written to his friend Constantijn Huygens about how he hoped to be able to avoid such faults and so to live for more than a century.[62]

Descartes was, of course, taking up an ancient and fundamental discussion in medicine about the relation of body and mind, the acknowledged foundation for preserving health and prolonging life. The famous ancient dictum to "know thyself," *nosce teipsum*—which Descartes took as his motto—was directed precisely at the intimate, complex, and dynamic relations between mind and body that were at the heart of both moral philosophy and medicine.[63] The only hope of living a good life in body as well as mind, Descartes agreed, lay in regulating one's physical and mental life properly. In this, he stood in a long tradition. At the end of the *Timaeus*, for example, Plato recommended the preservation of health in body and soul through a proper regimen that followed the good and due proportion.[64] The maintenance of health through this kind of "dietetics" concerned regulating with care not only one's food and drink but one's exercise, one's friendships, the situation of one's house, one's passions, and other matters of daily life (the Greek *diaita* meant a way of living or mode of life). Living according to a "diet" or regimen appropriate to each person meant to maintain or restore the balance (*krasis*) appropriate to each person, the foundation of health. Therefore, the aims of both the good life and the healthy life were, for most philosophers, as completely intermingled as Descartes's rational soul and body. To achieve both physical and moral goods, one needed properly to regulate mind, passions, and body. For others, one did this by the exercise (again) of right reason. As the learned Oxford scholar Robert Burton put it: "All Philosophers impute the miseries of the body to the soul, that should have governed it better by command of reason, and hath not done it."[65]

MEDICAL DISPUTES

Descartes's *Meditations* was, then, arguing for our ability to know the world, and for the use of natural reason rather than right reason to do it, even in the maintenance of health, but divided what we know about nature from the more doubtful subject of moral philosophy. It freed the study of nature from the vexed problems of contemporary religion, although it smacked of materialism. If there is any doubt about whether the implications of Descartes's *Meditations* were of medical interest, subsequent events should lay that doubt to rest. For following the publication of *Meditations* in 1641, Descartes became involved in a grave dispute in Utrecht about where his views were leading, a dispute that arose first not among the philosophers but between the physicians and the theologians.

Descartes's friend Reneri had begun teaching in his philosophy courses in

Utrecht, successfully and without objection, parts of the problem-solving *Essays* that accompanied the earlier *Discours*. Reneri soon acquired a colleague on the medical faculty, Henricus Regius, who was also an enthusiast of Descartes's *Essays*. Just two years younger than Descartes, Regius (born in 1598) had been educated in the same world of intellectual tumult. His father came from a distinguished Utrecht family, and like Descartes, Regius began his further education in law—at Franeker. He shifted to medical studies at Groningen and then, about the time Descartes came to the military camps of Breda, he matriculated in medicine at Leiden, where he would have encountered Professors Heurnius, Bontius, and Pauw. Sometime later, while Descartes was on his travels in German lands, Regius set out for studies in Italy, where he was robbed and forced to take service with the French army before finally making his way to Padua, where he took his medical doctorate in March 1623. As he was starting his medical career back in The Netherlands, he acted in 1625 as an unsalaried city physician at Utrecht and in 1630—not long after Descartes returned to the north—held the post of physician and rector of the grammar school at Naarden (on the Zuiderzee, north of Utrecht). The local clergy there already disliked his religious positions. In 1634, the same year Reneri moved to the Illustrious School in Utrecht with Descartes in tow, Regius returned to Utrecht, found a house in the same neighborhood as Reneri's, and married. He gave well-regarded private lessons in natural philosophy.[66] In July 1638, because of his successful teaching of the new philosophy to his private students—probably based on Descartes's *Essays*—Regius gained an appointment as a professor of medicine extraordinary (an honorary position) at the higher school in Utrecht, which had just been elevated to the status of a university; his position was regularized in the following March, when he joined the first professor of medicine, Willem van der Straten, as a full member of the faculty, charged with the teaching of botany and the principles of medicine.

Not long after Regius's appointment as professor extraordinary, in mid-August 1638, his neighbor and new colleague Reneri delivered a letter from him to Descartes asking to be accepted as "his disciple," to which Descartes responded with good wishes and an invitation to them both to visit.[67] Reneri was not well, however (he would die in mid-March 1639), so Regius visited Descartes on his own; the two got on very well. Regius took up the cause for Descartes by having his first pupil, Johannes Hayman, defend the circulation of the blood in a medical thesis delivered on 10 June 1640. But he was eager to give a fuller public statement of Descartes's ideas without upsetting the professor of philosophy who succeeded Reneri, Arnold Senguerd (who was an Aristotelian). Regius finally arranged this at the suggestion of the theologian and powerful exponent of Counter-Remonstrant Calvinism Gysbertus Voetius (or Voet) by using the time-honored tradition of presenting his views through the mouths of students who were defending their theses; if these were medical theses under his direction, Senguerd could hardly object.[68]

The result was three two-part disputations, collected and published in 1641 as *Physiologica sive cognitio sanitatis* ("Physiology, or the knowledge of health").[69] Regius treated the study of the body as something that could be approached without invoking any immaterial qualities: hot and cold were attributes of the motion or rest of the insensibly small material bodies that composed parts, as were hardness and softness, and so on. The only immaterial aspect of the human was the rational soul; everything else could be accounted for by matter, shape, spatial relationships, and motion. Living processes could be reduced to the study of the body's innate heat, and therefore the study of health and disease could be reduced to the study of how the innate heat set in motion certain ordinary bodily processes or failed to do so, with the circulation of the blood being the most fundamental process for stimulating the others. All the bodily actions, even generation, growth, and nutrition, are all produced by nature, without the need for the intervention of a rational soul. The whole is a careful elaboration of the body rooted in Cartesian principles, but with scarcely anything said about metaphysics and other such subjects, the collected theses give the impression of a rigorously materialistic system.[70]

A second series of disputations, beginning in November 1641, were even more aggressive in asserting a materialist view of the body. The third session ended in tumultuous shouting that continued even after the professors had left the room (although it should be said that thesis defenses were often raucous affairs). Apparently many of the theology students had come, and they were upset that the union between body and soul was said to be "accidental." In Aristotelian terms, the soul is the form of the body, and so of its essence; Regius was proposing that the soul was a circumstantial rather than essential attribute of being human, implying that the body could exist separately from soul, which was not, therefore, necessary for human life. This raised a host of sensitive theological problems, not least related to the question of the bodily resurrection. By mid-December, Regius was being attacked for defending the circulation of the blood, for his teaching in botany and medicine, and for the idea that humans are accidental beings (in a thesis that also attacked the Copernican system and anti-Aristotelianism in general). Only the intervention of the Utrecht burgomasters, who were responsible for the university, kept the arguments from becoming publicly personalized. In the days leading up to Christmas, however, Voetius singled out for condemnation Descartes's proof of the existence of God. Clearly, the argument that everything except God and the rational soul could be explained in materialistic terms had struck a raw nerve.[71] Descartes consulted with Regius and they prepared a counterattack, which resulted in the February *Responsio* of Regius. In it, Regius launched a general attack on Aristotelianism that his colleagues found highly offensive, and the work was immediately confiscated. This was followed by a condemnation of Regius and a prohibition against teaching the "new philosophy" (which was not enforced).[72]

By now Descartes was fully involved in attack and defense. A delay in printing the new, seventh set of objections and replies at the end of the second edition of the *Meditations*, which appeared at Amsterdam in the spring of 1642, gave him the chance to add to the book a long, self-justificatory "Letter to Dinet" (the provincial of the French Jesuits, who had once been his prefect at La Flèche). In this letter, Descartes included final comments that attacked Voetius by name, openly saying, "He is not a person of honesty and integrity."[73] Voetius and many others grew more and more bitter in their counterattacks, while Descartes and Regius grew even more vehement in defense of their views; the students were in an uproar at every public dispute where the issues came up, and the publications of the two sides grew increasingly ad hominem. Yet the views of Descartes and Regius were also growing apart. Descartes was truly taken aback by Regius's view that man was an "accident." Regius took seriously Descartes's argument that the material composition of our animal bodies gives rise to all the powers necessary for generation, growth, and nutrition, and even gave rise to the *anima*. He therefore disagreed with Descartes on the necessity of innate ideas in the intellect and with his proof for the existence of God. Philosophical materialism had been reintroduced among Italian philosophers who took up Averroist themes about double-truth, such as Cesare Cremonini, and Pietro Pomponazzi (who argued that even a rational soul needed to be material in order to think). French intellectuals such as Pierre Charron and Gabriel Naudé echoed the view that materialist explanations were sufficient for everything, although they conceded, when they had to, that revelation and the church taught about an immortal and rational soul. Such a materialist outlook left the intellect, which Descartes identified with the immortal soul, as something unnecessary for life. One could consider it superfluous in the study of animal bodies, including human bodies, and this might lead on to paying it no heed at all. In other words, Regius was, at least in his medical arguments, a monist and materialist, believing all things can be derived from a single source, in this case matter and motion. Despite laying the foundations for materialist arguments, Descartes remained a dualist, defining mind and body as essentially different although completely intermingled. Descartes saw Regius's materialist propositions as inherently atheistic and told him that as a Catholic he was forbidden from following him down that road.[74] To avoid becoming adversaries, they broke off relations for a while, and when they tried to converse again in 1645 they found themselves completely at odds about whether humans possessed innate ideas (Descartes was, of course, in favor, and Regius against). They never agreed on this, although by the late 1640s Descartes wrote favorably about Regius's integrity.[75]

THE INTERPRETATION OF THE PASSIONS

But in the midst of these battles, Descartes continued with his medical investigations, which led to yet further implications about human nature. From late 1641, he worked on the treatise that would be published in Amsterdam in 1644 as *Principia philosophiae* ("Principles of philosophy"). He had not had enough time to complete his studies on animals and plants, and on the human body, so he had to leave aside his plan to publish on these things, simply adding a few observations at the end "concerning the objects of the senses," while stressing that "everything that we clearly perceive is true; and this removes the doubts mentioned earlier."[76] His observations on the objects of the senses included a few comments on "internal sensations," or the passions, which he sometimes called "emotions." By the time of the publication of the *Principles*, he was deeply engrossed in analyzing the passions, and he arrived at some striking conclusions. He treated the passions as aspects of body that communicate to us how we can be happy and healthful: in other words, he rooted them in his physiological outlook rather than in moral philosophy.[77] He first took the usual line that reason needed to control them but later declared them all to be good; only a few people alive could manage to regulate them, and the rest of us should not worry about our health or virtue if we could not, so that we should embrace them.

Descartes's close consideration of the problematic relation between reason and the passions and his changes of view were based on his physiological knowledge coupled with the concerns pressed on him by the young Princess Palatine Elizabeth (born in 1618), one of the children of the ill-fated "Winter King," Prince Frederick of the Palatinate, and his wife, Elizabeth, sister of Charles I of England, the original "Queen of Hearts," then taking refuge in The Hague with Frederick Hendrik, Prince of Orange. Descartes had heard through mutual acquaintances in late 1642 that the young Elizabeth was reading his *Meditations*, and he managed an introduction, which led to a lifelong relationship.[78]

In a letter of 6 May 1643, Elizabeth asked Descartes what everyone wanted to know: he had clearly discussed how matters of soul and body had to be distinguished one from the other and considered according to different clear and distinct notions, but how did they interact, as he admitted at the end of the *Meditations* they did? More precisely, how could the soul—a thinking substance only—get the bodily spirits to exhibit voluntary actions?[79] Descartes replied by letter two weeks later, arguing that the soul had two aspects we could know: it thinks, and it acts on and is acted on by the body. "About the second I have said hardly anything," he confessed, since his first philosophical aim had been "to prove the distinction between the soul and the body, and to this end only the first was useful, and the second might have been harmful." He began by writing of cognition, which occurs by the application of primitive notions, and explained that we confuse matters when we use notions of how one body acts on another to think about how the soul acts on the body. Elizabeth wrote again to express her exasperation at not understanding Descartes's views, and he replied apologetically by noting that it was when one refrained from philosophy that one understood the union of soul and body most clearly. "It does not seem to me that the human mind is capable of forming a very distinct conception of both the distinction between the soul and the body and their union." The notion of a union of body and soul was something "everyone invariably experiences." He then more or less told her to forget about the problem: "feel free" to think what you want, he wrote. As he said to others he also said to her, you should understand "the principles of metaphysics" once during your lifetime, but thinking about them too long and hard would be "very harmful."[80] Rather than explain further, then, he begged off. Understandably, Descartes's answers did not satisfy Elizabeth.

A year later, however, he came back to a discussion of the relation between the soul and the body, not for philosophical but for medical considerations.[81] In doing so, Descartes was returning to a central point of his work. By this time Elizabeth was fondly calling him her favorite philosopher and her doctor. But she was not well and had decided on a course of diet and exercise, which Descartes approved. He clearly agreed with her that the cause of her ill health was the result of a troubled mind: amid the English civil wars and last years of the Thirty Years' War her family's fortunes, and her personal future, appeared increasingly dire. (Indeed, lacking a dowry, the princess would never find a marriage partner and ended her life as abbess of the Lutheran convent in Herford, Westphalia, a city that would become known for physicians who spoke out in favor of Cartesianism.) Descartes therefore told Elizabeth that to regain her health she had to control her imagination and senses through her intellect. "There is no doubt," he wrote, "that the soul has great power over the body, as is shown by the great bodily changes produced by anger, fear and other passions." As he explained:

> The soul guides the spirits into the places where they can be useful or harmful; however, it does not do this directly through its volition, but only by willing

> or thinking about something else. For our body is so constructed that certain movements in it follow naturally upon certain thoughts: as we see that blushes accompany shame, tears compassion, and laughter joy. I know no thought more proper for preserving health than a strong conviction and firm belief that the architecture of our bodies is so thoroughly sound that when we are well we cannot easily fall ill except through extraordinary excess or infectious air or some other external cause, while when we are ill we can easily recover by the unaided force of nature, especially when we are still young.[82]

Yet despite this constant advice to look on the bright side, Elizabeth continued to experience ill health, which Descartes attributed mainly to the steady bad news about her family's fortunes: although her brother Rupert made brilliant efforts on his behalf, her uncle Charles I of England had just been defeated at the Battle of Naseby. Elizabeth felt distresses of such a sort that "reason does not command us to oppose them directly or to try to remove them," wrote Descartes. "I know only one remedy for this: so far as possible to distract our imagination and senses from them [that is, these feelings of distress], and when obliged by prudence to consider them, to do so with our intellect alone." Descartes went on to distinguish between the intellect on one hand and the imagination and senses on the other. The imagination and senses governed the passions and affected the spirits and body; the intellect was separate and had the power to direct the imagination. Descartes had cured himself of ill health, he declared, by looking at things "from the most favorable angle."[83] In a following letter a few days later, Descartes tried to soothe his distressed patient further by sympathizing with her and stating, "The best minds are those in which the passions are most violent and act most strongly on their bodies." But following a night's sleep, one can "begin to restore one's mind to tranquillity" by concentrating on the best aspects of one's situation, "for no events are so disastrous . . . that they cannot be considered in some favorable light by a person of intelligence."[84]

In his advice to Elizabeth, Descartes was, of course, relying on the classical view that to maintain health one needs to know what is right and proper via reason, and then to act in accordance with it. To do this, one needed to be able to control the "passions," which keep getting in the way of reason. Perhaps study would help to distract her, he suggested.

Elizabeth agreed, and they began an epistolary conversation about *De vita beata* ("On the happy life") of the Stoic philosopher Seneca, which remained intense for several months and continued until the end of Descartes's life. It was a fateful choice for careful reading. Seneca is best known as a neo-Stoic, but he also discussed Epicurus's views with considerable sympathy. "I myself believe . . . that Epicurus' teaching is moral, upright and, if you look at it closely, austere," Seneca wrote.[85] Montaigne and Charron, both of whom Descartes had read, were among the many who followed Seneca in thinking that Stoicism and Epicureanism were perfectly in harmony.[86] It was Seneca who convinced Descartes's acquaintance and the most famous Epicurean philosopher of the period, Gassendi, that Epicurean views were not only moral but compatible with Christianity.[87] But Epicurus had outlined a system of natural philosophy that was completely materialistic, one in which the passions were all good, and "the good" was judged by pleasurable sensations rather than moral ends. So-called Epicurean "libertines," often well-educated gentlemen and noblemen, were therefore often accused of all kinds of immoral behavior, from irreligion to private sexual and gastronomic indulgences.[88]

In their discussion of Seneca, then, Descartes began by defending a position not unlike the Stoic one, in which reason should possess complete dominion over the passions for the sake of health and well-being.[89] One should employ the mind to discover by reason what should and should not be done in all circumstances, resolving to do as reason directs "without being diverted by . . . passions or appetites. Virtue, I believe, consists precisely in sticking firmly to this resolution"; and one should acknowledge that all goods that one does not possess are beyond one's power and so not worth thinking about. "So we must conclude that the greatest felicity of man depends on the right use of reason" and the controlling of the passions by this. Put another way, "happiness consists solely in contentment of mind . . . but in order to achieve contentment which is solid we need to pursue virtue—that is to say, to maintain a firm and constant will to bring about everything we judge to be the best, and to use all the power of our intellect in judging well." Elizabeth raised a critical objection, however: many people, including those who are ill, do not have the free use of their reason that this view of Seneca and Descartes assumed. He agreed that "what I said in general about every person should be taken to apply only to those who have the free use of their reason and in addition know the way that must be followed to reach such happiness." That is, some people do not know where true happiness lies, and others have a bodily indisposition that prevents them from acting freely. But he came back to the neo-Stoic view that the passions are vain imaginings, or distortions of reason—that is, errors of thought—so that "the true function of reason . . . is to examine and consider without passion" one's true good and to "subject one's passions to reason."[90]

Still not satisfied, however, in a letter of 13 September 1645, Elizabeth asked Descartes to give "a definition of the passions, in order to make them well known."[91] He began by approaching the problem physiologically, by discarding a number of common associations of the word "passion" and limiting his investigations to "the thoughts that come from some special agitation of the spirits, whose effects are felt as in the soul itself." He reported that he had begun to consider the passions in detail but soon began making excuses for not following up while instead diverting the conversation into a long discussion about free will, distinguishing carefully between the will (volition) and the passions of mind. He had not given up, however, and in order to comply with her request, Descartes returned to a further consideration of animal physiology. The disputes in Utrecht and elsewhere, provoked by the implications of his physiological ideas, were now boiling hot and required much of his attention, but he also remained committed to developing his physiology and medicine. Just after agreeing to reconsider his ideas for Elizabeth he told the Marquis of Newcastle that "the preservation of health has always been the principle end of my studies," and in 1646 he wrote to the French resident in Stockholm, Hector-Pierre Chanut, that because of this, "I have spent much more time" on medical topics than on moral philosophy and physics. To Chanut he conceded that he had not yet found ways to preserve life with certainty, so that he had decided only "not to fear death."[92] But he was obviously hard at his medical work and hopeful about the outcome.

By early 1646, he had drafted a work on the passions, which he sent to Elizabeth for her comment. (I believe that this was incorporated into his later book as "part III," as we shall see below.) In writing to Elizabeth about his new ideas, Descartes stressed the links between body and soul. He explained that the movements of the blood that accompany each passion were grounded in physical and physiological principles and that "our soul and our body" are very closely linked, but he also acknowledged that "the remedies against excessive passions are difficult to practise" and "insufficient to prevent bodily disorders." He still believed that such remedies might free the soul of being dominated by the passions and so enable it to exercise "free judgement." But now he declared that "it is only desires for evil or superfluous things that need controlling"; more striking still, he wrote that "it is better to be guided by experience in these matters than by reason."[93] Quite explosively, when writing a few months later to Chanut at Christina's court about how to present his philosophical views to the queen, Descartes declared that despite Chanut's expectations, "in examining the passions I have found almost all of them to be good, and to be so useful in this life that our soul would have no reason to wish to remain joined to its body even for one minute if it could not feel them."[94] When in late 1647 he attempted to secure the patronage of the queen—probably acting for Chanut as a political instrument of French influence during this period when the Sveo-Gallic alliance was threatened from antiroyalist libertines at her court—he sent Christina copies of his letters from 1645 to Elizabeth and the draft treatise on the passions he had written for her.[95] (Elizabeth considered this a dishonorable breech of personal loyalty, although

Descartes rather lamely explained that he was trying to gain support for her from Christina.) His new work may have helped to secure a new position for him, since he left for Christina's court in Stockholm in the autumn of 1649.

THE PASSIONS OF THE SOUL

Descartes had returned to working on his treatise on animals in 1648, but in November 1649, just three months before his death in Stockholm, the final version of his *Les passions de l'âme* ("The passions of the soul") appeared in print, dedicated to Elizabeth. It argued that the passions are the source of all goods in life: they are all good, and all the pleasures that are common to both soul and body, such as love, "depend entirely on the passions."[96] This was a formula associating good with pleasure, a large step beyond Aristotle's view that in some cases some of the passions can be good; it is almost unthinkable for Descartes's neo-Stoic predecessors who made war on the passions; and it went further than Barlaeus and other contemporaries toward explicitly making the passions into beneficial rather than harmful powers.

Many people other than physicians and philosophers had, of course, considered many of the passions to be good. Such views run rampant through the poetry and literature of the age. In Bredero's *The Spanish Brabanter*, for instance, the Dutch servant Robbeknol manages to wheedle a meal despite his master's bankruptcy, and begins act 3 by exclaiming, "'Tis said that he who drinks well, sleeps well, and who well sleeps, sins not; / And he who does no sin will certainly be blessed."[97] Even tongue-in-cheek, this is a good case for the sanctity of the bodily appetites. In a more serious vein is poetry such as that of the Muider Kring's host, Pieter Cornelisz. Hooft, who had returned from youthful travels in Italy with remarkable abilities for writing love sonnets. As an admirer put it, when he is at his best, Hooft "seems to dance on the surface of life like a skater on a deep frozen lake: one marvels at his airy grace while sensing hidden depths of passion."[98] One of his sonnets that Descartes himself would have appreciated sings:

My love, my love, thus spoke my love to me,
While on her delicate lips my lips were browsing.
Those words, too clear to be in need of glossing,
Entered my ears and stirred mysteriously
My inmost thoughts into tumultuous stress.
They did not trust the ear and at their pressure
I begged my dearest for a fuller measure
Of that confession, and she did confess.

Oh, bounty of the heart that overflows!
Entranced, each heart did th'other's heart imprison.
But when the morning star fled for the risen
Light of the sun, the sad truth too arose:
Oh, Gods, how close are things that are and seem!
How like the dream is life, like life the dream![99]

Such examples of the power of the passions—especially love—to show us our good can be multiplied many times over.

The young cavalier Descartes himself, of course, had taken a keen interest in poetry. Moreover, the treatise he gave to Beeckman before they parted begins by saying that the goal of music was to move the passions.[100] Early in the missing notebook in which Descartes recorded his dreams and thoughts is, according to the notes taken by Leibniz, a lengthy consideration of the passions: "In everyone's mind there are certain parts that excite strong passions, however lightly they are touched," noted the youngster.[101] Descartes would also have been aware of the contemporary current of French moral thought that focused on *amour-propre*, or self-love, as the stimulus for most human behavior.[102]

Descartes's book began, therefore, by noting that "the defects of the sciences we have from the ancients are nowhere more apparent than in their writings on the passions." Descartes continued to treat the definitions of soul and body separately, arguing that the passions act on the soul in the same way that objects made themselves known through sight. That is, as he had discussed in the *Meditations* and elsewhere, the soul is able to apprehend the world around it through the faculty of the imagination, which "grasps" information acquired from the senses. Descartes had, therefore, agreed with others that despite the indivisibility of the rational soul, it possessed more than one faculty (or "mode") that was intermingled with body. Now he was adding another mode to intellect, the purpose of which was to apprehend what is good for our bodies and lives: the passions. "The function of all the passions consists solely in this, that they dispose our soul to want the things which nature deems useful to us, and to persist in this volition." "The various perceptions or modes of knowledge present in us may be called passions," which can arise from either soul or body. But the passions are not only modes of knowledge, they are also (together with volition) attributes of the soul itself that allow the soul to act in the world.[103] In other words, passions arise from both mind and body, linking both and causing effects in each.

But because passions of soul are not the same as volition, they "cannot be directly aroused or suppressed by the action of our will." They can be modified "only indirectly through the representation of things which are usually joined

The classical view: the passions bound by divine grace and reason. Frontispiece verso of J. F. Senault, *The Use of Passions*, translated by Henry, Earl of Monmouth (1649). Courtesy of the Wellcome Library, London

with the passions we wish to have and opposed to the passions we wish to reject." Moreover, because of their relations to body, "the soul cannot readily change or suspend its passions." They "are nearly all accompanied by some disturbance which takes place in the heart and consequently also throughout the blood and the animal spirits."[104] Indeed, as the letters to Elizabeth indicate, Descartes's view places the passions at the center of the causes of disease. Fevers, for instance, were caused by putrefaction in the blood due to lack of the proper movement of the blood, which, Descartes was explaining, can be caused by the passions.[105] Much of the treatise on the passions is therefore focused on how the passions and bodily physiology are inseparably intertwined.

Descartes was concluding, then, not that we should arrange our passions to be in accord with our reason, but that we should arrange our minds to be in accordance with our passions, to desire what our natures wish us to desire. If we do so, we can compose our volition—which unlike the passions is subject to reason—to act in accordance with what is good and healthful for us. He responded accordingly when asked by Christina in 1647 to explain his view of the supreme good. Descartes first set aside both Christian revelation and the general good of all humankind to deal with the supreme good for each person. This "consists only in a firm will to do well and the contentment which this produces." "The goods of the body and fortune do not depend absolutely on us" but rather depend on the world; consequently, one of the greatest goods of the soul, to know the good, is also "often beyond our powers." The only good remaining to us, then, is to will what is good, which requires "judging what is best" by employing all the powers of our mind, and resolving to do it as best we can. Only by "the good use of free will" in this way could "the greatest and most solid contentment in life" be gained.[106] "Contentment" as the supreme good was, however, usually associated with the Epicureans rather than the Stoics (whose highest good was the honor and virtue accorded to acting reasonably). Remnants of his earlier, neo-Stoic, line of thought remained in his book: "Undoubtedly the strongest souls belong to those in whom the will by nature can most easily conquer the passions and stop the bodily movements which accompany them." There was even hope for the rest of us, since "even those who have the weakest souls could acquire absolute mastery over all their passions if we employed sufficient ingenuity in training and guiding them." Yet as long as someone lives according to what his passions tell us is best, "he will receive from this a satisfaction which has such power to make him happy that the most violent assaults of the passions will never have sufficient power to disturb the tranquility of his soul." Happiness and health flowed not from dominating one's life by reason but from living "in such a way that his conscience cannot reproach him for ever failing to do something he judges to be the best (which is what I here call 'pursuing virtue')."[107]

This last comment makes his aims quite clear: he was explicitly reshaping moral philosophy (the pursuit of virtue) into an argument about how to judge what is best for one's life, which depends in turn on knowing by experience how the passions bind mind and body. Seneca's work, too, spoke of steadfastness, generosity, and other qualities of character and judgment that come from living confidently according to what we judge best. It was a commentary on the theme of how happiness is not complicated or distant, for "you will find it just by knowing where to reach out your hand." This was because "the happy life is one that conforms to its own nature."[108] Descartes's work on the passions might be considered a treatise supporting Seneca but for one fundamental difference: Descartes thought he found the source of philosophy in nature, without the need to reach beyond it. In other words, Seneca had discussed "human nature" in terms of knowing virtue, whereas Descartes thought he could locate our nature in the way we are made, not needing to look beyond how we are composed for the true source of our natures. Indeed, he had even begun his *Principles of Philosophy* (1644) with a preface announcing that a proper moral philosophy would derive from an understanding of the composition of the natural world, including the human body.[109] Descartes's book on the passions is concerned, therefore, not with moral guidance but with physiological explanation and advice.

The naturalist's goals are doubly clear from the contents. It was the fullest exposition of Descartes's physiological views published during his lifetime. He began with a sketch of his mechanical physiological system, including remarks on the movement of the heart and blood, the production of animal spirits in the brain, the function of the pineal gland, the movement of the muscles, the abilities of the senses—which sometimes also cause "involuntary" responses in the body, as he called them—perception, and so on. In this physiology, moreover, Descartes explicitly moved away from almost all previous discussions of the passions, which root them primarily in the heart, instead also discussing the movements of the blood and spirits.[110] (This also moved him away from the powerful contemporary movement within the Catholic faith to venerate the sacred heart of Christ, which was furthered in the later seventeenth century by the Jesuits and the later-sanctified Margaret Mary Alacocque.)[111] The soul cannot easily overcome the stronger passions until "the disturbance of the blood and spirits has died down"; until then, it is capable only of checking the movements prompted by these disturbances. (Similarly, near the end of the book, he wrote about the "blood all in turmoil, just as if they had a fever.")[112] The second part of his book investigated the causes of the passions and analyzed the primary ones (wonder, love, hatred, joy, sadness, and desire). He also described how they prompt external signs, such as expressions in the eyes and face, color (blushing or pallor), trembling, listlessness, laughter, tears, weeping, groaning, and sighing.

The third and final part of Descartes's book was apparently written first, since it gives definitions of the passions, as Elizabeth had requested from him in September 1645. It is also the most aphoristic, explaining, for instance, "why those whom anger causes to flush are less to be feared than those whom it causes to grow pale." Here too, however, he underpinned his analysis with physiological explanation. For instance, paragraph 162 deals with "veneration." Descartes defines it as "an inclination of the soul not only to have esteem for the object it reveres but also to submit to it with some fear in order to try to gain its favour." After further comments on this passion, he concluded with a physiological comment: "The movement of the spirits which produces this passion is composed of that which produces wonder and that which produces fear (about which I will speak later)." If one follows this last suggestion, one quickly finds a succeeding passage on fear, which is defined as an excess of "timidity, wonder and anxiety" caused by surprise, which produces a "coldness" and a "disturbance and astonishment in the soul which deprives it of the power to resist the evils which it thinks lie close at hand." But the analysis of wonder he asks us to find occurs not later but earlier in the book, suggesting that the parts were printed in a reverse order from their composition. When one finds the section, in part 2, wonder is described as "the first of all the passions," with no opposite. It is caused both by an impression on the brain and by a movement in the spirits, but, uniquely among the passions, it causes no change in the blood or heart.[113]

In short, in his work on the passions Descartes comes close to giving us what in modern terms would be described as a biologically reductive analysis. As one recent commentator has put it, "In Cartesian psychosomatics, the feelings do not represent a primarily psychological occurrence expressing itself through the body, but rather a mirroring of bodily regulative processes that is useful for self-maintenance and survival."[114] By accepting our passions as good and authentic expressions of our true natures, and exercising our will so as to act in accordance with them, the mind will be so happy and tranquil that momentary disturbances will not cause ill health. Descartes still agreed with the ancients that a properly regulated life would yield health and happiness, but he does not speak of virtue other than as doing what one judges to be best from experience: he continued to shy away from invoking classical right reason as the source of knowledge. As he had written Elizabeth, "It is better to be guided by experience in these matters than by reason." Although the soul "can have pleasures of its own," the pleasures "common to it and the body depend entirely on the passions." Such pleasures, even a good conscience, depend on living in accordance with nature, the knowledge of which comes to the intellect from the passions. Even what other philosophers and theologians called "conscience" is thus the apprehension of acting in accordance with the true nature of our passions so that they are neither misused nor allowed to become excessive: "we see that they are all by nature good, and that we have nothing to avoid but their misuse or their excess." Natural reason can act to avoid these excesses and misuses by forethought and diligence. "But I must admit that there are few people who have sufficiently prepared themselves in this way for all the contingencies of life." And "no amount of human wisdom is capable of counteracting these movements [in the body] when we are not adequately prepared to do so."[115] In his considered view, then, mind and body were as one in the passions, while the intellect grasped sensory experiences by the faculty of imagination and, when it could, directed the body's voluntary actions by the faculty of volition.

If one were trying to find philosophical parallels, it would be possible to show that Descartes shares some views in common with the Aristotelianism of moderation, or with Renaissance neo-Platonism, which focused on love. One might also point to parallels with materialist Epicureanism and its goal of contentment. In such arguments, the relation between human reason and the passions surfaced again as one of the most pressing problems in moral philosophy.[116] Perhaps he was being persuaded of the coherence of this school of thought by his correspondent Pierre Gassendi.[117] More likely Seneca provided the bridge from an early neo-Stoicism to a later neo-Epicureanism. Moreover, in the 1620s, the problem of the relation between reason and the passions had moved Dutch commentators such as the Leiden philosopher Franco Burgersdijk to introduce into a discussion of Aristotelian ethics Plutarch's view of moral virtue as being rooted in the passions though still guided by reason.[118] Barlaeus had something similar to say in his inaugural oration of 1632. Dutch physicians were also raising doubts about the ability of reason to control the passions. In composing his regimen of health for the Indies, in the 1620s, Bontius had commented that "physicians have written much and variously about how to moderate [the passions]. But the motions of our mind (*anima*) are hardly within our control; I agree with Horace: 'control your passion or it will control you; check them, restrain these your bonds.' But who will not admit that what is easy for one is difficult for another of a different temperament? Therefore, setting down rules for the affects or passions of the mind is more a matter for a magician than for a natural philosopher."[119]

It should be emphasized, however, that Descartes is not explicit about his sources, for he thought of himself not as writing in a historical tradition but as breaking from it. Nor did he ever try his hand at producing a work of moral or political philosophy. What Descartes thought about his predecessors remains hidden: as he said of himself, *larvatus prodeo* ("I go masked"). The guidance offered by his natural reason aimed only at helping us to obtain health and physiological knowledge. When Descartes arrived at Queen Christina's court in later 1649, he therefore discovered that she found his work on the passions "mediocre." Her own interests dealt heavily with moral philosophy, and she was deeply engrossed in Greek studies with the erudite young Isaac Vossius. In her learned opinion Descartes offered nothing that could not be found in Plato or Sextus Empiricus (a skeptic). Consequently, the queen preferred his mathematics to his philosophy, meeting him only half a dozen times at most before his death. Many of her learned courtiers also made scornful comments about his views. He in turn had little time for their learned discussions about the ancients. Not surprisingly, then, when he died—after treating himself and refusing the doctors—one of many rumors held that he had been poisoned by the classicists.[120]

Descartes had begun doing philosophy with a proof for the existence of God, but during his many years in the Dutch Republic he became something of an empiricist. His arguments took their lead not from moral philosophy or religion but from what can be shown according to natural reason coupled with experience of the material world. His physiology remained silent on the subject of any way by which either divine or demonic spirits could affect the body, senses, or mind. Perhaps these were mysteries about which we cannot speak, but this means that he offered no guidance apart from urging us to live within ourselves, being attentive but wanting nothing more than what nature and destiny brings our way, and then doing our best. Then the passions themselves will lead us to happiness and health. There was nothing here for those who would judge his book on the passions according to his earliest work, nothing like his early proof of God or the attributes of the immortal soul and intellect that make it distinct from body. Even when it came to the passion he (like Seneca) considered indicative of the greatest souls—generosity—there is "nothing essential to true generosity." Or as Regius would have put it, generosity is an "accident" having in it nothing dependent on the love of God nor anything that makes it necessary for personal salvation. Indeed, one of the most astute recent commentators on his book on the passions notes that by comparison to his first works, Descartes seems to be in "retreat."[121] The contentment Descartes offered from the passionate union of mind and body never held out the promise of redemption. But it did promise more accurate knowledge of our nature, better health and longer life, liberty, and wealth.

"CARTESIANISM," THE BODY, AND POLITICAL LIBERTY

The implications of Descartes's views, whether published or communicated by conversation and letter, did not go unnoticed. By the mid-1640s the disputes over his physiology had spread to the university of Leiden, and by 1647 they had grown into a national controversy. The chief early "Cartesian" at Leiden, Adriaan Heereboord, saw his work in light of contemporary neo-Aristotelianism: that is, as a strongly modified but further plea for building natural philosophy on the evidence of the senses.[122] It was in keeping with recent trends, for following the purge of open supporters of the Remonstrants from the university in 1619, Franco Burgersdijk had taken over several of the chairs in philosophy, and he did his best to incorporate new information about nature into an Aristotelian framework of matter-form theory.[123] But after his death in 1635, no one of his stature was available to replace him. The university senate encouraged the curators to appoint a commission to look into philosophical instruction, which led to a decision to base teaching on Aristotle. In late 1644, a conservative Aristotelian, Adam Stuart (a Scot), gained appointment as the professor of natural philosophy with the backing of the faculty of theology, who together with the senior professor of theology, Jacobus Triglandius, and Jacobus Revius, a member of the college for students of theology (the Statencollege), launched an attack on Cartesianism. As a younger professor of logic who had been passed over for Stuart's position, however, Heereboord defended Cartesianism. Heereboord set out strong arguments about the good public life on the basis of the great importance of the passions.[124] The disputes grew bitter over the course of 1646–48, with Descartes's opponents making a mockery of his views, which prompted a letter from Descartes himself. In February 1648, following the outbreak of fisticuffs and hair-pulling at a disputation, the curators felt compelled to interrogate all the professors involved in these battles, severely reprimanded Stuart, Heereboord, and Revius, reiterated the importance of Aristotle, and tried to tamp down the increasingly heated battle by instituting various bureaucratic measures for better order, including ending the teaching of metaphysics; but when Stuart complained, the curators warned Stuart against his plans to raise the world against Descartes while quietly allowing private teaching of the Cartesian philosophy to continue. Skirmishes continued, with the students in particular hounding Stuart until his death in 1654; and the States of Holland were even forced to issue an edict in 1659 forbidding stamping and banging in lectures, orations, or disputations.[125]

Yet by the late 1640s, a number of academics, many of them physicians, were openly supporting Descartes's teachings. Given Descartes's own interests late in his life, it is not surprising that the chief early advocates for his outlook tended to see Cartesianism as a support for careful descriptive empiricism, including Johannes de Raey, Abraham Heidanus, Johannes Clauberg, Christopher Wittich, Lambert van Velthuysen, and Frans Burman. De Raey and Velthuysen had obtained medical degrees and so would have read Descartes with attention to his medical interests. Not all Dutch physicians adopted Descartes's positions, but many found his views helpful in moving physiological debate beyond Galenic and Aristotelian frameworks. After Sylvius joined the Leiden medical faculty in 1656, he and his students often spoke up in favor of Descartes's ideas about the material construction of the human body, although because of their work in chemistry and other subjects as well it would be unfair simply to call them Cartesians.[126] At the university of Louvain, Cartesianism was first openly advocated by members of the medical faculty, Plemp and Léonard-François Dinghens, although an investigation by the university in 1662 concluded by prohibiting such teaching and the teaching of philosophy outside the philosophical faculty.[127] Even Descartes's physics, when introduced by Jacques Rohault's *Traité de physique* of 1671, dropped his metaphysics, played up the experimentalist aspects of Descartes's views, and treated his knowledge as resulting in probabilistic rather than certain truths.[128]

When he died, Descartes left behind a treatise in French on human physiology, originally intended as a part of his abortive project *Le monde*. It had been shared with others in manuscript, such as Regius, and several copies seem to have been in circulation. Twelve years after Descartes's death, Florentius Schuyl collated and edited copies of it, translated the results into Latin, and published them as *De homine* ("On man," sometimes called "Description of the human body," 1662), after which Sylvius invited him to take a medical degree at Leiden. Only many years after his death, then, after many long arguments about his ideas, did Descartes's full views on human physiology become public.[129] In the early days, his medical outlook was being discovered mainly through his work on the passions.

But attempts to do what Descartes left unsaid—to ground moral and political philosophy on the physiology of the passions—had followed quickly. Early Dutch political theory, as elsewhere, generally remained rooted in ideas in which the goal of public virtue remained uppermost and even argued that this was obtained by controlling the passions through monarchical government. The humanist education of The Netherlands had been directed "towards developing civic virtues and preparing its pupils for a life of responsible leadership" just as much as elsewhere.[130] Orthodox Calvinists had come to hate anything that smacked of Cartesianism, while even the more liberal Arminians battled materialism, borrowing support from their English latitudinarian colleagues.[131] But for the libertine party in The Netherlands, Descartes was giving elaborate reasons for what Barlaeus had already noticed: all goods came from acting in accordance with nature rather than from trying to suppress the passions through the application of a rational doctrine or powerful sovereign. Drawing on the expressions of the original leaders of the Revolt, Barlaeus had argued that the chief reason for the flourishing state of the Republic was the pursuit of liberty, adding that liberty was not only good for the consciences of its citizens but the chief reason for their material prosperity. Descartes had given physiological and philosophical reasons for this: there was no higher reason by which one could act morally even if one wished, whereas following the passions allowed one to achieve health and other bodily goods, which was all the good we mortals have in our power to achieve. With the death of Frederick Hendrik in 1647, the Orangist alliance with the strict Calvinists had been revived under William II, but in 1650, as he was about to seize complete power, and with his army at the gates of Amsterdam, William suddenly died of smallpox. His death consolidated the hold of the so-called States Party, the republican and liberal group that held sway until the rise to the Stadholdership of William III in 1672. Members of the States Party—or the party of True Liberty, as they called themselves—developed a view of political economy that allowed far more freedom to the pursuit of the passions. The party's leader, Johann de Witt, saw himself as a skeptical realist who tried to live without illusions and without faith in the ability of humanity to save itself in this turbulent world of sin: as one historian has declared, he was "a Calvinist of an unmistakable neo-stoic type." And although De Witt himself acted with integrity, he knew that in general "nothing so much inspires men to love and affection as the feeling in the purse."[132]

Regenten such as De Witt took pride in empirical purity, in studying the world as it was rather than as people wished it to be. Pieter De la Court, and his brother, Johan, showed just how explosive the impact of Descartes's physiological arguments about the passions might be in such a context. The brothers De la Court produced some of the most remarkable statements of the century on capitalism and republicanism.[133] Well-educated Dutch merchants from Leiden, they developed their views into an explication of the political and economic ideas of the States Party.[134] As one historian has remarked, the brothers De la Court "regarded self-interest and passion as the basis of human conduct, but at the same time they developed the concept of the harmony of self-interests, possible only in a democratic community."[135] The clearest statement of their position was set out by Pieter De la Court, published as *Interest van Holland, ofte gronden van Hollands-welvaren* ("The interest of Holland, or the grounds of Holland's prosperity," 1662). De la Court began by noting that Holland had few natural resources, forcing its inhabitants to rely mainly on fishing, trading and some manufacturing rather than on agriculture. After estimating the number of people involved in various occupations, he concluded that "not the eighth part of the population of Holland can find their means of living [*nooddrust*] from their own land."[136] But he could show, by a brief economic history, why Amsterdam was a richer and larger trading center, and Holland a richer country, than had ever been seen before in the world: it was because of liberty. Liberty allowed people to worship as they pleased and to immigrate to the city and work in any productive capacity they wished, allowing them to pursue their interests freely, which produced wealth. He complained that the strict Calvinists were trying to limit freedom of conscience, that the freedom of the fisheries, trade, and crafts was also being narrowed by the growing power of institutions like the VOC, and that taxes were too heavy and therefore threatening to reduce trade. But when Holland had a thoroughly free government, compared to when the Prince of Orange and his allies ruled, the country flourished.

In arguing directly that the good state did not depend on a virtuous monarch, the brothers De la Court set themselves directly against those who wished to restore or enhance the power of the House of Orange.[137] In the same year, Pieter and Johan co-wrote *Consideratien van staat* ("Considerations of state," 1662) and *Politike discoursen* ("Political discourses," 1662), outlining an original republican theory and staking out positions on the most important issues of the day: the Stadholdership, the politics of trade, freedom of thought, the navy, the power of the clergy, and many others.[138] A few decades ago, the historian of political thought E. H. Kossmann noticed that with their many digressions on physiology and the passions, the theory of the brothers De la Court was "based on the latest psychology," that of Descartes's *Passions de l'âme*. (Pieter's son-in-law, Adriaan Heereboord, was the major defender of Descartes on the Leiden faculty.) They argued that a republic was the best form of government, yielding wealth and utility and fostering letters and sciences. They self-consciously sought to persuade the magistrates (*politici*) that they could establish this not on the theories of the schools but on experience and an analysis of the passions. In their view, the passions of individuals should be allowed to express themselves, and, providing that the political-economic system in which they operated was well ordered—rooted in the law of contracts—opposite passions would balance one another, yielding public harmony and tranquillity.[139] The De la Courts therefore held that "the public interest [is] the sum of individual interests" and that the true expression of the public interest was possible only in a democratically rooted and commercial republic.[140] They were not yet positing Adam Smith's hidden hand:

instead, the various self-interests of society had to be properly regulated through a fair legal system. But they saw that the prosperity of the state was based on setting individuals free to pursue their passions. As Pieter De la Court put it, "A good government is not where the subjects fare well or badly depending on the virtue or vice of the governors, but . . . where the fate of the governors necessarily depends on whether the governed fare well or badly."[141]

Benedictus Spinoza went even further. Although there is no direct evidence that he knew the De la Courts, their circles of acquaintances, at least, overlapped.[142] Born in Amsterdam in 1632 to one of the many Portuguese-Jewish families that had moved there to avoid persecution, Spinoza entered into the family business, mainly importing goods from the Mediterranean, in 1649. In the mid-1650s, the business suffered badly from the depredations of the English in the lead-up to the First Anglo-Dutch War. At the same time, he had been studying contemporary philosophy, including the work of Descartes. His views, which he expressed with a vehemence that deeply disturbed the Amsterdam rabbis, led to his expulsion from the Jewish community in 1656, which also meant the end of his life as a merchant. He had become a part of a group of freethinkers that included the former Jesuit Franciscus van den Enden, whose Latin school he attended. With support from his friends and a small income made from lens-grinding—at which he became extraordinarily skilled—Spinoza spent much of his time pursuing the implications of the new philosophy, publishing a work on Cartesianism in 1665, the *Tractatus theologico-politicus* in 1670, and the *Ethics* in 1675.[143]

There have been many clear and penetrating studies of Spinoza's philosophy, which became a touchstone for the so-called Radical Enlightenment.[144] For our purposes, it is simply important to note that it argued for an identity of mind and body in which an understanding of the passions was critical.[145] Like Thomas Hobbes and the Epicureans, Spinoza argued that "whatever is conducive to the common society of men, or, whatever brings it about that men live together in agreement, is useful, and, on the contrary, that is bad which induces discord in the state." He also adopted the view that "pleasure is not directly bad but good; but pain, on the contrary, is directly bad."[146] But he went further than even Descartes in arguing for the identity of mind and body in what he termed the *conatus*, "the striving, self-preserving and self-enhancing life-activity of a minded body." In his view, the passions played an essential part in the strivings of the conatus. Humans are creatures with needs "and of vital activity in the service of these needs. [Man's] whole psychic life is the mirror of this activity, or its double in consciousness. However, it is not that consciousness reflects these needs, but rather that consciousness *is* these needs themselves, under the attribute of thought."

Spinoza's is therefore a very strong identity theory, with an identity of "thought and emotion, thought and joy, thought and sorrow, thought and desire," so that "a change in the psychic character, or intensity, or quality of an emotion does not lead to a change in a bodily state; it is one."[147] With his identity theory of the passions, Spinoza also developed a republican political theory that had many similarities to the arguments of the De la Courts (perhaps even being deeply indebted to them),[148] stating that only in a republic could people live together harmoniously, keeping one another's bad qualities in check while acting on their passions.

It would be a mistake to think that the political materialism of the De la Courts, Spinoza, and others typified thinking in the Dutch Republic in the later seventeenth-century, since even the liberal regenten were sometimes offended and the Voetians were deeply angered by their views.[149] Moreover, the States Party collapsed in 1672, when Louis XIV invaded the country overland while the English waged war on the Dutch at sea, almost extinguishing the Republic. The orthodox Calvinists and the Orangists rose up against the Republican regenten and installed the young William III as head of the army and virtual head of state. The former leader of the Republic, De Witt, and his brother Cornelis, were brutally butchered during a riot in The Hague, strung up from hooks like hogs, with pieces of their bodies handed out or sold to members of the crowd. Spinoza, a great admirer of the De Witts, had to be locked in his room by his landlord to prevent him from posting a sign at the nearby scene of the incident (on which he had written *ultimi barbarorum*), fearing that Spinoza, too, would be torn apart.[150] After this horror, there is little evidence of any Dutch political treatises explicitly adopting the positive views of Spinoza or the De la Courts about allowing the passions freedom in a republic.[151]

But there are many hints of the persistence of these views, especially among the physicians: the Cartesian physician Cornelis Bontekoe, for instance, wrote that "during the union of soul and body . . . the soul is so greatly subject to the body that on the latter depends its wisdom and virtue."[152] For others, too—among them Caspar Barlaeus, René Descartes, the brothers De la Court, Spinoza—mind and body, God and Nature, commerce and wisdom, were part of a whole rather than separate. Early anatomists like Pauw had reminded their viewers that Adam and Eve had sinned by eating the fruit of the tree of the knowledge of good and evil, which had not only made us mortal but made us into conscious moral beings. A later generation took other lessons of the anatomists to heart, seeing the wonderful contrivances that made us—even our minds—as properly, perhaps entirely, bodily. Despite cruelty, fear, anger, and other powers that were so obviously destroying individual persons, nature was good, passions like love made life worth living, and prosperity resulted from allowing nature to take its rightful course. Individual persons might be debased coins, condemned to being melted down in hellish dissolution, but from their strivings the larger world grew wealthy, powerful, more pleasurable, and perhaps even healthier. In setting the knowledge of good and evil aside and acting as nature prompted, some people were hinting, the gates of paradise might be reopened during the period of mortal life, with health and prosperity in sight. It was a revolutionary view, perhaps as profound as Isaac Newton's later merger of earth and heaven into a single system of calculation. They would have understood perfectly their North American heirs a century later, whose rebellion against Britain was explained as a defense of life, liberty, and the pursuit of happiness.

7

Industry and Analysis

The conduct of life depends entirely on our senses, and since sight is the noblest and most comprehensive of the senses, inventions which serve to increase its power are undoubtedly among the most useful there can be.
—RENÉ DESCARTES, *The Optics*

However much speculation and theory excited argument, the development of knowledge in medicine and natural history depended on accurate description. But for both description and analysis, investigators also adopted information acquired from the latest technologies of production. The methods used by apothecaries, chemists, lens grinders, and others to produce objects were taken up opportunistically by naturalists. These methods sometimes also had unexpected consequences for the materials with which they worked, leading not only to more precise descriptive methods but to better tools of analysis. These methods in turn emerged from the activities of people engaged with the material world rather than from discourse and debate. The three examples below, from the 1650s through early 1670s, show how advances in anatomical knowledge came from new methods of preparing specimens, how a whole new world of life and form was discovered from the use of the single-lens microscope, and how new dietetic drinks such as tea were analyzed according to color indicators, which in turn emerged from the dyeing and bleaching industries. These projects all combined topics that are often separated by historians into different subjects, such as anatomy, chemistry, natural history, and microscopy. Yet not only do they show how intertwined investigative projects were, they also illustrate how commonly investigative techniques arose outside the walls of the universities. In the Dutch academy, enough investigators were open to natural knowledge from any source to quickly adopt new methods in furtherance of their projects. These in turn deeply affected conversations about the human body and nature, not only among scholars but among members of the broader public.

THE PRESERVATION OF BODIES

When it came to examining objects of natural history and anatomy, many signs of life quickly disappeared once something was collected. Once collected, plants and animals died and decayed, leaving only traces of their living form in the memory, and in the field one could observe carefully for a limited time. Memory could be improved, but it remained fallible. Drawing from life could preserve an image, but that was insufficient for allowing many subsequent investigations of the things depicted and required the viewer to trust in the artist. Apothecaries had developed various methods to preserve medicines as freshly as possible, mostly by placing them in tightly stoppered tin or ceramic jars. Chemical methods of preparation also yielded medicinals that kept their potency for long periods (unless they were in a volatile state), but the results fundamentally transformed the ingredients. Foods could be pickled in vinegar or immersed in sugar or salt, but this, too, changed their textures, appearances, tastes, and other attributes. Botany benefited from the invention of herbaria, in which dried specimens could be collected and studied at leisure, yet although this was satisfactory for most leaves and flowers, and for small, thin roots and stems, other parts of plants, especially fruits, could not be preserved this way, while colors and other attributes of the dried specimens were subject to fading.

Animal bodies could also be preserved to an extent by drying. Many insects, for instance, could be placed between pages of paper and dried. Many birds drawn "from life" for sixteenth-century illustrations were actually drawn from "mummified" specimens, in which the viscera were removed and the rest of the body was dried in an oven and sometimes salted—a method probably adapted from those used for the preparation of animal skins for clothing. By the 1620s, some people were using the method of removing the feathered skin and pulling it over an artificial body made of cloth, sometimes adding arsenic to the skin to keep down the subsequent pests. Although specimens prepared in these ways ordinarily did not last many years, the durability of feathers and their colors were sufficient to give the impression of a bird's appearance for some time, helping to generate an interest in ornithology in the sixteenth century. Some animal skins—most famously, those of crocodiles—could also be dried and prepared to look like the animal, and of course harder parts like hair, bone, teeth, and horn, and concretions formed in the body (such as bladder stones), survived well and could be collected.[1] Until the mid-seventeenth century, then, almost all collect-

The cabinet of natural history of Ole Worm: all dry specimens. Title page of Ole Worm, *Musei Wormiani historia* (1655). Courtesy of the Wellcome Library, London

ible examples of naturalia required desiccation to preserve them over time. Such methods were, however, useless for anyone who wished to examine the anatomy of plants or animals, which continued to require rapid investigation before decay set in.

One of the most frequently kept kinds of animal bodies found in the curiosity cabinets of the late sixteenth and early seventeenth centuries were mummified remains from Egypt; they were also considered to have powerful healing properties. Archaic Egyptian civilization held enormous interest for contemporary Europeans,[2] but so did their mummies, which were not simply dried but prepared and hence remained more lifelike in appearance. Given that the brain and viscera were extracted and the remaining skin and muscle were hardened by the preservative process, Egyptian mummies had little to offer the anatomist.[3] But they held useful powers for medicine. According to Karl Dannenfeldt, over many centuries embalmed or desiccated bodies acquired medicinal properties originally associated with bituminous products. In antiquity, the precious seepage of black rock-asphalt or pissasphalt from a mountain in Persia, locally called "mumiya," became particularly well regarded as a remedy. (The Christmas story, of course, involves magi from the East bearing myrrh and frankincense, two resins from "Arabia" that were also very valuable in medicine.) By the thirteenth century, however, the resinous, aromatic substance exuded from bodies found in Egyptian tombs was considered to be very similar. Because asphalt, along with myrrh and aloes, was said to be used by the Egyptians for embalming their dead, the true *mumia* could be found in the cavities of the head and body in the "mummies." It did not take long for the embalmed flesh itself—and even the wrappings—to be associated with the precious resin. Antonius Musa Brasavola's book of 1537 on medicinal simples defined "mumia as the remains of an embalmed body and the same as bitumen judiacum."[4] European demand for mummies became so high as a result that the Egyptian government outlawed the export of mummies, although a large contraband trade in both true and counterfeited mummy continued through the early modern period.[5]

By the sixteenth century, Europeans—especially those inclined toward iatrochemistry—considered mumia to be one of their most powerful medicines. Paracelsus described mumia as a force in living tissue that defended against invading "seeds" (*semina*) of disease. The textbook on chemistry by Joachim Tanckius, professor at Leipzig, simply stated that "mumia is the arcanum and secret of the microcosm."[6] The power of mumia could also be extracted from the flesh of the recently deceased. Oswald Croll believed that the best tincture of mumia was prepared from the flesh of a "red-haired man twenty-four years old, who had been hanged, broken on the wheel, or thrust-through, exposed to the air for a day and a night, then cut into small pieces or slices, sprinkled with a little powder of myrrh and aloes, soaked in spirits of wine, dried, soaked again, and dried." From this could be extracted a red tincture, "a quintessence, which could be used for cures of pestilence, venin, and pleurisy."[7] Andreas Tentzel's *Medicina diastatica* (1629) "was primarily devoted to mumia, of which he enlarged the scope and definition" by placing its potency in the life spirit that left the body upon death, proposing the "extraction of the mumia of the aerial body by interception of the dying breath."[8] It is no surprise, then, that Dutch physicians also set out recipes for making mumia.[9]

Recipes for making mumia were often included alongside receipts for embalming bodies, from which new techniques were developed that led to some breakthroughs in anatomy. By the late Middle Ages, it was customary to embalm the bodies of the wealthy and powerful, perhaps in imitation of the Egyptians, although more likely in imitation of the saintly. A body subject to no special care that did not decay after death might be the sign of a miracle, and therefore a sign that someone thought to be particularly holy in life had indeed become associated with the divine. Saintly bodies were more than merely dried: indeed, one of the attributes that investigators looked for in such cases was palpability, where parts of the body, such as a thigh, could be pressed by the fingers but then return to their original shape.[10] In other words, the miraculous body had to be lifelike in death. Similarly, perhaps it was to demonstrate their personal virtues that medieval royalty insisted on being embalmed after death. The physicians and apothecaries who did the work used some of the same methods they thought the Egyptians had used, employing myrrh, aloes, and other resins. As one commentator explained: in order to thwart the usual course of putrefaction, the Egyptians disemboweled the dead and repeatedly steeped them in bitumen and stuffed them with precious aromatics.[11] (Aromatic and oily resins were often placed under the rubric of "balsam," or in English, "balm," hence "embalm.") In the sixteenth century, for instance, the famous French royal surgeon Ambrose Paré used a recipe that was much imitated. The inner organs were removed from the body and deep gashes were cut into the limbs to drain the blood. The body was then washed with a sponge dipped in a solution of vinegar boiled with wormwood, aloes, coloquintida, salt, and alum (all known as preservatives). Various spices were then inserted into the gashes in the body and the abdomen, and the body was sewn up. It was then covered with "Turpentine melted with Oil of Chamomel and Roses; adding, if you please, some Chymical aromatick Oils," after which it was wrapped in cere-cloth and placed in a lead coffin well soldered and (again) filled with aromatic herbs.[12] For the embalming of Prince William of Orange (supervised by Forestus), many additional aromatics (supplied from Cluyt's shop in Delft) were used to wash out his emptied abdomen—including aloes, myrrh, absinthe, rosemary, the balsam herb, mint, salvia, lavender, marjoram, thyme, cumin, cloves, and nutmeg—following which his outer body was drenched in a solution of oil of lavender and oil of turpentine and smeared with a compound of wax, turpentine, and myrrh, before being covered with a cloth soaked through with turpentine.[13]

But a new and spectacular embalming process was developed in the mid-seventeenth that not only caused a public sensation but had unintended consequences for the study of human anatomy. The method was developed by Louis de Bils. It had some methods in common with Paré's and Forestus's, such as the use of oil of turpentine, aromatics (or balms), and a lead container. But De Bils found ways to preserve complete bodies and body parts in a lifelike state without the need to remove the inner organs or to drain the blood via deep incisions.

De Bils was neither a physician nor surgeon but lord of Coppensdamme and Bonem, both modest fiefs located in Flanders.[14] According to a later report of the French savant Samuel de Sorbière, De Bils (born about 1624) had begun dissecting at the age of thirteen when living in Rouen, and afterwards in Flanders and Rotterdam. Why he developed his interest and how he obtained bodies to dissect are unknown. His father and brothers were merchants, and he seems not to have had a classical education. Yet in 1646–47, when De Bils was living in Amsterdam with his young family, he knew two of the most important medics of the city, the surgeon Paul Barbette and the physician-chemist François dele Boë Sylvius, both of whom had keen interests in the new anatomy; a few years thereafter he had taken up residence in Sluis, a flourishing port in Dutch Zeeland not far from Middelburg, where he must have continued his anatomical studies. The next we know, in 1651 he gave the university of Leiden a number of preparations made at great expense, which the new professor of anatomy Joannes van Horne acknowledged with a written testimonial. Among them was a particularly remarkable specimen, described by Van Horne as "a dried human cadaver that appears to be freshly dead, the most worthy work" for display in the anatomical theater.[15] One can see how such a specimen crossed customary boundaries by Van Horne's juxtaposition of the words "dried" and "appears to be freshly dead": he did not quite know how to describe it. Shortly thereafter, Van Horne saw in The Hague another body "balsamed" by De Bils (*balsemen* remains the Dutch verb for the English "embalming"), in which the sinews and plump muscles appeared in the body as if it were alive. Apparently working alone, De Bils had found means to prepare human bodies so that they seemed lifelike rather than desiccated. His secret process was a fantastic new art.

De Bils was clearly experimenting with various expensive oils and resins such as myrrh when he engaged in "balsaming." Given the high prices of the imported balms, his experimental costs must have been enormous. According to what he wrote down in 1664, the process De Bils had developed was as follows: A tin box (*tinne kiste*) eight feet long by two and a half feet wide by three feet high was placed in a wooden box trimmed and caulked so as to let in no light and fixed with iron bands; in the top of the lid of the wooden box a trapdoor (*schuyve*) was cut that could be opened and completely sealed. The tin box would also be covered at the appropriate time with double wool blankets so that no light could enter. Into the tin box was introduced sixty pints of the very best rum, freshly made; fifty pints of Roman alum very finely ground; fifty pints of pepper very finely ground; one sack of salt finely ground, which must be poured in at this point; two hundred large glasses (*stoop*) of the very best Nantes brandy; and one hundred large glasses of the very best wine vinegar, all of which were well mixed in the tin box as quickly as possible so as not to let the power of the mixture get lost (*opdat de kracht niet te veel en verlighe van ditto substantie*). Twenty pounds of finely ground myrrh of the best kind and twenty pounds of the best finely ground aloes could also be added to the mixture. The corpse, wound about with a white linen sheet, was immediately dunked in this mixture, lying on and tied to a wooden platform (*stellinghe*) so that at least two feet of fluid covered the body. The boxes were closed for thirty days, except that three days after the body was put in the fluid, the mixture was well stirred, as it was twice more during the thirty-day period. Each time the fluid was stirred, the body was also taken out, unwrapped, washed in fresh brandy, flipped over to drain out any moisture via the mouth (being careful not to damage the hair or finger- and toenails), rewrapped in sheets, and replaced. After thirty days, the body was transferred to another box made like the first with a mixture of rum, pepper, alum, salt, brandy, and vinegar in the previous proportions, in which it was left for sixty days (with three stirrings and turnings). The above mixtures were for kings or others whose bodies were to be displayed in public. If this was not to be the case, the rum and alum could be left out of the first mixture and the spices had to be added, and in the second mixture no salt was added, nor rum or alum. Between the second and third soaking, the body was allowed to dry. The first box was in the meantime cleaned and filled with a third mixture, which excluded the rum, alum, and salt but included the myrrh and aloes; this mixture was stirred several times and the clear liquid that came to the surface was skimmed off. Then forty-four pounds of aloes, forty-four pounds of myrrh, twenty pounds of mace, twenty pounds of cloves, twenty pounds of cinnamon, twenty pounds of nutmeg (all of the best kind, finely ground), one quarter-pound of ambergris, one half-pound of black balsam, with one quarter-pound of oil of cinnamon were all blended and applied several times to the exterior of the body and allowed to dry. The body now rested in the third mixture for two months, being turned over periodically as before, and being washed and rinsed with the clear liquid skimmed off previously. If after all this the body fat had not completely dried up, the body would be placed in a small, tight stone room with two ovens burning low, one of which burned two pounds of mastix. After the body was thoroughly dried, the mixture of ambergris and other resins was again applied to the body. Once prepared, the specimen could be best kept in a tin box that let in no air.[16] Performed properly, this method turned mortal human flesh into a specimen that appeared as plump and fresh as in life, dead yet not subject to decay. It could be dissected and displayed to reveal any feature of human anatomy.

De Bils's personal affairs suffered badly in the early 1650s, yet he kept up his investigations. Two medical friends in Sluis, Drs. Abraham Parent and Laurens Jordaen, both of whom had studied at Padua, helped De Bils with his anatomical work and jointly published a pamphlet on De Bils's investigations on the anatomy of the inner ear. They both moved away in the mid-1650s, however, lessening De Bils's opportunities for anatomical study.[17] After the death of his father, De

Bils and his brothers, merchants in Rouen, became embroiled in various lawsuits against one another about the inheritance. Although he obtained the office of bailiff of Aardenburg, the pay was slight, and he seems not to have invested much energy in the position.[18] By 1657, De Bils was searching for new means of support. A physician in Brugges (not far from Sluis), Burchardus Wittenberg, wrote a short treatise praising De Bils's achievements highly and calling on a prince to support him, so that his work was not paid for out of his own pocket. Through an intermediary, De Bills tried to interest Professor Van Horne in working with him, but Van Horne balked, perhaps because of the probable expense. De Bils did finally get the financial support of a physician from Middelburg for his research and publication on the lymphatics. But this publication hit a raw nerve in Van Horne, who expressed complete surprise at De Bils's work. Van Horne quickly turned out a Latin translation of the book, though criticizing it at the same time.[19] But according to the historian G. A. Lindeboom, Van Horne "now applied himself to the making of fine anatomical preparations"[20]—a matter to which we will return in a moment.

Given his successes, the States General of the Dutch Republic issued an order on 9 August 1658 for the public provision of bodies to De Bils, while new translations into Dutch of anatomical works by Thomas Bartolinus and Paul Barbette allowed De Bils, who did not have good Latin, to study further. His friend Parent also published a notice again urging support for De Bils's work, which was so costly—especially the balsams. The city of Rotterdam, where De Bils had moved in order to be with Parent, set up an anatomical theater over the former English merchants' courthouse. De Bils used it for further studies on his secret method of dissecting and embalming and for the display of at least four dissected and embalmed cadavers.[21] A "sovereign power" (later rumored to be the king of Spain) did try to get him to sell his secret several times, but he refused in favor of setting up his show in Rotterdam, for which he charged an admission of one rijksdollar (two gilders). Despite the high entry fee, his display was heavily attended, by physicians and students as well as by the public, from ordinary people to ambassadors and princes. He also held public anatomical demonstrations. His special technique was to dissect bodies specially prepared so that no blood or other moisture was lost from them. For such theatrical presentations he charged even more.[22] As one later report had it, he had "found out a method of Anatomizing Bodies without effusion of Blood, after a new and unheard of manner, with an accurate separation, and manifest demonstration, of all the parts and little Vessels, thô they were less than Hairs."[23] An English visitor, the young Robert Boyle, was so enthusiastic about what he saw that he ventured into print for the first time in order to report on the events in Rotterdam and to urge the English government to purchase or to support work replicating De Bils's secret.[24]

But soon the academics turned against De Bils. After initially supporting him by encouraging students and others to attend his displays and demonstrations, Van Horne lashed out in writing: De Bils had become proud, telling the world that students learned more from him in half an hour than from Van Horne in two years. Van Horne decided that De Bils was a pretender, with neither learning nor proper behavior. The Amsterdam surgeon Barbette also turned against De Bils.[25] He, too, underlined De Bils's lack of academic education: "Philosophy, chemistry, astronomy, medicine, and daily practice" were absolutely necessary for understanding the workings of the body, but for two years the unlettered De Bils had pretended to be the great master overturning established learning. De Bils replied to these two sallies with his own pamphlet of March 1660, in which he ascribed Van Horne's and Barbette's criticisms to jealousy. In attacking them, however, he also attacked, and further alienated, learned physicians generally. The debate continued throughout De Bils's life (he died in 1669), for he had some supporters as well as antagonists in the Dutch Republic. But his one achievement that continued to be praised even by his strongest opponents was his balsaming of cadavers.[26]

De Bils promised to reveal the secret of his process for 120,000 gilders, although he was willing to part with two balsamed cadavers to one Duke Christiaan for a mere 16,000 gilders. At the end of 1661, it was rumored that he had sold his secret to a nobleman. This may well have been prompted by the attempts of Luis de Benavides Carillo of Toledo, Marques de Caracena, a follower of Don Jan of Austria, Stadholder of the Southern Netherlands, to purchase De Bils's collections for the university of Louvain. After inspection of some of De Bils's cadavers in November 1662, Gerard van Gutschoven, professor of mathematics and anatomy at Louvain, became quite enthusiastic about the possibility of obtaining De Bils's specimens. Don Jan therefore proposed to the States of Brabant that they purchase De Bils's cadavers and his secret method. By June 1663, eighteen articles had been drawn up by which De Bils agreed to provide Louvain with five cadavers and all his knowledge, including his secret embalming process. The method would be kept top secret: it would be written in duplicate in Latin, with a Dutch version for De Bils; the two Latin versions would be deposited in two strongboxes kept at two locations each sealed by two keys, one key to be held by the States of Brabant and the other by the professors of Louvain. Various other provisions ensured De Bils's oath that he had not revealed and would not otherwise report the secret to anyone else. In return, the states promised a payment of 22,000 Rijnsche gilders and a professorship salaried at 2,000 gilders per year, which would revert to his son after his death. De Bils would also establish an anatomy theater in Louvain and demonstrate there without charging admission. By October some changes were made to this draft contract, and Van Gutschoven began to learn De Bils's secret under his tutelage. At the same time, as word of the arrangement got out, powerful people began to insist in their wills that their bodies be embalmed according to De Bils's method.[27]

Finally, on 16 April 1664, De Bils's secret was handed over in writing and shown to Van Gutschoven, who was allowed eleven minutes in private to read it, after which he stated that he understood the methods of bloodless dissection and balsaming of bodies. By May, the five bodies De Bils owed the Louvain faculty were in hand, and the states paid out the 22,000 gilders. There was as yet no place prepared for the cadavers, and so they were placed in a basement of the library; after four hot months, they were laid out on tables under a roof with holes in it, which not only allowed rain and snow to damage books in the library but by 1666 caused signs of rot to appear in three of the five cadavers. This later became known to De Bils's opponents, who claimed it proved him a fraud. Yet despite continued financial difficulties, De Bils remained well regarded in Louvain and the southern netherlands: in early 1669 Flanders awarded him the benefice of a canon of 's Hertogenbosch and Sint Oedenrode and made him an honorary professor of anatomy at the Illustrious School. Also that year, several public demonstrations of his method were to be undertaken in the northern netherlands with the assistance of Tobias Andreae—but De Bils sickened and died.[28] A decade later, an English surgeon praised him as "the greatest Master of Ambalming in our Age." But rumor followed his name. He was reputed to have obtained a vast "Sum of Money" for his methods from the States of Brabant but was never allowed to take up the professorship at Louvain "upon this pretence, that *Bilsius* being of the Reformed Religion, it might in so doing lay itself obnoxious to an Excommunication from the see of *Rome*, according to the Statutes. There was great bickering about the matter, till at last *Bilsius*, loden with his Money, bid *Lovain* farewel, and returned to his United Provinces, where (as far as I can learn,) his famous Secrets were buried with him."[29]

NEW METHODS IN ANATOMY

Because De Bils had kept his method secret others had to guess at his means and experiment with possibilities themselves, resulting in a wave of new findings. In March 1661 it was rumored that a Dr. Hubertus of Leiden had discovered some of De Bils's secrets, and later in the year a story was circulating that

De Bils had sold his secret to a nobleman who passed it on to one Burrhus in Leiden—although nothing more is known.[30] Another student at Leiden, Theodorus Kerckring (later known as an astute anatomist and chemical physician), "is said to have invented" a means of "preserving dead bodies by covering them with varnish" or, in another version, by performing "experiments with liquefied amber to preserve corpses."[31] Another medical student, Gabriel Clauder, from Leipzig, was making a grand tour of Europe and England in 1660 and 1661 when he visited De Bils's exhibition and "applied his moistened finger to one of the bodies, and carrying it to his lips recognized the taste of salts. He started from this fact to attempt numerous researches, and succeeded in forming different compounds" of salt to perform the same function.[32]

But it was the Leiden professor of surgery, Van Horne, his colleague Sylvius, and their students, who developed their own methods in imitation of De Bils and thereby made a series of notable anatomical discoveries. The most remarkable results were obtained by Jan Swammerdam, who matriculated in medicine at Leiden in 1661; he quickly became one of the favorite pupils of Van Horne and Sylvius. One of the finest natural historical investigators of the seventeenth century, Jan Swammerdam was born in 1637 of a well-to-do family of Amsterdam.[33] His apothecary father was assembling a remarkable cabinet of curiosities, and as a youth Jan himself started collecting insects, by 1669 having acquired over twelve hundred specimens, a collection that eventually grew to three thousand.[34] Although his father intended him for the ministry, the younger Swammerdam was finally allowed to matriculate in the medical faculty at Leiden in 1661 (at the relatively late age of twenty-four), where he threw himself into the work of dissection and related studies with vigor. The findings made from the tremendously fine skill and ingenious investigative techniques he developed with some of his fellow students, such as Nicolas Steno and Regnier de Graaf, would amaze the savants of Europe. Working with Van Horne, for example, Swammerdam carried out pioneering work on muscles and respiration.[35] He vivisected and anatomized dogs and frogs at his rooms or occasionally at the home of Van Horne, who apparently underwrote the research expenses of his best students. His work on the contraction of muscles—showing that the Galenic idea that they increase in volume when contracted was wrong—was carried out with experimental skill and ingenuity. After a trip to France in 1664–65, where he gained a lifelong friend in Melchisédech Thévenot, who sponsored an active scientific academy, Swammerdam returned to Amsterdam and then to Leiden, where he took his medical doctorate in 1667 by defending a thesis on respiration that remains impressive. It showed him to be a physiological mechanist as well as an innovative searcher into nature's secrets.

But Swammerdam was also developing new methods for preserving animal bodies. After finishing his studies in Leiden he returned to Amsterdam, bringing with him an embalming method that seems to have been common to the Leiden investigators. Years later, when writing an introduction to a collection of Swammerdam's works (*Bybel der natuure/Biblia naturae*), Herman Boerhaave commented that "having gone through his courses [in medicine] with the most sudden and unexpected success, he immediately began to consider how the parts of the body prepared by dissection, could be preserved and kept in constant order and readiness for anatomical demonstrations; as such a discovery would free him not only from the trouble of repeated dissections, but likewise from the difficulty of obtaining fresh subjects, and the disagreeable necessity of inspecting such as were already putrefied."[36] But this is clearly too simple. As we have seen, De Bils had already become famous for his embalmed specimens, Van Horne was both impressed and angered by De Bils, and by at least 1661, when Swammerdam matriculated at Leiden, rumors were circulating that De Bils's method had been cracked. Around 1661–62, Swammerdam made a durable preparation of the thoracic duct, discovered by Van Horne in 1652.[37] More important, he seems to have worked out a new method of embalming. According to Justus Schrader —a slightly younger student of Van Horne's—Swammerdam's technique was to prepare a tin vessel large enough to receive an organ or even a small body. Into this was set a grate or screen resting two fingers' width above the bottom, on which the body was placed. Then oil of turpentine was poured in to a height of three fingers' breadth from the bottom. The vessel was covered tightly, leaving a small opening, and set aside for time to do its work. The very penetrating oil of turpentine entered the pores and replaced the fluids that caused fermentation and decay, which due to their weight descended through the screen to the bottom of the vessel while at the same time the volatile oils evaporated through the small opening in the top, leaving the specimen coated throughout with the hardened oil, which prevented it from decay. Different organs required longer or lesser times: an embryo took six months, a skeleton about two, the parenchymia of the heart three, a liver and a placenta one, a spleen ten days, and intestines a month.[38] A few other techniques helped prepare more complicated specimens.

It seems, then, that the investigators in Leiden had discovered oil of turpentine to be the critical ingredient. It was, after all, the chief ingredient in making generic balsam ("an aromatic oily or resinous medicinal preparation . . . of various substances dissolved in oil of turpentine").[39] The so-called turpentine commonly in use today, a product of fir and pine trees, has little relation to the substance called turpentine, or terebinth, in the seventeenth century. At that time the word still applied only to an exudation of the terebinth tree (now called *Pis-*

tacia terebinthus, or Chian turpentine). As John Goodyer explained in his 1655 edition of Dioscorides, the tree grew in "Arabia Petraea" as well as "Judea and in Syria & in Cyprus, & in Africa, & in the Islands called Cyclades." He also noted that "the Resina Terebinthina doth surpass all other rosins."[40] This is confirmed by the reports by one of the major sixteenth-century investigators, Antonio Musa Brasavola, who reported that "true terebinth was now imported in round lumps from Cyprus to Venice." Symphorien Champier, however, noted that "larch-tree resin was sold for terebinth."[41] In the later seventeenth century, the English military surgeon James Yonge further warned that "there is a base Turpentine-like substance commonly called Terebinth, brought from France, drawn from the Fir and other Trees, . . . which is no more the gum of the Turpentine tree, than Tar is." The proper oil or "spirit" of turpentine ("they being names promiscuously given to one and the same kinds of thing") was obtained after a slow distillation of the resin of the Terebinthus tree in a retort, which produced first a white, then a yellow, and finally a red oil, the last of which was the best. This oil of true turpentine, Yonge explained, "contain[s] in it the Balsam."[42]

With a few elaborations, the method developed by Swammerdam continued to be taught at the university of Leiden. For instance, Carel Maets (or De Maets, Dematius), who had been teaching experimental chemistry at Leiden since 1669, explained his private method of preserving bodies from at least 1674.[43] He elaborated the method in his *Chemia rationalis* of 1687: "After first removing the intestines, viscera, brain, and all other soft parts, it is then placed in a lead coffin [*cysta*] commodious enough for it, where it is soaked in clear oil of turpentine. After fourteen days, or when the oil has well penetrated all the parts of the muscles, remove it and wash it with spirits of wine, and put it in a place where it will dry." To preserve the soft tissues, they were first inflated and injected with lukewarm water so as to evacuate all the blood; then they were washed out with spirit of wine until no trace of blood remained, after which they were dried in appropriate shape and soaked in oil of turpentine.[44] Another former Leiden student, Stephen Blankaart, also wrote about the use of oil of turpentine for balsaming bodies.[45]

Experiments with oil of turpentine seem also to have led to developments across the Channel for using strong alcohol to preserve specimens. As one historian of chemistry noted, "Oil of turpentine was regarded as very similar to spirit of wine."[46] At a meeting of the Royal Society of London on 10 September of 1662, Robert Boyle—who had been so impressed by De Bils's methods—showed the society "a puppy in a certain liquor, in which it had been preserved during all the hot months of the summer, though in a broken and unsealed glass." It has sometimes been assumed that the liquor was alcohol, for a few months earlier the physician William Croone "produced two embryos of puppy-dogs, which he had kept eight days, and were put in spirit of wine in a glass-vial sealed hermetically."[47] One historian remarked about this that "it seems certain that Croone got the idea from Boyle, who was experimenting with alcohol before 1662, and who is quoted by Grew as the inventor of the method."[48] In 1663, Boyle's specimens of "a linnet and a little snake, preserved already four months, entrails and all, without any change in colour, in some spirit of wine," were to be found in the Royal Society's repository, although it is clear from his work diaries that Boyle was concerned about the shriveling and change of color of caterpillars and other specimens he tried to preserve in spirit of wine alone.[49] But Boyle seems to have obtained his best results by mixing spirit of turpentine and spirit of wine, for at a meeting of the Royal Society late in 1663, "Mr. Boyle suggested, that the oil of turpentine or spirit of wine is good to keep birds, and the first better than the second for that purpose." Two and a half months later he "observed, that he had a liquor compounded of spirit of wine and a little oil of turpentine, whereby . . . the bodies of animals or the parts thereof might be preserved. . . . He offered to preserve a hand and a larynx." And a week later he "presented the Society with a little bird, preserved for several months in oil of turpentine"; at a meeting yet a month later he suggested "that seeing [as] animals of remote parts have particular and considerable inward contrivances, some liquor, as spirit of turpentine, might be thought upon, and sent abroad for the preservation of internal parts, at least in smaller animals."[50] Indeed, one of the experiments most often repeated by Boyle was the action of oil of vitriol (sulfuric acid) distilled in a retort with turpentine, which yielded sulfur.[51] Such methods are spread more widely through the *Philosophical Transactions*. Moreover, the Italian anatomist Marcello Malpighi devised a method using fixatives and stains to view the blastoderm of a chick's egg, presumably learning "about the use of alcohol for the preservation of embryos from the letter which Boyle had published on 7 May 1666" in the *Philosophical Transactions*.[52] Aside from the expense and the tedium of periodically renewing the liquid, and the slightly imperfect results, the simplicity of suspending specimens in spirit of wine in a glass container made it an important new process. As one can tell from some of the Dutch methods, too, spirit of wine was also used with oil of turpentine to produce preserved specimens that could be handled.

The use of oil of turpentine and other materials for the preservation of anatomical specimens was rich in unintended consequences, however. The preservative techniques themselves allowed new matters to be discerned. Properly prepared with the oil, the tissues of bodies and organs remained supple but were toughened. Many vessels in the body are very thin and in their natural states are easily ruptured when examined. But after the vessels were treated with oil of turpentine, one could carefully insert other substances into them. Malpighi appears to have been the first to inject vessels with mercury in 1661 in order to see the tiny ramifications of their branches. But after treating with oil of turpentine, Swammerdam and his fellows were also able to inflate vessels with air, colored wax, and other substances. By these methods, they could examine the structure of the lung, the follicles of the human uterus, the ramifications of the vessels of the placenta, and so on.[53] For instance, on 21 January 1667, working with Van Horne on the human uterus, Swammerdam found the means to inject it with wax—a technique he developed, together with injections of air—filling out vessels that could not otherwise be discerned.[54] The work he did with Regnier de Graaf under Van Horne in the later 1660s established that what had been considered to be female "testicles" were ovaries containing eggs—which he considered to be more important in generation than the "liveliness" conveyed to the eggs by sperm.[55] In 1671 he published a fine engraving, dedicated to Dr. Tulp, that displayed details of the human uterus and followed this by publishing a written treatise together with the engraving dedicated to the Royal Society of London (to which De Graaf appealed for a decision on their priority dispute over the female organs of generation).[56]

Similarly, while living in rooms at his father's house rather than making a living by medical practice, and devoting virtually all his time to further investigations, Swammerdam expanded his own cabinet of naturalia, the centerpieces of which were a preserved child of one month of age and a lamb. His cabinet also contained a preparation of the lungs in which the trachea was filled with white wax even to the tiniest parts, the pulmonary artery with red wax, the pulmonary vein with rose wax, and the small orifices of the arteria bronchialis with a fire-red substance; he showed a liver similarly differentiated in balsam and wax.[57] (He also developed other techniques, for example showing that the human spinal marrow was composed of fibrous nerves by plunging the yet warm spinal vertebrae in cold water, leaving them there for twenty-four hours, and then carefully breaking off the bone to expose the marrow—which again had turned from an undifferentiated mass into tissues.)[58] What had begun as an attempt to preserve bodies from the process of decay had developed into a range of experimental techniques crucial to gaining anatomical knowledge.

It was Frederick Ruysch, a fellow student of Swammerdam's, who developed the Leiden methods to the highest pitch. Ruysch became perhaps the most innovative anatomist of the late seventeenth century. Trained as an apothecary, then matriculating at Leiden in 1664, then settling into medical practice in The Hague, in 1665 he published a work announcing his discovery of the valves of the lymphatic system (*Dilucidatio valvularum in vasis lymphaticis et lacteis*), which

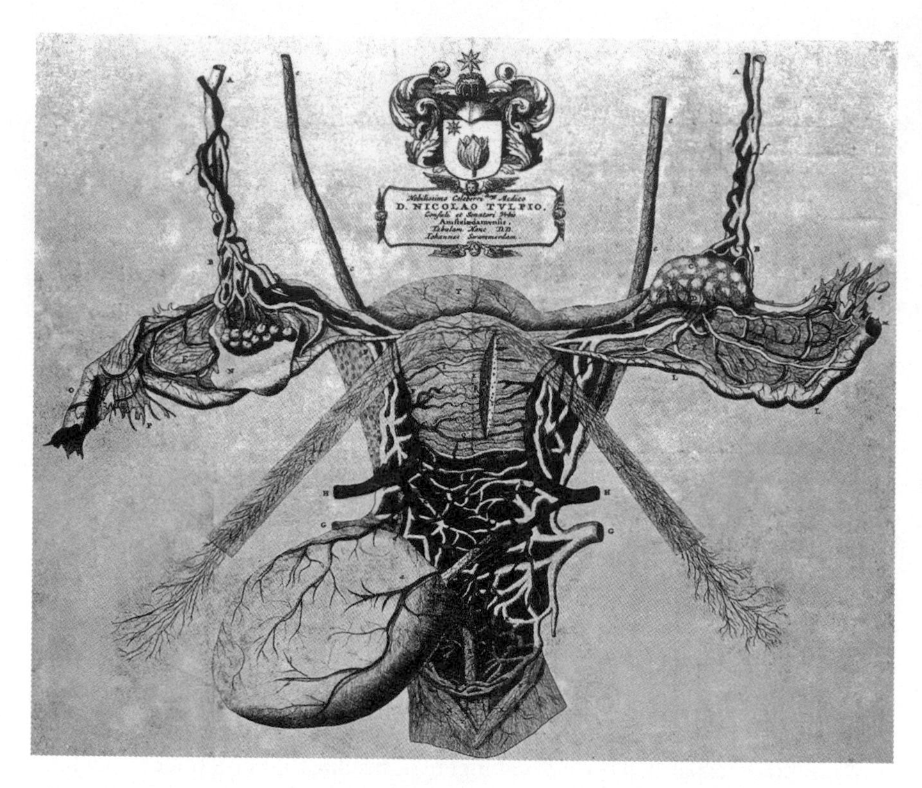

Female reproductive organs as depicted by Jan Swammerdam, published under the patronage of Tulp. From *Miraculum naturae sive uteri muliebris fabrica* (1672). Courtesy of the Wellcome Library, London

Swammerdam had also recently discovered but against which De Bils himself argued, leading to a heated pamphlet war. In 1666 he obtained the position of *praelector anatomiae* to the surgeons' guild in Amsterdam (following Joannes Deyman, who had succeeded Tulp) and soon became the anatomical examiner, and then instructor, for the Amsterdam midwives. In 1685 he also obtained the position of professor of botany in the Athenaeum and head of the botanical garden. But he became most famous for his cabinet. It was an extraordinary sight, full of strange fish and organs in bottles, and embalmed and preserved specimens, including humans, in lifelike poses and dress. The centerpieces of his displays were his thesaurii: dioramas of tiny human skeletons in poses memorializing the fleeting world of time by (for instance) playing violins made from hardened body parts, all standing among woods made from hardened arteries and veins and rocks made from bladder stones. Even as they drew the viewer's attention to mortality, the specimens represented permanence in the face of the forces of decay. When

One of Ruysch's "thesauri," made from fetal skeletons, kidney, gall, and bladder stones, and hardened body parts. A central figure looks toward heaven (the accompanying text says, singing "Ah fate, bitter fate") while accompanying itself on the violin; in front is a tiny skeleton holding a mayfly (the "one-day fly" made famous by Swammerdam); other figures also signify the brevity of life. Frederik Ruysch, *Thesaurus anatomicus* (1703). Courtesy of the Wellcome Library, London

Czar Peter visited Ruysch's cabinet, an embalmed baby lying in a cradle with glass eyes looked so lifelike and peaceful that he bent down to kiss the child. In 1717 the czar purchased this emblem of the new science, along with the rest of the specimens in the cabinet, for thirty thousand gilders—parts of the collection remain today in Saint Petersburg. (Ruysch had sold his collections before, afterward laboriously building up new ones.)[59]

The new cabinet of natural history: note the large cabinet with glass doors at the back of the room containing glass bottles with wet specimens; in front are other bottles, thesauri, and framed dry specimens, all arranged artistically. Title page of Frederik Ruysch, *Opera omnia* (1720). Courtesy of the Wellcome Library, London

If much of this work was the unexpected result of the public excitement about new embalming methods developed by a minor nobleman, other work was also furthered by well-to-do people outside the academy. It was, for instance, apparently the Amsterdam burgomaster Johannes Hudde who suggested to Swammerdam that he could investigate the vessels of the body more exactly by injecting different colors of wax into different veins and arteries.[60] This is yet another example of how keen some members of the regenten were to support his kind of investigations into nature.[61] Hudde came from a prominent commercial family in Amsterdam, his father having served as one of the Amsterdam members of the VOC's governing body, the Heren XVII. As a young man, Hudde attended lectures on mathematics at Leiden by Frans van Schooten, Jr., one of the best mathematicians of his day. Van Schooten's collected edition (1646) of the works of the French founder of modern algebra, François Viète, made Viète famous, and it was also Van Schooten who edited and translated into Latin Descartes's *Geometry* (1649), which he did working with Descartes himself. At least four sons of the regenten who studied with Van Schooten were mathematically capable enough to be mentioned by him in his *Exercitationes mathematicae* ("Mathematical exercises," 1657), including the famous Christiaan Huygens (son of Constantijn Huygens), Hendrik van Heuraet, Johan de Witt (the future grand pensionary of Holland), and Hudde.[62] These former students later kept up a steady correspondence about mathematics, being particularly interested in probability theory (especially actuarial tables), because of the dependence of Dutch public finance on lotteries and annuities: it was to Hudde that De Witt set out the important conclusions he reached that were later published as *Waedye van lyf-renten naer proportie van losrentien* ("The worth of life annuities compared to redemption bonds," 1671).[63] After he decided to devote himself to public service in the mid-1650s, Hudde made his way onto the Amsterdam vroedschap, on which he served for many years and from which he was chosen as a burgomaster an astonishing eighteen times before his death in 1704, being associated with the moderates. But he remained interested in the new science and its uses: for instance, he studied the philosophy of Spinoza with Spinoza himself, in 1670 used his influence to secure a teaching post in philosophy at Leiden for the Cartesian physician Burchardus de Volder (the first in The Netherlands to teach natural philosophy through the use of experimental demonstrations), and with Christiaan Huygens undertook a review for the States of Holland of the lower Rhijne (*Nederrijn*) and IJssel Rivers in order to plan how to prevent them from silting up.[64] Several important books by Amsterdam investigators were dedicated to Hudde, among them Gerard Blasius's important history of discoveries in anatomy.[65]

In his dedication, Blasius also mentions another powerful regent of Amsterdam and another friend of Swammerdam's, Coenraad Van Beuningen, one of the most important Dutch politicians of the later seventeenth century. Although as a member of the Amsterdam vroedschap he later worked with Gillis Valckenier to develop a middle-of-the-road group between the Calvinist-Orangist and the libertine factions, his association with the Rijnsburger Collegianten suggests that Van Beuningen was a bit of a freethinker himself, and he clearly took an interest in the new philosophy.[66] He apparently met Swammerdam in the spring of

1664, when Swammerdam traveled to France for study. A fellow medical student from Leiden who had also been pursuing fundamental work on anatomy under Sylvius and Van Horne, Niels Stensen (Steno), introduced Swammerdam to the private academy for the study of nature organized by Melchisédech Thévenot. At the time, Van Beuningen held the post of Dutch ambassador in Paris and was friendly with Thévenot, so the ambassador had occasion to be introduced to Swammerdam. Both Thévenot and Van Beuningen were impressed by Swammerdam's abilities and remained firm supporters of his work for the rest of his life.

With such support and his father's reluctant indulgence (allowing him to live at home on a small allowance), Swammerdam was able to continue his intensive anatomical investigations outside the academic world per se. He did travel to Leiden from time to time to work with Van Horne (who paid the expenses) until Van Horne's death in early 1670. Even more important, however, was his energetic participation in Amsterdam's Collegium Medicum Privatum ("Private College of Medicine"), an informal group who met at one another's houses from the mid-1660s into the early 1680s and carried out investigations into animal anatomy, publishing many of the results.[67] The group included several important Amsterdam medical investigators, such as Matthew Slade and Philippus Mattheus, and perhaps most notable, Gerard Blasius. Blasius had been born in Amsterdam in 1621 to an architect of the king of Denmark and attended Leiden, switching from philosophy to medicine during the Cartesian debates of the mid-1640s, taking his degree in 1648 with a thesis on the kidneys. He first practiced in the province of Zeeland before returning to Amsterdam. In 1660 he gained the position of city physician in Amsterdam, and at the urging of the inspectors of the Collegium Medicum he became the first medical professor of the Athenaeum as well, in 1666 managing to convert the post from *extraordinarius* into professor *ordinarius*. In 1670 he also became the librarian of the school; moreover, he performed anatomical demonstrations for the surgeons' guild and gave botanical lessons for the physicians and apothecaries in the garden of the city hospital, the Binnengasthuis.[68] He was a good analytical chemist and published on anatomy, including the anatomy of the nervous system, and medicine and medical practice more generally. He and Swammerdam were together apparently responsible for most of the published results (*Observationes*) of the investigations of their collegium, mostly anatomical work on worms, fishes, amphibians, reptiles, birds, and mammals, which has resulted in the claim that they were among the first comparative anatomists.[69] Swammerdam apparently also dissected human remains at the Binnengasthuis, where Slade held a position. After Slade's retirement in 1669, it was the regent Van Beuningen who obtained permission for Swammerdam to continue to anatomize the bodies of those who died there.[70]

But Swammerdam pursued many interests in addition to his studies on human anatomy, in which the mastery of technique continued to prove critical. Oil of turpentine was also the key ingredient in a new method Swammerdam used for preparing insects, for instance, which led to some of his most famous work. He was one of the first to see the anatomy of insects as something other than an almost undifferentiated jelly. He could do so in part because oil of turpentine not only preserved the bodies of insects (such as caterpillars) that could not simply be dried out but also turned the body fats of insects into a kind of lime, allowing them to be dissected and the lime carefully washed away, leaving their fibrous tissue exposed to the eye.[71] The dissections he accomplished with very fine scissors, tweezers, and mounting platforms of his own design. When the young Cosimo de' Medici traveled to the north in 1668 and paid a visit to the elder Swammerdam's famous cabinet, he saw the younger Swammerdam dissect a caterpillar to show how the wings of the future butterfly were already contained in its body. The demonstration held great importance in showing that metamorphosis was not an alchemical transformation of one kind of matter into another but rather an unfolding of parts already present. The grand duke was so impressed with the skill and novelty of Swammerdam's work that he offered him twelve thousand gilders for his collection of insects—an enormous sum—if he would bring it to Florence and enter his service, although Swammerdam turned down the offer.[72]

It was not long after the visit of Cosimo de' Medici that Swammerdam published (in Dutch) his magnificent, illustrated *Historia insectorum generalis, ofte, algemeene verhandeling van bloedeloose dierkens* ("General history of insects, or a general treatment of little bloodless animals," 1669).[73] In it, he made the revolutionary claim that all living creatures, even insects, develop from egg to adult through a series of material unfoldings rather than through internal transformations. Insects were neither generated spontaneously from fermenting and decaying matter, nor were they without an internal anatomy, nor were their various stages of life the result of qualitative changes in their substance—all proposals held to be true by Aristotle and by William Harvey, for example. Rather, both generation and metamorphosis resulted from the propagation and growth of parts already present in the egg.[74] He even found eggs within a caterpillar that itself had just emerged from an egg. He considered all the different membranes found in animal bodies, from insect wings to human skin, to be formed from minute vessels. In other words, he was able to give a materialist account—sometimes called a "mechanical" account—of the life of insects, with implications about how all the solid parts of animal bodies were formed.[75] He had even probed the sexual parts of several kinds of insects, shockingly discovering along the way that the chief bee in a hive possessed a uterus: it was not a king but a queen bee.

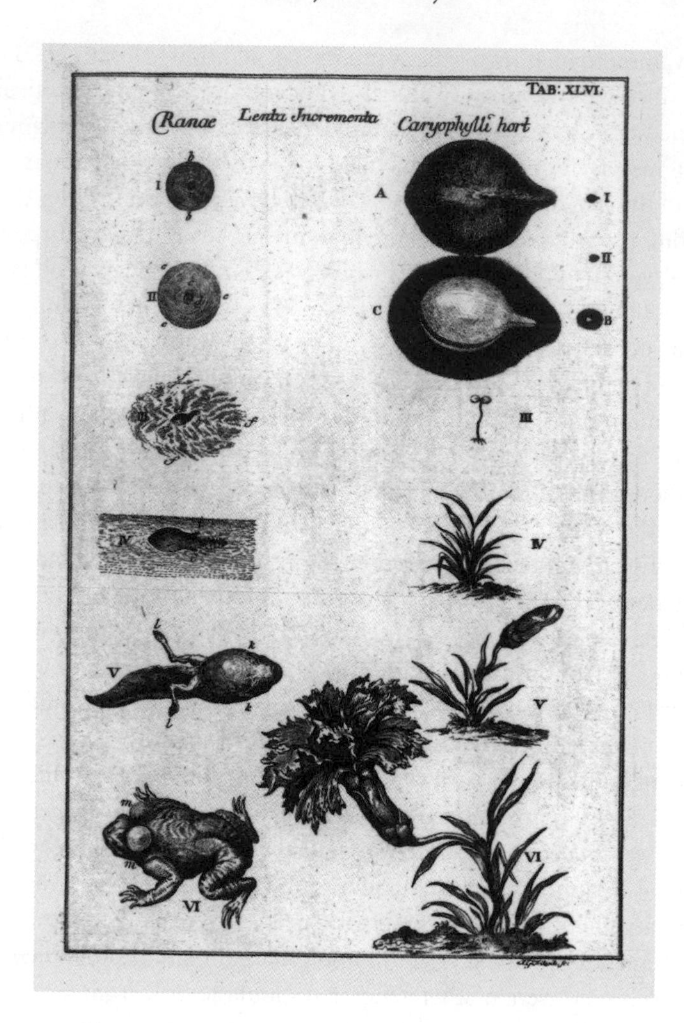

The development of a frog from an egg, paralleled by the carnation from a seed. From Jan Swammerdam, *Bibel der natur* (1752), table 46. Courtesy of the Wellcome Library, London

Yet he was explicit in noting that humankind could not probe to the level of the finest details and so would never have full and sufficient knowledge to understand how nature worked. When he dedicated the book to the burgomasters of Amsterdam, they were so pleased that they rewarded him with an honorarium of two hundred gilders.

EXAMINING THINGS MINUTELY

Although oil of turpentine, finely made tweezers and scissors, and a good magnifying glass were Swammerdam's first technical keys to exploring the anatomy of caterpillars and butterflies, his finest work depended on developing methods for using a new kind of simple microscope, made with a single-lens. Glass lenses to help with vision had been in use since at least the fourteenth century, presumably being used to magnify as well as to correct eyesight. The quality of cloth, for instance, was carefully graded by municipal officials in cloth-manufacturing towns before being offered for sale, and one of the most important ways for assessing quality was to "count" it (assess the density of the warp and weft per inch or similar unit of measure); this often involved the use of magnifying glasses.[76] As is now well known, the first documentation about someone putting two lenses together on either end of a tube in order to create greater magnification—a telescope—comes from a lens maker, Hans Lipperhey of Middelburg (a place known for its high-quality glass-making), who in September 1608 applied to the States of Holland for a patent on his device.[77] Prince Maurits attempted to keep the device top secret because of its obvious military advantages, but the news spread quickly, and in Venice, Galileo constructed his own after having seen or heard about it, making the most famous early use of it, while Johannes Kepler soon worked out the optical mathematics to make improvements thereafter. But it was another Dutch innovator, Cornelius Drebbel, who is associated with modifying a Keplerian telescope (using two convex lenses) no later than the early 1620s to examine small things up close—making a compound microscope. Again, Galileo also found out about the device and modified and manufactured his own, presenting examples to Prince Frederico Cesi and the Accademia dei Lincei. Cesi and a colleague, Francesco Stelluti, used it in examining bees for a work published in 1630; Gianbatisa Odierna published illustrations of his examination of the compound eye of a fly in 1644; and Pierre Borel printed a crude depiction of an antenna of a moth in 1656. The best of these early microscopes (Odierna's) magnified perhaps twenty or thirty times, most of them considerably less. Various improvements were attempted in the 1650s, with the most spectacular results in Robert Hooke's *Micrographia*, published at the end of 1664 and carrying the date of 1665.[78]

But someone unknown also invented a tiny, single-lens microscope with which it was possible to magnify small objects: hence the name, "flea glass." In 1637, Descartes mentioned them as though they were sold quite widely in The Netherlands for entertainment.[79] One method of making them was as a "blown bubble":

a tiny piece of glass was heated on the point of a needle, which formed into a little solid sphere of glass (although the pressure from the needle caused the sphere to be slightly deformed). This tiny glass ball could then be fixed to something, and when an object like a flea was placed close to it and examined through the sphere, its small parts appeared enlarged. In the preface to his *Micrographia*, Hooke described another method: a long, thin thread of glass was drawn out and then the tip was heated, which formed a bead, which could be removed from the stem by fine grinding. Hooke then fixed the glass beads with wax to a thin metal plate in which a needle hole had been pierced, which could be placed very close to the eye, magnifying objects to a much larger extent than any compound microscope. The ground-off place also created a flat deformation on the sphere, however, so Hooke later modified his method to simply turn the stem to one side, allowing an object to be viewed through the bead of glass, which now formed two convex surfaces.[80] Hooke found that these lenses magnified better than compound microscopes, with less chromatic aberration as well, but because they strained his eyes so badly he did not use them much.

Within a few months of the publication of Hooke's book—by April 1665—Christiaan Huygens had acquired a copy and was making extensive annotations in it while also corresponding with Hudde about the advantages of such simple microscopes.[81] Hudde had been making blown bubble lenses from at least 1663.[82] Swammerdam later gratefully acknowledged Hudde as the person who taught him how to make single-lens microscopes, probably sometime during the 1660s.[83] But as Swammerdam's work on insects went to press in 1669, he received a copy of Malpighi's book on the silkworm, in which Malpighi had painstakingly described the entire course of the creature's changing anatomy via microscopical investigations, causing Swammerdam to develop his own methods for trying to imitate and surpass Malpighi. He even found ways to dissect under a microscopic lens. It was trying work, causing inflamed eyes and fever, but it resulted in Swammerdam's gaining a reputation as the most skilled microscopical anatomist of all for several generations.[84]

It used to be asserted that the most famous user of the single-lens microscope, Antoni Leeuwenhoek, learned how to make them from Hudde.[85] More recently, the claim has been made that Leeuwenhoek found the method after seeing Hooke's *Micrographia* on a business trip to London in 1668, getting someone to translate the parts of the preface where Hooke explains how to make them. It is clear that the specimens sent and described in Leeuwenhoek's fourth letter to Henry Oldenburg of the Royal Society are modeled after Hooke's book, although that does not prove that Leeuwenhoek learned how to make a simple microscope from it.[86] It is plain, too, that Leeuwenhoek was a skilled glassblower, allowing

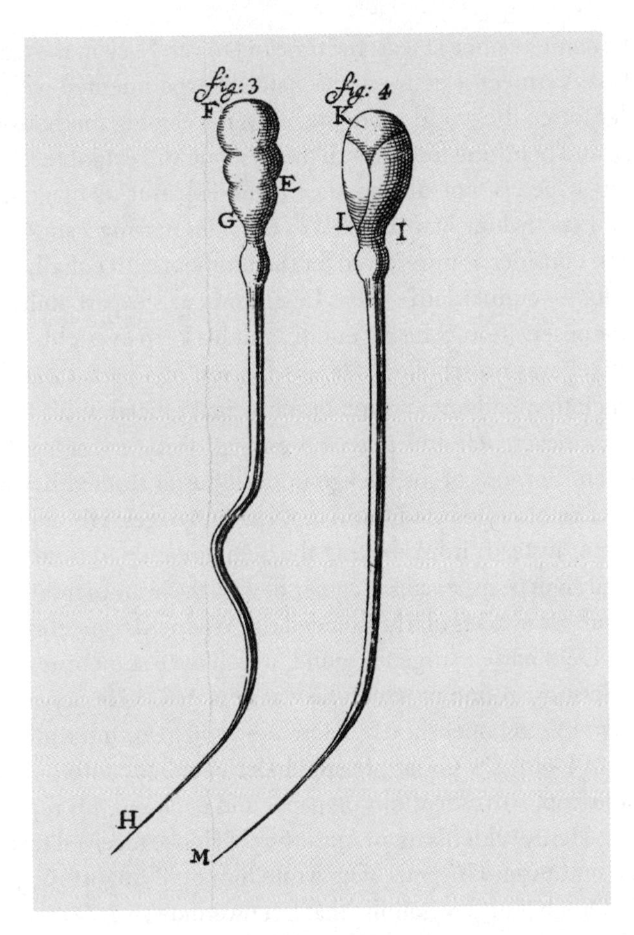

Two depictions of spermatozoa. Anthony Leeuwenhoek, *Anatomia seu interiora rerum* (1687). Courtesy of the Wellcome Library, London

him to make at least one of his lenses by creating a deformation on the sphere of a hot glass bubble.[87] Moreover, Leeuwenhoek lived in a city, Delft, whose inhabitants possessed great expertise in the handling of visual devices. He was baptized just four days apart (in October 1632) from his fellow townsman and painter Johannes Vermeer and later served as a executor of Vermeer's estate. Vermeer became famous for paintings that capture a moment in time in exact detail, in which he was helped by the use of a camera obscura: a darkened cabinet to which light was admitted by a small hole, on which a lens was placed that projected the scenes outside onto the opposite wall. Optical devices such as convex mirrors had been used by painters since at least the time of Jan van Eyck in the early fifteenth century.[88] Like Vermeer, Leeuwenhoek actively experimented with optical devices in order to see the world more precisely, developing methods for blowing and grinding and polishing lenses, with the best one still extant reaching a power of about ×266, which is "not far from the theoretical limit" of magnification with such a lens.[89] Exactly how he came first to consider making a single-lens microscope we must consider as unresolved for the moment. But equally important to his work, as for Swammerdam's, were Leeuwenhoek's expert ability to prepare specimens for observation,[90] his ingenuity, and his keen eyesight.

Leeuwenhoek was no scholar.[91] He was the son of a basket maker and grew to become a cloth merchant and minor municipal official in his native city, although he was never rich and powerful enough to be considered a member of the regenten. Because of his background, he never studied beyond primary school and hence never learned Latin or other languages. His opportunities for self-study came, instead, from making the acquaintance of a number of well-educated local figures, especially Regnier de Graaf, Swammerdam's fellow student and friend. He also regularly attended the Wednesday meetings of medical people at the Delft barber-surgeons' guild, which were sometimes the occasion for anatomy lessons, at one of which he was depicted.[92] He began to make his own single-lens microscopes shortly before 1673 and to examine all he could, exciting others in Delft. De Graaf pressured Leeuwenhoek into putting some of his early microscopic investigations on paper and enclosed this report in a letter of reference to Henry Oldenburg in April 1673.[93] Hooke received further reassurances from Constantijn Huygens, who wrote him on 8 August 1673 (during the Third Anglo-Dutch War), explaining that Leeuwenhoek was "a modest man, unlearned both in sciences and languages, but of his own nature exceedingly curious and industrious." After describing Leeuwenhoek's microscopical method, he returned to a portrayal of the man, saying that Hooke would "not be unpleased with confirmations of so diligent a searcher as this man is, though always modestly submitting his experiences and conceits about them to the censure and correction of the learned."[94] Oldenburg then invited him to communicate his observations regularly, which he did, in Dutch. Various members of the Royal Society translated them into English, and over the course of his long life (he died in 1723) 116 submissions appeared in the *Philosophical Transactions*.[95] His curiosity and technical ability for making discoveries amazed everyone and kept them interested. Leeuwenhoek is credited with careful investigations into a range of living creatures, with seeing the blood flow through the capillaries for the first time, and with discovering protozoa and spermatozoa, among other accomplishments.[96]

COLOR INDICATORS

The development of material techniques for probing phenomena can be noted even in something as seemingly distant as the promotion of tea drinking by Cornelis Bontekoe. Born in 1640, Bontekoe was apprenticed to a surgeon and later took a medical degree from Leiden, where he, like Swammerdam, studied with Sylvius and Van Horne.[97] Even after his graduation in 1667, his outspoken support for Cartesian materialism made him infamous among the more conservative members of the university, since he would attend public disputations and make his views known in what can only be described as confrontational arguments during the conservative academic retrenchment that followed the French invasion of 1672. He was a firm believer in the connection between plain speaking and truth telling, in medicine as well as philosophy, and he entered into many disputes with other physicians; in Leiden, however, the curators finally forbade him access to their buildings.[98] Two years after that Bontekoe published his *Tractat van het excellensie kruyd thee* ("Treatise on the excellence of the herb called tea," 1678), which in short order went through several editions, making him known to later generations as "the tea doctor."[99]

Tea had been briefly mentioned by Linschoten's work on the East Indies and was more thoroughly described in Jacobus Bontius's natural history manuscript of the early 1630s, which was supplemented with additional information for Willem Piso's publication of 1658. It would have been imported in small quantities at first, probably as part of the private trade driven by sailors, but by 1651–52 both Chinese tea and Japanese tea ("chiaa") were turning up at VOC auction sales.[100] The Portuguese court began drinking it about that time, and by the 1660s tea had been brought to the English court by Charles II's queen, Catherine of Braganza. Such examples no doubt helped to make tea popular in The Netherlands as well, although the chief advocates were former VOC employees who had returned from the East after becoming accustomed to tea. As late as 1685, the drink was still being introduced to even the highborn: Jesuits who had worked for the China mission found Queen Christina, then living in Rome, inquiring about the curious tea leaves drunk by the Chinese; the next day they brought her some of the leaves, along with implements for their preparation and some china cups to enjoy drinking it, so that she could taste real tea.[101]

Bontekoe believed, however, that neither "the Indians nor our people coming from the Indies" should be imitated "who, in this country, drink three or four cups of very strong and bitter tea sweetened with sugar. This habit is full of danger and a short way to the spoiling of the stomach, the blood and the brains."[102] His was instead a way of taking less strong tea in small but numerous cups, unsweetened, although adding milk was sometimes allowed. He also criticized the many vanities that already surrounded the preparation of tea (young women eager to make an impression drew his special ire), as well as the extravagance of some in their teapots, cups, saucers, and other paraphernalia. With mixed emotions, he remarked more generally on how Asian customs were influencing Dutch life: while recently "our work-masters have also shown their art in imitating the china and earthenware of that country" by making blue Delftware, "by means of a strange circulation, Holland is beginning to become Indian, even in dress, and may the Lord prevent it from becoming so in religion, which, if not in words, has certainly in the practical part of life, almost entirely become heathen again. But this carries me too far."[103] Whatever the implications of his views, Bontekoe was prepared to advocate powerfully not only tea but other medicines from the two Indies, such as opium, tobacco, chocolate, coffee, and "Jesuits bark."[104]

Bontekoe had his first experience with "tea-water" after a four-year period of agony from stone and gravel in his urine, which weakened him and caused him often to void bloody urine. "All the remedies I took stopped the bleeding but only at the cost of greatly increased pain, so that I preferred voiding blood day and night to suffering this pain, which was often unbearable and confined me to bed for a day or two." But probably while living in The Hague, where he is said to have practiced medicine from 1672, he had the good fortune to join a group of people who happened to drink tea plentifully, and "after a few weeks" of this, "I was rid of my pain and of the gravel and found with surprise that my urine was clear and not bloody, and never since that moment have I suffered from the one or the other."[105] He therefore wrote his book in Dutch for the good of his compatriots, for this was one of those "treatises which one not only may but should write, in a language which all can understand, because they are about things which everyone ought to know."[106]

Bontekoe's treatise began with a plea to his readers to lead their lives so as to preserve their health as best they could. Because of original sin humankind could not escape mortality, disease and pain, or the ignorance that hastens disease and death, while most people "hate or neglect" the knowledge that could best keep them well and preserve their life "and allow themselves to be led by their desires, and by the custom and example of those with whom they mix." But even though people were for the most part the cause of their own problems, "it is not beyond their power to prevent many of them, if only they would regulate their way of living, and that is what I believe doctors ought to work for most." He began by correcting their misapprehensions: the drinking of water was not harmful, as long as not too much cold was conveyed; tea does not dry up the body or make it thin or consumptive; tea did not "relax the stomach," a pseudo-problem in any case, according to the latest physiology; it neither produced bile nor harmed those suffering from bile; it never caused trembling of the limbs or the falling sickness; nor did it make men or women impotent or barren. Along the way he tried to put aside the fear of some that even if tea drinking resulted in "the country . . . [becoming] possibly full of healthy men and women," it "would be ruined by the loss of the taxes on wine and beer. It is not our business to show that the interests of the country are compatible with its health; this point lies outside the realm of medicine; apart from the fact that we neither intend nor hope that our nation will ever give up beer and wine and drink nothing but water." He then described the effects of tea on the mouth and throat, the stomach, the bowels, the blood, brain, eyes, and ears, the chest, belly, kidneys, and bladder. He concluded with a description of tea and how to prepare and drink it. "No one can ever, with any semblance of reason, complain that ten or twelve cups of tea-water, soon after dinner, do him any harm," while such a habit would in turn prevent many chronic complaints. When certain diseases gripped the body, it might be necessary to up the dosage to fifty or even two hundred cups a day: it was the finest (*deftigste*) medicine for "scurvy, gout, podraga and a hundred other complaints," including stone and gravel in the bladder, even curing dysentery, although it was not a panacea. In sum, tea warmed the blood without overheating it, moisturized, dried up consumptives without harm, diluted and "subtilized" thick, coarse blood, gave spirit to the blood without exciting it as did wine, beer, brandy, and so on, invigorated the weak, limp, and languid, "promotes the circulation of the blood and, when the blood begins to stop or coagulate, causes it to melt and continue its course at the rate necessary for life and health," tempered the "pungency and restrains acidity" in the blood, caused sweating, drove out poison, and cured fever. "In a word it is a pleasant remedy and a preservative for almost all defects of the blood and consequently for all that arises from these."[107]

Like many other physicians of his day, Bontekoe clearly highlights at length his experience with tea and its effects on disease. The collection of medical experience of the use of tea came not "from the mouths of this one and that," nor did he copy any data from various writers. His knowledge came from his own body and from his patients, he says, gathered together with care and caution.[108] But behind his explanations lies a set of concepts derived from his teacher, Sylvius, who in turn relied for his analysis on evidence derived from the local industries of bleaching and dyeing.

When it came to medical chemistry, Sylvius developed approaches that were thoroughly materialistic, despite previous lines of interpretation. The main successor to the spiritist Paracelsus was a nobleman from Brabant in the southern netherlands, Johan Baptista van Helmont, who also placed natural powers at the center of his theories.[109] After traveling widely, he returned to the southern low countries in 1605 and, despite having minor noble titles, practiced medicine, though without charge, until retiring to an estate at Vilvorde to conduct chemical investigations. He attacked the three principles of Paracelsus as much as the four-element theory of Aristotle, asserting that the causes of things could be discerned by the intellect (*mens*) only when it was inspired by grace and connected to the true ways of things via the labor of chemical experiments: what most people called "reason" was simply shared opinion. Objects of nature are soulful essences (*semina*) clothed in body, while humans, being the last creatures made by God, have in us a complete microcosm of the essences of the world. It is, then, the spiritual connection between us and them that yields real understanding. But at the same time he stressed that we cannot simply intuit these connections by meditation, needing the guidance of experiment to bring us illumination.[110] He therefore fully incorporated the commonly stated necessity to go find things out: for example, he showed that a plant in a pot will grow with the addition of nothing but water, indicating to him that water is the ultimate principle of nature.[111] He also taught that blood contained many of the essences of the body, so that bleeding was usually far more harmful than remedial, and that diseases are things-in-themselves rather than merely the absence of health due to bodily imbalances, which needed to be driven out by counterpoising essences contained in proper medicines.[112] Sylvius adopted many of Van Helmont's views about experimentalism and the matters of fact in chemical processes but discarded all his speculation about archeii and other immaterial powers. He also supported Harvey's views on the circulation of the blood and conducted expert work of his own in anatomy, especially on the brain.

Bontekoe clearly emerged from the Helmontian tradition as reshaped by Sylvius. He emphasized time and again that "blood is the treasure of life; it is owing to its warmth and circulation that all the members of the body move and live. Death is nothing but the stagnation of the blood. . . . To keep alive and well one ought to have a few pounds of blood, but it is more important that the blood one has should be thin enough, warm, subtle, full of spirit and activity."[113] Like other Helmontians, Bontekoe also argued against the common use of venesection. Like his teacher Sylvius, he considered many diseases to arise from an unhealthy state of the blood. Unhealthy blood was thick, slimy, cold, and acidic; healthy blood was "thin, warm, subtle, and full of spirit." For instance, consider what certain other drinks did to the blood: "It will not be hard to prove that, for our nation, wine contains too little water and too much acidity; that unmixed water, not to speak of other impurity, is too cold, that beer either inflames the blood too much or fills it with acidity and slime, and then the kidneys with gravel and the bowels with obstructions, that finally milk contains too much nourishment to be considered as a beverage and that whey and butter-milk are too raw and dirty and too useless to be drunk by human beings." Similarly, "wine, vinegar, butter, fat, oil and many more things . . . fill the stomach, the bowels, and especially the blood, with a sharp acid," and so produce bile.[114] Again, "dysentery is caused by a sharp acid exulcerating the bowels and fermenting the blood."

Such chemical analysis explains tea's good effects, too, for there is "no better remedy" for dysentery "than the alkali in tea, which gently neutralizes this acid and gradually stops the belly." More generally, "tea gives to the blood a volatile salt which is very subtle and almost entirely spirit; thus it makes the blood subtle and thin, and this is the best disposition that the blood can have." In cases of consumption, again, "the volatile salt and the subtle oil of tea are not among the worst remedies for taking acidity from the blood, tempering it and eliminating it through the pores and urine." To repeat, then, "tea warms, dilutes, and strengthens, and not only moistens the blood, but also tempers its sharpness, so that nature has nothing to equal it . . . in breaking the acidity in the blood and ejecting it by means of sweat and urine."[115] The critical ingredient in tea was its subtle oil and "volatile salt," also called an "alkali." As he promised to explain further in a forthcoming treatise, "I shall then prove that tea is an almost pure alkali volatile, a delicate, pure, subtle and volatile salt present in abundant quantity in the thin tubes of the strongly dried tea-leaves, which when soaked in water, easily melts and is rinsed out of the tubes by the hot, vapoury parts of the water." The evidence for this is as follows:

> Some well-water containing vitriol [copper sulfate], is immediately coloured brown, nay black by tea, likewise as tea-water poured on iron, at once turns to ink; or if vitriol melted in water is mixed with a little tea-water, it immediately becomes as black as ink. Seeing that ink can only be made by vitriol and alkali and that vitriol and tea-water immediately produce ink, it follows that tea contains an alkali, but a very subtle one, as tea exposed to the open air loses its force in a few hours, and even more because it immediately yields it to the hot water, which absorbs nothing so easily as a volatile alkali: I might still add that tea, soaked in vinegar, loses all its strength, the acid, the greatest enemy of an alkali, killing its opposite and annihilating the power of the salt in tea; this action again proves that tea contains salt, because this is annihilated by its opposite, acid.[116]

As Bontekoe put it in one of his later medical works, "People could tell if a body were salt or acid" by trying an experiment, in which an acid or alkali was applied and the results attended to; it was "the change of colors" and the precipitation of "the ground" that told which was which, with mercury being dissolved by acid, vitriol by water, "and so all the metals, vegetables, and animals can be properly examined."[117]

In this kind of analysis Bontekoe followed his teacher, Sylvius. He was the first to distinguish clearly fermentation from effervescence, the second of which was, in his view, the result of a mixing of a "spiritus acidus" and a "sal lixiviosum," or alkali. From such analyses, Sylvius developed his acids and alkali theory, which could—he believed—explain almost all diseases and suggest their remedies. In doing so, he divided the material substances of all things into fixed salts, acids, and "volatile salts" (alkalis), which reacted to one another materialistically, without the intervention of occult powers, causing things to happen in the body.[118]

Sylvius's theory about aqueous substances being either acid or alkaline had in turn emerged from new color tests—the first "litmus tests"—which emerged from the dyeing industry. As Bill Eamon has shown, most of the chemical color indicators were developed by painters and dyers.[119] As for dyeing, its art was ancient, revived in thirteenth-century Florence, and of keen interest to Europeans when visiting other parts of the world. As we have seen, indigo—used to dye cloth blue—was a major VOC import from Asia, while the very name of a place, Brazil, came to be associated with its chief export: brazilwood, used for dyeing.[120] A Central American import, cochineal—discovered by Swammerdam's microscope to be an insect—began to be brought back from New Spain as early as 1518 for use in dyeing. About 1630 Cornelis Drebbel, the famous Dutch inventor who spent much of his life at the English royal court, found how to combine cochineal with tin to obtain a bright scarlet color in wool. Just a year after Drebbel introduced his new method for making scarlet cloth, Edward Jorden, in his *Discourse on Naturall Bathes, and Mineral Waters* (1631) mentioned using Drebbel's scarlet cloth to distinguish waters according to their acidic or "salty" qualities (the one turning the cloth red, the other blue). Dyes therefore became of great interest to the European virtuosi: the first book brought out by the Royal Society of London, in 1662, was titled *An Apparatus to the History of the Commone Practices of Dyeing*, and Robert Boyle soon contributed *Experiments and Considerations Touching Colours* in 1664, in which he acknowledged dyers as important sources of his information. At the same time, in France, chief minister Jean-Baptiste Colbert spent much energy and money on promoting the art of dyeing and set many chemists to work on it.[121]

Bleaching was also important. And in northern Europe, the Dutch did most of the industrial dyeing and bleaching of cloth. The Dutch cloth industry centered on Leiden, in which, of course, Sylvius's medical faculty was located. Leiden had the second largest population in The Netherlands (about seventy thou-

Jacob van Ruisdael, *Bleaching Ground in the Countryside near Haarlem,* c. 1670.

sand in 1665), much of which constituted a large urban proletariat working in the cloth industry, which specialized in the new "bays," a fine and light wool fabric.[122] The annual cloth production of Leiden rose from 30,000 pieces in 1585 to more than 140,000 pieces in 1665.[123] The industry, moreover, involved not only weaving and spinning but dyeing and bleaching. The English, for example, exported to The Netherlands not only raw wool but what was then known as "Old Draperies": unfinished and undyed cloth to be finished—usually bleached or dyed—in the low countries. For dyeing, cloth was "fulled" (beaten and treated with alum, another import from Asia) to help the dye adhere to the fiber and then soaked in vats of dyestuffs. If cloth was not dyed, it was usually bleached. Bleaching, too, was an old art, but by the end of the Middle Ages, the Dutch had become renowned linen bleachers, with almost a monopoly on the trade in northern Europe until the eighteenth century. For bleaching, cloth was steeped in alkaline lye for several days, washed clean (a process called "bucking"), and spread on the grass for some weeks (called "crofting"), after which the process was repeated five or six times. To finish the bleaching, the linen was steeped in sour milk or buttermilk for several days, washed clean, and again crofted. To produce really white linen, months of repetitions—a whole summer—might be required.[124] The sandy fields covered with short grass behind the dunes near Haarlem were well suited for crofting, so that cloth was shipped to Haarlem in the spring not only from adjoining Leiden but from Germany, from Scotland and England, and even farther to be bleached. Any traveler going by canal boat from Leiden to Amsterdam would go via nearby Haarlem and its bleaching fields. It therefore seems no accident that the acid and alkali theory was developed in its strongest form by a professor of Leiden and used by one of his students to explain the beneficial effects of a new import from Asia.

PRESSING AGAINST THE LIMITS OF KNOWLEDGE

Thus, in anatomy, natural history, microscopy, medicine, and chemistry new discoveries were made by adopting cutting-edge methods of production to investigate the world. Hypotheses were built on the consequences, but for the most part it was recognized that such concepts were provisional. For instance, Swammerdam had clearly read Descartes among other moderns, offering an account of respiration in his medical thesis of 1667 that explained it by the pressure of the air entering the lungs when they expand rather than because of any innate principle of attraction. But the careful experimental work undertaken to demonstrate this view reflected the Cartesianism of the practical and empirical inquirer so familiar to medicine rather than that of the metaphysician. Two years after his thesis, in his *Historia*, Swammerdam declared, "Just as we cannot obtain true experience of all things, and have therefore no clear and distinct notion of the same, just so we should not foolishly imagine that we shall ever obtain through our reason true and real knowledge of the causes of things." "Our greatest wisdom" therefore lies only in gathering "a clear and distinct notion of the true manifestations or effects" of nature. Elsewhere in the same work he concluded, "There can be no stronger or more powerful reasons than those which are extracted from experience and practice, in which they must end."[125] As one historian has noted, "Swammerdam's conception of nature was based on rigorous order, a concept that precluded chance and corresponded with uniformity," but because he so often refused to give causal explanations for the ways of nature, his work "hardly suffice[s] to represent him as a mechanist."[126] So, like Pauw, Tulp, and other anatomists before him, Swammerdam consistently attributed the wonders of animal structure to God's creative ability without pretending to know the real causes. His teacher Sylvius is also known for frequently using such phrases as "it is possible," "we suppose," "we dare not in real earnest propose," and other phrases tied to probability rather than certainty.[127]

Not everyone took this tack. The uneducated Leeuwenhoek, for example, quickly obtained the self confidence of the applauded autodidact. He thoroughly trusted himself and his observations and his ability to draw conclusions from them even in the face of powerful disagreement, whether from the learned or high-ranking or from his fellow townspeople. For example, he stuck to his conviction that in generation the sperm was more important than the egg despite the opposite consensus among the learned physicians, and he held out against spontaneous generation in the face of heated opposition from ordinary folk.[128] Although he professed a clear and strong allegiance to empirical observation and placed little trust in academic learning, he also held strong religious convictions (at least later in his life) and adopted a number of simple mechanistic principles, mostly picked up from conversation with well-educated physicians. He listened carefully and sometimes changed his mind, but he usually did so because of better evidence in the form of observations or other matters of fact that he could verify himself rather than because of a better argument. As he learned more and more, he came to consider himself as belonging to the republic of letters and developed some pretensions. But Swammerdam was blunt about the consequences: "It is impossible to go into a discussion with Leeuwenhoek as he is biased and reasons in a very barbaric way, having no academic education."[129]

Of course, Swammerdam saw the world through his own preconceptions. He constantly reiterated the view that allowing for things in nature to occur by "chance" would lead to atheism. As one astute reader of him has remarked: "Underlying the urgency of Swammerdam's assault on the traditional concept of metamorphosis was in part, then, a broader concern to deny chance any place in the formation of living things."[130] One might go further, connecting this to Swammerdam's religious upbringing in a pious Calvinist household, and suggest that Swammerdam was not so much a preformationist as a predestinationist who considered God to work through his material creation. From about 1673 to 1676, he even came to be an avid follower of a religious mystic, Antoinette Bourignon.[131] One of his most impressive works was his book (1675) on the mayfly, which was thought to live for one day only and which he therefore took as a model of human mortality; in it, his careful descriptive work is interspersed with numerous poems to the Creator.[132] Or as he famously put it in a letter to Thévenot of 1678: "Herewith I offer you the Omnipotent Finger of God in the anatomy of a louse: wherein you will find miracles heaped on miracles and will see the wisdom of God clearly manifested in a minute point."[133]

More generally, many of the microscopical investigations of the 1660s and 1670s showed a general commitment to the corpuscular philosophy and "mechanism" in physiology. Indeed, it was tempting for many to think that because they could see the anatomical forms in little creatures, those forms must go "all the way down": that is, the fine structure of, say, the vessels of an insect must be based on other fine structures. For a while, Nicolaas Hartzoeker even thought that he could detect forms in the globulous part of spermatozoa, a homunculus, the early form of the later offspring.[134] The recognition that protozoa did not have observable inner anatomical structures came slowly, and only in the early nineteenth century would "cell theory" arise—the doctrine that all cells come from other cells, so that, for instance, bodily fibers are made of cells and not cells made of fibers.[135] Conceptual frameworks therefore had much to do with how observations were interpreted and even with what people thought they saw.

But it might also be put the other way around. As Catherine Wilson has written, the mechanical philosophy "was as remote and hypothetical in the programmatic writings of Descartes and Boyle as it had been in those of Lucretius. The grip that it held in the early modern period was related to the promise of actually being able to see, with the help of the microscope, tiny machines and invisible processes. Microscopical observation promised to give a deeper and more profound view of the world, but one that was based in experience rather than on thinking."[136] Or as Marianne Fournier has insisted, whatever their point of view, the work of all the microscopists "reveal[s] a deep commitment to the experimental method."[137] Who would have believed that rotifera could revive after years of desiccation, had the phenomenon not been observed? "With its monsters and unheard-of lives," the microscopic world "abounded no less than the distant Indies with strange and marvelous things."[138]

The promise of being able to see the smallest parts of the mechanisms on which life was built was, then, recognized as a dream rather than reality. But the reality of discovering new objects and phenomena continued to excite. Seeing live-looking anatomized bodies, the difference between the capillary veins and arteries in the human uterus, the fine structure of a caterpillar's anatomy or a snail's digestive system, strange new beasts in drinking water or the reproductive habits of aphids, even tea turning iron filings black as proof of its healthful effects, provided much grist for hypotheses, whatever they were. In the minds of many, then, the strange world of the Indies and the wonders of the invisible world were completely entangled. What it all meant was any mortal's guess. But one thing was sure: these new findings could not have been made unless a host of people had been aware of new techniques for finding things out, then opportunistically turning those techniques to analyzing objects.

8

Gardens of the Indies Transported

I do recall the day / This was a clover field, where thriving cattle lay. /
A garden, it looks now . . . , I'll be damned.

—CONSTANTINE HUYGENS, *Hofwyck*

Stimulated by a powerful combination of medical interests and enthusiasm for exotic garden specimens, Dutch investigations of the natural history of Asia and the southern Caribbean became renowned in the late seventeenth and early eighteenth centuries. These efforts took on additional seriousness after the government in Batavia issued instructions for obtaining local medicines, but they were supported by the continuing interest of well-placed people in The Netherlands for exotic plants and naturalia. The enthusiasm for gardens and cabinets, in which exotic specimens could be grown and shown, and the wealth committed to their establishment and expansion, remained one of the most potent reasons for finding things out and conveying them and information about them back to the home country. By the end of the seventeenth century, technical innovations even allowed tropical plants to be cultivated in the cold and dark north, while at the same time one or more huge, beautiful, and expensive folio volumes devoted to plants and animals were being produced each year for the collectors' market. Most of the specimens and information about them continued to be gathered from local people and scholars, who well knew their own environment; they were conveyed back to The Netherlands as objects and matters of fact, as they had been in Bontius's day. As in Bontius's case, too, many of the investigators, who endured personal suffering to find things out, came to hold the local people with whom they worked in high regard. The search for useful medicines and beautiful creatures had expanded greatly in scope since his day, but the main beneficiaries of this acquisition of goods and knowledge continued to be wealthy regenten and scholars.

SEARCHING FOR MEDICINES IN ASIA

The VOC's governing bodies took new initiatives in the late 1660s that resulted in a wide range of new work on the natural history of the Indies. Some of that work concerned medical investigations that seem to have been prompted by a physician who was employed as one of their merchants, Robert Padtbrugge. Padtbrugge's interest in the local medicines transformed the continued general observations about the wealth of medicinal substances found in Asia into a new policy of trying to use indigenous medicines as a substitute for the drugs imported from Europe.

Born in Paris in 1637, raised by his mother and stepfather in Amsterdam, and treated as a son by Isaac de la Peyrère—the infamous author of a work on the origin of humankind before the creation of Adam and Eve—he enrolled in medicine at Leiden in November 1661.[1] There he became friends with the younger Swammerdam, whom he may have helped with his experiments. He took his medical doctorate in 1663 under Van Horne, with a thesis on apoplexy.[2] But he decided to try his fortunes elsewhere, and so after marrying Catharina van Hoogeveen, a cousin of Johann de Witt, the leader of the States Party, he took service with the VOC as a junior merchant (*onderkoopman*). He departed The Netherlands in February 1664, arrived in Batavia in July, and in September moved on to Ceylon with a fleet under Rijklof van Goens. The main attraction of the island for the Dutch was its particularly rich natural resources—most importantly cinnamon, but also pepper, areca nuts (used in the practice of betel-chewing, which was widespread throughout South Asia), and other regional commodities, even elephants. The VOC had begun to seize some of its coastal cities from the Portuguese about two decades after setting up in Batavia (taking Batticaloa in 1638, Galle in 1640, and Negombo in 1644). By the end of the 1650s, Van Goens had expelled the Portuguese from the port cities of the island, securing a monopoly over the export of cinnamon not unlike that Jan Pietersz. Coen had secured over nutmeg, mace, and cloves in the Moluccas thirty years earlier, and he was then trying to finish removing them from the Malabar coast (what is now Kerala in southwest India).[3] For the next few years Padtbrugge served as a merchant in the Western Quarter of the VOC and as a member of the Council of Justice.[4]

Despite his other duties, however, Padtbrugge continued to take an interest in medicine and natural history, and it was he who seems to have provoked the

Heren XVII into action. Ceylon had already become noted for the riches of its medicinal herbs. Early in the Dutch presence on the island, one of the VOC's surgeons, Jan Carstens of Tonningen, is said to have "acquired a sound knowledge of local plants and herbs which were suitable in place of European drugs."[5] Another VOC surgeon on the island, Wouter Schouten (later well known for a published account of his voyages), commented in 1661 that "there are also intelligent lawyers, doctors, surgeons and barbers [here]." He criticized the medical practitioners for having "very little knowledge of anatomy, of things natural, unnatural, and contrary to nature, which ought to be the basis of their science. Thus, their principal knowledge rests upon experience." But it sufficed: "Their medicines consist of freshly plucked herbs and flowers, of which they know how to make decoctions, stupes, poultices and the like."[6] Padtbrugge followed up such observations. In early 1668, returning from conducting business at Basra, he presented the governor and council of Ceylon with reports on his investigations into diverse facts (*waerdige*) about minerals, herbs, medicines, and other things. One of these documents—a study made for the Heren XVII as well as for the government in Batavia—survives. It commented on the general environment and resources of Ceylon, including plentiful wood for ship masts, saltpeter for gunpowder, and a wealth of healthy and delicious herbs, roots, bulbs, fruits, and so forth. The plants were known to even the poorest inhabitants of the island, he wrote. Similarly, there are so many medicinals (*hulpmiddelen*) that a whole book should be written about them, but as a start he attached a list of thirty of the best medicinal plants and herbs of Ceylon. The island also possessed various excellent (*deftige*) hitherto unknown balsams and unusual gums. In short, the inhabitants had an abundance of good remedies at their easy disposal. The council wrote up a brief account of what Padtbrugge had told them in a letter to the Heren XVII dated 25 January 1668, describing him as a man well studied in medicine and the best doctor of Colombo (where the Dutch had established their headquarters).[7]

The observations of Padtbrugge had been made known at an opportune moment. In the early 1660s, the Heren XVII had begun trying to make the VOC settlements less dependent on supplies from The Netherlands. Pieter van Hoorn, one of the regenten of Amsterdam, arrived in 1663 as councilor-extraordinary to press for the development of Batavia and its environs more independently of home finance.[8] One of the consequences was the establishment in the same year of a "medical shop" (*medicinale winkel*) in the Castle of Batavia, which supplied all the medical chests to VOC factories and ships in the East or returning home. What Padtbrugge had seen were the implications in these combinations of interests and activities: with enough attention and energy, local medicinal plants could be supplied to the medical shop as a substitute for drugs sent out from Amsterdam, both cutting costs and furnishing good or even more effective remedies. Moreover, some of the Heren XVII were keenly interested in natural knowledge, most particularly Joan Huydecoper van Maarseveen (made a director in 1666). From at least 1669, the sergeant-major of Ceylon, Joan Bax, was sending his uncle, Huydecoper, shipments of natural history specimens from the island.[9] It must have been almost immediately on receipt of the January letter about Padtbrugge's views that the Heren XVII directed a letter of their own (dated 9 May) to the Council of the Indies in Batavia asking for a full investigation into the natural resources of Ceylon. A few months after receiving these instructions from the Heren XVII, the council in Batavia in turn sent a letter to Colombo requiring Padtbrugge, the VOC surgeons, and others to investigate these things and report to them.[10]

The delayed response in Batavia to the request of the Heren XVII was presumably to allow time for the new head of the medical shop, Andreas Cleyer, to develop a careful answer. Although not a physician, Cleyer was clever and ambitious, and he was now in a place where he could control most of the medical activities in Batavia and influence them almost everywhere the VOC operated. Cleyer had been born into a military family in 1634 in German Kassel.[11] It has been suggested that he had a sound education, perhaps even studying medicine for a short time at Marburg, although there is no evidence of his having taken a degree.[12] He had entered service with the VOC at least by 1661, when he shipped out to the East Indies as a "gentleman soldier" (*adelborst*). Where he served for the next few years is unknown (although he may have been among those sent to China in the wake of the fall of the VOC station on Taiwan), but he was at Batavia in 1664, apparently acting as a medical practitioner, for when his widowed aunt died, he is mentioned as a *medikus licentiaet*. He then inherited the property built up by his aunt's former husband, a lawyer and merchant in Batavia, which allowed him to expand his activities.[13] In a resolution of the governing council of Batavia of 15 December 1665, he was appointed to distill chemical medicines for the medical shop, for which he could charge 50 percent more than in The Netherlands while the Company supplied him what he needed for the work at half the usual cost. (Apparently such sweetheart contracts were not unusual in the wake of Van Hoorn's visitation.) He supplemented this income with a regular salary of sixty gilders per month plus expenses as head of the Batavian Latin school, which had about forty students. In May 1667, on the death of the head of the shop, Cleyer gained that appointment, too, as well as the title of physician to the castle and head of surgery (while stepping down from daily involvement with the school, although he became one of its curators). He held other administrative offices, such as membership in the college of commissioners to oversee marriages and small claims (Huwelijksche en Klein Zaken), joined the militia company that contained some of the most eminent men of the city, and from May 1672 served as one of the members of the city council for two years.

Cleyer had already requested drugs from the VOC station at the Cape of Good Hope.[14] A permanent settlement at Table Bay had been established there in 1652 under Jan Anthonisz van Riebeeck. Van Riebeeck also had a background combining medicine and business, starting out with the VOC as an assistant surgeon before gaining the rank of merchant and trading throughout East Asia. In the early 1650s he was called out of retirement to set up the Cape settlement as a way station where VOC ships could call to replenish their water and food halfway through their journeys to or from Batavia. To do this, Van Riebeeck both traded for the cattle raised by the local people, the Cape Khoikhoi people—then called "Hottentotts"[15]—and laid out a large garden, orchards, and woods, which were under the direction of Hendrik Hendricxsz Boom and his family. During its first decades, the main interest of the garden was in growing vegetable foods for the settlers and the ships calling there.[16] But by the mid-1660s, it was also growing local plants of botanical and medicinal interest, too, and sending them to gardeners back in The Netherlands.[17] It would eventually become the chief place for acclimatizing plants from throughout the Dutch East Indies. As early as 1668, therefore, Cleyer had written to the Cape inquiring about medicinal plants (although probably not yet in a search for exotic and new ones), and in 1669 he instigated the drafting of further letters from the governor and council in Batavia to the Cape commander, receiving in return garden seeds, artichoke plants, and medicinals for the Batavian medical shop. From then on, the Cape garden supplied many of the herbs for his shop, while it also sent medicinal plants and seeds to Colombo.[18]

Following Padtbrugge's report, then, Cleyer seized the additional opportunity. The letter of 24 April 1669 from the high government in Batavia to the government of Ceylon written with his involvement explained that the "doctor of the Castle of Batavia" had learned that some medicinal plants grew in large quantities on that island, "in particular colocynth apples in the Jaffnapatnam region and sarsaparilla near Caleture in the neighbourhood of Colombo." Cleyer wanted preparations of them, to see if they could be used as substitutes for medicines imported from The Netherlands. The high government at Batavia wrote: "Thus we shall be greatly pleased if you would recommend Doctor Robertus Padtbrugge (whom we take to be also a good herbalist) as well as the master surgeons and others who have knowledge of the medicinal herbs to speculate and satisfy their curiosity about these things in order to discover in Ceylon all that might be found to be a great comfort there."[19]

By the time the letter was written, however, Padtbrugge was on his way back to The Netherlands.[20] The governor of Ceylon, Van Goens, therefore requested that he be sent a couple of chemists to carry out the work requested—or perhaps he was merely getting after-the-fact approval since, probably in response to Padtbrugge's reports, he had already acquired two chemists "with good botanical knowledge," Abraham Goetjens and Willem de Witte.[21] Van Goens also requested a qualified physician, and after repeated requests he got the Heren XVII to appoint one for Ceylon (in November 1671): Paulus Hermann, born in Germany, well traveled in the medical faculties of Europe, with a medical doctorate from Padua and some months' work in the botanical garden at Leiden under tutelage of Charles Drelincourt. Stopping at the VOC station at Table Bay during the passage out, Hermann collected information, drawings, and a herbarium with the intention of publishing on Cape botany. Arriving in Ceylon in 1672, he studied the native flora with the help of local practitioners and "thoroughly studied the meaning of Singalese plant names."[22] Hermann was also appointed to the hospital in Colombo, although it was said of him after he left the island that "he took no good praise away with him from the soldiers and seamen that came under his hands." (It was also said that "he was a true tyrant over his slaves, with blows and whippings; he was . . . accused of killing a female slave whom he [buried] in the garden behind his house, and for some days [was] under arrest in his house, but was after set free.")[23] Because of his botanical studies on Ceylon, Hermann later obtained the position of professor of botany in the medical faculty at Leiden, where during the early 1680s he greatly expanded the exotic specimens held in the Leiden hortus.[24]

Letters generated by Cleyer prompted botanical work elsewhere, too. On the same day that the governor and council in Batavia wrote to Ceylon, another letter was sent to the governor and council of Coromandel asking for colocynth apples, nux vomica ("vomit nuts," from which strychnine would extracted 150 years later), and dried myrabolans, "in which Mr. Joachim Fijbeecq (also being a good herbalist) is to be recommended to add such medicinal herbs, which grow on the coast." Two months later, a similar letter went to the directorate of Bengal: "Since Mr. Jacob Frederik Strick Berts, first surgeon in Bengal (since he has long resided in the land of the Mogul and has roamed through it), must necessarily have obtained great experience and knowledge of many medicinal drugs and herbs in those lands, Your Honour will be well-advised to recommend and animate him to compose a list of all such things as he will think that they might profitably have been gathered in the said countries and sent hither so as to reduce the demands made on the home country."[25]

VAN REEDE AND THE BOTANY OF MALABAR

Such instructions from Batavia indirectly led to one of the great botanical works of the century, although produced not by a medical official but a military commander, Hendrik Adriaan van Reede tot Drakenstein. He was appointed commander for a new group of VOC factories along the Malabar coast of India not long after the letters instigated by Cleyer went out in 1669. Born in 1636, Van Reede was the younger son of a noble family from Utrecht, his father holding the post of "forester" of the mostly denuded province. It may have inspired the younger Van Reede to take an interest in plants. His father had also been a supporter of the Arminians, and although the son appears outwardly to have conformed to the orthodox branch of the Reformed faith, he was very tolerant of others. Most of his relatives were involved with the WIC, but Hendrik took service with the VOC. When the outbound ship he was aboard stopped at the Cape in the spring of 1657, he had a chance to see the five-year-old station and its expanding garden. Once he reached Ceylon, he served under Rijklof van Goens during the last campaign to control Ceylon (completed by Van Goens with the seizure of Colombo in 1656 and Jaffna in 1658). Following his success in Ceylon, Van Goens turned his attentions to the Malabar coast, culminating in the taking of Cochin in 1663, during which Van Reede stood out for his heroic actions, and in so doing acquired Van Goens's personal patronage. For the next two years Van Reede served as one of the councilors for the VOC administration of Malabar, as president of the council of Cochin, and as head of the daily administration of the city as "regedore maior" of the Rāja of Cochin, Vīra Kērala Varma, who would later support his botanical investigations. He went on, from 1665 to 1667, to serve as chief of Quilon, also on the Malabar coast. During this period, he was notably impressed by the flora of the woods and forests of Malabar; he also grew to be considered the Dutch administrator most knowledgeable about the people and resources of the region.[26] Van Reede moved back to Ceylon between 1667 and 1669 in a senior military rank and during most of 1669 fought off a siege of Tuticorin on the southern coast of India against a large army of the Neik of Madurai, leading to the Neik's defeat by Van Goens. But he returned to Malabar in 1670 as the commander of the region, a post he would hold until 1677. Van Reede continued to be active militarily and diplomatically, defeating the Zamorin of Calicut in 1671, turning back a French threat in 1672, and establishing the Union of Mouton in 1674, which settled local unrest. But because the VOC made Van Reede responsible to the council in Batavia rather than to Van Goens in Ceylon (as Van Goens thought was right), the two former colleagues became embroiled in a power struggle, with Van Goens finally forcing Van Reede's removal.

It was during his period as commander of Malabar—beginning not many months after the series of requests for local medicinal information had gone out from the council in Batavia—that Van Reede launched the large collaborative study that would result in the multivolume *Hortus Malabaricus.* The person who made the most thorough modern study of Van Reede and his work, Johann Heniger, concluded that it was probably because of the requests from Batavia that he had "the idea of an examination of the flora of Malabar."[27] Moreover, Van Reede's growing rivalry with Van Goens on Ceylon no doubt contributed to his desire to convince people that the vegetation of Malabar was equal or superior to that of Ceylon. Pepper was the chief product of Malabar, but the region also provided good sources of cinnamon (although "wild" rather than the domesticated variety of Ceylon), areca nut, various timbers, cardamom, products of coconut palm, and other botanicals, some of them medicinal. For instance, true calumba root, used widely in Europe, grew in Malabar rather than Ceylon but was imported to Europe via Colombo and consequently was called "Colombo" root and associated with Ceylon.[28]

The unrest in Malabar in 1673 had, however, temporarily dried up trade and crippled the crucial pepper harvest. Perhaps prompted by this financial exigency, or perhaps having enough leisure to act on previous orders or desires following the Treaty of Mouton, Van Reede followed Van Goens's example and had a chemical laboratory built in Cochin in 1674. As he later put it, he employed the chemist Paulus Meysner there to prepare "waters, oils, and salts" from the "famous medicinal herbs, fruits, and roots" of Malabar, "and to examine in what respect they are equal to or excel the European ones, in order to provide the Company's medical shop therewith and to avoid the annual expense of many ineffective leaves, roots, seeds, and ointments [being sent] from Batavia to Ceylon." Expanding on the theme in the preface to the third volume of the *Hortus Malabaricus,* "it would involve great profit for the Illustrious East India Company, which indeed would be able to save those expenses which it spends on transporting medicaments to that place. Indeed, it would be possible to use, at less expense and with greater profit, Indian medicaments . . . with the same, if not superior, virtues." He was especially pleased with the distillation of an oil from the roots of local cinnamon, which was used to good effect at the hospital at Cochin. But no doubt because of his rivalry with Van Goens in Ceylon, by a letter of 22 October 1675 the high government in Batavia prohibited Van Reede's further work, arguing that if the virtues of the oil produced from Malabar cinnamon became generally known, it would damage the market in Europe for the cultivated variety of Ceylon. Van Reede defended himself in a letter of 10 February 1676, pointing out that the Singalese of Ceylon had long known of the oil and used it in betel-chewing, and ignored this directive from the high council for the rest of his tenure in Malabar.[29]

Van Reede found an adviser on botanical matters in the person of Matthew of Saint Joseph, a friar of the order of the Discalced Carmelites and a keen medical botanist. As commander of Malabar, Van Reede had permitted Matthew to build two churches in the region in 1673, and in early 1674 he invited him to consult on the illnesses of some servants of the Company; the two got on well and began to work on a study of local plants. Matthew—originally called Pietro Foglia—was born about 1617 between Naples and Capua and had studied medicine at the university of Naples before entering orders in 1637.[30] In 1644, he was sent abroad as a missionary, working in Palestine in the mid-1640s under someone who was originally from The Netherlands, Coelestinus of Saint Liduina, originally Pieter van Gool, brother of Jacob van Gool (Golius), a professor of Oriental languages at Leiden and a botanist. At Golius's request, Matthew examined the plants of the Lebanon, before in 1648 he transferred to Basra—where he became convinced that the missionaries should concentrate on medical care, writing a book on medicine to assist their efforts—and then to other locations in South Asia. In all these places he continued his medico-botanical studies when he was able. In 1662, while resident in Malabar, Matthew asked Golius about the possible publication of his manuscripts, to which Golius agreed, persuading the Heren XVII to transport Matthew's manuscripts to The Netherlands. Since Matthew was frequently traveling about, however, the VOC's agents kept missing him. Much of his material finally reached Italy in 1673 through the hands of religious orders, parts of which appeared in print under the editorial direction of Giacomo Zanoni in 1675 as *Viridarium Orientale* ("Garden of the Orient"). It was no doubt because of Matthew's interests that the report of the Catholic mission in Malabar in 1672 paid close attention to the local flora and fauna.[31]

Matthew's studies seem to have provided the first template onto which Van Reede could map his developing interest in local botany. But like the good administrator he was, Van Reede also assembled an advisory board with a wide range of expertise. He called Paulus Hermann to Cochin from Ceylon in or about 1674, and Hermann caused Van Reede to revise Matthew's organization of the information he was collecting.[32] Van Reede also had at hand, in the person of a soldier named Antoni Jacobsz. Goetkint, someone from an Antwerp family of artists who could make careful drawings of plants. He found another fine draftsman in Marcelus Splinter, who came from a family of painters and sculptors of Utrecht; at least one or two other draftsmen were also employed by Van Reede. One of the ministers of Cochin, Johannes Casearius, undertook the work of putting the results of their studies into proper Latin, while the secretary of the VOC in Cochin, Christiaan Herman van Donep, served as a translator from Portuguese to Dutch and one Manuel Carneiro translated local languages into Portuguese. For information about the details of botany and medicine, Van Reede assembled over a dozen scholars, physicians, and botanists. They probably included some of the surgeons of the VOC hospital in Cochin, and the chemist and head of the local medical shop, Paulus Meysner. They certainly included a physician from Karapurram (south of Cochin), Itti Achudem, who came from the caste of Chogāns, a group deeply involved with the coconut industry and more generally very knowledgeable about local botany. Van Reede's consultants also included three scholars ("panditos") with an excellent knowledge of the regional medicinal botany: Ranga Botto, Vinaique Pandito, and Apu Botto, all of whom had fled Goa because of the increasingly intolerant Counter-Reformation there.[33] Many more were employed to gather specimens of the plants themselves. The local board members worked long hours every day for two years, discussing what plants should be noted, inspecting specimens before drawings were made, and answering questions, with their words being translated into Portuguese, then into Dutch, and finally into Latin. Van Reede attended many of the meetings, being particularly impressed with how the three panditos disputed amicably with one another, reciting from memory verses about plants from classical works.[34]

Because of the great importance of the work of local experts in the compiling of botanical information for Van Reede, it has been argued that the *Hortus* was "far from being inherently European," since it was "actually compilations of Middle Eastern and South Asian ethnobotany, organized on essentially non-European precepts." The historian who has taken this position, Richard Grove, also claimed that others "have failed to understand or identify the vital significance of the power of the Ezhava affinities within the text of the *Hortus Malabaricus*, with all that it implies for the assertion of the Ezhava classificatory system." Van Reede had, he believes, moved from a dependence on Matthew of Saint Joseph in 1673–74 "to a reliance on Malayali sources and, initially, to the professional expertise offered by three 'Brahmins,' . . . and a Malayali physician, Itti Achuden." Not only the descriptions of particular plants but the organization of the work as a whole shows, he says, in "the sequence of plants in *Hortus Malabaricus* that those which the Ezhava assumed to be related were treated successively, (even if the Europeans knew this to be contrary to their own classificatory system)," with the consequence that "the knowledge of the Ezhava has directly influenced the classification of *Hortus Malabaricus*."[35]

These claims for how local knowledge "directly influenced the classification" of the plants are asserted rather than demonstrated, however. The most careful student of Van Reede and the *Hortus*, Johannes Heniger, shows that the original organizational plan was drawn up before Van Reede assembled the group of local scholars and was based on Matthew of Saint Joseph's manuscripts; this was then altered according to Hermann's advice.[36] Moreover, when the first volumes were printed, his original organizational plan was also changed. Van Reede had wanted to publish one volume on trees, another on shrubs, and a third on herbs. But he sent his materials back to The Netherlands in batches. His publishers thought that the first shipment of manuscripts he sent was the complete set but considered it to be too short for three volumes. They therefore made up two equally sized volumes by adding some of his material on shrubs to that on trees for the first volume, and adding the material on herbs to the remaining material on shrubs to make up the second.[37] Moreover, whereas the first two volumes relied on the system of classification of Bauhin because that was preferred by the Leiden professor of botany Arnold Seyn, who helped with those volumes, later ones used John Ray's system, preferred by Van Reede's later collaborator, the Utrecht professor of botany Johannes Munnicks. The extent to which local knowledge from the Malabar coast—Ezhava, pandit, and otherwise—influenced the classifications in the *Hortus* therefore remains open to careful investigation.

There can be no question, however, that the information in the *Hortus* about the medicinal uses of the flora of Malabar was almost entirely due to Van Reede's consultations with local experts in these matters. Van Reede clearly approached the work without prejudice toward the various intellectual and religious traditions of his collaborators. This is obviously true of his sincere regard for the medical knowledge possessed by the local doctors he consulted. It is also true of his regard for the Catholic friar, Matthew of Saint Joseph. His initial conversations with Matthew even led Van Reede to recommend to the council in Batavia that the Carmelites be allowed to do missionary work in the area as long as they pledged their loyalty to the Company—a proposal eventually turned down. Moreover, the Reformed minister with whom he worked, Casearius, seems to have been as broad-minded as Van Reede. During his studies in Leiden between 1659 and 1661, Casearius had occupied the same house in Rijnsburg as Spinoza, and he then went on to Utrecht, where Regius was still propounding a Cartesian-influenced medical outlook. After completing his theology studies in 1665 Casaerius had difficulty in finding a living, so after marrying in 1667 he accepted an appointment with the VOC, arriving in Cochin in 1669. Van Reede got on well with him while criticizing the other clergyman of Cochin, Marcus Mazius, for his lack of Christian charity. But Van Reede's second in command, the upper-merchant Gelmer Vosburg, strongly attacked the influence of Matthew and Casaerius over Van Reede. The VOC's governor-general, Joan Maetsuyker (himself from a Catholic background), finally signed a letter of October

1675 banishing Matthew, while Casaerius was admonished to keep to his religious books and teachings. Van Reede protected Matthew by making him a paid adviser to the Company on local religious affairs (about which he knew a great deal). (At the same time, Matthew's work for Van Reede deeply upset the Carmelites, who tried to kidnap him in 1678; he was rescued by the soldier and draftsman Antoni Goetkint. In the end, the high government of Batavia allowed him to stay if he resigned from orders, but this was a step too far, and Matthew eventually returned to Surat, dying in 1691.)

In 1677, Van Goens finally forced Casaerius and Van Reede to leave Malabar for Batavia. Van Reede resided in Batavia during the months between May and November 1677, serving as councilor extraordinary on behalf of the Heren XVII and leading a debate on VOC policy for the Western Quarter. He wrote a long and detailed report about it, which he submitted to the Heren XVII after his return to The Netherlands.[38] It would deeply undermine the confidence of the VOC's directors in Van Goens.[39]

In Batavia, however, Van Reede continued work on his *Hortus* in conjunction with other naturalists, particularly a physician who had just returned from Japan, Willem ten Rhijne. Casaerius had fallen gravely ill during the trip to the capital city and died not long after their arrival. Van Reede therefore needed help, especially with medicine and Latin. Padtbrugge had just left. (During his second period in the East, he became head of the city of Jafnapatnam—about 1674—but as another of those who came into conflict with Van Goens, he was ordered home; probably intentionally missing the return fleet, however, he sailed to Batavia instead, where he resided from October 1675 until appointed governor of Ternate at the end of 1676.) Likewise, Van Reede had just missed Louys de Keyser, physician to the hospital in Batavia, who was then accompanying an embassy to China.[40] Herman Grimm, of Swedish origin, was a physician-botanist who was in Batavia as chief of the surgeons, after having first been posted to Ceylon in 1674. In 1678 he would be engaged as a "doctor chimicus" at sixty to ninety gilders a month, before repatriating back to Europe with his wife and family in October 1681. When Van Reede arrived in the city, however, Grimm was engaged in publishing at Batavia a work in Dutch on the medicines of Ceylon, the *Thesaurus medicus insulae Ceyloniae* ("The medical treasury of the island of Ceylon," 1677). (It was also translated into Latin and published in Amsterdam in 1679.) But as an advocate for Ceylon, Grimm was a protégé of Van Goens, and so Van Reede avoided him.[41] Cleyer himself was in Batavia, of course, and had laid out an extensive garden. But although he was a wealthy and powerful medical official, even had he wished to assist someone against whom Van Goens had set his face, he would have been of little use: he was in fact not an expert botanist,

Coconuts and trees, with a wild cinnamon tree depicted as the second from the right. Line engraving after J. Nieuhoff, *Remarkable Voyages and Travels to the East-Indies* (1682). Courtesy of the Wellcome Library, London

nor was his ability to write Latin any better than Van Reede's. Van Reede therefore found a collaborator in Ten Rhijne, who was independent of Van Goens. By the time he departed for home at the end of 1677, Van Reede trusted Ten Rhijne enough to leave copies of his text, drawings, and paintings in Ten Rhijne's possession.[42] In a letter of 1680, Ten Rhijne would write to a friend about "our Hortus Malabaricus" ("Horte nostro Malabarico"), and he was thanked in the preface to the third volume of the *Hortus Malabaricus* (1682).[43]

The first two volumes of the *Hortus*, based on material that had been sent back to The Netherlands before Van Reede left Malabar, appeared in 1678 and 1679, although because of the death in autumn 1678 of both Van Reede's publishers and one of his chief collaborators, Arnold Seyn, the rest of the work was put on hold for a while.[44] Several copies of the first volume were presented to the Heren XVII, who sent some of them on to the council in Batavia with a note saying that they were interested to know what might be done with the information in it for improving the medicines and medical supplies of the VOC, and suggesting that Hermann be sent from Ceylon to Malabar to head up the investigations. (Hermann escaped the appointed task by taking up the professorship in Leiden.) But the copies of the *Hortus* and its note made the new governor-general, Van Goens, furious. In his council's reply, they therefore told the Heren XVII in no uncertain terms that just as Ceylon's cinnamon was superior to that of Malabar, so was Ceylon medicinally superior to Malabar. They repeated that Ceylon was overflowing with rich medicinal resources and rather exaggerated Hermann's accomplishments, saying he had studied more than ten thousand of the island's herbs, shrubs, and roots, a great many of which had never been described by any author before. And they mentioned that Cleyer and Ten Rhijne were working together to make use of these botanicals for the VOC medicine chests, while at the same time Grimm was producing in the laboratory a number of chemical remedies from botanicals that grew in the woods and gardens of Java. They declared that they would soon need very few medicines to be sent from The Netherlands.[45]

THE DUTCH MARKET FOR RARITIES

But the quest for medicinals remained only a part of the reason why servants of the VOC invested time and energy in searching out the flora and fauna of Asia. They also supplied a growing market in The Netherlands for naturalia of all kinds, although this can only be glimpsed through the evidence of large enterprises. In 1676 and 1677, for instance, Joan Bax—who had been sending specimens from Ceylon back to his uncle, Director Huydecoper—collected plants and animals as a gift from the VOC for William III, gifts that ended up in the hands of Daniel Desmarets, William's court chaplain, who housed them in the menagerie and gardens of the prince at the palace of Honselaarsdijk. The gift was apparently in response to a request from the stadholder in September 1675 to collect for him from the East "all sorts of animals, birds, herbaria, cabinets, and other curiosities." In response, the Heren XVII sent out instructions to the various stations, including Ceylon, the Cape of Good Hope, and Malabar to collect birds, plants, bulbs, and seeds on behalf of William; Ceylon was also asked to send a couple of elephants, birds, and other tame animals, cinnamon and pepper trees, and other rare plants.[46] Van Goens sent the elephants in 1679. (Gifts of elephants and other animals were common enough between princes: "There came an Ambassador from the King of Bengal . . . who was to take to the Singelese King at Candy a Rhinoceros, two Horses, two Ibexes, one Buffalo-Bull and the Skin of another that had dy'd on the way, and one Dog," reported Johann von der Behr about one episode in 1646.)[47] The traffic in private specimens caused the Heren XVII to complain to the high government of Batavia in October 1677 that the last return fleet had been so "covered and obstructed in such a way with boxes, and that in such great numbers, as if they were whole gardens resulting in so great a weakening and damaging of the ship by all the weight on top that we were obliged to write off and prohibit herewith the sending of all those cuttings, trees, and plants, as well as the making of those gardens, and this now and then for the private use of those Chiefs."[48]

The eagerness of Dutch gardeners to acquire new specimen plants had clearly continued unabated. As elsewhere in Europe, elaborate rural pleasure gardens in The Netherlands were being constructed on a grand scale. The building of great garden estates by Maurits of Nassau in Brazil and at Cleves, for instance, had a considerable influence on the gardens of the princes of Orange.[49] The House of Orange came to possess six great estate gardens—at Honselaarsdijk, Huis ter Nieuburg, Naaldwijk, Rijswijkse Bos, Huis ten Bosch, and Noordeinde Palace—along with smaller gardens at their houses in Breda, Buren, IJsselstein, and Zuylensteyn. By the 1670s, some of the wealthiest merchants and regenten were also constructing similar monuments to living architecture. One listing of the chief estate gardens at the end of the century other than those of the House of Orange gives the following number: eight in North Holland, four in South Holland, three in Utrecht, three in Gelderland, and two in Overijssel. Despite the growing size of such garden estates, exotic plants remained important items of display, often being set out in urns to create special visual interest. At Honselaarsdijk, for instance, "the outstanding collection of exotic plants impressed foreign visitors."[50] The cities remained full of small private gardens at the back of houses, enclosed by walls, the wealthiest being ornate and entirely suitable for the display of showcase specimens. Indeed, "in all respects Dutch garden art in this period profited from flourishing horticulture."[51] The common conceit that by gardening one could re-create a bit of paradise by assembling examples of the variety of God's creation remained a constant throughout the horticultural literature of the day.[52] The chief work on the subject, Jan van der Groen's *Den nederlandtsen hovenier* ("The Dutch gardener," first published in 1669 with many subsequent editions), therefore calls Holland "'a beautiful arbour filled with all kinds of delights' where Dutch trade has assembled goods from all corners of the globe. Various gardens could therefore be compared, so to speak, with cabinets of nature or of art. As small open-air museums they accommodated rare plants from East and West, exotic birds, rare stones and shells from distant shores, topiary, statues, and fountains that, as a collection of naturalia and artificialia, provided a survey of the wonders of nature and of art."[53]

The botanical garden at Leiden also continued to make the acquisition of

Bird's-eye view of the garden estate of Heemstede. D. Stoopendaal after I. Moucheron. By permission of the Rijksmuseum, Amsterdam

exotics a priority. Two Italian diplomats to the peace conference at Nijmegen, visiting the hortus in 1677, remarked on the "infinite" number of plants from remote places and the truly noble rarities on display, which included not only strange aquatic and terrestrial animals from Asia but a lifelike prepared hippopotamus. Ten years later, a visiting French scholar reported that given the large numbers of items from Asia, the Leiden ambulacrum was called "le Cabinet des Indies."[54] Following his arrival from Ceylon in 1680, Paul Hermann in particular built up the botanical collections, traveling extensively in Europe to examine imports and to collect seeds and specimens. According to published catalogues, the number of species raised in the beds had grown from about 1,100 under Clusius in 1600 to 1,500 under Adolphus Vorstius at midcentury, while under Hermann, the number doubled to 3,000 in 1685. At the same time, the specimens of naturalia listed as on display in the ambulacrum attached to the garden also jumped, from about 110 curiosities in 1659 to 290 in 1680.[55]

In the midst of such enthusiasm, even the Amsterdam vroedschap established a new Hortus medicus in late 1682 on land beyond the precincts of the hospital, in a new eastern extension of the city called De Plantage, where it remains to this day. Although it retained close ties to the Collegium Medicum and the Collegium Chirurgicum (both of which paid substantial annual fees), and supported a professorship of medical botany at the Athenaeum (first held by Frederik Ruysch from 1685), the hortus was reestablished as an independent institution answerable to the city government via two "Commissarissen" and with the aim of collecting specimens of special botanical interest. Jan Commelin and Joan Huydecoper van Maarseveen served as its first commissioners. Huydecoper had been one of the Heren XVII of the VOC since 1666 and had long been using his influence within the company to collect exotics. He also had good contacts within the WIC, which he used to collect American plants. Commelin operated a large importing business in pharmaceutical commodities, supplying many apothecaries, the hospitals in Amsterdam and Gouda, and other retailers and institutions. After his death, he was succeeded by the physician Petrus Hotton; after Hotton departed for Leiden in 1695 (succeeding Hermann), Commelin's son, Caspar, took his place.[56] The first commissioners made great efforts to collect exotics, with Huydecoper in particular using his influence within the VOC to have ship captains and governors of various Asian factories send specimens and seeds directly to the Amsterdam hortus. The Amsterdam gardeners also attempted to acclimatize some of the most important commercial plants from the East Indies—such as cinnamon, camphor, and tea—and from the West Indies, such as pineapple. From there, the plants could be sent for cultivation in other parts of the Dutch world. Other directors of the VOC opposed this policy, however, "since dispersal of the areas of cultivation of valuable plants would weaken the VOC monopoly in the trade of these commodities." For this reason the orchard of camphor trees established at the Cape was cut down, and when Van Goens stopped there on his way back to The Netherlands he also made sure to destroy the cinnamon trees.[57]

Most of the new exotics came from Asia via the Cape's hortus, where many of them were acclimatized before transshipment to Europe. The first introductions of Cape plants into Dutch gardens went to the private grounds of Hieronymus Beverningk in 1663 and after, and to the hortus at the university of Leiden from 1668. Huydecoper's avid collecting of all kinds of exotic curiosities and plants from about the same time was conducted through his nephews Joan Bax van Herentals and Simon van der Stel, who both became governors of the Cape colony. By the mid-1680s, one visitor reported that it was "one of the loveliest and most curious gardens that ever I saw," with both European and exotic fruit trees, flowers, and herbs (tended by a large proportion of the five hundred slaves the VOC kept at the Cape).[58] Huydecoper urged Van der Stel to ship back only small amounts at a time until he became commissioner of the Amsterdam hortus in 1682, after which he tried to get Van der Stel to ship as much as possible. In 1683, he also commissioned copies of the drawings of Cape plants he was sent by Bax and Van der Stel, which he then sent to the merchant Jacob Breyne in Danzig. The grand pensionary of The Netherlands, Caspar Fagel, owned a botanical garden at his country house Leeuwenhorst (near Noordwijk), where he grew cinnamon, clove, and camphor trees first acclimatized at Cape and sent to him by Van der Stel. Fagel also requested and received permission to send a collector to the Cape in 1684 for acquiring interesting specimens for his personal use.[59]

Given such interests, people like Cleyer himself also invested large sums in the pursuit of horticulture in Batavia, no doubt to build a reputation with naturalists in Europe and with some of the governors of the VOC, but probably in part also for local prestige and even sincere curiosity. By the later half of the century, with the Dutch having dominated the area around the city to achieve local peace, Batavian residents were purchasing nearby grounds on which to establish gardens. Cleyer himself laid out a large botanical garden just outside the city that required the labor of about fifty slaves and hired a gardener, Georg Meister, who served him from 1678 to 1687.[60] Cleyer also sent botanical reports to Europe. His medical collaborator of the later 1670s, Ten Rhijne, was in communication with Jacob Breyne, who had received his training in commerce in The Netherlands and had become deeply interested in exotic botany while there. Breyne made contact by letter with many of the naturalists of the VOC, and Ten Rhijne had sent Breyne a branch of a camphor tree with leaves, flower, and even fruit from Japan in 1674; in 1677 he forwarded his collections and descriptions from South Africa and Japan, including an account of the cultivation and use of tea in Japan. Breyne published them in his *Exoticarum plantarum centuria prima* ("Exotic plants: The first hundred," 1678) with due credit to Ten Rhijne.[61]

Breyne also received news from Cleyer about an herb that seemed to work well against gout; he brought this to the attention of Christian Mentzel, court physician to Great Elector Frederick Wilhelm, who suffered from the ailment. Like Breyne, Mentzel was a member of the "Academia (or Collegium) Naturae Cu-

A branch of *Camphorifera japonica,* drawn from a specimen sent by Willem Ten Rhijne, from Jacob Breyne, *Exoticarum aliarumque* (1678). Courtesy of the Wellcome Library, London

riosorum," a group founded in 1652 among the physicians and naturalists of the German-speaking lands (which received royal protection in 1687, after which it became known as the "Leopoldina");[62] as a reward Mentzel or another member of the academia had Cleyer made a member, too, in 1678, under the classical nickname "Dioscorides."[63] Cleyer soon came to use Mentzel as his main conduit for sending information to Europe. On 25 November 1681, for instance, Cleyer sent Mentzel a treatise with ink drawings describing leprosy in Java, which his son, Johann Christian Mentzel, used as the basis for his medical thesis; it was in turn published in Latin in 1684 in the academia's *Ephemerides*. This reproduced as an etching the illustration sent by Cleyer (from an unnamed artist) of the face of a local woman of Batavia, an illustration so clear that modern physicians have praised it as clearly diagnostic and not to be surpassed for a century or more.[64] Following his two trips to Japan, Cleyer also sent the academia a number of descriptions of plants there, which were published under Cleyer's name although they appear to have originated with his gardener, Meister, who accompanied Cleyer on both his trips to Japan.[65] (Meister himself published a book on East Indian botany in 1692.)[66] Of the botanical drawings commissioned by Cleyer, 599 pages containing 1,060 pictures now reside in Berlin.[67]

Given such encouragement from Europe, in November 1681 Cleyer contracted the chemist Hendrik Claudius to go to the Cape to draw and paint Cape plants, start a herbarium, and to collect minerals, drugs, and other naturalia, all at Cleyer's expense.[68] Claudius arrived there on 16 February 1682 and got to work, soon being taken into the service of the local governor, Simon van der Stel, who had himself obtained a training in botany before sailing for the Cape in 1679.[69] In April 1685, Van Reede also arrived at the Cape again on his way east, having been called out of retirement by the Heren XVII to look into their affairs in Asia and root out corruption. He spent over a year at the Cape, from April 1685 to July 1686. He took a keen interest in the garden, in the orchards and woods, and in the ways and ideas of the Cape Khoikhoi people, since "as regards knowledge, brains, fairness, and reasonableness, in so far as required for their housekeeping and civil manner of government, they are not second to any other peoples"—he also tried to improve the conditions of the slaves and servants of the Company.[70] Much impressed with Claudius's work, Van Reede also kept his results—"He hath compleated two great Volumes in Folio of several Plants, which are drawn from life, and he hath made a Collection of all the kinds which he hath pasted to the Leaves of another Volume"—in his own rooms.[71] Moreover, the extraordinary representative of the Heren XVII authorized a four-month overland expedition to explore the copper mountains to the north (in Namaqualand).[72] Claudius served the expedition as secretary and draftsman, making watercolors and writing descriptions of the plants and animals along the way.[73]

Sharing his information would prove Claudius's undoing, however. Not long after the expedition, a group of French Jesuits stopped at the Cape colony (from 13 to 26 March 1686) on their way back to Europe from Siam (Thailand).[74] The group had been sent to the Far East by Louis XIV after one of the leading missionaries to China, Philippe Couplet, made a personal appeal to the king;[75] Couplet and Cleyer had been in close contact for some time;[76] so it was not surprising that someone who had worked with Cleyer and someone who had known Couplet would talk. Claudius freely communicated his knowledge with the Jesuits: "It is from him that we got all the knowledg[e] we have of that Country, of which he gave us a little Map made with his own hand, with some Figures of the Inhabitants of that Country, and of the rarest Animals, which are here inserted. The most remarkable things we learnt are what follow," wrote Guy Tachard in his 1686 book about the trip to Siam.[77] When a copy of the book came into Van der Stel's possession in late 1687, however, Claudius's action was held to be treasonable, and he was banished from the Cape; his replacement, Hendrik Bernard Oldenland, who had studied with Hermann, continued to help turn the Cape garden into an important research center.[78]

In the meantime, Van Reede himself had pushed on to Colombo, where he arrived on 13 October 1685 and stayed until early December. While he was there, interest in local botany was revived, for both medicinal and pleasurable ends. Aegidius Daelmans, a ship's doctor in service with the VOC, spent eighteen months on Ceylon beginning in 1687 (and then several years on the Coromandel coast) publishing a short interpretation of diseases and treatments in the Indies in 1687 titled *De nieuw hervormde geneeskonst, gebouwt op de gronden van het alcali en acidum* ("The new reformed medicine, based on the foundation of acids and alkali"). In it, he remarked that he was sent out to look for "new herbs and roots" on Ceylon useful to medicine, "of which I had already made such a large collection, that I presented Governor Pyl the pharmacopoeia of Ceylon, comprising both external and internal remedies, without there being any necessity for sending out for a single other from the Fatherland." As the title indicates, Daelmans followed the views of Sylvius (and Bontekoe) and favored the use of various chemical remedies, laudanum, and especially tea.[79] Additionally, from 1686 to 1689, several shipments of botanicals and even animals were sent to The Netherlands. Over the next several years Van Reede sent many plants back to the Amsterdam hortus himself, and in 1691, he issued an "instruction" to all the posts in the VOC's Western Quarters, requiring all kinds of plants to be collected along with a record of their names and sent to Ceylon, from where they would be shipped to The Netherlands, especially for the use of King William III and Jan Commelin in Amsterdam. "The government of Ceylon carried out the 1691 instructions for nearly one hundred years." When Cleyer's gardener, Meister, repatriated to The Netherlands in 1688, he carried with him a large number of plants from Cleyer's Batavian garden to the garden at the Cape colony (including a camphor tree and a tea shrub from Japan), and from the Cape he then transported a consignment from Van der Stel of seventeen cases of plants back to The Netherlands, where they were destined for the Amsterdam hortus and the gardens of William III and Fagel.[80]

No wonder, then, that after a visit to The Netherlands, William Sherard, a botanical enthusiast and Fellow of Saint John's College, Oxford, pseudonymously published a book titled *Paradisus Batavus* (Amsterdam, 1689): the "Dutch paradise."[81] Sherard had been drawn north from Paris, where he met the former physician of Ceylon and now professor of botany at Leiden, Hermann. Sherard had gone there in 1686 to study under Joseph Pitton de Tournefort, professor at the Jardin des Plantes, and early in the summer of 1688 Hermann had come to pay a call on Tournefort, who was preparing a catalogue of plants. Sherard and Hermann became friendly, and the Englishman returned with Hermann to The Netherlands later that summer. There he visited the private gardens of the Prince of Orange, of Hans Willem Bentinck at Sorgfliet, of Heironymus Beverning at Oud Teylinghe by Warmond, of Caspar Fagel at Leeuwenhorst, of Philips de Fline at Sparen-Hout by Haarlem, of Agnes Block at Vijverhof, and still others possessed by Simon van Beaumont, Frau Pullas, Daniel Desmarets, Johannes van Riedt, and others. He also visited the more public gardens in Leiden and Amsterdam. Just about every living thing named by Adam in the Garden of Eden was being rediscovered and renamed by the Dutch, or so it might have seemed to some imaginations.[82]

ACCLIMATIZATION

But the successful transportation of live plants over long distances by sailing ship took elaborate preparation and continuous attention. For instance, before he left for Jamaica to explore its natural history, the English physician Hans Sloane had instructions from Daniel Desmarets, chaplain and horticulturist to William III, about how to transport trees, shrubs, seeds, and bulbs compactly but without drying up, rotting, or sprouting. Trees and shrubs had to be uprooted with as much of their roots and surrounding soil intact as possible, with the whole wrapped into a ball of damp moss covering up to two feet of the stem. The different specimens were then laid in bundles and encased in a box, leaving several feet of free room at the top and sides. Milky plants could be transported in a bit of dry sand around their roots, again enclosed in dry moss, and encased in a simple wooden box; bulbous plants had first to be dried in the air away from the sun until faded and then also placed in dry moss in a wooden box; seeds had to be put dry into a well-closed box.[83] Notes from others included advice on how to poison the ship's rats with pieces of meat containing arsenic so that they would not gnaw anything green. In addition to the packing, baiting, and watching, so that the

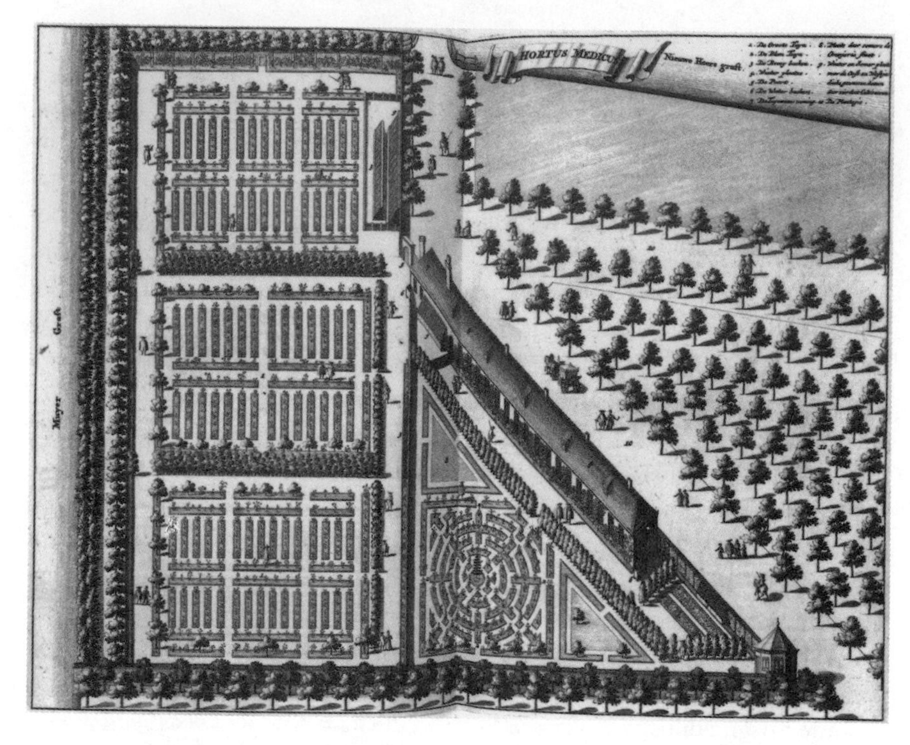

The Hortus Medicus in Amsterdam, showing the hothouse (on the diagonal) with glass frames leaning against it, from Caspar Commelin et al., *Beschryvinge van Amsterdam* (1693). Courtesy of the Wellcome Library, London

sailors did not harm the plants or saltwater soak them, keeping the roots of many of the plants moist with fresh water would have required the regular attendance of the gardener during the journey. Shipping animals was even more difficult. When Christopher Schweitzer's ship returned from Ceylon to The Netherlands in 1682, it departed with "a great many animals on Board, *viz.*, twelve Parrots, six Apes, two *Cacadus* of Ambona (they are white Birds as big as Pigeons, with a fine tuft on their Heads, and apter to learn to talk than the Parrots); we had a Crocodile an ell long, a young Deer and young Elk from *Bengal*, with lively white spots: All these dyed in two months time, except two of the Apes, *Cornelius* and *Margretha*."[84] When Johan Bitter returned to The Netherlands in 1688, the crew had fun secretly feeding his cassowary burning coals from the galley.[85]

Once in The Netherlands, many tropical plants had to be treated with additional care. From the late seventeenth century, the Leiden hortus's ambulacrum —with high windows facing south covered by rectangles of glass—served as a place into which tender plants could be moved in the winter. In the large estate gardens that followed a few decades later, "orangeries" were built—similar in design to the ambulacrum, with peat fires in braziers keeping the temperature a few degrees above freezing—where orange trees growing in large pots could be successfully wintered over.[86] In addition, in the second half of the seventeenth century, "hotframes with loose lights became fashionable in the Netherlands."[87] The soil inside the frames was sometimes mixed with tanbark to raise the temperature. Lean-to greenhouses also appeared, against south-facing walls, with roof lights at a steep angle to capture the low winter sun. Orangeries were improved with the construction in them of stoves with flues. But the most important improvement came in the 1680s with the development of true hothouses, heated by stoves that radiated their heat through a system of underground pipes, allowing the winter temperatures to be kept many degrees above freezing throughout the winter. They were very expensive: the "orangery" built by William III in 1680 at Honselaarsdijk cost thirty thousand gilders, probably because it was one of the first such hothouses. The medical gardens at Amsterdam and Leiden appear to have had such tropical houses around 1685 (the largest one in Amsterdam being heated by five stoves), and given the number of tropical plants in private hands from about the same time, private gardeners appear to have constructed similar buildings.[88]

With such means, many tropical plants began to flourish in Dutch gardens. In 1670, the wealthy Mennonite Agnes Block purchased a country residence at Loenen aan de Vecht from the estate of the Amsterdam councilor and sheriff Joris Backer and transformed it into Vijverhof, one of the era's great estate gardens.[89] Block was the first in Europe (in 1687) to grow a fruiting pineapple, a Brazilian native, having a medal struck on the occasion and commissioning the noted botanical artist Alida Withoos to depict it. (She also had Withoos paint her tea plant, possibly obtained from the Amsterdam garden, since Block and Commelin exchanged plants.) She must have possessed a hothouse to accomplish this feat, since the plant required temperatures no lower than about 10 degrees Celsius even in the winter.[90] In northern Groningen, the professor of botany Abraham Munting was able to raise bananas in his heated greenhouses, and it was even reported to the Royal Society of London in 1682 that he had been able to grow cinnamon and nutmeg trees.[91] One historian of horticulture has listed an impressive number of tropical plants grown in Caspar Fagel's estate garden Leeuwenhorst, near Noordwijkerhout. They "included baga (*Crinum zeylanicum*) and superb lily (*Gloriosa superba*) from Ceylon, the milkworts *Euphorbia antiquorum* and *E. nivulia* from India; Indian red sandal-wood (*Pterocarpus indicus*) and tanggooli (*Cassia javanica*) from Indonesia; Indian azalea (*Rhododendron indicum*), camphor wood (*Cinnamomum camphora*), and tea plant (*Camel-*

Oil painting of Agnes Block with her second husband, Sybrand de Flines, and two children, in her garden at Vijverhof, by J. Weenix (before 1697); note the pineapple depicted in the lower left. By permission of the Amsterdams Historisch Museum

lia sinensis) from Japan; *Dracaena fragrans* from tropical Africa and *Dracaena draco* from the Canary Islands; nasturtium (*Tropaeolum majus*) and pineapple (*Ananus comosus*) from Peru and Brazil; and the orchid *Brassavola nodosa* from the West Indies. From Japan came *Cycas revolute*."[92] Another assessment of lists and catalogues of plants cultivated in Dutch gardens late in the century has noted several species from Japan, twelve from Indonesia, many from Ceylon but none from Malabar, and a great many from the Cape. In addition, there were plants from the Mediterranean and North America and a few from the West Indies and Surinam.[93]

As Block's employment of an artist to depict some of her exotics also suggests, Dutch garden fanciers also created a rich pictorial record of their successes. For example, in 1685 William III asked the artist Stefan Cousijns to make colored drawings of the botanical collections at Leeuwenhorst. He stopped work in November 1688, perhaps because of Fagel's impending death, leaving thirty-five blank pages at the end of what was called the "Hortus Regius Honselaerdicensis," for after Fagel's death the collection of plants and illustrations was transferred to Honselaarsdijk.[94] The commissioners of the Amsterdam hortus, too, ordered watercolors to be made of the rare plants they collected and gathered them in the first eight volumes of the "Moninckx Atlas," named after the chief artist, Jan Moninckx. Other illustrators included Moninckx's daughter Maria, Alida Withoos, and Johanna Helena Herolt née Graff (daughter of the famous artist and naturalist Maria Sybilla Merian); later drawings were made by Jan Matthias Cok and Dorothea Storm née Kreps.[95] The illustrations were used for a four-part atlas in two volumes published by Jan Commelin and edited by the physician Fredrick Ruysch and the apothecary and botanist Frans Kiggelaer (*Horti medici Amstelodamensis rariorum plantarum historia*, 1697–1701). Huydecoper also had many of the drawings of plants made for him copied and sent on to Breyne in Danzig.[96] It was during this period, from 1682 to 1693, that the remaining ten volumes of Van Reede's *Hortus Malabaricus* appeared. There was enough public excitement about the plants of the Indies to support the publication of a huge, beautifully produced, erudite, and expensive volume, or set of volumes, about exotic botany every year or two.

RUMPHIUS AND MERIAN

The collection of natural rarities, and publications about them continued to be based on extraordinary work by people in both Indies. One of the most famous collector-authors lived on the far side of the Dutch East Indies, in the Spice Islands of the Moluccas: Georgius Everhardus Rumphius, later known as the "Pliny of the Indies" and "the blind seer of Ambon" (the second epithet arises from his decades of work after going blind).[97] A VOC merchant who came to love Ambon and the Moluccas, and with the support of his superiors was able to devote a great deal of energy to his studies, Rumphius is now best known for his *D'Amboinsche rariteitkamer* ("The Ambonese curiosity cabinet," 1705) and *Het Amboinsche kruid-boek* ("Ambonese herbal," in six volumes, 1741–50), both based on work he did in the 1660s and 1670s. They were two of the most impressive works on natural history published in the early eighteenth century. The result of years of labor, tragedy, and disappointment, they appeared only posthumously. Like Van Reede, Rumphius had support for his efforts from some of his superiors within the VOC, but nothing like a subsidy that would underwrite the costs of publication.

Rumphius's life history began not unlike Cleyer's. Born in Germany (probably around 1627 at or near Hanau—seven years before Cleyer), in 1652 Rumphius also signed on with the VOC as a "gentleman soldier" at about the same age, arrived in the Moluccas in 1654, and was soon promoted and put to work overseeing construction of defensive works and other structures. He must have accomplished all his assignments well and learned good Portuguese and Malay (the languages of trade), since after his first five-year hitch Rumphius transferred to the civilian branch of the VOC as a merchant (*onderkoopman*), which could only have come with the support of his superiors. Moreover, he secured the posting to Hila, which, because of the trade in cloves, was the second-most lucrative place within the territories of the factory of Ambon, which in turn was the richest of all the VOC factories. While living very well, he also became fond of the area, and apparently by the late 1650s—about the time that Piso's edition of Bontius appeared—he was beginning to compile a natural history of Ambon. Perhaps he was inspired by this edition, for one of the manuscripts he wrote (later lost) was a commentary on Bontius's natural history.[98]

By the time of his reappointment in 1663, Rumphius had clearly launched into an extensive attempt at a complete natural history of Ambon, and at the next renewal, in 1667, it was clear that his studies had become his paramount interest. From at least the late 1660s, he had explored the countryside and seashore with native-born friends, learning the local language to do so (as well as writing a now-lost dictionary of Malay). He established a household with a local woman, known only as Susanna, who helped him with his work. (He would name a rare orchid "Flos Susannae" after her, who first showed it to him.) As his recent biographer says, "One will find evidence of Rumphius' respect for the Indonesians in his work and in comments from his superiors. . . . He knew their food and medicine, their weapons and dress, their superstitions and stories. He was without a doubt the first (and for a long time the only) significant ethnographer of Indonesia." Moreover, he "often took the side of the local people against his powerful employer, nor did he scruple to lecture and criticize his European superiors. . . . One gets the feeling that he often was more comfortable with the indigenous population than with his compatriots." He wrote with great respect of "master" Iman Reti, for instance, who was "a Moorish Priest from Euro" who taught him the method for extracting oil from the wood of a tree. He may even have imbibed "the Indonesian ethic of liberality. He mentions it in his work time after time and always with approbation." His work also shows that Rumphius held views similar to those of Bontius and his Dutch contemporaries: he "preferred empirical evidence to theory" and "condemned man for trying to control nature. Nature is too immense [*improbus*] for our understanding, and it is vain to claim otherwise." At the same time, like Bontius, he attacked various "superstitions," such as "tapa" and other varieties of local religious expression, and the practice of magic.[99]

By early 1670, however, Rumphius was blind and by necessity had to move from the place he loved, Hila, to the town of Ambon. Yet he showed dogged determination to continue his projects. Because of the knowledge he acquired, and with the firm backing of Governor-General Maetsuyker, he was allowed to stay on with his rank and income as an adviser to the VOC. In his capacity of governor of Ternate and then of Ambon during the late 1670s and 1680s, Padtbrugge also firmly supported Rumphius, even securing a parcel of land for Rumphius's burial plot.[100] Rumphius had to abandon the text of the *Herbal* he had almost finished because he could not find anyone with whom he could work in Latin, but he began it again in Dutch. The color illustrations he had made remained, and he commissioned yet others. With the help of his family and servants, especially his son, Paul August, he also wrote two books on Ambon: *Generale lant-beschrijvinge van het Ambonsche gouvernement* ("General survey of the Ambonese possessions of the VOC"), completed in 1678, and *Ambonsche historie, sedert de eerste possessie van de O. I. Compagnie tot den jare 1664* ("History of Ambon, since its first possession by the VOC to the year 1664"), completed in 1679. Both were sent to the governing council in Batavia but never sent on to The Netherlands for publication because they contained too much confidential information.[101]

Despite these setbacks, Rumphius persisted. His knowledge of Ambonese natural history was becoming well known. From remarks made in the dedication to the Delft physician Hendrik D'Acquet in his *Ambonese Curiosity Cabinet* (see below), he seems to have been engaged in a substantial trade in naturalia from the region, especially in the many beautiful and unusual shells. He also sold most of his collection of curiosities to Duke Cosimo III de' Medici in 1682. As his reputation increased, Cleyer recommended him for membership in the Academia Naturae Curiosorum, and between 1683 and 1698 thirteen of Rumphius's "observationes" were published in their periodical, the *Miscellanae curiosa medico physica* of Germany.[102] A fire that burned down Ambon in 1687 destroyed his herbal, plus his other manuscripts, books, and collections. Cleyer sent him aid, in return for which Rumphius sent Cleyer a manuscript on Chinese pulse doctrine.[103] Fortunately, he had taken the precaution of storing partial copies of his most important works elsewhere, and with the help of his son Paul August and Philips van Eyck, a draftsman sent to him the next year by the VOC, he made up some of the losses. The first six books of his *Herbarium Amboinense* were sent to Batavia in 1690 and put aboard ship for The Netherlands in 1692, only to disappear again when the vessel carrying them sank. This time Rumphius had kept a clean copy. In 1696 they were again sent to Holland, accompanied by an additional three books, with the last three books following in 1697. As with Rumphius's first two manuscripts, however, this remarkable work was swallowed up by the archives of VOC. Only decades later was it rescued by the Amsterdam professor of botany Johannes Burman, who edited and published its six parts in four volumes between 1741 and 1755.[104] Even Rumphius's last work, *The Ambonese Curiosity Cabinet*, which had a better fate, appeared in print posthumously. Other parts of his unpublished work were incorporated into that of his friend and son-in-law, François Valentyn: Valentyn drew freely on Rumphius's work on animals, the *Amboinsch dierboek* (which has since disappeared), as well as Rumphius's illustrations and his *Lant-beschrijvinge*, for his own work on the Dutch East Indies, *Oud en Nieuw Oost-Indien* (1724–26).[105]

Even when Rumphius's *Ambonese Curiosity Cabinet* finally arrived in The Netherlands, the publisher complained about the state of the manuscript. In 1701, a year before Rumphius's death, it came into the hands of the physician and collector D'Acquet, a burgomaster of Delft. Rumphius had been corresponding with D'Acquet for many years and sending him specimens from the East, especially shells. Rumphius's manuscript was largely based on descriptions he made long ago, which he had gathered and sent on to D'Acquet in the hope that D'Acquet would find them worthy of publication. "Yr. Worship owns the majority" of the objects described "in your outstanding Cabinet." It was fortunate that he did, for in the publisher's dedication to D'Acquet, he complained that the manuscript passed on to him had been far from ready for the press. Many of the descriptions were incomplete "while other items were entirely lacking and which, without harming the order and design of the work, could not be left out." Moreover, "as far as the numerous drawings of Cockles and Shells were concerned, which the Author had promised us, or rather Yr. Worship, they never reached us." As a result, the publisher had to commission new illustrations of the missing items, which were done from "specimens from local Curiosity Cabinets." D'Acquet helped him to gain access to collections in private hands, and Simon Schynvoet, another collector, wrote the additional text and supplied the missing drawings.[106] Many of the illustrations for the book came from those D'Acquet already possessed.[107] Four years of additional work finally led to the book's publication.

In the same year that Rumphius's *Ambonese Curiosity Cabinet* appeared at the booksellers, another remarkable work was privately printed: *Metamorphosis insectorum Surinamensium ofte verandering der surinaamsche insecten* ("Metamorphosis of insects of Surinam," 1705).[108] It was written by another naturalist who was also an important painter and illustrator of natural history, Maria Sibylla

Merian.[109] She had returned to Amsterdam from a two-year expedition to Surinam, on the northern coast of South America, in 1701, the same year that Rumphius's manuscript reached Delft.

Merian was the daughter of the well-known engraver and artist Matthäus Merian the Elder of Frankfurt, and she received an apprenticeship in drawing and painting naturalia from her stepfather Jacob Marell. Among his teachers was Georg Flegel, one of those working in the tradition of Albrecht Dürer, and consequently Merian's work from as early as 1660—when she was thirteen—shows her painting in the style of Dürer. From the same early age she evidenced a keen interest in flowers, insects, and caterpillars. After marrying one of Marell's apprentices, Johann Graff, she moved with her husband to Nuremberg, where she set up her own business and took apprentices, selling cloth painted in flowers and experimenting with pigments to find the best ones to survive washing. Among other projects Merian pursued was also an attempt to find other caterpillars that could spin filaments strong enough to substitute for the silkworm. Very unusually, the beginning of her diary also shows her carefully following the life-cycles of the little creatures from egg to caterpillar, pupa, and butterfly, even taking an interest in their parasites. (This was almost a decade before Malpighi published on the silkworm or Redi described insects as arising from eggs rather than putrefaction). To make her observations, she raised the insects herself, discovering in the process which plants they fed on. Her first book (*Der paupen wunderbare verwandelung*, "Wonderful metamorphosis of caterpillars," 1679, with two subsequent volumes in 1683 and 1717) illustrated the life cycles of fifty of them and the plants with which they were found, "for the service of naturalists, artists, and garden lovers."[110] A year later she published a colored book of flowers, *Neues blumenbuch*.

In 1682, however, Merian returned to Frankfurt to care for her widowed mother, and three years later she took her mother and daughters to join her brother Caspar in the Labadist community at Wieuwerd, in Friesland. Through them she would find her way to Surinam. The Labadists were followers of the sectarian religious leader Jean de Labadie. Born at Bourg, near Bordeaux, and educated by Jesuits, he moved toward Jansenism and then to the Reformed communion until he was expelled from the pulpit in Middelburg by the local classis in 1669. Among those who became attracted to his life of constant piety and religious conversation was Anna Maria van Schurman, one of the most remarkable virtuosos of her age, defender of the ability of women to pursue intellectual quests, and friend of Princess Elizabeth. Fluent in Latin and other ancient languages, she had listened to many of the disputes about Cartesianism in Utrecht from behind a screen (so as not to disturb the male students), and Descartes himself had hoped that she might take up his views, but she did not break from the pietism of Voetius, adopting a view of nature that took on many of the new ideas but spoke of God as an "inner cause," the goal of our lives and thought. Her brother visited with Labadie for two months in 1662, and his high praise for the pastor brought Van Schurman into correspondence and then conversation with him; in 1668 she gave up her worldly goods to join Labadie's community. Her "autobiographical" *Eucleria* (1673) became perhaps the most profound statement of the aims of the Labadists.[111] After various tribulations in Amsterdam—the libertine regenten, including Johann de Witt, did not much like the sect—the Labadists took refuge with Descartes's former patient, Princess Elizabeth, now abbess of Herford, in 1670; they moved to tolerant Altona (just north of Hamburg) when the French threatened Herford in 1672; and in 1675, with war from Sweden threatening and Labadie having died, they settled in Wieuwerd on a large estate surrounding the manor house of Walta-slot. Because of an inheritance, three sisters of the Sommelsdyck family who were part of the Labadist community were owed large sums of money to be paid them by their brother, Cornelis van Aerssen van Sommelsdyck; he turned the place over to them in turn for reducing by half their financial entitlements. Here in Wieuwerd, too, another member, Hendrik van Deventer, set up a chemical laboratory in which he produced soap, chemical salts, and various pills for the treatment of fevers. The sale of these pills, which became quite popular and known as "Labadie pills," helped support the community. By the time Merian joined the community, Van Schurman had died, but since believers could be freed of their marital bonds to unbelievers, her husband was turned away when he tried to join her, eventually returning to Nuremberg to file for divorce. Sometime during this period she was inspired by a collection of large and brightly colored butterflies from Surinam sent to Walta-slot by its former owner and now governor of Surinam, Van Sommelsdyck.[112]

Merian remained at Wieuwerd until her mother died in 1691. (The Labadist community at Wieuwerd began to fall apart after Van Deventer stopped contributing to the community of goods of all the members in 1692, instead investing his large medical income in his family.) She and her daughters then left for Amsterdam, where in 1692 her eldest daughter, Johanna, married another ex-Labadist who had complained loudly about the sect's practice of mortifying the flesh.[113] In Amsterdam, Merian again supported herself by selling colored fabrics and by preparing and selling paints for artists. She became friends with Caspar Commelin, who employed Johanna to paint plants for the "Moninckx Atlas," and she visited the remarkable collections of naturalia possessed by Nicolaas and Jonas Witsen, Frederic Ruysch, and Levinus Vincent.[114] But she also developed the ambition of traveling to Surinam both to collect additional specimens for such liefhebbers and to paint its insects. In 1699, with the help of the directors of the West India Company, she followed Johanna and her son-in-law to Surinam, with her other daughter, Dorothea Maria Henrice, in hand.[115] At the time, Surinam had about one hundred mostly privately owned sugar plantations, with a European population of approximately one thousand, worked by about ten thousand African slaves; there was an uneasy peace with the Caribs and Arawaks, who controlled the woods.[116] Among other places she visited was the plantation La Providence, still owned by the Labadists (although none of them worked it anymore), and throughout her journeys she conversed with the slaves and native Americans while growing increasingly contemptuous of the planters, who could not understand why she was wasting her time on anything but sugarcane and its products. After two years, suffering from the climate and probably stricken with malaria, she returned to Amsterdam with her daughter Dorothea (Johanna remained behind for a time). The specimens she had collected (in bottles of strong brandy, between sheets of paper, and in boxes) were displayed in the great town hall of the city. Many of them, and the paintings she had completed in Surinam, she sold.

With her abilities as an artist and an expert in making paints and dyes, Merian also earned an income from coloring the copperplates of many works of natural history. Publishers did not yet have the technical means to print color plates, but they sometimes advertised that coloring could be added to plates afterward for an additional price. The problem then was ensuring the accuracy of the colors of the things represented. For instance, the most important Dutch author of a book on insects in the decades before Swammerdam, Johannes Goedaert, added a foreword to his *Metamorphosis naturalis ofte historische beschrijvinghe van . . . wormen* ("Natural metamorphosis, or an historical account of caterpillars" in three volumes, 1662, 1667, and 1669) offering to do the coloring himself if the buyer so desired, an implicit guarantee of authenticity.[117] Merian apparently contracted with publishers to do something similar. Many of the engravings for Rumphius's *Amboinese Curiosity Cabinet*, tinted by watercolors, ended up in Merian's hands, for her use as templates when customers wanted illustrations in their copies colored in.[118] (Following the discovery in the 1970s in Petersburg of Merian's manuscripts, which contain fifty-four colored illustrations of specimens in Rumphius's book, it has sometimes been claimed that she was responsible for drawing many of the originals for Rumphius's work, but that remains improbable.)[119]

Merian took full advantage of these several commercial methods when publishing her own book, which contained sixty folio-size engravings of the naturalia of Surinam with a facing page of text for each: the *Metamorphosis insectorum* of 1705. Commelin helped her to find proper Latin names for her creatures. The book is most famous for illustrating the insects (again often including their eggs and caterpillars, and sometimes their pupae, as well as the adult form) placed on the plants on which they were commonly found. But she included illustrations of other animals as well, including lizards, snakes, and spiders. In many of them, the proportional sizes of the creatures is not correct, some appearing too large or too small in relation to the other animals depicted on the page, although all are arresting and "lifelike" images.[120] Yet as one commentator has noted, "Some of these resemble designs for embroidery, particularly the style known as *peinture à l'anguile*"—she may have had that market in mind, too. To produce the books, Merian seems to have contracted with the printer herself, since she made sure that they were done on the highest-quality paper in fine copy, and then made them available for purchase at her residence (on the Kerkstraat, between Leidschestraat and Spiegelstraat), either with or without the plates colored. When orders were made for the coloring, she may have had her daughter Dorothea's help, but the best first editions now extant are thought to have been colored by Merian. To carry on the family business after her death, a printing of a second edition, enlarged by twelve plates, was overseen by her daughter Johanna and published by the bookseller J. Oosterwyk of Amsterdam, while her other daughter, Dorothea, published the third volume of her illustrated work on insects.[121] Merian also knew that the original drawings from which the plates were made would fetch the most money, so three sets of "originals" are known today: one partial set in St. Petersburg, one in the British Library (Sloane collection), and one in the royal collections at Windsor. But even the printed editions were very expensive, since it is said that Czar Peter purchased two volumes of her work in 1717 for three thousand gilders. He was so impressed that he also obtained the services of Dorothea, who moved to Saint Petersburg as a court painter.

The production of the books of Van Reede, Rumphius, and Merian reminds us of the extent to which no book is simply the product of an author's mind. Particularly in enterprises such as natural history and medicine, enormous numbers of people needed to be consulted, from those who collected specimens in the field or commented on the uses of natural things to friends and family, supporters and correspondents, paymasters and printers, engravers and colorists. The networks of expert correspondents could reach from the far side of the East Indies to The Netherlands, England, Germany, and Italy. Both "authors" and "readers" were constantly checking with one another and their acquaintances to ensure the accuracy of the written descriptions, images, and colors. By the late seventeenth century, moreover, many of the steps in compiling and transmitting information had been commercialized, although personal favor and reputation still counted

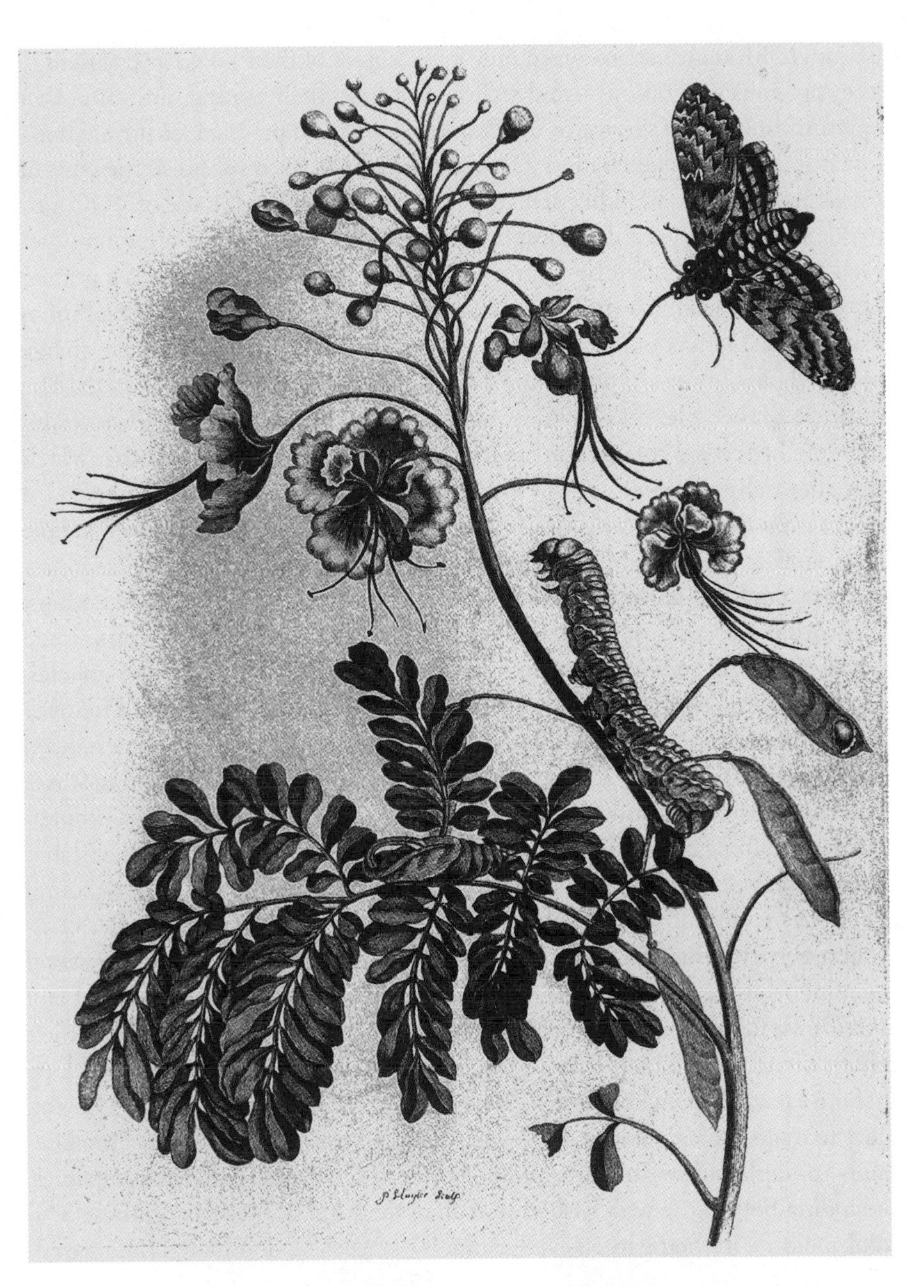

Flos pavonis, from Marie Sibylla Merian, *Dissertatio de generatione et metamorphosibus insectorum Surinamensium* (1726). Courtesy of the Wellcome Library, London

for much. Most authors received only a few copies of their books as payment—as in previous generations—and so lived not from their writing but from other enterprises, whether service in the Company, medical practice, or private trade in naturalia. Van Reede had wealth and power enough to subsidize the costs of research and publication himself, although he no doubt made use of VOC paymasters whenever he could. Rumphius depended on outside support, sometimes even from the VOC, but he never had the clout to hold in his hands a printed book resulting from his work. In such cases as Merian's, however, the author as craftsperson (with enough credit to support the upfront costs) might take charge of the whole business of producing a book herself and reap all the rewards. The directors of the WIC and VOC would have been pleased to see this successful example of vertical integration, as it would now be called, such as that which they themselves pursued.

But even though commerce supported the infrastructure that moved people and manuscripts around the globe, these instances also remind us that business interests could sometimes also conflict with exchange. With rare exceptions, the work in natural history and medicine is best seen as a part of private trade within the VOC and Surinam company than as a part of their business policies. Van Goens disliked the touting of things Malabar because they seemed to make Ceylon seem less important; he certainly would rather cut down acclimatized trees than let the spread of their cultivation threaten the VOC's monopoly over fine spices; Merian found the planters of Surinam to be brutal dullards, uninterested in anything but coining money from the production of sugar by slave labor despite the wonders of the world around them; Claudius could even be found guilty of treason for sharing information about the natural history of the Cape with a French Jesuit; Rumphius's manuscripts were generally treated as secret documents in his day. Without the business enterprises that were the East and West India Companies, none of their work would have been possible, but their efforts often came at high cost to body, spirit, and purse, costs they bore in hope of future rewards from liefhebbers. Nevertheless, the continued attention they all gave to matters of fact and utility—often covered by a gloss in praise of the Creator—underlines how the values of the businessmen in charge of the companies, including both those who wished to reduce their medical costs and those who took pleasure in their gardens and cabinets of curiosity, dominated their world.

9

TRANSLATING WHAT WORKS

The Medicine of East Asia

> Once two parties have clapped hands over a deal neither retracts an inch from his promise. In the general run of business deals . . . promises are deferred and irksome lawsuits follow; but here contracts whose surety is as uncertain as the clouds of a fickle sky are never broken. Within the agreed time, regardless of gain or loss, sales or purchases are honoured. The great merchants of Ōsaka, the foremost in Japan, are great in spirit, too, and such are their methods of business.
>
> —IHARA SAIKAKU, *The Japanese Family Storehouse* (1688)

The VOC sent Dr. Willem ten Rhijne to Japan not on their initiative or his or even as a part of the policy to adopt Asian medicinals within the VOC domains or to collect exotics for the liefhebbers in The Netherlands. He took ship because of a request from the Japanese government. In 1667, reiterated on 1 April 1668, the government of the reigning shōgun, Tokugawa Ietsuna, asked the VOC to send them a physician with botanical and chemical experience.[1] By the mid-1670s, Ten Rhijne, who obtained the appointment, found himself deep in challenging conversations about the medicine of Europe and East Asia. From his work in Japan, Europeans would for the first time have an account of the practice of acupuncture; for their part, the Japanese would learn more about chemical medicines and why Europeans did not distinguish between the yin and the yang pulse.

JAPANESE KNOWLEDGE OF EUROPEAN MEDICINE

The Japanese had long been aware of European medical practices. In the early sixteenth century, the system of "tribute trade" governed by the Chinese empire had been replaced by private trade, bringing Chinese merchants to Japan and Japanese merchants themselves to many places in East and Southeast Asia.[2] The growth of private trade offered opportunities for piracy and for the entry of new brokers, however, and so Portuguese merchants soon found their own way to Japan. The first three of them arrived at Tanegashima (just south of the large island of Kyushu) in 1543, when the pirate Chinese junk they were aboard was blown there by a storm. From about 1555, the Portuguese operations in East Asia were organized from Macao in southern China. Given the growing distrust between the Japanese and the Chinese at sea, the Portuguese managed to become brokers between the two, exchanging Chinese silk for Japanese silver at a profit of perhaps 200 to 400 percent.[3] New mining and smelting technology brought from China in the later sixteenth century allowed the Japanese to increase considerably their output of silver. With monetization of the Chinese economy, not even this could satisfy Chinese demand for the metal, so after the Spaniards arrived in the Philippines in the later 1560s, they sent ships laden with Peruvian silver across the Pacific for exchange with Chinese merchants.[4] (In all likelihood even greater amounts of this precious metal arrived in China by the Pacific route than through Indian Ocean routes from Europe.)[5] But the Portuguese came to conduct their business with Japan by sending only one large ship a year, which traded for a time before catching the winds back to Macao. It was therefore not through merchants but through missionaries that Europeans learned most about Japan and through whom the Japanese in turn learned most about matters European.

The first Jesuit missionary to Japan, Francis Xavier, decided to go there after meeting three Japanese men in Malacca who had fled from Kagoshima, a city on Kyushu, and he instructed them in Portuguese and baptized them. They accompanied a small party led by Xavier to Japan, arriving in 1549 in the bay of their hometown aboard a Chinese junk. The local daimyō (lord) received them well in expectation that merchants would follow, although these preferred to trade up the coast of Kyushu at Hirado instead. By the time Xavier left in 1551, Portuguese missionaries and merchants were becoming well established on Kyushu. He left behind not only many converts but foundations in several parts of the island and in Yamaguchi, just across the straits on the main island of Honshu. In order to attract Portuguese trade, another lord, Sumitada, ruler of Omura, near Hirado, gave land and residences to the Jesuits, forcibly converted his vassals, and in 1567 built a Christian church at a small fishing village called Nagasaki. It worked splendidly, making Sumitada one of the most important lords on Kyushu, with others soon imitating him, including the person emerging as the most important leader in war-torn Japan, Nobunaga, who needed a counterweight to his Buddhist enemies. Over the next decades, the Jesuit missions prospered. In the early 1590s, following the union of Portuguese and Spanish crowns, the Spaniards also introduced Franciscan and Dominican missionaries to Japan.[6]

With the missionaries also came European medical practices. One of the first Europeans to practice medicine in Japan was Luís de Almeida, who arrived with the Jesuits. Born into a "New Christian" family of Lisbon, and well educated (although it is not known where), Almeida practiced as both physician and surgeon. When he traveled to the Indies as a merchant, however, he apparently sailed on the same ship as a group of Jesuit missionaries and decided to continue on with them to Japan, where about 1555 he took up residence as a lay brother at one of the foundations established by Xavier, at Funai (now Oita) in the principality of Bungo, on the east side of Kyushu. There he started to practice medicine on the local people with the permission of the daimyō, and his activities soon established the basis for a large medical compound with various wards. He was especially successful in his surgical treatments, dealing with gunshot wounds—the Japanese had only recently adopted the use of firearms from the Europeans—and with something called "cander," a common Japanese diagnosis that indicated a swelling or carbuncle. Almeida even set up a system of training practitioners, which by 1558 had become something like a medical school in which students (both Jesuits and Japanese) were given both a theoretical and a practical education.[7] When the Franciscans entered the country, they, too, set up hospitals, at least two of them in Kyoto.[8] By the 1590s, many medical interchanges were visible. One of Almeida's students, a converted bonze given the name of Paul de Tonomine, had first been educated in Japanese medicine, and he came to take charge of medicine at the Jesuit foundation in Funai.[9] In the last quarter of the sixteenth century, a Portuguese surgeon moved in the other direction, adopted the habits of the Japanese, took the name of Keyu, and ended up practicing in the great city of Ōsaka.[10] Perhaps most influentially, Kurisaki Dōki, trained in surgery in the Philippines (after having been put aboard a ship in 1590 at the age of eight), returned to Nagasaki in 1617 and there established a school of surgery of his own.[11]

By then, however, the Japanese government was becoming increasingly suspicious about European influences. While the great lord Nobunaga, like many of the local daimyōs, had found Christianity and trade to be useful in building state power, his successor, Hideyoshi, developed other views. In consolidating his control over all of Japan Hideyoshi turned his attention to Kyushu and nearby areas in the mid-1580s, and found that Christianity was being used to build a state within a state with the Jesuits as its chief ministers. Moreover, because of internecine feuds between the Jesuits and Franciscans after the arrival of the Franciscans, he discovered much about the operations and plans of the two societies. By 1597, Hideyoshi had ordered the Jesuits out of Japan, burned churches and executed a few Christians, and forbade any feudal chiefs to practice the new religion. His death shortly thereafter prevented the strict enforcement of these edicts. His successor, the first Tokugawa shōgun, Ieyasu, at first had no quarrel with the Christians but also soon came to distrust them, suspecting Christian converts of supporting his chief rival for power, Hideyori. He issued a proclamation in 1614 banning Christianity and all missionaries, destroying all churches, and compelling all Japanese converts to renounce the faith. After his death in 1616, his son and successor, Hidetada, issued an even more severe edict against Christianity, making it a capital offense for any Japanese to practice the religion or to have any relations with Christian missionaries. Many hundreds, perhaps thousands, became martyrs, enduring horrifying agonies. The persecutions became even worse under his son and successor, Iemitsu, especially on Kyushu, at Nagasaki in particular.[12]

Into this atmosphere of growing suspicion the Dutch made their way. Their first representative arrived in 1600. An English pilot named Will Adams was the senior surviving officer of the *Liefde*, one of the early ships out of Rotterdam that attempted to reach the Spice Islands by sailing west; the vessel was blown by storms to the coast of Funai and the remaining crew was on the verge of starvation when the Japanese boarded—only a few could even stand up. Adams made himself useful to the authorities and rose to become the master shipbuilder for the Tokugawa government, the *Bakufu* (literally the "tent government" of the military leader). He also emerged as an adviser to Ieyasu about foreigners and the world beyond Japan. Through Adams, and with an exchange of letters between the shōgun and the Prince of Orange that recognized each other as equals, the newly founded VOC gained the right to trade with Japan in 1605. It established a factory in Hirado in 1609 under the leadership of Jacques Specx (who stayed until 1621—and later became one of Bontius's acquaintances on Java), and together Specx and his friend Adams managed to build good Japanese-Dutch relations. The English East India Company also set up in Hirado, in 1613, but with less able leadership pulled out after ten years. In the meantime, until the Bakufu put an end to it in 1621, both companies used Hirado as a privateering base to raid Portuguese and Spanish shipping.[13]

But in an atmosphere of growing distrust, the Bakufu closed off Japan from most of the rest of the world. The Spaniards were considered to be the main worry and were expelled in 1624. In 1635 Japanese merchants themselves were limited to traveling to Korea and the Ryukyu Islands; in 1636 any Japanese who had been further abroad was forbidden to return. To avoid additional uncontrolled exchanges, foreigners in Japan were banned from learning Japanese, with all discussions, whether about buying and selling or anything else, falling into the hands of a select group of translators who reported to the government. A rising of Christians near Nagasaki in 1637, the Shimabara revolt, provoked in part by punishing taxation, led to the insurgents being forced behind the walls of a town and their subsequent massacre, the final suppression of any public sign of Christianity. The Portuguese were blamed for instigating the revolt and were expelled in 1638. When envoys were sent from Macao to explain that they had had no connection to the revolt, they were beheaded, as were fifty-three shipmates, just thirteen members of the party being sent back to tell the tale.

These events also had consequences for the VOC. During the revolt, the VOC supported the Bakufu militarily, and afterward it was rumored by their enemies that they also took part in the annual ceremony imposed on the inhabitants of Nagasaki, the "Efumi," which compelled all residents to trample on portraits of Christ or the Virgin and Child. (The Dutch always claimed that they were exempt from this practice, although they took every precaution to scour their ships for any tokens of Christianity, even flags with crosses on them, when in sight of Japan, hiding them away until they had departed.) But the expulsion of the Portuguese left the merchants of Nagasaki in desperate straits, and they petitioned the shōgun to have the Dutch removed from Hirado to Nagasaki, which they were compelled to do in 1641.[14] There a small party of VOC officials was allowed to remain resident throughout the year on the man-made island in the harbor called Deshima. The name derived from the word for "fan," since it was indeed built in the shape of a fan, about 200 feet wide, 560 feet long on the side facing the city, and 700 feet long on the harbor side (about 15, 700 square yards, or about 32 acres in total). A landing gate on the west side could be opened to allow the unloading of VOC ships by small boats. The whole was surrounded by a high wooden wall topped by double rows of iron spikes; warning signs were posted on the water sides to keep all boats away, and guards patrolled the island hourly. Here the Dutch built (at their own expense) living quarters, warehouses, a garden, and quarters for their Japanese overseers, guards, and interpreters; they paid rent to their hosts and also paid for fresh water, which flowed from the city through a bamboo pipe. A single stone bridge connected the island to the shore, and all foot traffic came under the careful eye of Japanese sentries stationed at each end. With permission, Japanese could come and go—no doubt many of them what the Dutch called *dwarskijkers*, or spies working for the inspectors of the Bakufu (the *metsuke*)—but the Dutch were confined to the island except in exceptional cases.[15] The city was one of five under direct Bakufu control, with two *bugyō* (or "governors," as the Dutch called them) governing the city, one living in the shōgun's capital of Edo (now Tokyo), the other in Nagasaki; each

Island of Deshima, c. 1670. The garden in the upper left was further developed into a botanical garden over time. Courtesy of the Wellcome Library, London

alternated place with the other every six months or so. As a further check on the bugyōs, another official of the Bakufu was stationed in Nagasaki, the *daikwan*, who ranked below but was independent of them. The merchant guilds of the city (organized into the *Kaishō*) oversaw daily affairs. Here on Deshima, in full view of Nagasaki proper, the VOC officials supervised the unloading of the ships that visited yearly and saw to the auctions of goods to Japanese merchants. In winter the head of the factory—again under conditions of virtual captivity—would pay an annual visit to the shōgun at his seat in Edo.

Yet despite the difficulties imposed by the policies of *sakoku* ("closed country"), some Japanese remained interested in European knowledge, especially astronomy and medicine. The campaign against Christianity meant that studying European texts could be risky, but some scholars persisted. Hayashi Kichizaemon, who taught Western astronomy, was executed in 1646 on suspicion of being a Christian. He probably composed the basis for the *Kenkon bensetsu* ("Western cosmography with critical commentaries"), which not only explained Western calculatory astronomy but European methods of horoscopic astrology and astrological medicine based on the correspondences of the macrocosm-microcosm. The public credit for the composition of this work, however, goes to Christovao Ferreira, a Jesuit missionary who under severe torture had renounced Christianity. He took the name of Chūan Sawano and became an interrogator of others and an adviser to the Bakufu on how to deal with foreigners and their knowledge. At the end of his life, about 1650, he put the finishing touches on the *Kenkon bensetsu*, which has a probably fictional preface saying that it was derived from a text taken from Jesuit missionaries who were trying to smuggle themselves into Japan in 1643 and was later turned over to Chūan a few years later for translation. (The work is not taken from any known publication, however, hence the hypothesis that it was stitched together from several sources by Kichizaemon.) The preface says that Chūan "translated" it into Japanese words, which he wrote down phonetically using the Roman alphabet. A few years later, around 1656, an official translator, Kichibei Nishi, read it aloud while another, who was also a neo-Confucian scholar and physician, Genshō Mukai, wrote it down in Japanese characters and provided a commentary. It is said that Kichibei was a pupil and son-in-law of Chūan,[16] while Genshō had studied with Kichizaemon, although he remained an orthodox neo-Confucian and believer in Shinto. Genshō was scathing about the absence of proper philosophical underpinnings to the work, as he understood them, explaining to his Japanese readers that Westerners "are ingenious only in techniques that deal with appearances and utility, but are ignorant about metaphysical matters and go astray in their theory of heaven and hell. Since they do not comprehend the significance of *li-ch'i* or *yin-yang*, their theory of material phenomena is vulgar and unrefined. But this vulgarity appeals all the more to the ignorant populace, and stupefies them." In other words, what appealed to Genshō were the "universally valid astronomical measurements of the West," which he freely praised for their descriptive accuracy and ingenuity while also considering "the emphasis on appearances for their own sakes trivial and vulgar."[17]

Such comments indicate the continuing power in Japan of classical Chinese concepts, which often held up the unification of moral, natural, and medical cosmologies as the ideal. But there were many incentives for creating a more independent Japanese view of nature that could incorporate elements of European learning. Under the rule of the Tokugawa shōguns, Japan experienced a long period of peace, with growing prosperity for most everyone except the samurai.[18] The younger sons of displaced samurai often tried to make a career as physicians, teachers, and scholar-advisers to the powerful. If all went well, and one became a physician and adviser to a great lord like a daimyō, then perhaps one could even recover one's samurai status. Moreover, during the middle of the century, the country got its direction from a "mixed group of senior officials and advisers, none of whom enjoyed a clearly superior status with well-defined authority," which meant that they had to attempt to cooperate in keeping the government intact, therefore being open to new ways if these allowed agreement.[19] Many scholars were therefore at work to create a culture that would substitute learning for raw military power. In their efforts, many attempts were made to create an ideology that was more uniquely Japanese, with the efforts of previous decades to establish a pristine and robust Confucianism giving way to sharp and sometimes violent debates about the proper form and content of learning.[20]

Many well-placed Japanese scholar-physicians were therefore seeking to adapt the principles of Chinese teachings to their own ends, and in the process they were intrigued by the possibility of using European medical knowledge in recasting some of their learning.[21] Over the previous centuries, a number of Japanese authors had studied and translated many important Chinese medical texts.[22] They did not do so aimlessly, of course. From at least the fifteenth century, a number of such works showed evidence of adapting the principles of Chinese medicine to Japanese circumstances. For instance, one of the most important Japanese medical authors of the sixteenth century was Manase Dōsan, who composed the *Kirigami* (parts of which were set down in 1542, the rest between 1566 and 1581). He had studied the great Confucian and Daoist texts at the celebrated school of Ashikaga under a master (Tashiro Sanki) who had spent twelve years in China in the late fifteenth century. Manase was therefore brought up on the medicine of yin and yang as expounded by such medieval Chinese authors as Zhu Danxi and Li Gao (also known as Li Dongyuan or Li Mingzhi).[23] But in his own writing he separated the domains of medicine and religion, distancing medicine from Buddhism and stressing diagnosis and therapy over cosmology. Although Manase was well known to the Portuguese and in turn took an interest in their medicine, his works show no explicit trace of Western influence. Nevertheless, the kinds of practical interests found in his work led to a reconsideration of the pharmacopoeia in the early seventeenth century, which remained based on Chinese medicaments but incorporated some European elements.[24]

The same scholars who studied European astronomy therefore also gave attention to European medicine. Chūan, Kichibei, and Genshō turned to the Dutch on Deshima, continuing to admire their descriptive accuracy and ingenuity without appreciating their philosophical principles. Chūan visited the Dutch on Deshima at least twice in July 1648 to inquire about medicines and to observe an operation on the leg of a Japanese official's servant so that he could perform the same, and in October he smuggled in an *eenhoorn*, only to be told by a ship apothecary that it was the horn of a rhinoceros rather than of a unicorn. At his death he left behind a work on Portuguese and Spanish surgery (*Nanban-ryū geka hidensho*), which was later printed (1696) and reprinted (1705) under the title of *Oranda-ryū geka shinan* ("Lessons in surgery of the Dutch school").[25] It is said that Chūan taught Kichibei learned European medicine.[26]

From studies such as these, a critical group emerged in the persons of the translators. When the Dutch factory was moved from Hirado to Nagasaki in 1641, eight Japanese interpreters went with them, and they were soon joined by two others already living in Nagasaki, Bada Seiuemon (who had been an interpreter to the Chinese and switched to being a Dutch one), and Kichibei (who had been serving as an interpreter to the Portuguese). Another was appointed in 1643. This group was further regulated and expanded in 1656, when further interpreters were added (over a period to 1672) and the classification of "senior interpreter" (*ōtsūji*) and "junior interpreter" (*kotsūji*) was introduced. About twenty families came to form a kind of guild or joint-stock company, sharing the proceeds and handing down their business to sons, with one or two inspector interpreters overseeing their activities.[27] The VOC officers frequently complained about the inadequacies of the translators' knowledge of Dutch, but many transactions, however awkward, were managed daily. Most work concerned business exchanges, but clearly some translators used their developing abilities to discover more about what the "red-haired" barbarians knew more generally.

With such gathering interest, then, the visit of a VOC surgeon to Edo in 1650 set off a small storm of curiosity about Dutch medicine and surgery. Caspar Schamberger had been born in 1623 in Leipzig, where he apprenticed in surgery before gaining a great deal of experience in the Thirty Years' War.[28] After joining the VOC, he was sent to Deshima, from where he accompanied a special envoy, Andries Frisius, on a diplomatic journey to the seat of the shōgun. (The Dutch had signed a treaty with the hated Portuguese, and to show his displeasure the shōgun had refused the annual gifts from the VOC.) Shōgun Ietsuna was ill when they arrived in Edo, however, and they had to spend some months waiting for an audience. (Frisius eventually persuaded the Bakufu that the treaty was of no consequence.) While in Edo, some of the high-level officials of the court sought out Schamberger for treatment of various complaints and thought him so successful that he was asked to stay on for another six months to teach his surgical knowledge, leading to the establishment of a school of "Caspar-style surgery" (*Kasuparu-ryū geka*). An anonymous work (*Kōmō-geka*, "Red-hair surgery") described his methods of using ointments, plasters, and poultices; Inomata Dembei, an interpreter who studied with him, wrote two others with the help of six other interpreters: *Kaspar-ryū-i-sho* ("Medical book of the Caspar school") and *Oranda-geka-sho* ("Book of Dutch surgery"). One of Dembei's pupils became particularly important: Kawaguchi Ryōan conveyed Caspar's methods to Kyoto, northern Honshu, and Shikoku within two decades, and published another work on Caspar's methods (*Caspar dempō*) in 1661.[29]

Following Schamberger's visit to Edo, one of the governors of Nagasaki, Inoue Chikugo-no-kami, ordered the VOC to deliver to him works on anatomy in Portuguese and an herbal (he was eventually supplied with a copy of Rembert Dodoens's *Herbarius oft cruidt-boeck*), and in 1654 Inoue oversaw the translation into Portuguese of several Dutch anatomical works.[30] In the same year, Genshō was ordered to compile a work on the surgical methods of the Dutch after he had studied with the VOC surgeon Hans Jonson (writing *Kōmō-ryū-geka-hōmō*, "Secret outlines of red-haired school surgery"); fifteen years later, he composed a thirteen-volume treatise on medical botany.[31] Moreover, Schamberger's successors on Deshima were afterward constantly bothered by visitors, often the personal physicians of daimyōs, trying to learn about Dutch medicine and surgery. One of them, Hatano Gentō, as he was leaving for Edo in the late 1650s, asked for a certificate from the surgeons on Deshima testifying to his training with them, and several similar certificates were issued in later years, some of which survive.[32] One was issued to Arashiyama Hoan, physician to the lord of Hirado; in turn, one of his pupils, Katsuragawa Hochiku, started a medical tradition called the Katsuragawa School that incorporated Dutch methods. One of the signatories to Hoan's certificate, the surgeon Daniel Busch, who served on Deshima from 1662 to 1666, was authorized by officials in Nagasaki to visit patients in the city in order to give them medical treatment.[33] Another certificate was awarded to the interpreter Kichibei in 1668; he then went to Edo to practice medicine and surgery, developing a school called "Nishi-method" (*Nishi ryu*) that combined Chinese, Portuguese, and Dutch practices.[34] Studying with the Dutch surgeons continued to result in texts, as well: Narabayashi Chinzan (known as Shingohei) was a pupil of Willem Hoffman (the chief surgeon on Deshima from 1671 to 1675) before going on to become a full-time practitioner and teacher of Western medicine himself and originating one of the most famous families of physician-translators. In 1706, he published a six-volume work called *Kōi geka sōden* ("An explanation of Dutch surgery"); the third volume, on surgical procedures, is clearly based on the work of Ambrose Paré.[35] The Japanese were more skeptical about Dutch medicines than about their surgical techniques. The fourth volume of the work begins with an "introductory remark" explaining that the Dutch therapeutic procedures described are abbreviated because they were of little significance.[36]

But in the late 1660s, many in Japan also exhibited a lively interest in learning about aspects of Dutch medicine beyond surgery. One of those Schamberger had successfully treated for gout in Edo was Inaba Masanori, daimyō of Odawara, lord of Mino, and *rōjyū* (a senior councilor or "elder" in the shōgun's government). Inaba's interest in Western medicine became clear at the same time that Kichibei was studying on Deshima: the daily account of the head of the VOC factory (the *Dagh-register*) of 1668 reports Inaba ordering several medical books from the Dutch in 1668, including an anatomical work ("Spiegel der anathomie") and a botanical one (*Hortus Eystettensis*).[37] Moreover, in November 1667 the bugyō of Nagasaki had already asked the head of the VOC factory to send him a physician—not merely a surgeon. The message was repeated in 1668 during the annual visit of by the Dutch station chief to Edo, and on a third occasion shortly thereafter. In the third instance, the Japanese translator Sinosie left the Dutch with the impression that the shōgun wanted them to send him a personal physician, although that may not have been intended.[38] Daimyō Inaba, who wished to obtain a Dutch physician for his own needs, seems to have been behind this request.[39]

A DUTCH PHYSICIAN IN JAPAN

Not until 1672 did the Heren XVII finally deal with the request "from the Bugyō of Nagasaki" to send "a good physician who knows medical plants well."[40] They had not responded promptly to the initial request, and in 1672, moreover, they were suffering from the invasion of The Netherlands, which would certainly have provided many reasons for inaction. But the period from 1672 to 1674 also saw a change in how commercial exchanges between the Dutch and the Japanese were managed, causing a large fall in the income from the factory in Deshima. The VOC had previously auctioned items to the highest bidder, mostly taking silver in exchange (which was in turn used to facilitate other exchanges in Asia). This led to huge profits: in 1671, the Company made a 65 percent profit (amounting to over 1.5 million gilders) in the Japan trade alone. But the Bakufu began to worry about the amount of silver flowing out of Japan. In 1672, therefore, they raised the exchange rate on silver and imposed a "municipal sale" (*Shihō Shōbai*, which the Dutch called the *taxatie handel*), in which Japanese officials decided on the value of the items for sale. These actions cut the profits made at Deshima by slightly more than half (to about 30 percent). The head of the factory blamed the new system on Inaba.[41] Perhaps, then, the Heren XVII were attempting to mollify the Bakufu, and especially Inaba, when they advertised the post on 9 May 1672.

The chamber of Amsterdam considered five physicians for the posting to Japan between October 1672 and February 1673.[42] Three are otherwise unknown,[43] and two had taken medical doctorates. One, Adriaan van der Poel, had been born about 1642, lived and practiced in The Hague, and took his medical degree at Leiden with a thesis on angina after being in residence for a little over a week.[44] The other was Willem ten Rhijne, much younger but clearly well qualified and unmarried, and so free to make his way in Japan. Ten Rhijne had been born in 1649,[45] studied at the Athenaeum in his home city of Deventer (in 1665), then at the university of Franeker (in 1666), and finally at Leiden, where he matriculated in March 1668.[46] Sylvius was then developing his theory about acids and alkalis in the cause and treatment of disease, and Ten Rhijne became deeply imbued with the values of both chemical and Hippocratic medicine, which was evident in the thesis he presented in the next year under Sylvius's direction, demonstrating how Hippocrates had anticipated the chemists in some of his ideas.[47] (One of his professors, the editor of Descartes, Florentius Schuyl, soon took the same position in a publication, which Ten Rhijne considered to be plagiarism, though they probably both owed their proposition to Sylvius.) He got to know his fellow student Swammerdam well and made a trusted friend in Frederick Ruysch. Ten Rhijne also became a fine botanist. Like many Dutch medical students, he took his medical degree at the cheaper university of Angers, in France, in a private ceremony on 14 July 1670; it must also have been about this time that he visited the Hôtel-Dieu in Paris before returning to Amsterdam and joining the Collegium Medicum.[48] But with the war of 1672 having almost destroyed the country, the young doctor sought out the job in Japan. Ten Rhijne later mentioned Pieter van Dam, one of the Heren XVII, as a patron: perhaps he recruited him.[49] The chamber awarded him the appointment on 6 February 1673. He was granted the rank and salary of a merchant and, while awaiting transit to the East, at the chamber's request, improved his skills in the glass-blowing methods necessary for making equipment for chemical work.[50]

Ten Rhijne sailed for Batavia in June of 1673 aboard the *Ternate*, which carried 152 sailors, 106 soldiers, 5 women, and 10 children. The ship made it safely to the way station on the Cape by mid-October; there Ten Rhijne took the opportunity to investigate the local natural history.[51] The herbarium he collected there, along with drawings, was later sent to Jacob Breyne in Danzig; he also observed the customs and language of the indigenous people. After a month-long layover, the ship sailed to Batavia, arriving after a total of six months at sea near the end of January 1674, not long past Ten Rhijne's twenty-fifth birthday. During the second stage of his journey, Ten Rhijne almost died of a pestilential fever—almost half the ship's company did—but he found his way back to health through the plentiful use of Monsiegneur Charras's "Trochischi de viperis" (pastiles of viper-flesh), a remedy he later commonly prescribed to others. He then spent the late winter and spring in Batavia before sailing to Japan. Among other activities, in March 1674, in the anatomical theater that had been built in Batavia just one year earlier, he delivered an address on the antiquity and dignity of chemistry and botany.[52]

One of the people with whom he had serious conversations about the kind of medical practice he might expect in Japan was one of the Reformed ministers of Batavia, Hermann Busschoff. Busschoff had come to take a keen interest in the Asian practice of moxibustion and explained it to Ten Rhijne. He had arrived in Batavia almost twenty years earlier and served on Taiwan (where he had not been able to convert the locals to Christianity because he found the Sinkang dialect they spoke incomprehensible) before returning to Batavia in 1657. At the beginning of the next year he became one of the five Reformed ministers of the city. Throughout his time in the East he suffered from bouts of ill health, including gout.[53] But he had recently allowed his wife, Elizabeth Bargeus, to persuade him to let an "Indian doctress," as he called her, treat him according to her methods. She burned moxa in about twenty places on his feet and knees over the course of a half hour, which relieved his pains. This moxa was "a very soft and woolly substance" made from "a certain dried Herb" (which he could not identify), prepared according to a method kept secret by the Chinese and Japanese, by which the Chinese in particular "drive a good trade with it." The moxa was made into "a little pellet . . . scarce of the bigness of a small white pea, at one end somewhat sharp, and at the other end flat; and this they put with the flat end on the place where the Burning is to be made, setting fire to the upper sharp end by some small Aromatick sticks." It combusted quickly, leaving only a bit of oil on the skin.[54] Busschoff, apparently at the urging of Ten Rhijne and possibly with his help, finished a manuscript on this therapy and sent it back to Utrecht, where it was published and led to a lively European debate about moxa and moxibustion. In turn Busschoff urged Ten Rhijne to find out more about moxibustion and the additional practice of acupuncture when he arrived in Japan.

Ten Rhijne shipped out to Japan with the party of Martinus Caesaer, the official appointed to become the new head of the factory at Deshima, on 20 June 1674, arriving at Deshima on the last day of July. On Thursday, 2 August, the official still in charge noted the arrival of the distinguished doctor "with all his special knowledge" for medicine and other business, who had been sent by the governors of the VOC to serve the "kingdom" of Japan. The translators had already discovered his arrival, asking many long questions and returning several times, being especially interested in the rarities that he had brought along, such as musk, blood coral, and amber.[55] Two months later, the buygō himself visited the doctor.[56] One of the buygōs, Ushigome Chuzaemon, had asked a series of medical questions during the last trading period, and these were put to Ten Rhijne. On 12 November, two officers of Chūzaemon and some physicians came to collect the answers, but Caesaer thought that because they could not comprehend Dutch, they had no idea whether the explanations given by the translators were correct. He further thought that the translators' knowledge of medicine was insufficient either to understand Ten Rhijne or to put his knowledge to good use. The delegation apparently also brought along medical texts written in Chinese, but since Ten Rhijne could not read Chinese, the lack of understanding persisted, despite Ten Rhijne's long and painstaking efforts to communicate. Two days later, Caesaer wrote that Ten Rhijne and the senior surgeon, Willem Hoffman, continued to do their best every day to answer the questions put by the translators, although they obviously continued to think that their interlocutors understood them poorly.[57]

The difficulties of finding a Japanese vocabulary for Ten Rhijne's language were great. Especially when it came to the explanatory principles of medicine rather than the demonstrative methods of surgery, long linguistic histories and assumptions about the body and nature's powers lay behind the outlooks of all the parties involved, principles that could not be translated without additional long and difficult discussion. Even with direct borrowing, in which both sides simply wrote down approximations of the sounds made by the other party in reference to particulars, such as medical simples, it was hard to make the exchange of knowledge work. Nevertheless, the learned interpreters with whom Ten Rhijne interacted had a much greater understanding of the Dutch language and Western medicine than he at first realized. Chief among them was Shōdayū Motogi. Although Ten Rhijne initially doubted Shōdayū's facility with Dutch, he came to recognize in him a person with a good knowledge of medicine and a keen mind and in a later publication mentioned him with appreciation. Another interpreter Ten Rhijne later praised was Iwanaga Sōko, a pupil of Genshō (who had been among the leaders in studying European astronomy and surgery). Iwanaga was staying on the island of Deshima in the company of other physicians, including one of the four chief physicians of Nagasaki, Yanagi Nyotaku.[58]

Iwanaga had apparently been preparing for a much larger exercise, for a few days later, on 17 November, the translators presented Ten Rhijne with another 160 questions at the order of another bugyō, Okano Magokurō. Ten Rhijne was ordered to work with the translators in preparing the answers within a month, so that they could take them along on the annual journey to Edo.[59] The text of 165 questions and answers survives and was published in 1976.[60] The extant text is in the form of a chapter in the first volume of a work compiled by Katsuragawa Hochiku, the *Oranda yakuō zasshū* ("A collection of various Dutch medical recipes"). Katsuragawa was later a founder of a school of medicine and in turn a pupil of one of those who had been granted a Dutch certificate in surgery, Arashiyama Hoan. Katsuragawa described his text as "a record of pathology, pharmacology and so forth from questions posed by Iwanaga Sōko on behalf of the government, to a Dutch physician. Although some of the details are uncertain, everything is written down."[61] Credit was given to Iwanaga at the end, with the comment that eight translators worked on the answers.[62] Iwanaga must have prepared the questions for Ten Rhijne in a way that he thought Ten Rhijne could answer—so a fair amount of preparation had already been undertaken even before the translators began work.

Most of the questions concerned how Europeans dealt with particular medical conditions and the composition of the medicines recommended. But many assumptions lay behind them. The first question, for instance, asked why in diagnosis the Dutch felt only the radial pulse on the left hand.[63] In answer, Ten Rhijne explained that they felt either the right or the left wrist, regarding the two pulses as identical. The translators went on to record him as saying, "This is because the pulse is the circulation of blood, which comes from the heart." He probably did not say that the pulse is the same as the circulation of the blood but rather that is was an effect of it. But the translators went on to record, more correctly, that "since the root of the pulse is the heart, feeling the pulse on the right or left wrist yields the same result." The next question plunged deeper into problems of commensurability, asking how the Dutch discriminated between the yin-type and yang-type (or warm natured and cold natured) symptoms of carbuncles.[64] Not only did Iwanaga's questions probe how Europeans distinguished yin and yang symptoms—which they did not, of course but often with regard to "carbuncles," another conceptual morass. According to a dictionary of 1847, the Japanese word for this condition can also be translated as "aposteme" (French), "swelling" (English), or "ettergeswell" (Dutch, for a swelling with pus, or an abscess).[65] This was not a particularly important concern of medical practice in The Netherlands or elsewhere in Europe at the time (although apostemes had been the focus of much attention in the medieval period), but an intense interest in "carbuncles" on the part of Iwanaga is manifest in the many questions he asked about them. The interpreters seem to have thought of a "carbuncle" as something like a sign of a general cause of diseases. A century before, Almeida had been treating a common Japanese diagnosis called *cander,* which indicated a "swelling or carbuncle."[66] One might speculate that the translators were concerned to find out how the Europeans discerned the places where excess qi built up in the body, which came over as something like a "swelling" under the skin.[67] Japanese works of the surgical "Caspar School" had already found a word for abscesses: books written by Inomata Dembei (an interpreter who studied with Schamberger) are, for instance, said to discuss "the origin of purulence and the use of plasters with respect to the treatment of wounds."[68] And indeed, many of the remedies elicited from Ten Rhijne describe the use of plasters, ointments, and other external applications. Ten Rhijne in turn described Japanese medicine as having a concern for "whirlpools of somewhat deeper blood [that] lie concealed in fleshy areas."[69] Given the training of many of the interpreters in Dutch surgery, then, they expected to discuss their interest about one of the most general causes of disease in terms of the "purulence" of the Caspar School, with their concern for congealed qi coming over as an interest in "carbuncles."

For his own part, Ten Rhijne seems to have tried to reach an understanding with his interrogators, responding sympathetically to questions about "carbuncles" with his own account of the general causes of disease. In response to the second question, for instance, the interpreters took him to be saying that "the Dutch distinguish four different causes of carbuncles: 1. those caused by blood; 2. those caused by mental depression; 3. those caused by (disorder of) the liver; 4. those caused by moisture." One might redescribe these as diseases of the blood, passions, digestion, and environment. It would appear, then, that Ten Rhijne was trying to tell his interlocutors about his own theories, strongly influenced in turn by the views of his master, Sylvius, giving first place to chronic and disabling problems arising from thick and acidic blood before moving on to the non-naturals. For example, in his answer to question five (about how to administer internal remedies to carbuncles that persisted), they took him to be distinguishing "carbuncles involving 'muddy wetness' as a result of accumulated moisture within and 'muddy blood' as a result of accumulated blood accompanied by pain." They took the answer to the tenth question to be: "With fever, the blood of the whole body becomes muddy. Muddiness of the blood is a sign of carbuncles."[70] This "muddy blood" is reminiscent of Sylvius's view of how the serum of acidic blood thickened and gave rise to various diseases, including fevers and chronic illnesses. To get even this far despite fundamental misunderstandings suggests long and earnest conversation. In the end, however, perhaps the clearest transmissions from this exchange were the lists of drugs. Indeed, Ten Rhijne was to continue the conversations about natural history and botanicals throughout his visit.

For his part, Ten Rhijne also asked questions of the translators and physicians, and they shared something of their medical views with him, particularly in explaining their practice of moxibustion and acupuncture, which he later conveyed to Europe in Latin.[71] "The jealous Japanese," he wrote, were reluctant to share the mysteries of their art with anyone, particularly foreigners, concealing their books, "like most sacred treasures," in chests. But various Chinese medical texts, including authentic diagrams of the channels and points, had come into the hands of someone at the VOC station on Deshima. These had been neglected until Ten Rhijne arrived and took possession of them, and because of working with well-educated Japanese physicians on the prepared questions, he had the opportunity of consulting a Japanese physician who understood Chinese (Iwanaga) in order to have the text and pictures explained. Unfortunately, much of the original had to go untranslated because of the limited knowledge of Dutch possessed by the interpreter, Shōdayū. In Ten Rhijne's view, he spoke only "faltering Dutch in half-words and fragmentary expressions," although he also possessed "more experience in medical matters than all the other interpreters—but he was also more cunning."[72] Ten Rhijne managed to get Iwanaga to render

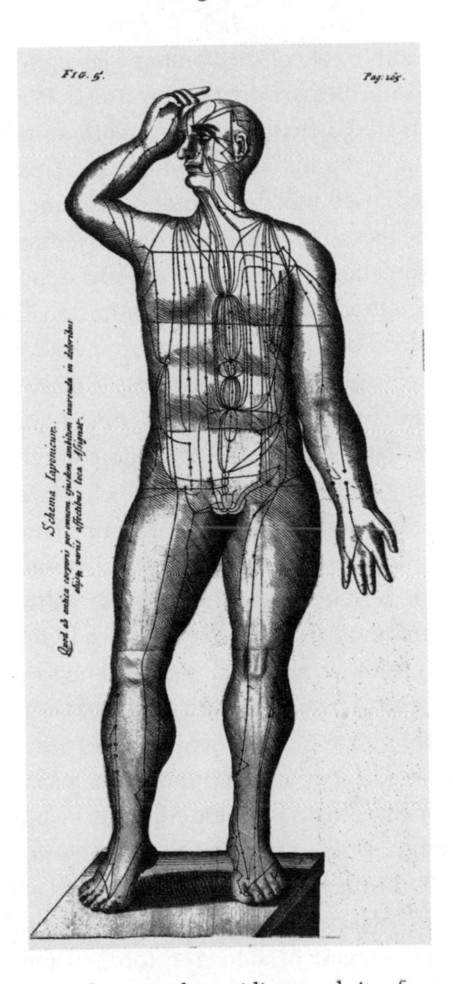

Chart of Japanese figure with meridians and sites for acupuncture and moxa, from Willem ten Rhijne, *Dissertatio de arthritide* (1683). Courtesy of the Wellcome Library, London

the Chinese into Japanese, and Shōdayū the Japanese into Dutch, which Ten Rhijne then turned into Latin. Most of what he conveyed from the Chinese text gave an account of the history and practice of medicine in East Asia.[73] He then had the various diagrams synthesized into four representations, two Japanese and two Chinese, each pair showing the back and front of the body.[74] The points for inserting the needle were marked with green dots, the ones for moxibustion with red ones. He also had many other figures showing "the lateral parts," which he would publish if they were thought to be worth the expense, he later told his readers.

Ten Rhijne wrote admiringly of the ability of Japanese practitioners. Knowing where to insert the needles or burn the moxa so as to avoid injury took great knowledge, so he compared them to the captains of ships who can find their destinations without going aground.[75] The Chinese and Japanese, particularly the Japanese, he explained, used moxibustion and acupuncture to treat most diseases. At the same time they detested phlebotomy as letting out healthy as well as diseased blood, therefore shortening life.[76] As a consequence, he argued, they had developed these two methods of venting impurities and "winds" (which caused the pains) from the blood. He thought that their methods were in accord with those of Hippocrates, who urged physicians to use cautery to remove winds from the blood so as to relieve pain.[77] Although the Chinese and Japanese were ignorant of anatomy, they "have perhaps devoted more effort over many centuries to learning and teaching with very great care the circulation of the blood, than have European physicians, individually or as a group. They base the foundation of their entire medicine upon the rules of this circulation." This they taught by "precepts and rules" derived from Chinese sources.[78] Although he thought that the lines in the diagrams might have been developed fortuitously, the points undoubtedly indicated places where blood vessels joined in "tiny foundations of blood which are also seats of pain due to the assault of tainted winds." From such considerations Ten Rhijne assembled a treatise of his own on the use of needles in European surgery and in acupuncture, and on the practice of acupuncture.[79]

Ten Rhijne's exchanges with his Japanese colleagues paid dividends. After their month of work together, which must have been exhausting for all, the bugyō himself visited Ten Rhijne on 16 December. He was met at the end of the bridge to Deshima by Governor Caesaer, Ten Rhijne, and other officials, who escorted him to the East India Company headquarters for food and drink. The Japanese were reportedly quick to give Ten Rhijne many more compliments than he could return. They asked how cold the weather was in Holland, heard how the Dutch wore two or three layers of clothing in the winter, and so on. The goodwill Ten Rhijne was generating for the Company was clear enough that on 21 December Caesaer decided that, despite the expense, he should make provision for Ten Rhijne to go with the small group that would journey to Edo. Ten Rhijne's usefulness to the medical interpreters also allowed him to gain some local privileges. On 7 February 1675, for instance, two of the best of them (Shōdayū and Yokoyama Yozauemon—or Yokoyama Yozaburō—who was known to the Dutch as "Brasman"), accompanied him and a Dutch apothecary into Nagasaki to visit a pharmacy, where they discussed various drugs. They continued their discussions about drugs over the next two days.[80]

On 12 February, Ten Rhijne set out with the rest of a party of six Dutch officials on the annual journey to the shōgun's seat at Edo. The Dutch governor and a few accompanying persons (usually four in all) were required to make this month-long journey each year, with a large train of Japanese guards and officials as escort (paid by the VOC). At Edo they presented the shōgun and his servants with gifts, a report on the outer world, and a promise of continued good behavior in exchange for their right to trade with the Japanese, sometimes also being asked to sing, dance, or otherwise make fools of themselves.[81]

The day after his arrival in Edo on 15 March, Ten Rhijne was visited by Kichibei Nishi, the former translator and student of European astronomy and medicine, who had now taken the name "Genpo" and from 1673 had been a physician to the Bakufu. He came on instructions from the bugyō and bearing gifts from his master, for which Ten Rhijne expressed thanks. Governor Caesaer proposed that Ten Rhijne be sent to the court to give instruction on drugs and medicine, and Genpo agreed to convey the message and to visit often. The next day, they had a chance to review the answers to the medical questions Ten Rhijne had been asked in Nagasaki. On the following day, Genpo and a physician to Rōjū Inaba visited Ten Rhijne and asked a few questions about the breast cancer of a woman. The cancer had been discussed on earlier visits (on 23 April 1673 and 30 April 1674, on the first occasion the woman having been examined by the surgeon Hoffman), and although she was still alive, the cancer now covered her bosom. Genpo came again on the following day, and on the next, 20 March, Ten Rhijne and the senior surgeon, Hoffman, received word to prepare for visiting the lord of Izu, Matudaira Izu no Kami. There they saw a boy of fifteen or sixteen years who looked like skin and bone and was unable to walk because of dreadful abscesses in his right leg that caused him to suffer severe fevers. Ten Rhijne examined him and thought that he was still curable using some of his own methods, but the boy refused to try Dutch medicine and was left to await death. Two days later, on 23 March, the two Dutch medical practitioners paid a call on Rōjū Hotta Bizen no Kami (lord of Bizen), who complained of pains in his finger- and toenails. After returning to their quarters, Ten Rhijne prepared medicines and sent them to him. By now, however, arrangements were under way to return to Nagasaki, and Ten Rhijne was among the Dutch who were summoned to the bugyō to take their leave; on the same day (31 March) Genpo visited them and dined with Caesaer and Ten Rhijne, asking further questions about the best methods to treat breast cancer, which Ten Rhijne explained carefully.[82] The next day, one of the lords who ranked just beneath the shōgun himself, Mito Chūnagon Mitsukuni, sent medicinal samples to Ten Rhijne to ask about their names—to which he replied in writing. A group of Japanese physicians visited him the day after, asking more questions (probably to complete the text of the 165 questions and answers), to which he patiently replied while they wrote down his answers. On 6 April the Dutch finally departed Edo, arriving back on Deshima on 14 May.[83]

After returning, Ten Rhijne was still consulted but was under much less pressure than before. There were no more formal written questions, and only a few entries mentioning him are to be found in the station's daily record. About two weeks after taking up residence on Deshima again, an assistant translator came from the bugyō to get the same medicines that had been given to the Rōjū Hotta in Edo, by which he had received benefit; a month later, a chief translator, Tominaga Ichirobei ("Mierobe"), asked for medicines for one of the lords. Ten Rhijne was consulted at the end of June by the translator Shōdayū about a woman who had lost her senses while suffering from an intense fever following childbirth (probably puerperal fever); Ten Rhijne explained that he could give advice only after examining her, and the bugyō gave permission. After seeing her, Ten Rhijne thought that not all was lost and sent her some medicines, hoping that if the cure were successful it would increase the reputation of Dutch medicine. It was also noted on 15 November that Ten Rhijne was ordered to prepare oil of cloves, which would be sent to Edo.[84]

During this period Ten Rhijne continued to help those interested in learning about Dutch medicine. On 5 December, the chief translator, Tominaga Ichirobei, took an anatomy book, the *Opera Spiegel* (a work requested by Inaba in 1668—see above), from the governor's chest, asking whether he could have it whenever he wished (the answer was "yes"). He returned with the book on 10 January 1676, having taken it to the bugyō and then discussed it with physicians (some of them in Edo), telling Ten Rhijne that they had many questions and wished to discuss it carefully with him. Indeed, three physicians arrived on 17 December to question him. They also got a chance to drink "distilled water," probably the new medical concoction invented by Ten Rhijne's teacher, Sylvius: genever, or gin.[85] The book in question was a copy of one of the best summaries of anatomy then available. It was a large, thick, and well-illustrated folio of the collected work of Adriaan van den Spiegel, edited by Johannes Antonides van der Linden, the professor of botany and anatomy at Franeker, and printed by the great Amsterdam publisher Johannes Blaeu. Spiegel had studied medicine at Leiden at the end of the sixteenth century before going on to Bologna and Padua, where he ended as professor of anatomy and surgery, dying in 1625. His two most important works were published posthumously: *De formato foetu* ("On the formation of the human fetus," 1626) and *De humani corporis fabrica libri decem* ("On the fabric of the human body in ten books," edited by his student Daniel Bucretius in 1627). The first was supported by a large number of well-drawn full-page copper-plate engravings, the second by almost one hundred more. The *Opera*, dedicated to a number of prominent regenten and directors of the VOC, not only collected these works and plates together but supplemented them with works by Caspar Aseli on the lacteal vessels, William Harvey's *De motu*, Johannes Walaeus on the motion of the chyle and blood, and Spiegel on semitertian fevers, gout, worms in the body, and botany.[86] One of the chief translators with whom Ten Rhijne worked, Shōdayū Motogi, is known to have composed a short book on anatomy (called "Anatomical charts of Holland"), the first Western-style anatomical treatise in Japanese and a full century ahead of the more famous work of Sugita Genpaku (the *Tahel Anatomia*);[87] apparently Shōdayū was one of the three physician-translators who came to question Ten Rhijne about Spiegel's work.

About a month later, Ten Rhijne again made the journey to Edo with the governor, Johannes Camphuis, departing on 27 February and arriving on 14 April.[88] Genpo came to visit on the seventeenth. Two days later an envoy from a senior lord, Aoki Tōmi no Kami, came to ask about containers for distilled waters (possibly stills), and Ten Rhijne explained how to work them.[89] By now, however, Ten Rhijne was fed up with being made so little use of. When he had taken the position, he had expected to serve the shōgun. Instead, he was being politely fobbed off by the Bakufu. He wrote a long letter to Governor Camphuis, recounting his situation as he understood it. He explained that when he arrived, he had been set some questions by Bugyō Ushigome Chūzaemon, for which he was thanked. The other bugyō, Okana Magokuro, ascribed to a different medical theory, but he, too, seemed satisfied since he did not ask further questions. On reaching Edo the first time, he had also been investigated by representatives of the shōgun—by Genpo and the physician to the Rōjū Inaba—and Bugyō Chūzaemon ordered him to see patients, particularly the woman with advanced breast cancer and the young man with an advanced case of an abscess in his leg. Yet the results apparently did not satisfy the Japanese, since all had gone silent since then. Ten Rhijne was upset that his professional ability was being judged by the success of such cases. Someone unnamed seemed to be spreading lies and innuendo, he thought, blocking his chances. Although he loved liberty like life itself, as all freeborn people did, he was willing to become a slave if necessary to help the interests of the Company. He therefore placed himself in Camphuis's hands, asking him to use what influence he had to break the silence by explaining the situation to the person behind the scenes or threaten to take the case to the shōgun's council.[90]

A few days later, Camphuis recorded that he had earnestly requested of the translator "Brasman" (Yokoyama Yozaburō) that he bring up Ten Rhijne's case. Camphuis wanted it made clear that Ten Rhijne was not a mere physician but had been expressly sent from The Netherlands by the Heren XVII at the request of the shōgun and his council.[91] Genpo visited again on 22 April, getting advice about how to treat pains in the foot, and three days after that, Shōdayū arrived on behalf of the buygō to ask about the use of some drugs that he brought along. They also had a further discussion about Ten Rhijne's situation. Two days later, on 27 April, Ten Rhijne was himself present at the audience between Camphuis and Shōgun Tokugawa Ietsuna, but nothing came of it.

Camphuis and Ten Rhijne's position may have taken the Japanese by surprise. They may never have thought that their requests to have the VOC send a physician would be interpreted as a request to place a VOC doctor in the shōgun's household. That not only presumed the superiority of Dutch medicine, or at least its better efficacy in some cases, but would have opened the doors to new kinds of intrigue that would not have been welcome. Nevertheless, once Camphuis made his view plain, they seem to have wanted to show how much they respected the doctor—although without granting him the appointment he sought. On the day following the audience, a messenger from the head of the shōgun's council, the Tairō Ii Hakibe no Kami Naozumi, came to express his gratitude to Camphuis for paying some of the expenses of his physician, who had been sent to study medicine with Ten Rhijne; and on the same day Ten Rhijne was invited to visit the stables of the shōgun to give advice about them. (The VOC had sent a gift of two Persian horses in 1668, which were bred with local horses to produce superior stock for the shōgun.)[92] Camphuis and a high Japanese official also had a long meeting on the next day (29 April), in which the matter of Dr. Ten Rhijne was discussed again. Camphuis once more explained that the doctor had more than proven his excellence and had been sent out from The Netherlands after the Heren XVII had received a request (*eÿsen*) from the Shōgun and his council nine years before. He also explained the recent efforts to have Ten Rhijne's situation presented. Genpo visited again on the last day of the month, and two days later a Japanese physician visited in the afternoon. He and the Dutch delegation held discussions in the castle of the bugyō, where he showed them large amounts of various medical simples and questioned Ten Rhijne about them, receiving comments on about half of them before he had to leave. After the Dutch returned to their inn, a message of thanks came from Genpo through Brasman, and two personal physicians of Rōjū Inaba came to seek further information about some of the drugs recommended for him. Each day after, Ten Rhijne continued to have high-ranking Japanese physicians coming to ask questions: on 4 May, for instance, an envoy from Bugyō Okana Magouō, together with Brasman, asked about balsams and oils they had been wondering about for ten years. The next day, on the eve of their departure, Ten Rhijne even had a visit from two of Inaba's physicians, and then from the shōgun's physician, who brought presents of dried fish and rice wine. The Japanese cooks and translators who served the Dutch seemed impressed: they considered such gifts unprecedented. But when the Dutch delegation left for Nagasaki on 6 May, Ten Rhijne went with them.[93] He had been consulted widely and deeply, and by some of the most eminent physicians in Japan. They seem to have been impressed, and they offered him tokens of esteem, although since we do not have their side of the story it is impossible to know whether Ten Rhijne satisfied them in their further tests. But he did not obtain the position he thought he had come to claim. Perhaps it had all been a misunderstanding.

The rest of Ten Rhijne's time in Japan left little record. He arrived at Deshima on 7 June 1676. Three weeks later, the officers of the Company were visited by the bugyō, whom they hosted for dinner, sharing some of their distilled waters with him. At the gathering the bugyō asked Ten Rhijne about remedies for the heart. He answered well, it was recorded, probably explaining about Sylvius's medicinal liquor, which was considered a cordial. But there is no further record of his engagements. He set sail on 27 October, arriving in Batavia on 13 December 1676.[94]

The misunderstanding about Ten Rhijne's appointment created further mistrust between the VOC and the Bakufu. The message received by the Heren XVII was that the doctor had been permitted to leave Japan after having willingly and civilly offered medical instruction to them, but he had been sent away with no recompense or thanks.[95] Their further interpretation of events was that the bugyōs had requested them with great vigor to send "an expert master in medicine" who was an excellent botanist, and they had sent Dr. Ten Rhijne, a man of much study and capability. But the Japanese "made very little use of him, and treated him like an ordinary surgeon," and opposed him to the point where they let him depart "without extending him any gratitude" for his efforts. This provoked a litany of complaints from the station chief on Deshima to the shōgun about the translators.[96] It all suggests that they blamed the senior translator, Brasman (Yokoyama Yozaburō), for the wasted effort.

PUBLISHING ON ASIAN MEDICINE IN EUROPE

Because Ten Rhijne had been sent out for the sole purpose of going to Japan and serving the shōgun, the officers of the VOC seem to have had little idea about what to do with him once he was back in Batavia.[97] He was elected a deacon of the church on 12 January 1677 and made one of two regents of the leper's hospice on 28 July in place of a merchant who had been sent on a mission to China. Over the next two years he may have helped to create a sense among members of the city council that a new institution was needed, since they decided in May 1679 to begin construction of a new leper house on the island of Purmerend.[98] For his part, he launched proceedings to recover the back pay he thought he was owed.

But between 1677 and 1679, Ten Rhijne made further use of his time by involving himself deeply in further studies of naturalia. He helped Van Reede with his botany of Malabar. He also composed several works in Latin and sent them to his friend Frederick Ruysch for publication in Amsterdam.[99] Moreover, Ten Rhijne assisted Cleyer with his studies of Asian medical plants and practices. Cleyer, of course, held almost all of the medical positions in Batavia himself. Only a few weeks before Ten Rhijne returned to Batavia at the end of 1676, Cleyer added the post of apothecary to the chief hospital (on very favorable conditions) to his list of appointments, and in 1681 he gained the post of "inspector" of the lepers. It has been somewhat improbably suggested that he felt threatened by the presence of a well-educated physician like Ten Rhijne.[100] From his view, had Ten Rhijne any hope of advancement, he would have needed Cleyer's support. A report of February 1679 from the governing council in Batavia to the Heren XVII therefore mentions the botanical work the two of them were doing together, prompted by their receipt of the first volume on Van Reede's *Hortus Malabaricus*, which had been sent by the Heren XVII together with a note asking about the medicines and medical supplies of the VOC. In the council's reply, they mentioned that Cleyer and Ten Rhijne were collaborating to make use of these botanicals for the VOC medicine chests.[101]

There are also clear indications that between 1677 and 1679 Ten Rhijne assisted Cleyer in the medical work for which the Cleyer became famous: the publication of information about Chinese pulse doctrine. The work originated from Cleyer's correspondence with the Jesuit mission in China. The Dutch had little prolonged direct contact with China. An expeditionary fleet sent by the VOC in 1622 to capture Macao from the Portuguese failed in its main objective but went on to establish a base on the Pescadores Islands; in 1624, Ming officials forced them to move from the Pescadores (which they considered part of China) to the harbor of Tayouan (currently called An-p'ing) on Taiwan, to which China made no claim. After 1633, when a VOC fleet was defeated by Admiral Zheng Zhilong off Chinmen (Quemoy), they gave up hope of forcing direct trade with China by conquest. But Taiwan gave the VOC a source of gold in exchange for Japanese silver—gold being necessary to purchase textiles on the Coromandel coast[102]—and from the island the VOC could raid Portuguese and Spanish shipping while trading with the many Chinese and Japanese merchants who visited their factory. Many Chinese merchants even settled there, especially after refugees began to pile in during the 1640s as Qing forces wrested coastal China from the Ming. The VOC soon also became concerned with establishing territorial dominion over the aboriginal inhabitants and eventually (by 1642) managed to expel the Spaniards from settlements they, too, had set up on the island.[103] While the Company sent several embassies to China in the course of the century in the hope of obtaining trading rights, they came to little. The one in 1656 came closest to success, only to be blocked by the influence of Adam Schall, a Jesuit then in residence as an astronomer at the imperial court.[104] Finally, the leader of a semi-independent Chinese trading company, Zheng Chenggong, called "Coxinga" by the Dutch, at the head of a large fleet and army, eventually forced the surrender of the VOC's fortress in 1662 and expelled them from Taiwan.[105] As with Japan until its closed period, then, most of what Europeans learned about China came not from merchants, especially not from the Dutch, but from Jesuit missionaries.

But most of what they learned of Chinese medicine came into Cleyer's hands before being sent on and published. Cleyer's connection to the Jesuit mission was conducted mainly through Philippe Couplet. Couplet was a speaker of Dutch, having been born in Mechelen. He was among a group of three young men from the low countries who had been inspired to volunteer for the China mission after a visit to Louvain in 1654 by one who had been there, Martino Martini. Perhaps Cleyer and Couplet first met in Amsterdam during the religiously tolerant mid-1650s, when the young missionaries lived there for some weeks while awaiting a ship for Lisbon. More likely they met in China during the later 1650s or early 1660s. Couplet served in the southeast, including Fuzhou, where in October 1662, some months after Taiwan had fallen to Coxinga, a Dutch fleet arrived to try to make an alliance with the new dynasty of the Qing (Zheng having opposed them, too) and to open trade, and for the next two years the Dutch and the Qing were on good terms.[106] Couplet welcomed the Dutch to Fuzhou and for some time acted as an informant for them about Chinese politics and trade (and by letter sent his kind regards to the playwright Vondel and the printer Johannes Blaeu in Amsterdam).[107] Perhaps he met Cleyer in Fuzhou, since Cleyer shipped out with the VOC on 9 January 1662, and as a soldier he might well have been sent directly on to China to support the efforts against Coxinga and his son.[108] (In November 1663 the VOC in alliance with Qing forces expelled Zheng from Quemoy and Amoy, forcing him to withdraw to Taiwan in 1664.) It is equally clear that by 1666 Couplet and Cleyer knew each other: when in 1664 the new emperor, Kangxi, under pressure from neo-Confucians and those who feared foreign involvement in the country, cracked down on Christians and arrested Schall and two other prominent Jesuits, the rest of the missionaries were first brought to Beijing and then, in 1666, taken to Guangzhou (Canton), where a Dutch dele-

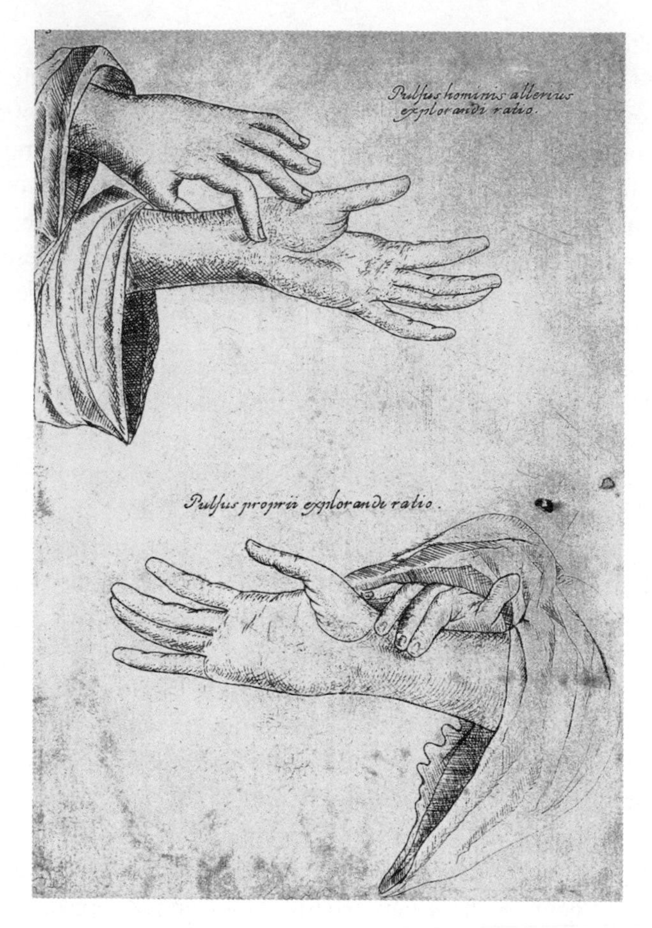

Pulse taking, from Andreas Cleyer, *Specimen medicinae Sinicae* (1682). Courtesy of the Wellcome Library, London.

gation on its way to Beijing crossed their path. In a letter to the delegation about their situation, Couplet sent his best regards to two of "his special friends," one of whom was Cleyer.[109] By that time, Couplet had himself become deeply interested in Chinese medicine.[110]

Couplet and Cleyer continued to write one another, with Couplet sending him information and Cleyer returning newspapers, medicines, money, and requests for Chinese medical works, particularly being interested in manuscripts on Chinese "pulse-method."[111] (The services Cleyer provided to Couplet were extended in 1670, with Cleyer handling the mail that the Jesuits in East Asia wanted to send on to Europe, collecting it in Batavia and forwarding it as a

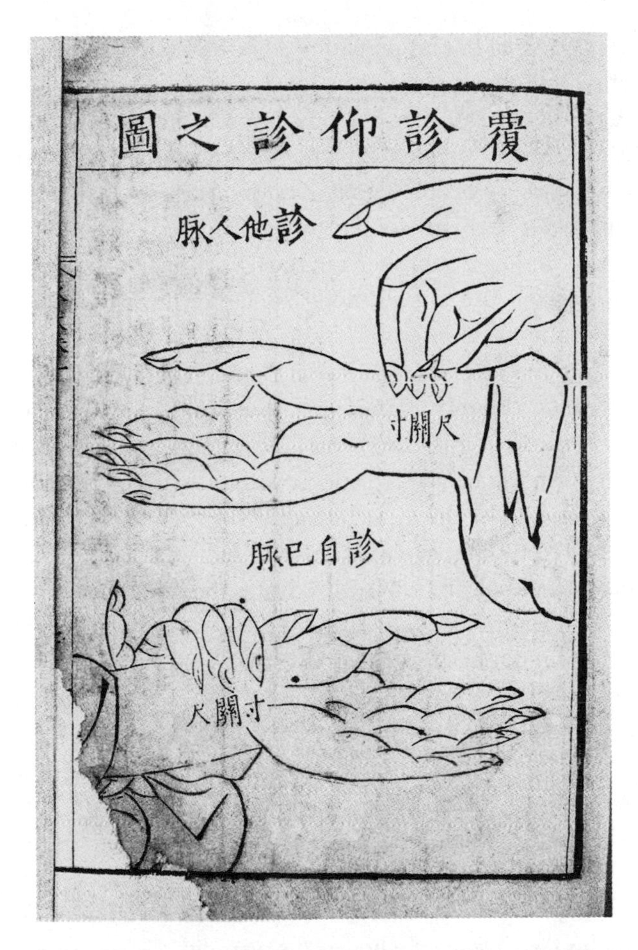

Pulse taking, from *Tuzhu Wang Shuhe Maijue* (Wang Shuhe's secrets of the pulse). Woodcut illustration from an edition of 1554. Courtesy of the Wellcome Library, London

package to the printers Blaeu or Clement Verschoor in Amsterdam or Balthasar Moretus in Antwerp, who in turn opened the packets and sent on the individual letters.)[112] Among the materials that Cleyer obtained from Couplet were translations and analyses of Chinese medicine compiled by Michael Boym, a Polish-born member of the China mission: Boym's work largely introduced Europeans to the learned medicine of China.[113] Boym had met up with Couplet in 1658 in Siam (now Thailand), when both were on their way to China, Boym returning after traveling in Europe as an ambassador for the Ming. During those years, he had published a short work on Chinese plants and, in an account of his work in

China, given notice that he was preparing a book on Chinese medicine and the Chinese method of touching the pulse for diagnosis and prognosis.[114] In Siam, Boym gave Couplet a treatise on the pulse which, with a packet of letters and possibly a list of Chinese medicinal plants, he forwarded to Governor-General Maetsuyker in Batavia (probably through the good offices of Jan de Ryck, head of the Company's factory in Siam) for printing in Europe. The treatise on pulse disappeared without finding a publisher, but the list of medicinals later circulated in manuscript among the Jesuits and among the Dutch, for Cleyer and Rumphius, and possibly Ten Rhijne, all of whom received support from Maetsuyker, apparently possessed copies.[115] In the years after Boym's death in 1659, Couplet collected at least some of Boym's manuscripts, as well as further translations of Chinese medical works done by himself and his brethren, and sold them to Cleyer.

It is likely that Ten Rhijne assisted Cleyer with the work on the medical botany of China first collected by Boym, as he had helped Van Reede with the botany of Malabar. But Ten Rhijne seems also to have helped Cleyer with other manuscripts he had obtained from Couplet, editing them and polishing the Latin, since Cleyer had no particular knowledge of East Asian medical practices and his ability to write Latin was as shaky as Van Reede's. The resulting work, the *Specimen medicinae Sinicae, sive opuscula medica ad mentem Sinensium* ("The pattern of Chinese medicine, or medical work according to Chinese thought"), was published in 1682 at Frankfurt with the help of the physician Sebastian Scheffer, with Cleyer's name on the title page as editor. Because the printer died before completing the project, the intended preface, dedication, and frontispiece were missing from the final copy, and various copies were bound containing different sections or the sections in different order.[116] (Moreover, since Boym's name did not appear in the *Specimen* whereas his own did, Cleyer came to carry the accusation of plagiarism—in print from 1730—although since the 1930s several scholars have given him the benefit of the doubt.)[117]

The title page advertised the *Specimen* as divided into six parts (*opuscula*). But in fact it contained only some of the parts mentioned, together with a list of simples that did not appear on the title page at all.[118] The first section contained four parts, amounting to a translation of works attributed to Wang Shuhe (whose name was transliterated as *Vám Xó Hó*)—which modern commentators consider most likely to have been a late medieval version of the *Mo Chüeh* ("Sphygmological instructions")[119]—and of the legendary Yellow Emperor, Huangdi (Romanized in the text as *Hoam ty*). The second section consisted of "Medicamenta simplicia, quae à Chinensibus ad usum medicum adhibentur" ("Medicinal simples, which are possessed by the Chinese for use in medicine"). This treatise is composed of 289 entries, each giving a brief account of the qualities of the

Illustration intended for the title page of Cleyer, *Specimen medicinae Sinicae* (1682), showing a Chinese physician taking the pulse. By permission of the Staatsbibliothek zu Berlin

simples and their uses for various conditions; in all likelihood, this was the botanical list of Boym's that had been circulating in manuscript.[120] As we have already seen, it is very likely that Ten Rhijne had assisted Cleyer to put this section in order. The third part had several subsections: a treatise on pulses collected by an "erudite European," four chapters that would be reprinted in Boym's *Clavis medica* in 1686 (to which we will turn in a moment), parts of letters, and additional pieces of medical information, including anatomical tables showing the channels and points for use in acupuncture and moxibustion. The fourth section was a short work on diagnostics according to the appearance of the tongue, with thirty-seven diagrams. The first, third, and fourth sections are almost certainly based on Boym's manuscripts.[121] The Latin style of part 1 is different from the third, however, and Cleyer later stated that he had the help of a learned European in assembling the fragments of part 3 into a whole:[122] this unnamed person was no doubt also Ten Rhijne.[123] For all its difficulties, the *Specimen* became one of the foundational texts through which Europeans glimpsed Chinese medical practices, becoming an inspiration for John Floyer's studies of the pulse, for example.[124]

By the time the manuscript was on its way back to Europe, however, Ten Rhijne and Cleyer had fallen out. In a letter to friends in Amsterdam in March 1681, Ten Rhijne protested that Cleyer had been elbowing him out of the way for some time in affairs about the Chinese work on the pulse, even denigrating his reputation among acquaintances in Europe, although he could prove from Cleyer's letters that it was Cleyer who had broken their working agreement (*stipulatie*). The naked truth was that Ten Rhijne had compiled all the works on the pulse and written a commentary on them without anyone else's help, but the copy of the work Cleyer had sent back to The Netherlands to be printed had a part missing. The times were too delicate for him to unfold the whole story, he sighed, so he quickly changed the subject.[125] But no doubt he meant that Cleyer had suppressed his commentary on Chinese pulse-taking, since he also wrote to Henry Oldenburg in 1681 about having sent a copy of a treatise on the pulse to friends in Amsterdam (see below). Two years later he wrote that the matter was still too *dangereus* for him to explain, but he repeated that he had compiled the parts collected by Couplet into the text that became the *Specimen*.[126] A month later (21 March 1683) he commented in a letter to his friend Johannes Groenevelt (asking him to pass on his letter to the Royal Society of London) that "The Treatise about pulses is put out in Germany by Andreas Cleyer (one that has done no more than purchase a Translation by force of money) which yet without an Interpreter can do Little Good in Europe. I have Taken some pains about it but there are Mysteries not yet to be revealed which yet I hope the Learned world Shall some time partake of."[127] Later in 1683, he complained directly to the Royal Society of London that the treatise on how the Chinese touched the pulse, with commentary, which had not been fully completed, had appeared in an edition under another name despite the agreement that had been drawn up in all good faith and foreknowledge. He had expected that his views and efforts would be publicly acknowledged, as had been agreed in advance, although political and domestic affairs had conspired against this for the moment. But the Jesuit Couplet, who was deep in these affairs and had recently returned to Europe, promised to explain all in due course.[128]

In a "PS" to his letter of March 1681, however, Ten Rhijne noted that a more complete copy of the Chinese work on pulse and other works were being sent to Amsterdam by Cleyer.[129] These parts were synthetic accounts of Chinese medicine rather than translations, as in the *Specimen*. Mentzel was eager to have the Academia publish this second work, too.[130] They found an outlet through the Academia's *Miscellanae*, the work appearing in 1686 under the title *Clavis medica ad Chinarum doctrinam de pulsibus* ("Medical key to the Chinese doctrine on the pulse").[131] The *Clavis* was composed of two major treatises—one of which might just possibly be the discussion on the pulse Ten Rhijne claimed to have written—plus three prefaces written by Boym about his work on the pulse and the treatise on medical simples (the first and last dated from Siam in 1658) with the addition of an "annotation" to the preface on simples by someone unnamed (again, possibly by Ten Rhijne). The work was indeed the "key" to the first book and to Chinese medicine more generally, for rather than trying to translate Chinese works, it explained carefully, in terms a European audience would understand, the principles that lay behind their practices.[132] Mentzel also made sure this time that the proper persons got the credit they deserved, as he understood it: the title page credited Boym as author, Cleyer as the collector of the works, and Couplet as bringing the works to Europe and purging it of errors. (Couplet had stopped in Batavia on his return to Europe for some time in early 1682, speaking there with both Cleyer and Ten Rhijne. Ten Rhijne commented to his friends in Amsterdam that Couplet was affable and sociable, and would be a good colleague with whom to share their work, but that they should remember that he was a Jesuit.)[133] Cleyer made certain, however, that Ten Rhijne's name was not mentioned. Instead, Cleyer continued to slander him: according to a letter to Scheffer, Cleyer had sent his work to the Academia Naturae Curiosorum for publication because Ten Rhijne and his "slaves" (*mancipiis*) in Amsterdam had suppressed the printing of Cleyer's work there.[134]

Ten Rhijne's sense that Cleyer did not deal honestly had foundation beyond their personal differences. Shortly after meeting Couplet early in 1682, Cleyer went to Japan himself as a merchant, carrying the title of "third head of the factory" (*Derde Opperhoofd van Japan*). There had never been a third chief before, strongly suggesting that the trip was organized by Cleyer's powerful friends on the Batavian council for his personal gain. After he left for Japan in 1682, however, the Council of the Indies inspected his books and found that he had left debts of thirty thousand gilders, which caused them to reassert their supervision of the Castle dispensary by making the Castle surgeon the manager of it (rather than the Castle physician), while contracting out the post of city apothecary.[135] Despite these financial problems, after arriving back in Batavia in March 1684, Cleyer built a grand house later reported to have been the largest, the most magnificent, and by far the most expensive in Batavia. A year later, he obtained the post of merchant in Japan again. When he arrived at Deshima at the end of August 1685, he found that the Japanese had changed the trading system again: they had raised the valuations of the goods brought by the Dutch and allowed them to be sold at open auction, but placed a limit of 140,000 gilders (340,000 tael) on what the Company could take out of Japan each year. Cleyer responded by engaging in smuggling. When the Nagasaki officials discovered this, many of Cleyer's Japanese associates were executed (some by crucifixion) and Cleyer was expelled with orders never to return, on pain of death; he departed on 19 December 1686, afterward being rumored to have arrived in Batavia in possession of sacks of gold coins. Despite an inquiry launched by the Heren XVII, the government in Batavia made excuses for Cleyer and he was never punished.[136]

Ten Rhijne's dispute with Cleyer seems not to have affected his own reputation with the Council of the Indies, for they appointed him to the important committee of six sent to the west coast of Sumatra to look into the gold and silver mines of Silida in 1679. These mines had been worked without much profit for about ten years, but the Heren XVII decided to send a party of more than sixty miners there to get better results. Within two months Ten Rhijne returned to Batavia as the group's representative to present their 390-page report about how small the mines were and the troubles they were having. Additional officers were sent to the mines in May 1680, among them Ten Rhijne and Dr. Hermanus Grimm. When he returned to Sumatra, Ten Rhijne found that serious fevers had laid low many of those employed in the work.[137]

But Ten Rhijne did not let his argument with Cleyer stop him publishing on his own. He had been joined on Sumatra by his younger brother, a VOC surgeon, who brought him news from his friends in Amsterdam and a gift of the two volumes of the voyages of Jean-Baptiste Tavernier in the Indies. After returning from the mines on Sumatra, he continued his studies. He was, for instance, learning Arabic and asking for copies of Erpenius's grammar and the "Lexicon polyglotton" or any other lexicon to be sent to him, as well as any publications of the "collegium curiosorum" (what we would call scientific academies) of Germany, France, England, Denmark, Italy, and elsewhere.[138] He insisted, moreover, on chasing up the earlier works he had written. In the later 1670s, he had sent a number of manuscripts to his friend from student days at Leiden, Frederick Ruysch, intending them for publication. But he heard little back from Ruysch or his other friends, and he began to suspect that the publishers were not interested (whether because of Cleyer's influence or because they simply did not seem marketable is unclear).[139]

In July 1681, therefore, Ten Rhijne wrote to Henry Oldenburg, the secretary of the Royal Society. His former teacher Sylvius had been in communication with Oldenburg, as had former friends of his student days, such as Regnier de Graaf, and Ten Rhijne must have heard that the English virtuosi were open to communicating with people like himself. He told Oldenburg that he had various observations which he would be willing to communicate and that he had sent a treatise to Holland on the use of moxa and acupuncture in Japan and the use of pulse diagnosis by the Chinese, which he would like to be printed in England.[140] Ten Rhijne's letter arrived in London about five months after it was written, and since Oldenburg had recently died, it was brought to a meeting of the Royal Society by Theodore Haak on 18 January 1681/2. The contents of the letter caused considerable discussion. Ten Rhijne had mentioned that the moxa was made from artemisia, and Frederick Slare affirmed that he had made "a kind of moxa of the fibres of mugwort," the common artemisia in England, when he had dried them well and blown away the dust. Thomas Henshaw thought that the virtue of moxa was in the burning and cauterizing alone, but Robert Hooke believed that the plant itself might have some "peculiar virtue" in its substance, perhaps in its "solid oil." Ten Rhijne's remarks about Chinese knowledge of the pulse also led the president, Sir Christopher Wren, to note that the Chinese "were extremely curious about feeling the pulse of the patient" not only in the wrist "but in divers other parts of the body, by which they pretended to make great discoveries of the disease." Even the ancients probably knew more about the use of the pulse than modern European physicians, and perhaps Galen's remarks on pulse diagnosis were not fully understood. Hooke then suggested that in feeling the pulse one might discern the different states of the parts within the body. Wren thought that the pulse in the arteries was distinct from the beating of the heart and that since the arteries were covered by three coats of muscles, the motion of the different parts of the pulse might indeed yield important information.[141]

The conversation indicates clearly that the English virtuosi were well aware of information about Asian medicine filtering back through the Dutch. Ten Rhijne had first encountered the practice of moxibustion when he met Busschoff in Batavia, and he had encouraged him to write up his experiences. Busschoff sent the resulting short treatise and samples of the moxa to his son (a lawyer) back in Utrecht. His son published his father's book in Dutch in 1674 and apparently began to sell the moxa that his father sent to him. Constantijn Huygens, secretary to the Prince of Orange, was among those who noticed, and wrote to Henry Oldenburg at the Royal Society, sending along a copy of the book. Huygens also struck up a correspondence with Busschoff in the Indies,[142] while recommending the use of moxibustion for the gout to the English ambassador in The Hague, Sir William Temple, who also experienced great relief from it.[143] The Dutch microscopist Antoni Leeuwenhoek quickly turned his attention to investigating moxa itself with his little lenses, sending a report on it to the Royal Society.[144] At the request of the Royal Society, therefore, a translation of Busschoff's treatise was published in 1676. A German edition appeared at Breslau in 1677.[145] The interest among the virtuosi in moxibustion also soon led Thomas Sydenham to mention moxibustion favorably in his book on the gout.[146]

Brief comments on acupuncture had led to interest in that subject, too. Willem Piso's 1658 edition of Bontius's natural history contained the following observation: "I will now set out to tell what happens in Japan, which surpasses those wonders [*miracula*]. In cases of chronic head-ache, or obstruction of the liver and spleen, or in cases of pleurisy, they pierce the organs mentioned with a silver or bronze pin [*stylo*] not much thicker than the string of a lute [*cythrarum*], slowly and gently being pushed through the said organs until the pin comes out the other side, which I also saw done in Java."[147] This report is not in Bontius's original manuscript and was therefore probably one of the later additions made by Piso from someone else's report.[148] In 1669 Henry Oldenburg had translated and published in the *Philosophical Transactions* "Some Observations Concerning Iapan, made by an Ingenious person, that hath many years resided in that Country," which took the form of twenty observations; number eight commented that "There are many Medicinal waters, and Hott-Springs there, which the Inhabitants use in their distempers. They have particular Medicines; but they let no Blood. They make much use of Causticks, by applying upon some nerve or other the powder of *Artemisia* or *Mugwort*, and Cotton, which they set on fire. They always drink their liquors warme."[149] Almost nothing was known of Chinese methods of touching the pulse, only that the Chinese spent much time and effort in the examination.

The virtuosi therefore asked Secretary Francis Aston to follow up this letter energetically. He should not only write back to Ten Rhijne at once but look into the contact Ten Rhijne suggested, an old friend from his hometown and medical Leiden now living in London, Johannes Groenevelt, to find out if the treatises could be procured. He and John Houghton did so within days.[150] Aston was urged to hasten his reply, which he did, inviting Ten Rhijne to send the society any further observations he had about the medicine or natural history of Asia and giving him a list of the subjects they were interested in. In the meantime, Groenevelt wrote to an Amsterdam friend he and Ten Rhijne had in common, Caspar Sibelius, asking that the treatise be sent over, and it was delivered at the end of

March.[151] Aston reported this to the society at the meeting on 5 April, and by the next weekly meeting he also reported on the progress he had made with reviewing the manuscript and seeing to its printing.[152] By the end of June, Groenevelt could report to Sibelius that the Royal Society planned to print Ten Rhijne's treatise and asked to have Ten Rhijne's picture sent over, so that it could be engraved for the frontispiece.[153] The book finally appeared a year later, in May 1683.[154]

The manuscript may have been more than the society had bargained for: the final printed Latin and Dutch text took up 334 pages. It contained not only Ten Rhijne's work on acupuncture but almost all his work other than the commentary on Chinese knowledge of the pulse (which had possibly appeared in the *Clavis*): an introduction by Busschoff, a general discussion of the gout (including a section on its cure by caustics, among them moxibustion); four Japanese diagrams showing the points to which the moxa and the acupuncture needles ought to be applied, together with a short description of Japanese medical practice; a discussion of acupuncture; an account of a pestilential fever that had struck Ten Rhijne's ship when he was sailing to Asia; and three assorted essays.[155] A long summary of the work quickly appeared in the Royal Society's *Philosophical Transactions*, which began with the theory Ten Rhijne had adapted from his Japanese colleagues: "This Author treating of the *Gout*, . . . asserts Flatus or Wind included between the *Periosteum* and the bone to be the genuine producer of those intolerable Pains . . . and that all the method of cure ought to tend toward the dispelling those *Flatus*."[156]

The publication piqued the interest of many physicians in Europe. A translation into Dutch was published not long after,[157] and the English physician John Locke, in exile in The Netherlands for political reasons, gave a copy (probably of the edition published in London) to one of Ten Rhijne's friends who had helped in getting it published, Caspar Sibelius. Sibelius wrote Locke a letter noting that Ten Rhijne had confirmed that artemisia was indeed a proper substitute for moxa and that he, Sibelius, often used it successfully in "cauterizing parts that are painful but not inflamed" as well as in cases of gout. Two years later, Sibelius copied out for Locke part of a letter of Ten Rhijne, which explained that "Cauterisation with moxa can without hesitation be applied more often than once, though the Japanese are immoderate in this respect. I myself . . . commonly employ it three times, unless the trouble is persistent and the cause of it deep-seated and obscure. I shall produce my Notes on Treatment by Cautery at the proper time and place; I am too distracted by other things just now."[158] The interest in acupuncture also grew in Europe.[159] More than forty years after Ten Rhijne's publication, the noted German surgical encyclopedia of Lorenz Heister included a section on acupuncture with an illustration of the needle taken from Ten Rhijne's book.

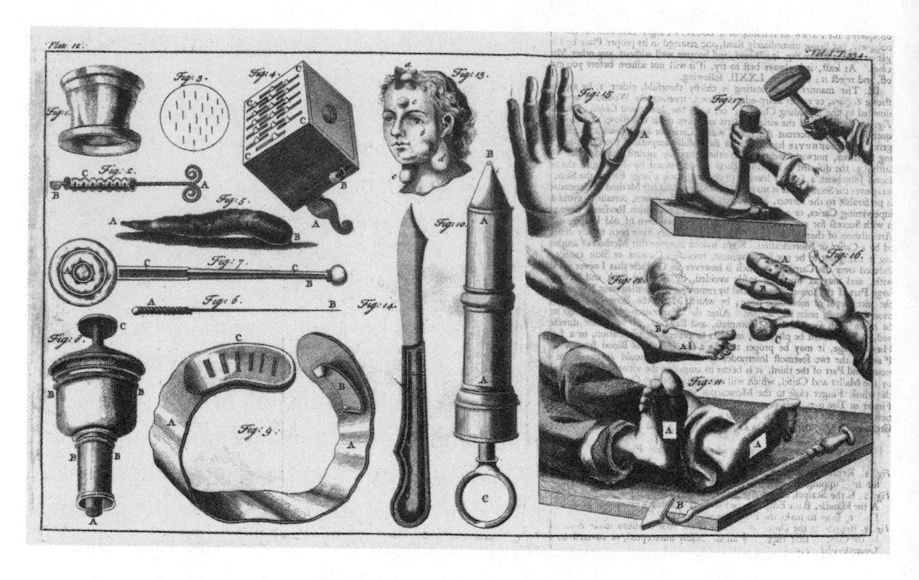

Acupuncture needle and hammer (figs. 6 and 7), depicted among surgical instruments in Lorenz Heister, A *General System of Surgery* (1743). Courtesy of the Wellcome Library, London

Heister had his doubts, however: "One wonders how such clever nations can esteem these remedies so highly," he sighed.[160]

Not long after Ten Rhijne's book appeared, moreover, another practitioner in service with the VOC visited Japan and confirmed many of Ten Rhijne's observations. Engelbert Kaempfer's work became more widely known in the eighteenth century than Ten Rhijne's, while it took a similar approach to explaining the practices of acupuncture and moxibustion, and to explaining why they worked. Kaempfer, from Lemgo in Germany, joined a Swedish legation to Persia in 1683 that was attempting to establish trading relations, but after various disappointments he joined the VOC in 1685, going on to Bandar Abbas as the physician to the factory there until 1688.[161] Having proven his worth, Kaempfer then moved to Batavia, and from there he took up a two-year posting to Deshima from autumn 1690 until autumn 1692. He then returned to Europe and in November 1693 took a medical doctorate at Leiden on ten exotic matters of fact he had observed (*Decadem observationum exoticarum*), such as a worm that infests the people around Hormuz and much of the Persian Gulf called the *dracunculus*, the mumia of Persia called "Mouminahi," and acupuncture and moxibustion. Acupuncture he termed "a Japanese cure for colic," which the Japanese called *senki;* it arose from a gas of the intestines, mesentery, omentum, peritoneum, and muscles of the abdomen, which the use of the needle allowed to pass out of the body, giving relief. Moxa, he confirmed, was "a Chinese and Japanese substance used in cautery," which worked by dissolving the "impacted matter of the illness" or by drawing it out "dissolved into gas."[162] Kaempfer returned to his native city in 1694 before in 1698 becoming physician-in-ordinary to Friedrich Adolf, Count of Lippe. His reputation was established by the book he published in 1712, *Amoenitatum exoticarum politco-physico-medicarum fasciculi* V ("Exotic politico-physico-medical pleasantries in five sections"), which contained many important observations about Persia and Japan at the time he lived there, with fascicle 3 a slightly expanded version of his medical thesis, including his commentaries on acupuncture and moxibustion.[163]

As for Ten Rhijne: despite his anger at Cleyer, he retained his positions in Batavia. When he married in December 1681, he was titled a member of the Council of Justice and "Physician to the VOC."[164] He took his place on the Council of Justice in March 1681—which kept him very busy in subsequent years—and a few months later, in August, he took up an appointment as one of the three physicians of the new leprosarium on the offshore island of Purmerend (along with Cleyer, Louis de Keijser, and two surgeons). In that capacity, he became the chief inspector of leprosy in the city, ferreting out the victims and ordering them to the island (unless they could support themselves under a form of house arrest), where they would be cared for until they died. The disease constituted a serious problem for all the city's inhabitants. Ten Rhijne served as the chief diagnostician, deciding whether various manifestations of unhealthy skin constituted leprosy or other problems, such as syphilis, and among the cases of leprosy, of which kind. Sometime during his first five years as an "external regent" of the leprosarium, Ten Rhijne wrote a treatise on Asiatic leprosy (*Asiatise melaatsheid*), which was published in Amsterdam in 1687. It shows his careful attention to discriminating among varieties of the disease.[165] By 1690, about 170 inmates were being cared for in the hospital per year, including about 7 "Europeans and Creoles" and 164 "Natives and Chinese"; about the same total number (but a larger proportion of Europeans) were isolated in their homes. Ten Rhijne clearly remained quite open-minded about using Asian methods of treatment, for he invited a practitioner from the island of Makassar whom he thought had cured the disease in some cases to try out his treatments on Purmerend. But although the trials were begun on 14 August 1691, doctor "Care" or "Sara" Jagera died on 5 November, ending the testing before a clear outcome could be determined.[166] Ten Rhijne also became a curator of the gymnasium of Batavia in 1684, and the account of the Cape he wrote based upon his stopover there in 1673 was printed at Schaffhausen in Switzerland in 1686 (it was also later printed in English in

1732, translated perhaps by John Locke). It probably appeared through the good offices of a physician who was part of his circle of acquaintances, Johan Jacob Wepfer; the most famous part of it was Ten Rhijne's early account of the genitalia of "Hottentot" women, which would be cited for several generations.[167] He also continued to serve on the Court of Justice, becoming one of those embroiled in the scandalous dispute between the merchant Johan Bitter and his wife Cornelis Cnoll (daughter of one of the chief Dutch merchants in Japan, Cornelis van Nijnroode, and his Japanese concubine Surishia), which caused Bitter to accuse Ten Rhijne of being a drunk.[168] He died in Batavia in 1700.

The publication of books on the medicine and natural history of East Asia was the outcome of countless human relationships, in Ten Rhijne's case spanning the globe from Amsterdam to Batavia, to Deshima and Edo, cities in China, and back to Amsterdam, Frankfurt, and London. It involved famous and anonymous people: Ten Rhijne's boyhood friends, his teachers, his patrons, his momentary friends and rivals, the governors of the East India Company, a minister's wife in Batavia and the minister's son in Utrecht, an anonymous local woman practitioner, the secretary of the Prince of Orange, the English ambassador in The Netherlands, Dutch chiefs of the trading post at Deshima, the shōgun, the governors of Nagasaki, a great many Japanese medical scholars and translators, Polish and Flemish Jesuits in China, unknown Chinese practitioners, famous physicians like Thomas Sydenham and John Locke, virtuosi in London and Amsterdam, Danzig, and Berlin, and printers and booksellers—to say nothing of merchant vessels plying the seas. Similar networks were also deeply involved in interpreting the results of these apparent matters of fact and deciding on how much they could be trusted and used. Human agency was at work on the part of all of the participants, not just the European intellectuals. The issue of the "authorship" of many early modern books, and certainly of the development of new medical practices, is therefore very much open to interpretation.

But the examples of how Ten Rhijne, Cleyer, and others tried to understand and interpret the medical practices and texts they encountered in Asia are also instructive. They were of course eager to find out what other practices might work. To do so, they tried things out for themselves and often decided to adopt the new methods. Describing the medicines and instruments, and how to use them, was only part of their work, however, for they also wanted to be able to give explanations about how the practices worked, presumably to reassure readers. But these explanations were framed in terms of their own understanding of the fundamentals of nature, translated into materialistic language. Words like "pneuma," "spirit," "gas," or even "impacted matter" might refer back to something like Chinese *qi*; but even when it came to explain such a concept in a work like the

Clavis, it was done by materialistic metaphors and similes. None translated the East Asian view perfectly. The same also applied in those instances where there is evidence of how others interpreted European ways, such as how the Japanese translators and physicians tried to understand Ten Rhijne's views on precipitates in the serum of the blood in phrases that come back to us as "muddy blood." Many things crossed from one language to another, but they tended to be those of the world of objectivity: of plain nouns, adjectives, and verbs that refer to what the senses tell us. Abstract concepts crossed much less readily, and then only in a confused manner. Misunderstandings and suggestions could be creative, as when John Floyer's imagination was captured by the idea that the Chinese knew so much more about the pulse than Europeans and in working this out invented the pulse watch. But counting the pulse accurately was in no way similar to the complicated diagnostic nuances that Chinese practitioners drew from their touch. Culture certainly made translating the whys and wherefores as understood by one group extraordinarily difficult. But it was no barrier to useful goods or the business of how to do something. Global trade encouraged materialistic exchanges.

10

The Refusal to Speculate

Sticking to Simple Things

And now I seem to hear the philosophers disagreeing with me. But the true unhappiness, they say, is to be engrossed in folly, to err, to be deceived, not to know. Nay, this is to live as a man. . . . Nothing can be called unhappy if it fulfills its own nature.

—ERASMUS, *The Praise of Folly*

The last three decades of the seventeenth century and well into the eighteenth century formed a period of conservative reaction in The Netherlands. The country had been destroyed in the *rampjaar,* the "disaster year" of 1672, and neither the economy nor the nation's previous exuberant self-confidence was ever the same. Many blamed the problems on those who had governed before the invasion: the States Party, with its Francophilia, supposed nonchalance about national security, and liberal views in religion and philosophy. The religious activists in the Reformed church saw themselves as the bulwark against the continuing threat of Catholic imperialism led by France's Louis XIV and against subversion through excessive intellectualism and even atheism among their compatriots. In the new order, the young William of Orange had seized the office of stadholder, rewarded the organizers of the mob who butchered the former chief of the States Party, Johann de Witt, and his brother Cornelis, and purged municipal, provincial, and national governments in favor of his supporters. Although he generally placed politics before religious doctrine, William took good care not to offend the religious conservatives who so strongly supported him and the military buildup against their nation's Catholic enemies. The country remained deeply split between the conservatives and the mostly libertine regenten of Amsterdam and their allies in other cities, who sometimes called themselves the party of True Freedom. Natural philosophers and physicians remained on the front lines in the often bitter debates about God and nature, often correctly being identified as carriers of materialist ideas of various sorts and sometimes loudly condemned for undermining morality and public order.

DIFFICULTIES OF THE NEW PHILOSOPHY

The implications of the new order for those who supported seemingly abstruse positions such as Spinozism, Cartesianism, Cocceianism, and related points of view, were quickly clear, even for professors at Leiden, since people cared about such ideas. William exerted far more influence over affairs at the university than he possessed by right—he even came to be called the "Chief of the Curators" (*Opper-Curator*), although the stadholder did not legally hold a position on the governing board—with virtually all professorial appointments passing under his hand for approval. In the next few years, as his enemies in the field were pushed back and he had more time to give his attention to local affairs, he wanted intellectual debate stifled in favor of a moral consensus. In 1676, a conservative theologian whom William chose on four occasions to govern the university's academic affairs as rector magnificus, Frederik Spanheim, came to him with a list of propositions that he considered to be deeply in error. Spanheim had already been supported by the curators in his 1673 condemnation of the Cartesian views of Professor Theodoor Craanen (who was subsequently shifted from the teaching of philosophy to medicine), and in 1675 the curators disciplined two doctors of theology who had behaved badly toward Spanheim during a disputation; the conservatives had even been physically assaulted. William agreed that Spanheim's paper listed things to which all people who believed in the Lord God—not only Christians—took offense and had the curators prohibit the promulgation of such propositions in lecture or print. When several professors at whom these steps were aimed objected, even publishing pamphlets to rally their allies, the curators acted. The almost-eighty-year-old theologian Abraham Heidanus took the fall for them all and was dismissed from his post.[1] The theological faculty remained in the hands of Voetians (who followed the lead of Descartes's most famous opponent at Utrecht) until 1689. For a while, students and other interested parties, such as Cornelis Bontekoe, supported Heidanus and his fellows by disrupting the lectures of Spanheim and his allies, and Cartesianism and other such views continued to be discussed in private lessons and in the medical faculty. But conservatives had won the struggle for what could be argued in public.[2]

In the country at large, too, sympathizers of Cartesianism and other tendencies within the new philosophical movement continued to meet, to read and discuss, and sometimes to publish their views, but they were usually attacked as soon as they went public.[3] An example is the case of Balthasar Bekker, a Calvinist clergyman hounded from office in the early 1690s. During his studies at the universities of Groningen and Franeker during the 1650s, at the height of the controversies over the views of Descartes, Bekker became committed to the new philosophy. At the time of the greatest public acceptance of Cartesianism, in 1668, he published a work about it arguing a standard Cartesian position: that it could never be a threat to religion because the foundations of philosophy and religion were different. But his adult catechism of two years later caused bitter controversy for taking positions aligned with the biblical criticism of the Leiden professor of theology Joannes Coccejus, who emphasized a theology rooted in the notion of God's covenant with humanity, separating theology from philosophy and relying on new critical methods of the interpretation of scripture. (Because the Cocceians rejected Aristotelian terminology in the interpretation of scripture, they could freely support new philosophical views, such as Cartesianism.) After various moves he settled in Amsterdam in 1679, and in the early 1680s he took part in an excited controversy over the appearance of several comets. Like Pierre Bayle of Rotterdam and other supporters of the new philosophy, Bekker published a pamphlet arguing that comets were not portents or indications of anything supernatural. By the late 1680s, he was working on a huge project to review all history since biblical times in order to prove that the powers of supposed witches were mainly the result of superstition. He took the Cartesian position that matter and spirit mixed only in humanity, so the Devil (a creature without a physical body) could not intervene in nature. Only God possessed the power to alter its course, and only good angels could carry out God's will when affecting the human spirit was required. Biblical passages were clear enough that the Devil existed, but God kept him locked up in Hell. When other parts of scripture spoke of him and other evil spirits, they did so figuratively, not literally. Consequently, "there would be no magic at all if men did not believe magic exists."[4]

Despite advance warning, his *Betoverde weereld* ("The world bewitched," published in four volumes from 1691 to 1693), caused a furor when the first two volumes appeared, with religious traditionalists accusing him of being an agent of atheism in advocating such a benign and naturalistic view of the world. The Amsterdam consistory—split between conservative Voetians and Cartesio-Cocceians—debated at length what to do, and after meeting with Bekker for some months, they composed a ten-page document of "clarifications" of his positions, in effect "correcting" them in a more agreeable direction, to which he put his hand. This was published at the beginning of 1692. But it merely led to further controversy, with the North Holland Synod condemning his first two volumes in July 1692 and declaring Bekker unfit to hold the office of preacher, even debarring him from communion. Rather than back down, he responded angrily, forcing the Amsterdam burgomasters to keep the peace by suspending him from office, although they continued paying his salary. In return, Bekker published the second two volumes of his work. It appeared in many Dutch and foreign editions, becoming a classic of the early Enlightenment among those who were not believers in the traditional religions. In 1694, William III tried to depoliticize such disputes by getting the States of Holland to do what Amsterdam had been doing: to grant Voetians and Cocceians equal share in all material affairs of the church, including appointments to the ministry.[5] The policy helped to tamp matters down. By the early eighteenth century, intellectual divisions within the Reformed church were no longer as vehement or as significant as in previous generations.

Nevertheless, the perceived threat from the views of such people continued to be felt, as it was throughout Europe in the later seventeenth and early eighteenth centuries. "Cartesianism" was bad enough, but many were termed "Spinozists." It was easy to cast Spinoza in the role of scapegoat, since his most famous slogan was *Deus sive natura* ("God or nature"), equating God with nature, and so undermining the common religious principles that God was above nature, created the world for humankind's use, endowed us with immortal souls that hoped for heavenly redemption, pronounced commandments that required our obedience if we were to obtain salvation, and sent a continuing revelation supported by the evidence of miracles that superseded anything that could be learned from the laws of nature. During the mid-1670s, at the same time that the supporters of Cartesianism were being silenced in Leiden, Spinoza himself was being watched so closely in The Hague by the local synod and its political allies that he spent his last eighteen months in virtual seclusion (he died in February 1677). Nevertheless his associates managed secretly to publish the last of his works in 1678, to a huge public outcry. By the 1680s, a number of Dutch philosophers and theologians were deeply concerned about a widespread, informal, underground movement in which his "gangrenous" ideas seemed to lead the way, contaminating not only many well-educated libertines but also the common people. Spinoza's name became intimately associated with what one recent historian has called the Radical Enlightenment.[6]

In the face of such threats, some Cartesians retreated. A good example is the physician Bernard Nieuwentijt.[7] Born twenty years after Bekker, in the small village of Westgraftdijk in North Holland in 1654, son of a Reformed minister, Nieuwentijt entered Leiden in February 1675 to study medicine, but he was soon expelled for bad behavior, probably for his actions in favor of the Cartesians during those heady days of trying to fight off the conservative crackdown at the university. He enrolled at Utrecht instead, to study both medicine and law, and took his medical degree in 1678 with a thesis (*De obstructione*) that exhibited a fondness for Cartesian corpuscularianism. After settling into practice in his hometown and gaining a solid reputation, he obtained the post of city physician at the larger nearby town of Purmerend (just north of Amsterdam) in 1682. Two years later, he married Eva Moens, the well-to-do widow of a sea captain, and in the same year he gained election to the city council. By the early 1700s, he stood among the local regenten, becoming a burgomaster in 1702 and eight times thereafter. He proved an able mathematician, too: in the late 1680s and early 1690s, he developed an approach to calculating infinitesimals, engaging in controversy over Leibniz's interpretation of the new calculus in works of 1694 and 1696. At some point, however, he decided that his Cartesianism had led him astray into an outlook verging on the atheistic and came back to a defense of religion. The resulting multivolume work was a renunciation of his previous philosophical positions while preserving his abiding interest in factual investigation: *Het regt gebruik der wereltbeschouwingen ter overtuiginge van ongodisten en ongelovingen aangetoont* ("The right use of the contemplation of the world for the convincing of atheists and unbelievers," 1714 and many subsequent editions).[8] Far from a work of theological doctrine, it did not attack other religious positions, only atheists and unbelievers, particularly Spinoza and his followers. It became one of the most popular books of the period, a powerful work of natural theology (or "physico-theology"), proving the existence of an omnipotent and benevolent God from the order, harmony, and providence of Nature.

The basic argument that God's presence could be demonstrated in his creation was ancient, supported by Augustine, and given modern dress in the work of Raymond Sebond, Montaigne's treatise on Sebond, works of literature such as the estate-garden poetry of Cats and Huygens, works on insects by Goedaert, Swammerdam, and Merian, and most famously of all at the time, the long essay by the English naturalist and minister John Ray, *The Wisdom of God Manifested in the Works of Creation* (1691). Nieuwentijt found that he could use the evidence of the new science he liked so well to refute the metaphysical speculations of the Cartesians and Spinozists. Using examples from many authors, as well as from his own experimentation with air pump and lenses, much of it done in an experimental "College" he set up with other burgers of Purmerend, he took the position that basic experiences (*grond-ondervindingen*) lead to generalizations about nature that need to be confirmed by experiment (*proef-ondervindingen*); when they were, certainty followed. In this sense, he thought of his mathematical work as empirically derived as well. Deductive methods could show us nothing for certain except the ingenuity of the human mind. He also thought that he could prove the validity of a literal interpretation of scripture (versus the Cocceians) by showing that remarks in it prefigured some of the discoveries of his own day, meaning that superhuman knowledge (revelation) contributed to its composition. Furthermore, he argued that belief, as when we trust the reports of other people's experiments, was also critical to the growth of true knowledge; in this way, too, belief in the Bible is also trustworthy.[9] In laying out the details of creation in encyclopedic detail, Nieuwentijt also believed that he was demonstrating that the wonders of nature could not have been formed by chance alone. He therefore reintroduced teleology: this was evidence of creation with a plan and purpose, showing God to be all-wise, benevolent, and almighty (while acknowledging that God's promise of deliverance and his mercy could be proved only from nature). His *Gronden van zekerheid* ("Grounds of certainty," 1720) published two years after his death, made a further argumentative case for how real certainty could be derived only from the empirical approach.

THE EDUCATION OF HERMAN BOERHAAVE

Others moved in the opposite direction, from a religious view of nature to one that stressed natural reason. Such a man was the most famous medical professor ever associated with Leiden, Herman Boerhaave, who grew into one of the chief advocates of his day for giving close attention to descriptive facts rather than speculating on their ultimate causes. As a student in philosophy and theology in the late 1680s and early 1690s, he was drawn into the intense debates of his day about the underlying order of the cosmos and at first advocated the view that the moral implications of any particular philosophy of nature should indicate whether its physical principles were correct. But although he never abandoned personal prayer, he gradually came to rely not on metaphysics but on observation and experience as the best guide to how things really were in the world, although that would not lead to any understanding of God. As a professor, he would teach his many students that speculation about the causes of things was unnecessary, even misleading.

As a youth intending to go into the ministry, Boerhaave began his public career allied with what might be called the moderate-conservatives of the Reformed church, but after he failed to advance in the church and took up medical teaching instead, he moved to what might be called the moderate-radicals of natural philosophy. His father, Jacobus, was a minister in the Reformed church, being associated with Jacobus Triglandus, Jr., grandson of the Triglandus who together with Jacobus Revius had led the attack on Descartes at Leiden in the 1640s. As a fellow member of the faculty of theology, the younger Triglandus supported Spanheim's efforts to bring peace by prohibiting the teaching of Cartesianism.

Jacobus Boerhaave made certain that Herman became adept at Latin and Greek, while the younger Triglandus gave the youngster private lessons in Hebrew and Chaldean.[10] The boy matriculated as an undergraduate in philosophy and theology at Leiden in 1683 at the age of fourteen, not long after his father's death, supported by a scholarship for the training of clergymen. In the passages later assembled as his autobiographical notes, the "Commentariolus," Boerhaave wrote of his debt to the professor of peripatetic philosophy (Aristotelianism), Wolferd Senguerd.[11] Senguerd seems to have been thought to be a safe pair of hands, having been appointed in 1675, at the height of the intellectual controversies, because he was considered a philosophical traditionalist. But he quickly developed methods of argument that cut the ground out from under the feet of most metaphysical concepts, while also supporting the experimental natural philosophy introduced by his colleague Burchard de Volder in 1675 as a way to gain more certainty about the operations of nature. In 1679 Senguerd and Johan Joosten van Musschenbroek even constructed a new kind of air pump that became the model for most later instruments designed to experiment with air.[12] At the time Boerhaave became his pupil, Senguerd was between publishing a book on natural philosophy (the *Philosophia naturalis*, 1680) that defended the vacuum and atomism as presented in the form of Christian Epicureanism by the French philosopher Pierre Gassendi, and a compilation of disputations on various experimental investigations (*Inquisitiones experimentales*, 1690).[13] Senguerd taught the youngster "dialectic, metaphysics, the use of the globe, and politics," while also presiding over his first five public disputations.[14]

Nevertheless, in 1689, a professor of classical languages, Jacobus Gronovius, arranged for Boerhaave—still a student—to give a public oration in the great auditorium of the university that attacked Gassendi, and so by implication Senguerd, for thinking that Epicurus had anything to offer. For this speech, the curators awarded Boerhaave a gold medal with a special inscription on it: they at least approved of his views, and he may even have been speaking for them.[15] The oration was a critique of the new understanding of Epicurus emerging from contemporary studies without really taking on board the evidence for it, supported by common pieties and concluding on a patriotic note. If we take him at his word, Boerhaave was completely disgusted with the views he thought to have been held by Epicurus. History was replete with examples of absurd ideas that once seemed brilliant, he said, singling out for his own time two unnamed authors—presumably Hobbes and Spinoza—"who hold horrible opinions on God and the human soul." Similarly, the materialist doctrines of Epicurus were not noble, as Gassendi had falsely argued, but led to gluttony, sexual indulgence, and all kinds of debauchery and sin. After excoriating Epicurus's ethics, he moved on to his natural philosophy: "And because his ethics originate in his physical doctrine, we should consider his teaching on this subject." Epicurus's thoroughgoing materialism upset him deeply: "the scourge of liberal arts and culture asserts that the mind is a mass composed of a varying structure of atoms." The doctrine had recently been "recalled from the realm of Hell by most sinful men," he reiterated, going on to imply once more that he meant the unmentionable Hobbes and Spinoza. Such people taught that nothing like an immortal soul exists, only matter; hence, they did not "hesitate to teach that perpetual happiness consists in progressing from one desire to the next." But this was "a heinous and sinful opinion that robs us of all that is divine and eminently desirable!" The "sanctuary of pleasure" in such doctrines turns out to be a "temple of Pain." Fortunately, at Leiden "the Sacred Pledge of the True Religion is guarded against those evildoers" with the help of "the Fathers of the Republic and the professors" who, he trusted, would be "preserved, long and happily, for the salvation and protection of this jeopardized Republic, of our religion, on all sides attacked, and of the lofty arts and sciences, by the Supreme God Whom we humbly adore."[16] Boerhaave had put his finger on the pulse of the times, and it beat in righteous and patriotic fear of immoral materialism.

The crucial passage in his lecture had been his move from a consideration of Epicurean ethics to Epicurean physics. After showing that the ethics were horrifying, he had made the claim that they originated "in his physical doctrine," which must therefore be equally bad. That is, Boerhaave was inviting his audience to measure the natural philosophy of materialism not by the standards and proofs of natural philosophy itself but by its consequences for the moral life. He was, in other words, invoking the principle of right reason, which he would have absorbed from his classical education. Since "the good" contained in it "the true," anything that went against the teachings of goodness for human life must also be untrue.

Yet mysteries about the episode remain. By attacking Gassendi's natural philosophy, Boerhaave was publicly turning his back on his former professor of philosophy.[17] His "Commentariolus" mentions how he had become "charmed by the enticement" (*dulcedine prolectante*) of geometry and trigonometry in 1687, perhaps implying a drift toward Cartesianism already.[18] In any case, he seems to have been moving away from Senguerd toward his colleague the former Mennonite and supporter of Cartesian metaphysics De Volder, who had been one of Sylvius's pupils and took his degree with a thesis attacking Aristotelianism.[19] De Volder would in turn be Boerhaave's promoter for his doctorate in philosophy (defended in December 1690). In these early stages, Boerhaave apparently thought the Cartesian distinction between mind and body to be a principle from which he could launch an attack on philosophies that were monist and placed knowledge of bodies first, since his doctoral thesis was on the subject of the distinction between mind and body (*De distinctione mentis a corpore*), which repeated the severe criticism of the monist views of Epicurus, Hobbes, and Spinoza.[20] Perhaps as he was moving toward De Volder's views in the late 1680s he had felt compelled to reenter the old debate between Descartes and Gassendi.[21] Yet De Volder was not simply a semi-Cartesian: he was one of Spinoza's acquaintances and remained a discrete but steady defender of the power of his views.[22] Moreover, one of the professors of medicine for whom Boerhaave later expressed the greatest respect, Lucas Schacht, not only affirmed Cartesian principles but "defended propositions that would not have been rejected by the young Spinoza."[23] It may be, then, that in inviting Boerhaave to deliver the oration, Gronovius was able to put words in the student's mouth with which he already disagreed, hoping that by committing himself in public Boerhaave would return to the fold privately as well.

In any case, this oration was the last occasion on which Boerhaave implicitly invoked right reason in making judgments about natural philosophy. In later years he took a position similar to Descartes's own, commenting only on truths that could be known through the senses and natural reason. One cause may have been the influence of his teachers, especially De Volder. But in later years, Boerhaave also remembered having developed a growing contempt for all debates based on arguments about first principles. He had worked his way through the Church Fathers in chronological order from Clement of Rome onward, being impressed by the "simplicity" of the early authors and regretting how the scholastics, even the Cartesians, had later corrupted theology with "subtleties" and philosophical debates. From these came bitter disputes that contrasted with the fundamental goal of peace with God and one's fellows. In short, he learned that metaphysics was a poor foundation for a true understanding of the simple teachings of the sacred writings. Indeed, as he later put it, he planned to take a doctorate in theology with an oration on the subject of why unschooled (*ab indoctis*) men had made so many converts to Christianity in the beginning, whereas the highly educated moderns made so few.[24]

The shift in young Boerhaave's interests away from metaphysics was bolstered by other studies. To support himself after graduation, from the summer of 1690 to the end of 1691, he worked for the university library cataloguing the remarkable collection of books and manuscripts purchased from the estate of the "libertine" Isaac Vossius, which for the first time made many of the "moderns" publicly available to the university community.[25] At the same time, his mentor and patron, Jan van den Berg, who was secretary to the board of curators, urged Boerhaave to study medicine, perhaps as a means of making a living.[26] In taking up the study of medicine, Boerhaave would have been continuing to work along the lines of his former professor De Volder. Perhaps under his influence, too, Boerhaave also taught mathematics to "most able young men" (*lectissimis juvenibus*). But in his own mind he clearly intended to study medicine only for the knowledge of nature, not for clinical practice. He read the medical authors as he had tackled the theologians: in chronological order beginning with Hippocrates up to the "English Hippocrates," Thomas Sydenham, with whom he was particularly impressed. His ready access to the books of Vossius, who had lived in England, made this possible. Boerhaave also studied anatomy books and attended the public anatomical lectures (then being given by Antonius Nuck), but he did not attend the lectures of the medical professors aside from a few—presumably on the foundations of medicine—by Charles Drelincourt. Nor did he ever go to Paulus Hermann's lectures on botany and materia medica, even though he read Hermann's *Flora* and spent hours in the Leiden hortus and botanizing in the field.[27] He also opened the bodies of animals and cultivated chemistry by day and night. His method, then, was to read critically and to supplement this with physical demonstrations and chemical experiments. It would have been the best introduction to the natural sciences of the day available to an autodidact. He was confident enough in his knowledge to decide that he should take a degree in medicine, although he continued to plan to take a doctorate in theology.[28] In 1693 he traveled to Harderwijk to receive a medical degree, which he did in July with a thesis on "The Utility of Examining Signs of Disease in the Excrement of the Sick" (*De utilitate explorandorum in aegris excrementorum ut signorum*).[29]

Boerhaave's continued hopes for a position in the church were, however, almost immediately dashed by an occasion on which someone inferred that he had imbibed dangerous views. Looking back on the incident not long before his death in 1737, Boerhaave wrote somewhat cryptically that "upon returning to Leiden from the university in Gelderland, he was innocently ruined when someone did not form the right opinion of him, from which occasion he foresaw obstacles to his desire for a ministry in the church."[30] Especially since he was returning as a newly minted medical doctor, there would have been echoes in his conversation of sympathies for contemporary natural philosophy, and of course physicians had been in the forefront of supporting a variety of materialist positions, even making Cartesianism into one in the 1650s. Moreover, physicians in Leiden had just introduced Newtonian natural philosophy, which for some years also attracted Boerhaave. One of the Scottish exiles in The Netherlands during the 1680s, Sir James Dalrymple, had become friendly with the theologian (and Boerhaave's teacher) Trigland, and on Dalrymple's advice, and no doubt with Trigland's support (and with William III's religious adviser Bishop Gilbert Burnett downplaying any doubts that the prince may have had), the curators appointed another Scotsman, Archibald Pitcairne, to a position on the medical faculty in 1691. Pitcairne was one of the early enthusiasts for applying Newton's principles of natural philosophy to medicine, and he encouraged many of the medical students to defend theses that took a favorable view of Newtonianism. But in the summer of 1693—about the time Boerhaave took his doctorate in Harderwijk—Pitcairne returned to Edinburgh for a holiday, there married for a second time, and never returned to Leiden, to the fury of the curators.[31] Perhaps Pitcairne did not find Leiden to his liking: as a Scottish Tory, he was, after all, politically and religiously opposed to William III, and a number of Newtonian principles would have caused concern to the religious conservatives. Newtonianism included an enthusiasm for mathematics and Copernicanism (as with Descartes), atomism and the void (as with Epicurus and Gassendi), the invocation of mechanical principles to account for all changes in nature, and above all for experimentation rather than metaphysical speculation as the fundamental method of inquiry into nature, differing from Cartesianism and Epicureanism mainly in allowing for a creator God to be distinct from yet present in the system of nature.[32] In the climate of the day, however, such views could easily be interpreted as favoring the position that held nature to be the only thing worthy of study.

One of Boerhaave's colleagues therefore later explained away the difficulties he encountered on his return to Leiden as a huge, almost comical, misunderstanding. At the funeral oration he delivered for Boerhaave in November 1738, Albert Schultens told a story. Having just taken his medical doctorate at Harderwijck, traveling back to Leiden in a canal boat, Boerhaave overheard several fellow passengers condemning Spinoza. We can infer that Boerhaave made the common mistake of trying to turn what was a political conversation into an academic debate, for according to Schultens he asked his fellow passengers if they had ever actually read anything by Spinoza. In doing so, he would have implied that he was familiar with Spinoza's work, and apparently the conversation stopped dead. But Schultens also reports that someone who had been listening with favor to the previous discussion asked the name of this newly minted doctor and carefully wrote it down in his notebook. After Boerhaave and the passengers got out at Leiden, a rumor began circulating that he was a secret Spinozist. "Oh, what a false accusation! that betrayed you with such a plague," commented his memorialist. This was the episode that reputedly ended Boerhaave's chances of advancement in the church.[33] Perhaps, given Boerhaave's newfound enthusiasm for the study of natural bodies, the accusations leveled at him came closer to the mark than he or his friends later wished to admit. Yet given Boerhaave's later comment about his innocence and the importance of the story in Schultens's account of his life (no doubt heard from Boerhaave's lips), it is clear that he was deeply wounded by the event, and perhaps it further sensitized him to the dangers of judging a person's qualities by the remarks they uttered, for he would in future be very careful about such matters.

Almost nothing is known of Boerhaave's life during the next decade except that he remained in Leiden. Without a future in the church, Boerhaave turned to the practice of medicine. When invited by a nobleman close to William III to move to The Hague, Boerhaave declined, apparently not wishing to be involved in the games of dissimulation present at court and council. Remaining in Leiden, he later said, left him free to pursue his studies as he liked without needing to pretend to believe what he did not or deny what he did. He busied himself with visiting patients (something he had no previous experience of), tutoring pupils in mathematics, conducting chemical experiments, and reading medical and religious works.[34] As for the medical faculty, during the 1690s, the curators and William III often could not agree on new appointments, and it was run down. Jacob Le Mort had taken over the chemistry teaching after the death of De Maets in 1690, but William III blocked the attempt to appoint him as a professor. In 1693, following Pitcairne's departure and Nuck's death, only two professors remained, and after various attempts to attract other eminent physicians to the vacant posts failed, the curators settled for Frederik Dekkers (a local physician in Leiden and an uninspiring teacher) and Govert Bidloo (an expert anatomist but also physician to the king and hence mostly absent from Leiden). When Hermann died in 1695, he was replaced by one of Ruysch's assistants from the botanical garden in Amsterdam, Petrus Hotton, another worthy but not especially distinguished selection. Drelincourt's death in 1697 gutted the faculty. The economic problems of the period exacerbated the situation: student enrollments declined and a financial crisis arose.[35]

When Boerhaave reemerged on the scene, he did so to bolster student numbers in the medical faculty. He obtained an appointment as a lector in Leiden's medical faculty in 1701. This was a position inferior to a professorship, and at about a quarter of the salary, with the duties of reading the classical texts with the students to familiarize them with the standard sources. In tipping this position to him, his patrons clearly continued to consider his greatest strength to be his knowledge of texts and philosophy. But this time he took good care to avoid philosophical and theological controversy.

He used the occasion of his appointment to deliver an oration recommending the study of Hippocrates, hero of those who lauded the power of clinical description. He now spoke as an advocate for careful descriptive investigations into diseases and their bodily origins and for pursuing these matters of fact rather than speculations about ultimate causes. He argued firmly against the implications of philosophical right reason for knowing natural things: nothing should be judged by its goodness, but rather the mind should be cleared of all a priori judgments for the matters of fact to enter it fully and correctly.[36] The oration began with an attack on "science" that was "based on theoretical speculation without any reference to practical experience." Toward the end of the oration, he returned to the theme. Those who start with "general principles of nature, like matter, movement, and the shape of corpuscles," and then move on to "analytical thinking and using a seemingly plausible reasoning" that they apply to the facts, inventing rules for healing from their speculations, are basing medicine on "figments of their imaginations," yielding only disappointment to themselves and harm to their patients. To put medicine right rather required the physician to emulate Hippocrates in following Nature as "the sole guide when investigating facts." To do this, one had to be "free of all sectarianism, unfettered by any preconceived ideas, [and] devoid of all leanings towards prejudice," which would allow the investigator to "merely learn, accept, and relate what he actually sees." Hippocrates himself simply gave "a lucid description" of a disease and then a method of treatment that was "thoroughly tested out and confirmed by favourable results in many cases." In recent years, Thomas Sydenham in England had properly revived this Hippocratic method. The concentration on descriptive detail should be bolstered by a "simple style which sets forth the subject matter briefly and clearly" when reporting the facts. The facts alone should "determine the argument and not the other way round."[37]

Boerhaave's advocacy of starting with the facts and only then using natural reason to determine their meaning continued for the rest of his career, although he did sometimes offer further philosophical support for this position. He did so, for instance, in another oration, in 1703, "On the Usefulness of the Mechanical Method in Medicine." His teaching had been so successful that he began to attract offers from other universities, which led the Leiden curators to offer him an increase in his salary and the promise of the first vacant medical professorship, a promise commemorated with this oration. In it, he praised mechanics as an "almost superhuman science" for the power it gave humankind to move any mass based on a few simple and certain principles. Yet in this science, too, one had to begin with "the effects to be observed in each body" by means of sense perception. Once a multitude of facts had been discovered about each body, then arguments of a geometrical kind "will discover much more than would have been possible with the aid of sense alone." From both methods, it was clear "that the human body is in its nature the same as the whole of the Universe which is open

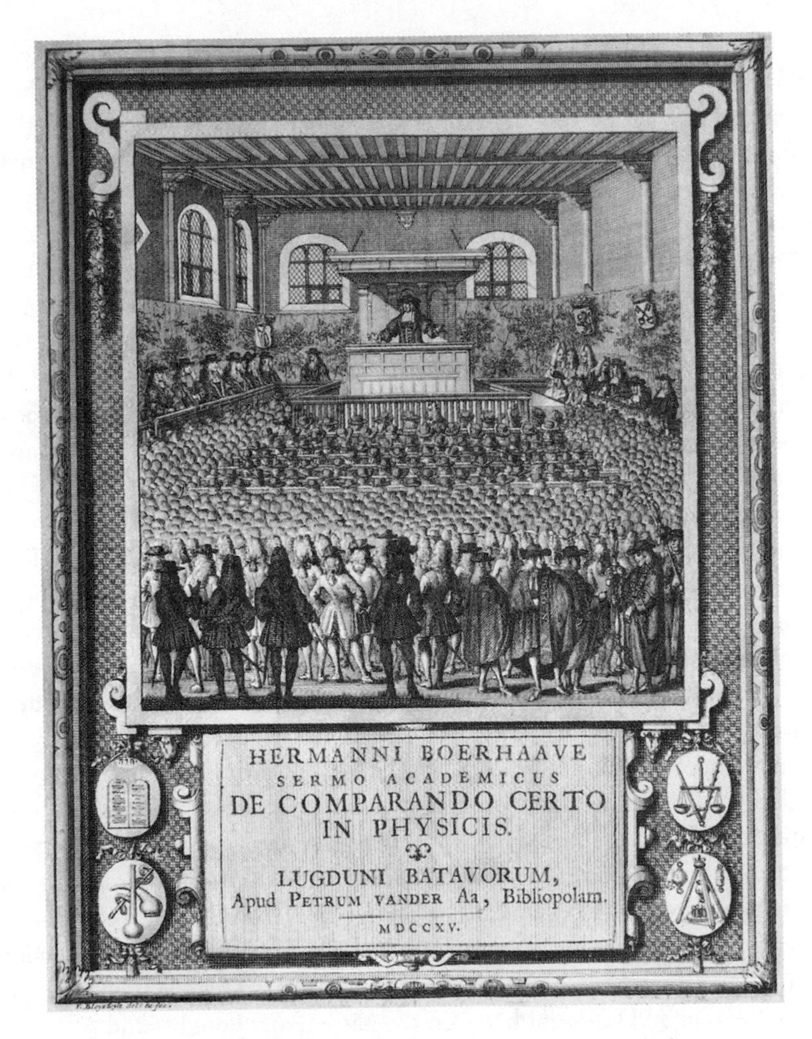

Boerhaave lecturing in the main hall of the university. Title page of Herman Boerhaave, *De comparando certo in physicis* (1715). Courtesy of the Wellcome Library, London

to our view." He then described some of the many facts that had been found out about the human body by means of observation, microscopical studies, the ligation of vessels in living animals, the injection of mercury into dead bodies, the careful observation of diseases, and comparative anatomy. Using such methods, Malpighi and others had shown that the body was composed of both simple vessels and "mechanical parts" such as glands and muscles. Hippocrates himself had discovered that the body was composed of only these two categories. Because it was liquids that flowed through the channels of the simple vessels, physicians also needed to learn from hydraulics and chemistry how to study them. From such investigations new treatments might even be devised. On such a view there was no need to invoke innate powers, much less the intervention of supernatural forces, to explain all that is necessary, even for life itself.[38]

Putting right reason aside might be allowed to physicians, who were after all concerned with treating bodies rather than guiding souls. There was one common objection to the mechanical account offered by Boerhaave, however: the ability of the mind to affect the body. Yet whoever made such a remark was assuming that "mind" was distinct from body, and so composed of something else. Boerhaave therefore asked a question of his own: "Who has been able to locate the secret of this marvellous relationship in any constitutive part of either mind or body?" In other words, no dualist had yet offered a satisfactory account of how mind and body interacted. That was because the sources of such things are not open to investigation by the senses, and so beyond our comprehension. He remained convinced that metaphysical speculations were both useless and pointless for solving such problems, for even if one assumed that the mind could control the body, "as soon as the capacity of thinking influences our body, every effect it brings about therein is wholly corporeal." It did not, therefore, matter if the first cause was corporeal or not. The physician needed only to perceive the corporeal effect "and to scrutinize and guide it" toward health. That was sufficient: the physician does not need to know ultimate causes, only bodily effects.[39] Boerhaave made the same point near the beginning of his textbook on medicine, the *Institutes* (*Institutiones medicae*, first edition 1708) by remarking that "Man is composed of a Body and Mind, united to each other," but "the Nature of these are very different, and . . . therefore, each has a life, Actions and Affections differing from the other." As almost everyone agreed, "There is such a reciprocal Connection and Consent between the particular Thoughts and Affections of the Mind and the Body, that a Change in one always produces a Change in the other, and the reverse." In an almost exact paraphrase of Descartes, he concluded: "We cannot understand or explain the Manner in which the Body and Mind reciprocally act upon each other from any consideration of their Nature separate; we can only remark by Observation their Effects upon each other, without explaining them."[40]

The content of Boerhaave's soon-to-be famous medical text therefore differed greatly from previous ones. He continued the basic arrangement by dividing the medical institutes into five parts: the elements of nature and the natural functions of the body based thereon; pathology; semiotics; hygiene; and therapeutics. What changed most substantially between Boerhaave's textbook and previous ones such as Fernel's, however, was the content of the first part, commonly called "physiology." Boerhaave said nothing about the four elements or humors, about formal or final causes, about faculties or powers, about natural, animal, or nervous spirits, or about any of the rest of classical teaching about fundamental principles. He even—significantly—left out talk about reason (or mind) and the passions. Instead, he explained all that was necessary for a physician to know in terms of observable solids and fluids alone. He dwelt on this experimental physiology, treating the other four parts of the institutes (pathology, semiotics, hygiene, and therapeutics) briefly but also materialistically, stressing the *what* and *how* of medicine rather than the *why*. After a short review of the history of medicine, for instance, he drew the following the lesson for the student readers of his textbook: "From hence it appears that the Art of Physic was anciently established by a faithful Collection of Facts observed, whose Effects were afterwards explained, and their Causes assigned by the Assistance of Reason; the first carried Conviction along with it, and is indisputable; nothing being more certain than Demonstration from Experience, but the latter [that is, causal reasoning] is more dubious and uncertain." His heroes, Hippocrates and Harvey, advanced knowledge farthest because they had paid close attention to "demonstration from experience," whereas Galen had corrupted ancient medical teachings with his elaborate reasoning. To discover truth in physic, all knowledge had to be based on sense experience, which meant accounting only for those things "which are purely material in the human Body, with mechanical and physical Experiments." Or, as he repeated, discussions of first causes "are neither possible, useful, or necessary to be investigated by a Physician." The physician should therefore confine himself to materialistic explanations only, based as they were on the demonstrations of anatomy, chemistry, and mechanics via the natural and experimental philosophy.[41]

For the practicing physician, the outlook Boerhaave held also simplified treatment. Not only did he applaud the Hippocratic attention to describing clinical cases in all their exact detail, he believed that particular symptoms pointed to particular physiological causes. The disease and its cause would consequently be the same in anyone found to be suffering from the same symptoms. Treatment would then be to modify the corporeal cause of the disease, usually by medicine, sometimes with an accompanying regimen; this treatment would be similar in all suffering the same effects. Age, gender, social class, and other variables might make someone more or less susceptible to particular errant physiological processes. But whether king or queen, duke or peasant, merchant or servant, anyone with one kind of a disease should be treated for it in the same way as anyone else. Such an outlook stood directly against the classical view that because each person had a different temperament and lived in different circumstances, the cause of illness in each patient had a singular origin, often requiring a different treatment. When Boerhaave published his textbook on treatment of disease, then, he titled it *Aphorisms* (1709), following the example of Hippocrates in more than title by setting down only observations of disease and successful treatment. Again, his theme was *how,* not *why:* observation and experience, not speculating about causes or connecting modes of treatment to long explanations of why they worked.

Despite his emphasis on matters of fact and natural reason alone, Boerhaave allowed room in his thinking for an immortal soul and God's lawful design of humankind and the rest of creation—but exactly what he thought about the divine remained private.[42] It is unlikely that Boerhaave was a thoroughgoing philosophical materialist in the sense of considering all knowledge to be the result of the movement of material bodies acting on others, some of which we call "ideas." Late in his career, Boerhaave wrote of a Hippocratic *impetum faciens* or *enormon,* a faculty of the soul that bound soul and body together. But as one historian has put it, even this "was for the physician really not accessible." Human life might have "a vitalistic beginning" in God's creation but from then on could be treated entirely according to mechanical principles.[43] It was later said that he prayed every morning, and in his last written words he referred to being about to leave the earth at the command of the "Almighty" (*Ominpotens*).[44] But for him, at least in his public declarations after 1700, there was no way to gather evidence about God's nature, will, or plan via the five senses, so such things were unknowable. This was a long way from the conservative theological and philosophical position where he began as a student.

Boerhaave's insistence on a matter-of-fact approach can clearly be seen in Boerhaave's first work as a professor of medicine. The opportunity came when in 1709, on the death of the professor of botany, he claimed Hotton's chair in the medical faculty (he had been promised the first vacancy). Because he had no reputation for knowing botany, however, it took the learned world by surprise. It may even have taken him by surprise, for upon assuming the chair, Boerhaave delivered an oration on simplicity in medicine, a nice rhetorical play on the notion of medical simples, and advocating many of the same themes raised in previous orations, but containing almost no suggestion of a close acquaintance with, much less an agenda for, the subject he was appointed to teach.[45] There had been more qualified candidates for the post. As soon as news of Hotton's death spread, the well-known Caspar Commelin was given an increase in his salary to keep him in Amsterdam, there were rumors that Johann Philipp Breyne of Danzig (who had, for instance, published botanical descriptions not only by himself but by Ten Rhijne, Rumphius, Cleyer, and others) might be the best replace-

The Leiden botanical garden in the early eighteenth century.
Title page of Herman Boerhaave, *Index plantarum* (1710).
Courtesy of the Wellcome Library, London

ment, and Johann Jacob Scheuchzer of Zurich—well connected with German and English naturalists, a collector of naturalia, and an expert on alpine plants—was being pushed for the post by some members of the curatorial board, most particularly by the curator and diplomat Pieter Valkenier. But Boerhaave's long-standing mentor, Jan van den Berg, steered the appointment to him in a close vote. From the outside, it was clear that local interests had prevailed over the international interests of the botanical community.[46]

Determined not to let Leiden's reputation worsen, however, Boerhaave did as he preached and worked hard to master the descriptive details of the science. He took great pains to compile an account of the contents of the Leiden hortus, quickly bringing out a catalogue (the *Index planatarum*, 1710) containing about 3,700 plants, many described for the first time. Though it has been called "an indifferent piece of work," in terms of its botanical significance—Boerhaave himself wrote that its standard was not as high as he wished, although as high as he was able—it is an impressive achievement for someone previously untrained in anything more than the elements of botany.[47] Although he had the ambition to rearrange the layout of the plants to accord with the system of Joseph Pitton de Tournefort, he did no more than introduce some new plants along the borders, leaving the beds as arranged by Hermann and modified by Hotton, illustrating the carpological system (a classification based on the seeds and fruit) of Robert Morrison. But following in his immediate predecessors' footsteps, he soon turned his attention to the flowers. As the English physician Hans Sloane wrote in 1690, following the publication of Hermann's *Paradisus Batavus* in 1689, "people begin to find fault with all methods hitherto, and to arrange them only by the flower," although "how long this may last I cannot tell." William Sherard not only edited that work by Hermann but put Boerhaave in touch with Sébastian Valliant, leading Boerhaave to promote Vaillant's view that the parts of flowers could be compared to the sexual organs in humans and animals, a system that Carl Linnaeus (who studied with Boerhaave in 1735) would develop in its most influential form. Although he considered the collection of naturalia in the ambulacrum to have decayed into ruin, Boerhaave also did his best to build it up by collecting minerals, corals, animals, and other specimens of naturalia. To do so he asked the curators to seek support from the VOC, WIC, and admiralty boards. He also deposited his own collections, in two green cases, in the ambulacrum, although specifying to the university that they remained his. Most important, he immediately wrote to botanists throughout Europe for seeds and specimens, building up a wide network of colleagues.[48]

BASING A MORAL ORDER ON MEDICAL MATERIALISM

Boerhaave's silence on ultimate causes may have been one of the reasons why he seemed such a great teacher to so many students. They might be Calvinist, Anglican, Lutheran, Catholic, Jansenist, Jewish, Mennonite, atheist, or anything else and yet never be concerned about his religious orientation. Boerhaave's critics took note: in his *Apologia pro vera et saniore philosophia* ("Defense of the true and sound philosophy," 1718), the Franeker professor R. Andala accused Boerhaave of being a skeptic and an atheist.[49] The English physician-surgeon Daniel Turner attacked Boerhaave's *Treatise on the Venereal Disease* (1729) for being merely empirical because it dealt with treating the symptoms rather than the cause.[50] Boerhaave's refusal to speculate in public can even help to make sense of the praise bestowed on him by the most famous—or infamous—materialist of the eighteenth century: Julien Offray de la Mettrie. Before he became a materialist, La Mettrie seemed embarrassed that Boerhaave could be interpreted as irreligious. In his translation of Boerhaave's *Institutes* in 1739 and 1740, La Mettrie included a "Vie de Monsieur Herman Boerhaave," in which he dealt directly with the question of Boerhaave's silence on first causes. After quoting long passages from Boerhaave's oration "On Servitude as the Physician's Glory" (1730), which describes the need for the physician to understand and cooperate with nature in order to heal the patient, La Mettrie explained that it would be crude of him to comment on it "in order to make it appear that the learned professor would not recognize other causes of everything that occurs in the world." Boerhaave recognized that laws are "imprinted on each part of the human body," and in invoking "Nature" he should not be suspected of irreligion. His very moral life stood as proof that he contemplated the future life.[51] Ten years later, however, La Mettrie had completed an eight-volume translation with commentary on the *Institutes*, in which his notes interpreted Boerhaave as a materialist.[52] It may even have been through puzzling out Boerhaave's meanings on the relation between brain and mind that led La Mettrie into explicit materialism himself.[53]

For some medical students of Boerhaave's generation, however, it was not enough to refuse to comment on the relation between the good and the true: they wished to derive an understanding of human behavior from natural truths. For this, continuing discussions about how nature prompted actions—an understanding of the passions—remained critical. The views of one of Boerhaave's contemporaries who studied medicine at Leiden, Bernard Mandeville, were much more radical. (Because he moved to London and wrote in the common tongue there, he is often mistakenly considered to be English.) Mandeville believed that the careful study of nature could provide a guide for human behavior, although this had nothing to do with old-fashioned moral philosophy and everything to do with the proper pursuit of the passions within a system regulated by the "clever politician," who would ensure that the public benefited. His work, like Boerhaave's and even Nieuwentijt's and Bekker's, signals how religious and political debate was being driven by the findings of medicine and natural history, even in a period of conservative reaction.

Mandeville was born just over a year after Boerhaave into a family of doctor-magistrates. On his father's side were three generations of medical doctors who also served on various political offices in the city of Nijmegen. His mother's family also served their city, her father being a naval captain of the admiralty board of Rotterdam, the city to which they moved three years before Bernard's birth in 1670. Shortly after the family's move there, his father became city physician, a board member of the city hospital, an alderman of the adjoining jurisdiction of Schieland, and a lieutenant in the militia.[54] It is certain that the Mandevilles were deeply supportive of the anticlerical and republican values of the States Party against the order of Calvinist orthodoxy and princely influence that came to dominate not long after Bernard's birth. The Mandevilles were, for instance, well acquainted with Adriaan Paets, one of the former leaders of the States Party in Rotterdam, a lawyer who had been a member of the city council and ambassador to Spain, and a strong supporter of religious and political liberty.[55] Given his various civic duties, Bernard's father would have come into close contact with Paets and others of his circle, while his maternal grandfather, after whom Bernard was named, had served with Paets on the admiralty board.[56] Among his many accomplishments, Paets had helped to form the Collegiant movement in Rotterdam. Beginning near Leiden in Rijnsburg in the 1620s and 1630s, the Collegiants composed a fluid group of anticlerical religious dissenters who met for the purposes of religious self-education and free discussion. By the time the young Paets joined them in the middle of the century, their millenarian tendencies were declining, moving them increasingly toward strong support for more secular rationalism; many like Paets were even touched by Socinianism. Although they met in discussion groups, like the English Quakers (who also had a large community in Rotterdam), Paets's Collegiants were firmly opposed to the Quaker religion of personal inspiration; Paets himself had written a treatise against revelation.[57] Rather, Paets shared many ideas with the Rotterdam merchant and Collegiant Jan Bredenburg, who in 1672 circulated a manuscript that strongly defended causal determinism and reason as a better guide than scripture, causing a great debate among the Collegiants; Bredenburg soon also exhibited an acquaintance with Spinoza's philosophy.[58] Paets also knew Spinoza personally, and he used his Rotterdam friendship with Pieter de Groot to make contact with the powerful republican theorist Pieter De la Court.[59] Before his death in 1686, Paets did much to bring the values of the party of True Freedom to prominence in Rotterdam and to keep them alive during a period of Orangist revival.

The influence of older States Party leaders like Paets had been felt not only in Mandeville's political upbringing but in his formal education. As a burgomaster of Rotterdam in 1681, Paets had been instrumental in bringing two French

Protestant intellectuals to Rotterdam after Louis XIV destroyed the Protestant academy in Sedan: Pierre Jurieu and Pierre Bayle.[60] They made it possible to transform the local Latin school into the Illustrious School in which Mandeville enrolled. He studied there before Jurieu and Bayle fell out with each other.[61] But by the time Mandeville left in 1685 (the year of the Revocation of Nantes, which brought Huguenot refugees north by the thousands), Jurieu had begun to think of himself as the chief spokesman for Calvinist orthodoxy against Bayle's pleas for toleration and the defeat of "superstition." From Jurieu, Mandeville would have had the Calvinist position rammed home: that salvation could not be obtained through human reason or will. Mandeville's commentators have often remarked on his high standards for virtue. As Thomas Horne has put it, "Mandeville maintains throughout his work a rigorous moral standard, [but one] which man is unable to live up to."[62] From Bayle, Mandeville would have absorbed suspicion about human motives, a sharp intellectual skepticism, and possibly the view that moral virtue had little or nothing to do with whether people called themselves Christians.

In addition to being reared on the political outlook of the party of True Freedom and taught by the likes of Bayle and Jurieu, Mandeville became deeply imbued with philosophical Cartesianism and empirical science through his medical studies.[63] In October 1685, the month of the Revocation, the fifteen-year-old Mandeville went off to study at the university of Leiden, matriculating in the faculty of philosophy the following September.[64] The disputation he defended in 1689 at the end of his philosophy training, *De brutorum operationibus* ("On the activity of animals"), took the view that animals lack sensibility—an extreme "Cartesian" position.[65] He then moved into the medical faculty just before its difficulties of the 1690s. Drelincourt was aging but still an excellent medical scholar; Nuck was launching an impressive range of anatomical demonstrations; Hermann taught botany at the university and clinical medicine in the city's Caeciliagasthuis; Dematius offered courses on chemistry until his death in 1690; the brilliant clinician Schacht had just died. They all stressed the importance of performing detailed investigations into nature and the utility of such studies for repairing the health of fallen humanity. After two years, Mandeville successfully defended a medical thesis—just three months after Boerhaave, whom he must have met as a student—with the title of *De chylosi vitiata* ("Of corrupt chylification"). He argued, along lines that would also be familiar to Boerhaave and other Leiden teachers and pupils, that corrupt chylification, or bad digestion, caused the illnesses of hypochondria and hysteria. The "Stomachiack Ferment" of digestion occurs when the stomach receives animal spirits through the hollow nerves lining it. These spirits are composed of both fine and gross particles.

For proper fermentation, a balance of the two is necessary. But the brain also consumes the finer particles in thinking, while the grosser are consumed by the body in exercise. If a person does too much thinking or gets too little exercise, the gross particles accumulate in the stomach and make the digestion "sour."[66] The roots of Mandeville's physiology can be traced to both Paracelsianism and learned anatomy, to his teacher Nuck's work on the lymphatic system, to Van Helmont's interest in the subject of digestion and fermentation, to Cartesian corpuscularianism, to Sylvius's acid and alkali theories, and to the general doctrine of vascular secretion that Boerhaave later codified.

Amid his medical studies, Mandeville and his father became embroiled in a bitter political affair in his hometown: the Costerman Riot and its aftermath. The conservatives in Rotterdam flexed their muscles even further after the Prince of Orange became the king of England in 1689. The Mandevilles opposed them, taking the side of those who attempted to oust the city's Orangist bailiff (*baljuw*), Jacob van Zuijlen van Nievelt, who ran Rotterdam on behalf of William III. He was known to many as a corrupt and domineering official who brought the stricter Calvinist clergy into a network of informers and political supporters. Van Zuijlen had seen to it that some young, well-to-do militiamen were caught one night on watch refreshing themselves with a cask of wine on which duty had not been paid, but during the ensuing scuffle one of Van Zuijlen's taxmen died by a sword blow. A militiaman, Costerman, was accused of dealing the fatal stroke (perhaps falsely) and executed.[67] A group of people who hated Van Zuijlen as an officious, rapacious, and extortionate tyrant and religious hypocrite thereafter gathered to plot his downfall. Bernard Mandeville and his father were among them. On 5 October 1690, egged on by satirical verse broadsides against Van Zuijlen—one of which was probably written by Bernard himself—a crowd gathered in front of the bailiff's house, blasted in the front door with a cannon, and systematically ransacked the place, even pulling down the building's stone facade. The following day, the city government dismissed Van Zuijlen. A court case against the bailiff for various crimes in office followed. Mandeville's party looked set to win its case against him when William III shifted it to the High Court of Holland and Zeeland, where he controlled the process, obtaining an acquittal for the bailiff. The Rotterdam city council was also forced to pay a large fine. William then expelled from the city government any men sympathetic to the old States Party ideas and introduced his own clients, including strict Calvinists and the reinstated bailiff. Van Zuijlen also took revenge on his opponents. At the beginning of 1693, his enemies, such as the elder Mandeville, were banished from the city; the party of True Freedom in Rotterdam was finally put to rest.[68]

By that time, Bernard had completed his medical doctorate and took himself off to London, where he set up a medical practice.[69] It would be another decade before he ventured into print again. When he did so, it was with a poem supporting an antiestablishment cause, a work of Latin praising an older Dutch medical graduate of Leiden (and Willem ten Rhijne's friend) Joannes Groenevelt, who was being tried by the College of Physicians for malpractice: "A whole Collegiate Troop, a weak-brained Host, / Their nervy Prowess all no more than Boast." Groenevelt, Mandeville believed, had discovered the true and beneficial use of cantharides and camphor to treat urinary problems, whereas the officers of the college had never explored useful medicines empirically. Instead, they were foolishly and ineffectually relying on old and conservative medical views while hindering innovators like Groenevelt.[70] In his anger at the college censors, Mandeville articulated views common to a wide range of other London medical practitioners.

He began to supplement his income from medical practice with work as a translator, in 1703 and 1704 producing an English edition of Jean de la Fontaine's *Fables*, a collection of *Aesop's Fables*, and a long poem in imitation of the French author of burlesques, Paul Scarron, titled *Typhon*. In 1705, he anonymously published another satirical poem poking fun at the establishment, the work that would eventually bring him infamy: "The Grumbling Hive: or, Knaves Turn'd Honest." It began by describing "A Spacious Hive well stockt with Bees, / That liv'd in Luxury and Ease," and went on to praise the kingdom's law and military might, science and industry, system of government, and all the other glories of the hive. They were busy bees, "Millions endeavouring to supply / Each other's Lust and Vanity," though many were knaves, who "With downright Working, cunningly / Convert to their own Use the Labour / Of their good-natur'd heedless Neighbour." In fact, "All Trades and Places knew some Cheat, / No Calling was without Deceit," be it law, medicine, the church, the army, the ministers of government, and all. Justice had dropped her scales more than once in favor of the rich and powerful in order to punish the poor, honest, and hardworking. Still, while "every Part was full of Vice, / Yet [was] the whole Mass a Paradise; / . . . Such were the Blessings of that State; / Their Crimes conspir'd to make them Great." In other words, "Virtue . . . Made Friends with Vice." That the general good might arise from individual corruption was accomplished by clever administration: "This was the State's Craft, that maintain'd / The Whole of which each Part complain'd." Avarice, prodigality, luxury, pride, envy, vanity, folly, fickleness, and inconstancy employed millions and encouraged material improvement. "Thus Vice nurs'd Ingenuity, / Which join'd with Time and Industry, / Had carry'd Life's Conveniencies / . . . / To such a Height, the very Poor / Liv'd better than the Rich before."[71]

Nevertheless, despite their thriving, the bees complained about the viciousness of their hive and, though aware of their own cheating, shouted for honesty and virtue in everyone else. In the end, moved by indignation, Jove "rid / The bawling Hive of Fraud," which caused prices to collapse, law courts to fall silent, good doctors to disburse themselves throughout the country and cultivate authentic cures, lazy clergy to resign and the rest to care for the poor, and government ministers to live on their salaries instead of the bribes and perquisites of office; fine clothes, horses, carriages, and houses were sold off, the army was brought back from foreign posts and used to fight only when necessary, and so on. Of course, this led to an economic collapse, with the consequent end of arts and sciences; the population declined, the hive's territory decreased, and their enemies pressed them on all sides, causing great loss of life, and the remaining bees toiled long and hard for necessities alone. The moral: "Then leave Complaints: Fools only strive / To make a Great an Honest Hive." Just as the grapevine runs to wood and chokes out other plants when wild but produces fine wine when "tied and cut," "So Vice is beneficial found, / When it's by Justice lopt and bound."

Two arguments were implicit in "The Grumbling Hive." One was that "reason" consists only in natural reason, not right reason, an argument commonly accepted by many physicians and philosophers, including Boerhaave. The proper craft of the state was founded not on judging what was correct morally but by material goods, which can be calculated according to natural reason. The second was more radical, that the passions govern human nature, a point on which Boerhaave was silent. In other words, Mandeville continued to advocate the views of the republican States Party, the anti-Orangist party of True Freedom, and Bayle: allow people to do as they chose within a well-regulated set of rules governed by clever "States' Craft," and the generality would flourish. Disguised as an ephemeral work of amusement, the poem drew no significant public attention. Nine years later, Mandeville added some prose explanations to his verses in order to make the serious intentions of his criticisms plain, resulting in the first edition of a book with the title *The Fable of the Bees: Or, Private Vices, Publick Benefits*, published about a month before the death of Queen Anne in the summer of 1714. Although there was a second printing of the *Fable* in late 1714, it remained anonymous and little remarked, at least publicly. Beneath the surface, however, Mandeville's outlook was entirely consistent with a truly materialist point of view.

Mandeville's views were developed more clearly in other writing, where he advocated an epistemology that favored empirical investigations over reasoned debate. For instance, in a contribution of 1709 to *The Female Tatler*, Mandeville declared that in medicine "university learning is irrelevant to curing patients."[72] And in his *The Virgin Unmask'd* of the same year, the elderly aunt constantly explodes her niece's fond hopes and long chains of reasons by introducing the real experiences of life. In his *Treatise of the Hypochondriack and Hysterick Passions* (1711) Mandeville elaborated on the implications of experience in medicine. Mandeville's stand-in is named Philopirio, or "a Lover of Experience, which I shall always profess to be."[73] The "useful part" of physic is "attain'd by an almost everlasting Attendance on the Sick, unwearied Patience, and judicious as well as diligent Observation." Hippocrates the close observer was his hero. "'Tis Observation, plain Observation, without descanting or reasoning upon it," he repeated, "that makes the Art"; once again, "it is the Observations, and not Reasons, that constitute the Art." Philipirio refuses to indulge in hypotheses or to "reason about the Causes and Seat of the Distemper"; he defends the ancient sect of "empirics" against the criticisms of Galen and others; he is for setting down observations systematically but is against "the Speculative part of Physick"; he also mocked those who had taken up the recent fashion of trying to explain medicine mathematically. Medicine was to be learned not from reading and reasoning but from experience. This also meant that a physician ought to specialize in treating one disease only, so as to accumulate as much experience with it as possible. On the other hand, the well-to-do patient, named Misomedon (a name apparently meaning "a hater of guardians") is described as "a Man of Learning, that had made Physick his particular Study."[74] Misomedon thinks he knows everything about health and disease from his intensive reading. In this case the patient, not the doctor, is the pedant: the doctor knows his texts, but they are interpreted in light of his medical experience, whereas the sickly gentleman has interpreted his experiences in light of his reading. In getting this backwards and trying to cure himself with bookish knowledge without clinical experience, Misomedon has only made his disease worse.

But if experience rather than explanations were at the heart of medicine, the doctor would not be able to demonstrate his ability to patients by uttering the right words. Many things could be known that the insufficiency of language made impossible to convey to others exactly. "There are no Words in any Language for an hundreth part of all the minute Differences in many things that yet are obvious and easily perceptible to the Skilful." For instance, the painting Philopirio saw in Misomedon's house seemed to be an original Van Dyke, but Philopirio could not explain precisely why he thought it so. Mandeville's solution is to claim that patients, too, must rely on experience rather than words: if a physician speaks plainly and directly, and his prognoses prove true, then "you may trust him with the rest." The doctor's protagonist, Misomedon, is finally convinced: treating the sick is to reasoning about medicine as war is to fencing. The one demonstrates courage and ability, the other only skill.[75] Rather than help patients through reasoned counsel, then, Mandeville offered expert help in manipulating material nature to achieve benefits. A great many medical practitioners of the day were promoting the same outlook: judge us by our fruits, not by our speech or appearance.

But Mandeville also took account of the power of the passions to dominate even natural reason. For if the physician treated patients not by offering advice—by appealing to reason—but only by offering expert treatment based on experience with nature, what happens if the patient refuses to accept the physician's expertise, as in the case of Misomedon? Mandeville's answer is to appeal to the patient's passions. His medical treatise therefore began with a discussion of pride, and toward the end he remarked that in the age of Augustus virtue was perfectly understood although vice abounded.[76] Philopirio weans his patients from excessive somatic concerns by speaking wittily but honestly. The dialogue itself becomes a therapeutic tool. Philopirio plays on Misomedon's pride, finally cajoling him into following the doctor's advice rather than his own reasonings.[77] Mandeville can offer his patient health only by working with the greatest passion (pride) through flattery and witty conversation, coaxing him into relying on a doctor's sound advice. Based on his experience (and grounding in Dutch medical theory), Philopirio gets Misomedon to embark on a regimen of diet and exercise, baths, vomits, and strengthening medicines, which bring him back to health. Like the regenten dealing with "State's Craft," then, passions properly regulated by the physician bring material benefits.

Mandeville applied this outlook to the physicians themselves. The beginning of therapy lay not in the physician's right reason but in his passions, the search for reputation and gain.[78] The physician should therefore be allowed to make appeals to the public about what he could cure, since the physician's passions could be satisfied by developing better treatments that would in turn yield a good reputation and so a good income. Mandeville himself was not shy about advertising his expertise. His treatise on hypochondria and hysteria, for instance, advised readers to contact him through the bookseller. Mandeville vigorously denied that such publicity constituted quackery, arguing instead that when practitioners skilled in a particular disease made themselves known to the public, patients as well as practitioners benefited: "If a Regular Physician writing of a Distemper, the Cure of which he particularly professes, after a manner never yet attempted, be a *Quack*, because besides his Design of being instructive and doing Good to others, he has likewise an aim of making himself more known by it than he was before, then I am one."[79] But he was not a quack because he obtained good results. The difference between the quacks and the good practitioners lay not in a virtuous character, then, but in true ability to intervene in nature effectively.

Such medical views promoted what might properly be called the profession of clinical medicine. They had to play a complicated game, attacking on two fronts at once, distancing themselves from empirics and rationalists alike. In favoring experience, they did not support simple empirics. In Mandeville's treatise on hysteria, for instance, Philopirio attacks the reliance on medicines of Misomedon's wife, Polytheca, and her daughter, when they instead should be changing the way in which they lived. Nevertheless he has his character Philopirio mix his own drugs for his patients, just like an empiric. In his case, however, it was to insure their purity and saving his patients the money an apothecary would have charged. A medical education made a difference, while mere empirics had none; what a Hippocratic education endowed the student with was previous experience, the tried and trusted knowledge of how to describe and identify diseases correctly, and how best to cure them. In making the clever doctor's expert observation and experience the root of medical knowledge, Mandeville intended to dismiss the meddling of learned persons who reason about medicine without proper experience, or who think that they can judge the best outcomes by some sort of moral calculus, whether these pseudo-rationalists be either medical conservatives or patients who read too much. One needed to begin with experience (including the experience conveyed in words through descriptive truths), and then merely add natural reason to decide how to apply such knowledge. This was the basis of the "clinical" or "experimental" medicine of the period, which intended to give the doctor *expertise* beyond that of both patients and overly-rationalist physicians.

Other clinicians would have approved much of what Mandeville had to say. By the 1720s Boerhaave, for instance, was flourishing as a professor inculcating just such expertise into his students. He had added to his professorship of botany the chair of medicine and the task of giving clinical lectures in the Caeciliagasthuis in 1714 (although until the 1730s he seems to have done so mainly for his private pupils rather than the medical students generally),[80] and in 1718 he took over the chair of chemistry as well, for which he produced another textbook again emphasizing experience and experiment (explicating more than 220 experiments, most dealing with plant chemistry).[81] After amassing an outstanding reputation as a teacher and clinician, and a large estate, he resigned the chairs of botany and chemistry in 1729. He continued to refuse to enter into controversial philosophical subjects, remaining committed to the advocacy of experience with things and natural reason.

But at the same time, because Mandeville was led to think that natural reason is the outcome of the passions that moved our thoughts—or to put it another way, that what we think of as "reason" is post-hoc rationalizing about our experience, passions, and interests—he became a political controversialist. In 1720 he issued *Free Thoughts on Religion, the Church, and National Happiness.* Then, in the spring of 1723, during the corrupt but effective Whig ministry of Robert Walpole, he issued yet another edition of the *Fable*, with an expansion of the prose remarks and the appending of two essays, *An Essay on Charity and Charity-Schools* and *A Search into the Nature of Society*, which immediately drew fire from the forces of public order. The Middlesex grand jury brought charges against the publisher of the *Fable* and several of "Cato's" journalistic letters attacking the clergy and the charity schools; in late July, an open letter to "Lord C,"[82] attacking Cato and the *Fable*, was printed in the *London Journal.* Mandeville replied in the same paper with a vindication of his book, printed and circulated this defense among the public, and in early 1724 brought out another edition of the *Fable* to which the presentment, the letter to Lord C, and his defense were appended. He soon also published two bitter works of social commentary, *A Modest Defense of Publick Stews* (1724) and *An Enquiry into the Causes of the Frequent Executions at Tyburn* (1725). Further printings of his *Fable* also appeared in these years, and he published *Part II* of the *Fable* in 1728.

It was, then, his attack on the charity school movement that made the seriousness of Mandeville's position clear, leading to charges being laid before the grand jury and Mandeville's own increasingly outrageous counterblasts against the forces of so-called public morality. "These violent Accusations and the great Clamour [were] every where raised against the [*Fable*], by Governors, Masters, and other Champions of Charity-Schools," he declared. The essay in question began with a meditation on charity, or "that sincere Love we have for our selves . . . transferr'd pure and unmix'd to others, not tied to us by Bonds of Friendship or Consanguinity." The charity schools, he thought, were not really founded on such true acts of self-denial, but the ambition to gain honor and public respect, such as the recent magnificent gifts in the will of the very wealthy Dr. John Radcliffe: "Pride and Vanity have built more Hospitals than all the Virtues together," Mandeville declared. Moreover, such gifts often did more harm than good: "Charity, where it is too extensive, seldom fails of promoting Sloth and Idleness, and is good for little in the Commonwealth but to breed Drones and destroy Industry." He urged that while the helpless needed relief, most of those seeking charity mainly needed to be put to work. As for the charity schools themselves, the notion that they would make the poor more moral was a fantasy. The causes of crime and incivility were many but did not stem from ignorance: in fact, the worst criminals were the most cunning and knowing. "Craft has a greater Hand in making Rogues than Stupidity, and Vice in general is no where more predominant than where Arts and Sciences flourish"; indeed, the converse was also true, for "Ignorance is . . . counted to be the Mother of Devotion, and it is certain that we shall find Innocence and Honesty no where more general than among the most illiterate, the poor Country People."[83] Although the number of professors in the practical subjects at the universities ought to be increased (Oxford and Cambridge mainly being full of useless clerics), schools are not the place to improve children's morals. The same passions of pride, emulation, and love of glory motivated the rogue and the honest soldier. What made the difference was not education but circumstance.

Mandeville's "defense" even more sharply claimed that the passions governed human conduct. His analysis "describes the Nature and Symptoms of human Passions, detects their Forces and Disguises; and traces Self-love in its darkest Recesses; I might safely add, beyond any other System of Ethics." He was merely "search[ing] into the real Causes of Things." As a consequence, he was demonstrating that the most important political danger lay in giving way to those who thought that people could be made more moral through reasoning with them. Those who advocated such a position were the worst hypocrites and should be told "to look at home, and by examining their own Consciences, be made asham'd of always railing at what they are more or less guilty of themselves." Although the benefits of a great and luxurious nation inevitably brought "Inconveniences, which no Government upon Earth can remedy . . . private Vices by the dextrous Management of a skilful Politician, may be turn'd into publick Benefits."[84] The true politician, led by personal ambitions rather than disinterested charity, achieved public goods by playing on the real mechanisms of human action generally, the "private vices" (or passions).

The presentment to the grand jury confirmed the point by directing its ire toward attempting to argue with him: the securing of the Protestant succession depended on remaining on the side of the Almighty and hence required "the Suppression of Blasphemy and Profaneness." Mandeville had not only attacked the church and its clergy, the universities and schools, and "recommend[ed] Luxury, Avarice, Pride, and all kind of Vices." But in addition he was castigated as a determinist, "affirm[ing] an absolute *Fate*," which denied the "Government of the Almighty in the World."[85] In short, his opponents thought, Mandeville's arguments about human nature led to admitting that nature determined our fates, leading to atheism and the promotion of vice, with all their attendant political dangers.

The world Mandeville described, to which so many people objected, was both beautiful and terrible. Medicine progressed because its investigators took the materialist foundation of the human condition as their object of study and paid close attention to the descriptive particulars. These showed not only the frame of life, disease, and death but also how those natural powers called the passions could determine our "mental" experiences: our desires and even our thoughts. Better treatments of disease were being discovered via careful investigation of the natural flora and fauna, all kinds of new diagnoses and treatments that just might have good effect, and carefully recorded clinical experiment, all because of physicians' desire to increase their personal reputations for effective treatment and hence their incomes, which would enable them better to satisfy their passions. If asked to look around at developments in natural history, Mandeville would have been able to point to curiosity as a passion, but even more to the passions exhibited in grand estate gardens, small flower gardens containing exotics, and cabinets of natural rarities that showed the owner's connections with the strange, far-off places that brought wealth from desired goods. These forms of knowledge, too, were rooted in the careful description of things, in all their precise detail, and in the technical means for finding, growing, and preserving their bodies. The demonstration of wealth and expertise embodied in such collections was not merely a sign of high status—which many people naturally craved—but also a sign of the kinds of difficult skills they acquired to reach the top: consensus-building and a steady ability to live up to promises made, the ability to muster resources enough to hold out against or bring down any enemy, and above all the ability to find out and manage information. The people who ruled the country from which he originated had become rich and powerful not from the expropriation of land but from their ability to develop and master the latest techniques for exchanging things and the tangible signs of things, such as bills of exchange and account books. It was, then, not simply their passions but the placement of passions within a system of political relations that shifted history in certain directions, one sign of which was material betterment for many of those absorbed into the rapidly growing monetary economy. Another sign was the increase in knowledge about objects, the objectivity of natural matters of fact.

Most people, of course, preferred to read the previous two centuries as exhibiting a dualistic trajectory: better knowledge and a decrease in "superstition" came because of the rise of the new and experimental science and philosophical enlightenment, with a growing material economy merely providing the means to sustain the lives of those who wished to devote themselves to advancing thought. Boerhaave, for instance, accepted that the best way to know what physicians needed was to act as if knowledge depended only on what could be known through the senses—that is, what could be known about material things—and in turn to how natural reason could order such knowledge and draw out its implications from simple axioms. It was unnecessary to search into the foundation of things, which could only lead to confusion and enmity between warring metaphysical positions. And there were those like Nieuwentijt who saw change as a sign of God's providence. They preferred to think that they went about their work from humble and noble motives, including the search for signs of God's will, not from such base passions as the desire for riches on this earth or to get ahead or dominate others, even if that happened to be one of the outcomes.

Yet Mandeville, Boerhaave, and Nieuwentijt agreed that empirical knowledge of nature should be cultivated, not because it made one more virtuous but because it made one more expert. Knowledge worked, producing material results. As a doctor, Mandeville was better than others not because he was more virtuous than the rest of fallen humankind but because he had studied the material world deeply in one aspect. He had gained more knowledge of the body and the passions, so that he could work material benefits for his patients. The expert doctor had become like his "skilful Politician," who by "dextrous Management" could turn private vices into public benefits. Other contemporaries also defended views that produced all things necessary for human life and health from nature in the absence of God or the soul, such as Friedrich Hoffmann at the university of Halle. Most went far in their speculations, defending corpuscularianism, Leibnizian motive forces, and other physical hypotheses, which gave some hint about their view of the relation between God and nature. Hoffmann's colleague Georg Stahl found hypotheses about the physical world incomplete without a theory of how reason and God could control the corporeal.[86] But Boerhaave did not. Nor did Georgio Baglivi, of the Collegio della sapienza in Rome. Baglivi, for instance, had urged: "The two chief Pillars of Physick, are Reason and Observation: But Observation is the Thread to which Reason must point. . . . From what has been said, 'tis manifest, that not only the Original of Medicine, but whatever solid Knowledge 'tis entitutled to, is chiefly deriv'd from Experience."[87] Boerhaave not only mentioned Baglivi with praise, he also co-opted other favorers of the new philosophy into his work, such as Francis Bacon, Robert Boyle, Isaac Newton, and Thomas Sydenham. Those he sought to emulate all wrote of reason as a supplement to observation, experience, and experiment, rather than the other way around. The new and experimental physician was not superior to the empiric because he used reason and the empiric did not but because of the superior experience he had gathered from any and all sources. And he was superior to the dogmatist because he refused to speculate beyond the observable. In both instances, what drove him to be better than the others was his pride in his ability to get things right even when it upset the preachers of virtue.

11

Conclusions and Comparisons

For the thoughts of mortal men are miserable, and our devices are but uncertain.

—*Book of Wisdom* 9:14

Medicine and natural history clearly emerged as the big science of the early modern period, not only in the Dutch Republic but throughout Europe. For more than 150 years, from Clusius to Boerhaave, Dutch intellectuals found various ways to say that they hoped to describe and explain nature according to what could be known about it through the use of the five senses supplemented by reason. Objective facts not only could be deployed for the sake of utility, however, but could also take on the attributes of taste and discernment. The powers and pleasures of accurate descriptive knowledge were not only praised by the chief advocates of the new philosophy, they were also noted as the chief concern of European science by others, such as Japanese scholars who interested themselves in such matters. Many advocates also acknowledged that because this kind of knowledge was rooted in bodily experience, its source lay not in abstract reason but in the passions. Advocates for the old ways of knowing therefore worried that the new science was providing explanations that necessarily led away from the knowledge of the good that was innate in the immortal soul, therefore leading to disorder and atheism. But the new philosophy nevertheless continued to excite interest, garner public attention, and establish itself near the centers of power.

What people valued in this new philosophy was consonant with the values embedded in commerce. Merchants took a deep interest in natural facts because they were essential to business. But so did many other people concerned with bodily experience, particularly medical practitioners. The new philosophy was of course not about buying and selling per se. Yet the ways of life associated with commerce that increasingly dominated Europe focused attention on the objects of nature. The gathering of facts not only created great excitement but depended on enormous investments of time, energy, expertise, experience, and money. It also required collaborative work. Even what seem like minor bits of information came into being through the labors of large networks. Armies of people therefore worked to gather new and old information and to sort out the true from the false, which was no simple task. To put it in terms that modern historians of science will understand: in recent years we have had many excellent studies of the *production* of knowledge, but we also need to note that values of systems of *accumulation* and, particularly, of *exchange*, also changed the kind of knowledge produced. Facts had the advantage of being easily communicated from person to person, penetrating cultural borders without—at first appearance at least—altering deeply felt assumptions about the world, whereas concepts and theories about nature were deeply imbued with local cultural values, such as religious or philosophical outlooks, and were less easily exchanged.[1] It was no accident, then, that the so-called Scientific Revolution occurred at the same time as the development of the first global economy. That world linked the silver mines of Peru to China as well as Europe, the sugar plantations of the Caribbean and the nutmeg-growing regions of Southeast Asia to slave labor as well as to luxury consumables, and a wealth of new information circulating in coffeehouses and lecture halls to books and the natural objects available in European gardens, cabinets of curiosity, and anatomy theaters.

Why did investigating empirical details seem satisfying when in earlier generations such efforts had manifestly been condemned as the cause of error and sin? The explanation lies not in better concepts but in shifting priorities. As commercial cities and the financial capital they produced became ever more important for the larger political systems of which they were a part, the values of the urban merchants, including their intellectual values, were increasingly dominant throughout society. This view surfaced clearly in Constantijn Huygens's long paean to his country estate, *Hofwyck*. In this country-house poem (*hofdicht*) Huygens praised his garden in various ways, following the main theme laid down by some of the Roman poets and most importantly by Lipsius in the 1580s: the garden was a retreat from affairs where the learned person could recreate the mind and exercise the body. But some of the power of the poem arises from Huygens's honesty in knowing that the purity of country life was an idealization. Toward the end, a country-born "driveller" accuses him of spending a fortune in converting useful grassland into a ground for mere pleasure and luxury. He responds to this criticism with the usual defense, that a busy man of affairs not only deserved but required such a place of "moderation, hospitality, knowledge, wisdom, piety, tolerance, and merriment," and he adds that his money was honestly earned. But he does not win the debate outright, nor does he find that the country is simply virtuous and the city simply deceitful. This greatest of country-house poems therefore concludes that "the city is a nice place, too."[2] He well knew that his very escape from the world of affairs depended on the commercial ventures of it.

What was occurring in the Dutch Republic was therefore happening elsewhere in Europe as well. For instance, the Republic of Venice became renowned for fostering the new science at its university in Padua during the peak of its economic and political power, when the senate appointed leading figures as professors despite the opinions of the old guard. It was there that antischolastic philosophy flourished, there that Vesalius conducted his famous anatomical studies, there that Galileo carried out his early work, cheek by jowl with the shipbuilders of the arsenal. Galileo is also known for his later residence at the Medici court in Florence, another place governed by merchants and former merchants, although not quite as independent of the Catholic Church as he might have liked. But even in places like Naples and Rome, the new philosophy flourished under the patronage of wealthy men with deep interests in the material world, and again, it was mainly medicine and natural history that captured their attention.[3] So, too, the new philosophy was furthered in Spain by merchants and nobles who were engaged in expanding commerce;[4] in German-speaking lands, it made inroads in great centers of commerce such as Augsburg, Hamburg, and Rostock, as well as at princely courts trying to further the commercial development of their lands. Indeed, the program of economic development among small and medium-sized principalities, which came to be called "cameralism," made powerful use of the concept of the body politic and focused attention on the development of local natural resources.[5] In Sweden, similar concerns led Linnaeus to develop his simple methods for allowing students and laypeople to describe useful plants on field trips, resulting in the binomial system.[6]

In France, the monarchy tried to stimulate the further development of the country in the wake of its religious wars, and as it did so, the chief minister, Cardinal Richelieu, placed men trained in the new medicine and natural history in charge of the garden he founded in Paris—in which he also encouraged the teaching of chemistry—and the institution he founded for furthering both commerce and science in the capital city, the Office of Address and academy of Théophraste Renaudot.[7] Informal societies of people keen to study the new science also began to gather in the mid-seventeenth century, not only around the letter-writer Marin Mersenne but most especially around Henri-Louis Habert de Montmor and Melchisédech Thévenot, at whose meetings participants could discuss and witness, even engage in, demonstrations. Louis XIV's chief minister,

Colbert, founded the Academy of Sciences in 1666 as part of his plans to advance the material interests of France despite the reluctance of the university of Paris to engage in the new enterprises; he put the academicians to work on mapping the kingdom, developing plans for dredging rivers and harbors, and investigating natural history. The academy he chartered in Caen shortly thereafter also pursued medical and natural historical subjects.[8] In England, Scotland, and Ireland, the development of the new science was also stimulated by the utilitarian and tasteful interests of city and court, with many of the earliest "scientists" being such men as the gardeners Tradescant father and son or the physicians William Gilbert, William Harvey, and Sir Thomas Browne. The Hartlib Circle and the Oxford Philosophical Society of midcentury were dominated by concerns for medical and natural historical subjects, as were the Royal Society of London of 1660 and the later scientific societies founded in Dublin and Edinburgh.[9] No wonder that the largest group of members by far within the early Royal Society consisted of the physicians, apothecaries, and others concerned with medicine and natural history, while the watchword for the group as a whole was "utility." Nor should it come as a surprise to find that the vast majority of early modern "scientists" throughout Europe were deeply interested in working with bodily objects and thinking about their consequences.[10]

But the physicians, virtuosi, leifhebbers, and savants in all these countries did not simply exemplify commonalities, they worked to establish them. They traveled and got to know one another, often exchanging letters, and sometimes books and specimens, for years afterward.[11] There were, for example, many personal contacts between Dutch and Italian virtuosi despite religious and political differences, and the same could be said for the Dutch and Iberians; the contacts between the Dutch and the Russians and Scandinavians were important; those between the Dutch and the French were very strong; the relationships between the Dutch and Germans were even more powerful; and their connections with the English, Scots, and Irish extraordinarily firm, despite three wars. Dutch innovators such as Cornelis Drebbel found careers in England during the Revolt, while in turn royalist exiles from the English civil wars such as Robert Moray turned their hands to chemistry and similar pursuits in the Dutch Republic before obtaining the charter for the Royal Society from his fellow chemist in exile, King Charles II. Looked at from the perspective of any nation, the to-ing and fro-ing of men of learning along well-established trade routes is conspicuous. Because of people's ability to exchange descriptive information and generalizations, the new science could lay claim to being a universal method of investigation, even when those participating in it hesitated or disagreed about its conceptual foundations. The best means to discern the truth was to communicate with others, and the cities were interconnected by dense webs of information exchange. The encounters of similar-minded people throughout Europe were therefore frequent and sustained, allowing the development of intellectual "movements."

But the matters of fact that such virtuosi valued so highly were collected not from themselves alone but eagerly sought from a wide range of other kinds of people. As Robert Boyle put it in his early work of enthusiasm, *Of the Usefulnesse of Experimental Naturall Philosophy* (much of it based on hoped-for improvements in medicine):

> Nor should we onely expect some improvements to the *Therapeutical* part of Physick, from the writings of so ingenious a People as the *Chineses;* but probably the knowledge of Physitians might not be inconsiderably increased, if Men were a little more curious to take notice of the Observations and Experiments, suggested partly by the practice of Midwives, Barbers, Old Women, Empericks, and the rest of that illiterate crue, that presume to meddle with Physick among our selves; and partly by the *Indians* and other barbarous Nations, without excepting the People of such part of Europe it self, where the generality of Men is so illiterate and poor as to live without Physitians. For where Physick is practised by Persons that never studyed the Art of it in Schools or Books, many things are wont to be rashly done, which though perhaps prejudicial, or even fatal to those on whom they were tryed, may afford very good Hints to the learned and Judicious Observer: Besides, where the Practitioners of Physick are altogether illiterate, there oftentime Specifics, may be best met with.[12]

As Boyle and his contemporaries knew so well, the medical and scientific revolution of the early modern period involved hosts of people all over the globe, of all kinds of social ranks, backgrounds, and training, who were grubbing around for facts—just the simple, curious, unexpected facts—about which "Judicious Observers" like himself were trying to ascertain their truth, utility, and moral value. The virtuosi were going out and collecting information from others, accumulating it, and exchanging it with one another.

The intellectual capital in which the virtuosi dealt, then, was based on the knowledge of acquaintance (*kennen*) rather than of knowledge of causal explanation (*weten*). The passion for exacting natural detail so evident in the work of the anatomists and natural historians has its counterpart in the precision of descriptive astronomy, mechanics and engineering, and mathematics. To the eyes of the twentieth and twenty-first centuries, an age of theoretical physics, virtual reality, and religious revival, the notion that science begins with material facts instead of theories sometimes seems simple-minded. But placing a high value on getting the facts right before generalizing about their meaning remains necessary for judging worldly events quickly and accurately and is the foundation on which the modern information economy still rests. Even in their speculative theories, then, investigators were self-consciously seeking "secondary" causes, explanations about how things happened rather than the reasons why.

But even though people throughout the trading networks that extended over the globe contributed information, and in merchant cities everywhere the knowledge of nature was being accumulated and exchanged, the intellectual capital in which they dealt required a willingness of people to invest in it by devoting their time, attention, energies, and income to the acquisition and exchange of objects and information. It sometimes indirectly brought monetary benefit, such as when Clusius or Sylvius accepted offers as unusually highly paid professors, or when De Bils sold his secret to the States of Brabant or Ruysch sold his cabinet to the czar of Russia. But even in such cases, it took many years of enormous effort and investment for a relatively small return. For most, like the younger Swammerdam, their investigations into nature ate up far more time and income than they returned. Financially speaking, then, the accumulation of natural objects and objective knowledge was an activity into which surplus wealth was sunk. This reinvestment reflected the continuing value these investigators placed on knowledge of the material world. The development of the new philosophy therefore depended not only on personal commitment of many kinds of people but sometimes also on the patronage of the directors of the VOC and similarly wealthy merchants, or that of the Prince of Orange and his like elsewhere in Europe.[13] Many of the patrons were themselves engaged in studies of nature—or at least in the collection of objects, specimens, and books about it—and appreciated the difficult investigations of those who were discovering new things for human benefit, edification, and pleasure. Princes acted as patrons out of the same appreciation for objective knowledge shown by the wealthy merchants and by those investing their own income and effort. The new philosophy arose not from codes of aristocratic honor but from the objective values inculcated by commerce.

The same seems to have been true of the translators in Japan, who were also involved with commerce. No doubt the merchants of Ōsaka, Fuzhou, Cochin, and elsewhere in Asia also placed a high value on objective natural knowledge. But in such places, merchants had much less status than priests, mandarins, or warriors, and had almost nothing to do with government. The kinds of knowledge they valued most highly could therefore hardly become dominant. There was nothing wrong with the minds, knowledge, or mentality of people outside Europe; but for most of them, as for many in Europe itself, the highest forms of natural knowledge explained why things were, turning their attention to causes rather than precise description. As the neo-Confucian scholar and physician Genshō

Mukai remarked in his study of the astronomical knowledge of Europeans, they "are ingenious only in techniques that deal with appearances and utility, but are ignorant about metaphysical matters and go astray in their theory of heaven and hell."[14] The priests and ministers of Europe opposed to Descartes or Galileo could not have put it better.

Shifts in the valuing of one kind of knowledge about nature or another did not come without a fight, then: or better, not without constant fighting. Not only were there holdouts against considering knowledge of objects to be important, but there were constant battles about its implications. Yet its importance for solving real and pressing problems of material life, bringing satisfaction to the senses, and comforting the mind in the knowledge that this was how the natural world really was could not be denied by those engaged in the active life. The new science did not always simply knock at the door of wisdom and ask to be seated at the table with other manifestations of knowledge. Instead, it sometimes arrived in the form of the very things that were consumed at the meal: new foods and medicines that made the body feel strong or exotic flowers that brought delight to eye and nose. Its apparent utility to the work of preserving life and restoring health was a particularly strong argument in its favor. Trading companies were certainly not established for the disinterested pursuit of knowledge, yet their servants sometimes spurred on the often Herculean investigations of nature that began to tie together the practical information of all the people of the world. The beginnings of a global science occurred during the period of the rise of a global economy. Surely that was no coincidence.

Notes

1. Worldly Goods and the Transformations of Objectivity

1. Although I acknowledge that *science* is an anachronistic term, I use it here as a shorthand for "natural knowledge," which encompasses such subjects as natural philosophy, natural history, medicine, and technology.
2. Motley, *Rise of the Dutch Republic*; Motley, *History of the United Netherlands*.
3. For a recent example, see Boorstin, *Discoverers*.
4. Sarton, *History of Science*; Hooykaas, "Rise of Modern Science," 471–72.
5. On the problems in this approach, see Hatfield, "Metaphysics and the New Science."
6. For a fine study, see Golinski, *Making Natural Knowledge*.
7. Even Shapin, *Social History of Truth*, is an attempt to establish how "scientific knowledge" was given credibility: e.g., xvi.
8. Kranenburg, *Zeevisscherij*; Bloch, "Whaling"; Bruijn, "Fisheries."
9. Smith, *Complete Works*, 1:330–31; I owe this reference to Andrew Wear.
10. Steele, *Mediaeval Lore*, 94–96. For an account of the ancient tradition of writing about the wonders of the East, see Campbell, *Witness and the Other World*; Rossi-Reder, "Wonders of the Beast."
11. *Travels*, 248.
12. Fischel, "Spice Trade"; Lane, "Pepper Prices"; Wake, "Volume of European Spice Imports"; Keay, *Spice Route*.
13. Deerr, *History of Sugar*, 12–72.
14. Quoted in Brothwell and Brothwell, *Food in Antiquity*, 83.
15. Deerr, *History of Sugar*, 73–95.
16. Lombard, "Questions," 180.
17. Subrahmanyam and Thomaz, "Evolution of Empire." For recent narratives, see Corn, *Scents of Eden*, 3–106, and Keay, *Spice Route*.
18. Quoted in Jardine, *Worldly Goods*, 290.
19. There is a large body of literature on the Portuguese ventures; one of the best succinct accounts is Lach, *Asia in the Making of Europe*, 1:91–103.

20. Lane, *Venice*, 286–94; Lane, "Mediterranean Spice Trade"; Magalhães Godinho, "Le Repli"; Wake, "Changing Pattern"; Wee, "Structural Changes"; Chaudhuri, *Asia before Europe.*
21. Quoted in Jardine, *Worldly Goods*, 290.
22. Wee, *Growth of Antwerp Market*, 2:124–40. On the Portuguese spice trade, see Lach, *Asia in the Making of Europe*, 1:119–26.
23. See, e.g., Ramsay, *City of London*, 7–32, who sees London as a "satellite city" of it (33–80).
24. Verlinden, *Beginning of Colonization*, 17–32; Deerr, *History of Sugar*, 19.
25. Mintz, *Sweetness and Power*, 31–33.
26. Stols, "Expansion of the Sugar Market," 237–38.
27. Deerr, *History of Sugar*, 453; Bulbeck et al., *Southeast Asian Exports*, 107–41.
28. De Vries, *European Urbanization*, 158–60.
29. Quoted in Marnef, *Antwerp*, 3.
30. See, e.g., Goldthwaite, *Wealth;* Jardine, *Worldly Goods;* Brotton, *Renaissance Bazaar;* Welch, "Art of Expenditure"; Hollingsworth, *Cardinal's Hat.*
31. De Vries, "Connecting Europe and Asia," 82.
32. Momigliano, "Ancient History."
33. Goldthwaite, *Wealth*, 255.
34. Goldthwaite, *Wealth*, 248.
35. I admit to using the word heuristically here rather than historically, since in English, at least, this usage of *taste* became commonplace only late in the seventeenth and early in the eighteenth centuries. Yet there is no doubt that the social values later captured by the phrase "good taste" were already evident in the fifteenth century.
36. Gadamer, *Truth and Method*, 35, italics in original. See also Bourdieu, *Distinction.*
37. Gadamer, *Truth and Method*, 36.
38. For views of how this applied to medicine, see Bylebyl, "*De Motu Cordis*"; Cunningham, "Fabricius and the 'Aristotle Project'"; and French, "Languages of Harvey."
39. Daston, "Marvelous Facts"; Daston, "Baconian Facts"; Harrison, "Curiosity."
40. Kenny, *Uses of Curiosity*, 2, 5, 15, 8, 432. For a sample of the "concept-based studies," see Whitaker, "Culture of Curiosity"; Bynum, "Wonder"; Daston and Park, *Wonders;* and Benedict, *Curiosity.*
41. The definitions are taken from the *Oxford English Dictionary.*
42. Shapiro, "Law and Science"; Shapiro, *Probability and Certainty;* Shapiro, *Culture of Fact.*
43. See esp. Alpers, *Art of Describing;* Melion, *Shaping the Netherlandish Canon;* and Swan, *Art, Science, and Witchcraft*, 36–40.
44. Kaufmann, *Mastery of Nature;* Smith, *Body of the Artisan.* For medieval precedents, Hutchinson, "Attitudes toward Nature." For overviews emphasizing the lack of interest in living examples on the part of medieval clerics, see Stannard, "Natural History," and Clark and McMunn, *Beasts and Birds.* For a more positive assessment, see Salisbury, *Medieval World of Nature*, and Salisbury, *Beast Within.*
45. Silver and Smith, "Splendor in the Grass."
46. There are many excellent works on the history of Flemish and Dutch art that stress its naturalism or realism. See esp. Smith, *Body of the Artisan;* Kiers and Tissink, *Golden*

Age of Dutch Art; Brown, *Dutch Paintings;* and Alpers, *Art of Describing.* For a view that stresses the emblematic content, see Jongh, *Questions of Meaning.*

47. Daston and Galison, "Image of Objectivity"; Galison, "Judgement against Objectivity," quotation on 328.
48. Paolo Sarpi, *History of the Council of Trent,* trans. Brent (1620), 8:799.
49. Solomon, *Objectivity;* but also note the objection of anachronism by Zagorin, "Bacon's Concept of Objectivity."
50. *Greek-English Lexicon;* also Pomata and Siraisi, *Historia.* For an argument that the value placed on empiricism arose from historical studies, see Seifert, *Cognitio Historica.*
51. Thorndike, *History of Magic,* 1:41–99; Beagon, *Roman Nature;* French, *Ancient Natural History.*
52. Albertus Magnus, *On Animals.*
53. Grafton, "Availability," 787.
54. Nutton, "'Prisci Dissectionum Professores,'" esp. 113; see also Bylebyl, "Medicine, Philosophy, and Humanism," and Nutton, "Hellenism Postponed." On the difficulty of identifying Greek botanical names even today, see Lloyd, *Ambitions of Curiosity,* 110–11.
55. Nauert, "Humanists"; French, "Pliny and Renaissance"; Grafton, *Bring Out Your Dead,* 2–10; Ogilvie, *Science of Describing,* 30–34, 121–33.
56. Eamon, *Science and Secrets,* 269–300.
57. Foust, *Rhubard.*
58. Barrera, "Local and Global."
59. He lectured on Hippocrates' Greek *Prognostics* in 1537.
60. Auerbach, *Mimesis,* 281. See also Bakhtin, *Rabelais.*
61. The practice of giving *consilia* is said to have originated with Taddeo Alderotti, who imitated the lawyers in many ways: Siraisi, *Taddeo,* 270–302. See also Cook, "Good Advice."
62. See esp. Lonie, "'Paris Hippocratics'"; also Smith, *Hippocratic Tradition.*
63. Jones, "Life and Works of Fabricius."
64. Bosman-Jelgersma, *Foreest;* Nutton, "Pieter van Foreest"; Bosman-Jelgersma, *Petrus Forestus Medicus.*
65. Pomata, "Menstruating Men," 114.
66. Tribby, "Cooking (with) Clio and Cleo"; Watson, *Theriac and Mithridatium.*
67. Palmer, "Pharmacy in Venice," 110; Palmer, "Medical Botany"; also Bylebyl, "School of Padua."
68. Hunt, "Garden in Venice."
69. The northern European languages adopted variants on the Germanic root for enclosed space, becoming the Gothic *gart,* the Frankish *jardin,* and later the English *garden* and *yard,* and the Dutch *gaard.* The Dutch also, however, evolved the word *tuin,* probably deriving from the particle *dún,* which when occurring in place-names meant a fortified space: De Vries, *Nederlands etymologisch woordenboek.*
70. Ergun and Iskender, "Gardens of the Topkapi Palace"; Garcia Sánchez and López y López, "Botanic Gardens in Muslim Spain"; Hobhouse, *Plants in Garden History.*
71. For an overview, see Masson, *Italian Gardens.*

72. For an overview, see Stuart, *Plants*, 11–25.
73. Heniger, "Eerste Reis," 30.
74. Terwen-Dionisius, "Date and Design."
75. On Ghini (35–36) and more generally on Renaissance medical faculties and botany, see Reeds, *Botany*; and Engelhardt, "Luca Ghini."
76. Terwen-Dionisius, "Date and Design."
77. This is a point I owe to a former doctoral student, Stephen Eardley. See also Peitz, "Problem of the Fetish," and Daston, "Speechless."
78. Honour, *European Vision*.
79. For an argument about how collecting developed from religious interests, see Pomian, *Collectors and Curiosities*; and Lugli, *Naturalia et mirabilia*, 93–121.
80. For an overview, see MacGregor, "Collectors and Collections"; Kaufmann, *Mastery of Nature*, 174–94; Bredekamp, *Lure of Antiquity*; Daston and Park, *Wonders*; and Findlen, *Possessing Nature*.
81. Olmi, "Science-Honour-Metaphor"; Olmi, "From the Marvellous to the Commonplace."
82. Meadow, "Merchants and Marvels"; Evans, *Rudolph*, 247; MacGregor, "Collectors and Collections," 74.
83. MacGregor, "Collectors and Collections," 76; Helms, "Essay on Objects."
84. Meadow, "Merchants and Marvels," 190–95; Bredekamp, *Lure of Antiquity*, 28–31; for Quickelberg, see Lindeboom, *Dutch Medical Biography*, 1579–80.
85. MacGregor, "Collectors and Collections," 73.
86. For more on Aldrovandi, see Findlen, *Possessing Nature*, 17–31.
87. See esp. Olmi, "Science-Honour-Metaphor," 6–7; also Daston and Park, *Wonders*, 154–58.
88. On the private trade of VOC surgeons, see Bruijn, "Ship's Surgeons," 243–48, 278–85.
89. Schulz, "Notes on the History of Collecting," 206–9.
90. See the letter of Borganrutio Borgarucci reporting on Calceolari's collection in Orta, *Dell'historia de i semplici aromati*, 348–52.
91. In the low countries, e.g., apothecaries had a monopoly on the sale of sugar in the medieval period but lost it later in the fifteenth century: Backer, *Farmacie te Gent*, 25.
92. Laughran, "Medicating with 'Scruples,'" 96–97; Bénézet, *Pharmacie et médicament*, esp. 351; DeLancey, "Dragonsblood."
93. Jean Fernel, *Methodo medendi*, quoted in Reeds, *Botany*, 25–26.
94. On Cordus, see Dannenfeldt, "Wittenberg Botanists," 229–36.
95. It was imitated the following year with the publication of the German herbal published by Peter Schoeffer, *Gart der Gesundheit*.
96. Wellisch, "Conrad Gessner: A Bio-Bibliography," 159–60.
97. MacGregor, *Tradescant's Rarities*; Jones, *Turner*.
98. Boxer, *Two Pioneers*; Varey, Chabrán, and Weiner, *Searching for the Secrets*.
99. Beagon, *Roman Nature*.
100. Augustine, *Ennaratio in Psalmum XLV*, 6–7, quoted in Jorink, "Het boeck der natuere," 21; see also Berkel, *Citaten uit het boek*.

101. See esp. Bono, *Word of God*, 48–84.
102. Stannard, "Natural History"; Steneck, *Science and Creation*.
103. Shank, *"Unless You Believe,"* 139–200.
104. Carreras y Artau and Carreras y Artau, *Historia de la filosofía*, 2:104–5, 107, 109, 114, 118, 120, 151, 157–59.
105. Montaigne, *Complete Works*, 404.
106. Montaigne, *Complete Works*, 454.
107. Montaigne, *Complete Works*, 151–52.
108. Erasmus, *Praise of Folly*.
109. For brief remarks and suggestions for further reading, see Carlino, *Books of the Body*, 178–80; also Burney, "Viewing Bodies."
110. See esp. Park, "Criminal and the Saintly Body"; Park, "Relics of a Fertile Heart"; Brownstein, "Cultures of Anatomy"; and Park's forthcoming book, *Visible Women: Gender, Generation, and the Origins of Human Dissection*.
111. For overviews, see Keele, "Leonardo da Vinci"; Galluzzi, "Art and Artifice"; Schultz, *Art and Anatomy*; Carlino, *Books of the Body*, 64–65; and, with reference to the northern netherlands, Heckscher, *Rembrandt's Anatomy*.
112. See esp. Bynum, *Resurrection*, and Vidal, "Brains, Bodies, Selves, and Science."
113. Lorch, "Epicurean in Valla's *On Pleasure*."
114. Rupp, "Matters of Life and Death," 268–69; for a notable example of public ill-will against the anatomists in eighteenth-century England, see Linebaugh, "Tyburn Riot against the Surgeons."
115. Carlino, *Books of the Body*, 213–25.
116. For this and other such remedies, see Brockbank, "Sovereign Remedies."
117. Black, *Folk-Medicine*.
118. Cathy Stewart originally drew my attention to executioners and medicine: she has been studying their "magical" powers generally in the early modern Germanic world. For the Netherlands, see Spierenburg, *Spectacle of Suffering*, 30–32, and Huberts, *Beul en z'n werk*, 172–77.
119. For recent studies, see French, *Dissection and Vivisection*, and esp. Carlino, *Books of the Body*, and Brownstein, "Cultures of Anatomy."
120. Bylebyl, "School of Padua"; French, "Berengario"; Nutton, "'Prisci Dissectionum Professores'"; section 1 of Cunningham, *Anatomical Renaissance*.
121. Pagden, *European Encounters*, 51–87; Sawday, *Body Emblazoned*, 1–15.
122. Siraisi, *Clock and Mirror*, 9.
123. Carrillo, "Voyages and Visions."
124. Carlino, *Books of the Body*, 204–5.
125. The story is recounted in O'Malley, *Vesalius*, 64.
126. Carlino, *Books of the Body*, 172–75.
127. Carlino, *Books of the Body*, 27–32, quotation on 30.
128. Francis Bacon, *Novum organum*, bk. 1, aphorism 70.
129. Sloane, *Voyage to Madera, Etc.*, sig. Bv.
130. Hacking, "Participant Irrealist," 283, responding to Latour and Woolgar, *Laboratory Life*.

2. AN INFORMATION ECONOMY

1. Simmel, *On Individuality*, 45–46, 51; these passages are translated from his *Philosophie des Geldes*.
2. Kaye, *Economy and Nature*, 37–55.
3. Marx is quoted in Proctor, "Anti-Agate," 381. For provocative thoughts on the problem of money and morality, see Parry and Block, *Money and Morality*, 1–32; Le Goff, "Merchant's Time"; and Kaye, *Economy and Nature*, 80–87; see also the classic essay by Thompson, "Moral Economy of the English Crowd." Ancient thinkers could also praise the virtues of exchange, however: see Vivenza, "Renaissance Cicero."
4. Simmel, *On Individuality*, 47, 43.
5. Throsby, *Economics and Culture*, 14; a touchstone for this point of view is Sahlins, *Culture and Practical Reason*.
6. Mukerji, *From Graven Images;* Gadamer, *Truth and Method*, 35–42; Bourdieu, *Distinction;* Reddy, *Money and Liberty;* Thomas, *Entangled Objects;* Hoskins, *Biographical Objects*.
7. Ofek, *Second Nature*.
8. Appadurai, *Social Life of Things*, 29.
9. Weschler, *Boggs*, 22.
10. In recent years, there has been a spate of important works on the history of the passions. See, e.g., Levi, *French Moralists;* Hirschman, *Passions and the Interests;* Stearns and Stearns, *Emotion and Social Change;* Chartier, *Passions of the Renaissance;* MacDonald, *"Fearefull Estate of Francis Spira";* Ruggiero, *Binding Passions;* James, *Passion and Action;* and Reddy, *Navigation of Feeling*.
11. Acts 1:3, KJV: "To whom also he shewed himself alive after his passion by many infallible proofs."
12. James 5:17.
13. Francis Bacon, *Thoughts on Natural Things*, in *Works* 10:296–97 (quoted in Wilson, *Invisible World*, 52). Also, in an early draft of his *Opticks*, Isaac Newton used the phrase "fits or passions" to describe the causes of actions: Shapiro, *Fits, Passions and Paroxysms*, 180, 182.
14. Park, "Organic Soul"; for a fine contemporary description of the parts of the soul and the place of the passions therein, Burton, *Anatomy of Melancholy*, "The Anatomy of the Soul": 1.1.2.5–11.
15. For an excellent overview, see Park and Kessler, "Psychology."
16. Shapin, *Scientific Revolution*, 13; emphasis in the original.
17. Thompson, "Moral Economy of the English Crowd."
18. Daston, "Moral Economy," 4.
19. Daston, "Baconian Facts," 357–58; see also Daston and Park, *Wonders*.
20. Gadamer, *Truth and Method*, 9–19. For my own part, I would say that impartiality, public-spiritedness, and a general generosity of character are very important attributes for both scientists and public servants, whether they are seeking to gain knowledge or to give advice; disinterestedness, by contrast, suggests an aloofness from worldly affairs that can lead to terrifying consequences.
21. Hirschman, *Passions and the Interests*.

22. On the importance of migration in the period, see, e.g., Wrigley, "Simple Model of London's Importance"; De Vries, *European Urbanization*, 175–249; Lucassen, *Migrant Labour*; and Moch, *Moving Europeans*, 22–60.
23. Elsner and Rubiés, "Voyages and Visions," esp. 8–20.
24. Gregory, "Charron," 91.
25. Quotation from his *Discours*, in Descartes, *CSM*, 1:119; on his definition of reason in the same treatise, see 1:140.
26. Sassen, "Reis van Mersenne."
27. For a good overview of the "methodising of travel in the sixteenth century," see Stagl, *History of Curiosity*, 47–94; more generally, see Maczak, *Travel in Early Modern Europe*.
28. For a transcript of his advice, see Read, *Mr. Secretary Walsingham*, 18–20.
29. Francis Bacon, "Of Travel," in Warhaft, *Bacon*, 90.
30. Stoye, *English Travellers*, 13.
31. Williams, "Voyages and Visions."
32. Pollmann, *Religious Choice*, 14–15; also Lindeman, Scherf, and Dekker, *Reisverslagen; and* Dekker, "Dutch Travel Journals."
33. Chaney, *Grand Tour*; Black, *British and the Grand Tour*; Stoye, *English Travellers*; Black, *British Abroad*; Chaney, *Evolution of the Grand Tour*; Dolan, *Exploring*; Dolan, *Ladies*.
34. Granovetter, "Strength of Weak Ties," 1366, 1367–68, 1376, 1378.
35. Milroy and Milroy, "Social Network and Social Class," 1; also Milroy and Margrain, "Vernacular Language Loyalty"; Milroy and Milroy, "Linguistic Change"; Milroy, *Language and Social Networks*.
36. Cook and Lux, "Closed Circles?"; Goldgar, *Impolite Learning*; Stegeman, "How to Set Up a Correspondence."
37. For an account of the geography of the Beurs of Antwerp and of Amsterdam, see Calabi, *Market and the City*, 64–75.
38. See esp. Buchan, *Frozen Desire*.
39. The ability to make one kind of theory "commensurable" with another has been critical to much of the work in the history and philosophy of science in recent decades. See esp. Kuhn, *Structure of Scientific Revolutions*. See also Feyerabend, *Science in a Free Society*; Hacking, *Historical Ontology*; Hadden, *On the Shoulders*; and Kaye, *Economy and Nature*.
40. Mueller, *Venetian Money Market*; Munro, "Bullionism"; Swetz, *Capitalism and Arithmetic*, 257–97.
41. Kaye, *Economy and Nature*.
42. Sargent and Velde, *Big Problem of Small Change*, 102–8.
43. McCusker and Gravesteijn, *Beginnings of Commercial and Financial Journalism*, 23–24, 29.
44. De Vries and Van der Woude, *First Modern Economy*, 691–92.
45. For a clear statement of the importance of information in economic systems, see Mokyr, *Gifts of Athena*; with regard to South Asia, Bayly, *Empire and Information*; and very generally but provocatively Vermeij, *Nature*.
46. Schneider, *Nederlandse krant*, 18–45.

47. Hunt and Murray, *History of Business*, 207–13.
48. See esp. Muldrew, *Economy of Obligation*.
49. Klein and Veluwenkamp, "Role of the Entrepreneur," 41.
50. Dekker, *Lachen*; Dekker, *Humour*.
51. Bredero, *Spanish Brabanter*, 47, 78.
52. Schama, *Embarrassment of Riches*. It should be said, however, that there is little evidence of the Dutch being embarrassed by their riches per se.
53. Latour and Woolgar, *Laboratory Life*, 192, 198, 240.
54. Smith, *Business of Alchemy*, 20–22, 25, 28, 101–2, 131–40.
55. Shapin, *Social History of Truth*; for more general reflections, see O'Neill, *Question of Trust*.
56. See, e.g., Biagioli, *Galileo Courtier*.
57. Wessels, *Roman-Dutch Law*, 218–19.
58. See esp. North and Thomas, *Rise of the Western World*, esp. 132–145.
59. Bossy, "Moral Arithmetic."
60. Landes, *Revolution in Time*, 72–76.
61. On musical horology, see Gouk, *Music*, 202–4.
62. Elias, *Time*, 115. See also Iliffe, "Masculine Birth of Time."
63. North and Thomas, *Rise of the Western World*, 139.
64. Bontekoe, *Thee*, 127.
65. Jones, *Ancients and Moderns*.
66. Shapin, "House of Experiment."
67. See esp. De Vries and Van der Woude, *Nederland*, 1500–1815 (edited and translated as De Vries and Van der Woude, *First Modern Economy*), and Israel, *Dutch Primacy*. On urban specialization, see Lesger, "Intraregional Trade and the Port System in Holland, 1400–1700."
68. De Vries, "On the Modernity of the Dutch Republic"; De Vries and Van der Woude, *First Modern Economy*.
69. Gelderboom and Jonker, "Completing a Financial Revolution," 20. It should be noted, however, that other commercial cities in the region, such as Hamburg, Köln, London, Rouen, and La Rochelle benefited from the exodus from Antwerp, perhaps just as much as Amsterdam.
70. Israel, *Dutch Primacy*, 43–48.
71. Deerr, *History of Sugar*, 453.
72. Israel, *Dutch Republic*, 319.
73. I follow here the persuasive account of 't Hart, *Making of a Bourgeois State*, esp. 187–215, and 't Hart, "Freedom and Restrictions," who emphasizes the decentralized nature of the Dutch state rather than its "strong" characteristics, which are argued by Schama, *Embarrassment of Riches*, and Israel, *Dutch Primacy*. See also the account in Glete, *War and the State*, 140–73, which argues that the power of the Dutch state lay in its ability to support "complex organisations," and Lachmann, *Capitalists*, 158–70.
74. Bonfield, "Affective Families"; MacFarlane, *Culture of Capitalism*; Haks, "Family Structure"; Demaitre, "Domesticity."

75. Roberts, *Military Revolution*; Oestreich, *Neostoicism*, 76–89; Parker, *Military Revolution*; 't Hart, *Making of a Bourgeois State*, 34–39; for an alternative view, see Black, *Military Revolution?*
76. Tracy, *Financial Revolution.*
77. Hunt and Murray, *History of Business*, 208.
78. Hacking, *Emergence of Probability*, 92–98, 111–18.
79. De Vries, *Dutch Rural Economy*; Zanden, "Economic Growth in the Golden Age"; Soltow and Zanden, *Income and Wealth Inequality.*
80. 't Hart, *Making of a Bourgeois State*, 122–23.
81. It has been noted that the VOC and later trading companies like it were "distinctively Western," although the same economic historian thinks that their financial influence in giving the European economy an edge over the Asian was important only over a period of a couple of centuries: Pomeranz, *Great Divergence*, 198.
82. Subrahmanyam and Thomaz, "Evolution of Empire." For recent narratives, see Corn, *Scents of Eden*, 3–106, and Keay, *Spice Route.*
83. Pearson, *Spices*, xxiii.
84. For an account of the Chinese traders, see Gungwu, "Merchants."
85. For the legal framework of the early VOC, see Wessels, *Roman-Dutch Law*, 227–28.
86. In addition to specific references below, see Gaastra, *Geschiedenis van de VOC*, and Zandvliet, *Dutch Encounter.*
87. Terpstra, *Opkomst der westerkwartieren.*
88. Gøbel, "Danish Companies."
89. This was famously done through the influence of an English pilot in Dutch service who became one of the shōgun's trusted servants, Will Adams. For a narrative, see Milton, *Samurai William.*
90. For this and the next paragraph, see esp. Bruijn et al., *Dutch-Asiatic Shipping*, 5–11, and Gelderboom and Jonker, "Completing a Financial Revolution."
91. Leupe, "Letter Transport Overland."
92. For a study of how the English East India Company put in place formal methods of compiling and communicating information between Asia and England, see Ogborn, "Streynsham Master."
93. Bruijn et al., *Dutch Asiatic Shipping*, 11–55.
94. Wittop Koning, *Handel in geneesmiddelen*, 21, 30–31.
95. Glamann, *Dutch-Asiatic Trade*, 15–22, quotation on 22.
96. Glamann, *Dutch-Asiatic Trade*, 18. On coffee, see Bulbeck et al., *Southeast Asian Exports*, 142–78.
97. Latham and Matthews, *Diary of Pepys*, 6:300.
98. Parker, *World for a Marketplace*, 57.
99. De Vries, "Connecting Europe and Asia," 62.
100. Haitsma Mulier, "Language of Seventeenth-Century Republicanism," 182; and more generally, Kossmann, *Politieke theorie in Nederland*; Haitsma Mulier, *Myth of Venice*; Gelderen, "Machiavellian Moment and the Dutch Revolt"; and Gelderen, *Political Thought.*
101. Frijhoff, "Amsterdamse Athenaeum," 40–41.

102. For other contemporary precedents, see Barlaeus, *Marchand philosophe*, 41–56.
103. Plato, *Dialogues*, 3:256.
104. *Politica* 1.7–13, in Aristotle, *Works*, 1255b–1260b.
105. For an analysis, Barlaeus, *Marchand philosophe*, 81–90, where Secretan argues for Stoic influence on Coornhert, although in my view Aristotle predominates.
106. De Vries and Van der Woude, *First Modern Economy*, 137. On Coornhert more generally, including his 1587 Dutch-language *Zedekunst* ("The moral art"), a response to Lipsius's *De constantia* that argues for personal virtue rather than religious discipline, see Israel, *Dutch Republic*, esp. 567; Ten Brink, *Coornhert*; and Hamilton, *Family of Love*, 102–7.
107. Barlaeus, *Marchand philosophe*, 91–95.
108. Quoted from Schmidt, *Innocence Abroad*, 196. On how greed was targeted as the reason for the early problems in Brazil, see 284.
109. Levi, *French Moralists*, 225–33; Keohane, *Philosophy and the State*, 183–202.
110. Quotation from Tuck, "Grotius and Selden," 509; see also Tuck, "'Modern' Theory of Natural Law," and Tuck, *Philosophy and Government*.
111. Barlaeus, *Marchand philosophe*, 96–98.
112. Here and below, my account of Barlaeus's address is based on the critical edition of Barlaeus, *Marchand philosophe*.
113. Of the many important studies on the theme, see esp. Baron, *Crisis of the Early Italian Renaissance*, and Pocock, *Machiavellian Moment*.
114. Bury, *Idea of Progress*; Martines, *Power and Imagination*, 197–99.
115. Quoted from the English translation, Le Roy, *Of the Interchangeable Course*, from title of chap. 4, 130v.
116. Stimson, "Amateurs of Science"; Houghton, "English Virtuoso."
117. The second edition of Henry Peacham's *Compleat Gentleman*, of 1637, explained that the term *virtuoso* (then new to the English language) was equivalent to the Dutch *liefhebber*. See Woodall, "Pursuit of Virtue," 14.
118. Demiriz, "Tulips in Ottoman," 57.
119. Deschamps, "Belon"; Delaunay, "Belon."
120. Earlier in the century, Suleiman had captured Belgrade and even unsuccessfully besieged Vienna; Busbecq attempted to negotiate an end to the horrifying conflict in Hungary, which was concluded by a peace treaty in 1562. On Busbecq, see the essays in *Busbequius* and the introduction to Busbecq, *Life and Letters of Busbecq*.
121. Opsomer, "Quackelbeen."
122. Wijnands, "Commercium botanicum," 75; Stuart, *Plants*, 11–25.
123. Opsomer, "Plantes envoyées par Quackelbeen."
124. Dumon, "Betekenis van de Busbecq"; Busbecq, *Life and Letters of Busbecq*, 1:417–18.
125. Busbecq, *Life and Letters of Busbecq*, 1:108, 107.
126. Segal, "Tulip Portrayed," 10.
127. Hunger, *De L'Escluse*, 108; Dumon, "Betekenis van de Busbecq," 34.
128. Pavord, *Tulip*, 6, 141–43, 173–74.
129. Both Pieter Pauw and Adriaan Pauw had the same grandfather: Adriaen Pauw, senior. Among the children of Adriaen senior were Pieter Pauw, senior, who was Professor

Pieter Pauw's father, and Reinier Pauw, who was Adriaan Pauw's father. See the biographical entries in Aa, *Biographisch woordenboek.*

130. Jong, "Netherlandish Hesperidies," 15, 25.
131. Tjessinga, *Adriaan Pauw,* 15.
132. For a thoughtful study on the love of flowers as a cultural rather than "natural" attribute, see Goody, *Culture of Flowers.*
133. See Goldgar, "Nature as Art," and her forthcoming study of the tulip craze.
134. Dash, *Tulipomania,* 89–91.
135. Hingston Quiggin, *Survey of Primitive Money.*
136. Dash, *Tulipomania,* 162–74; Pavord, *Tulip,* 137–77.

3. REFORMATIONS TEMPERED

1. Vermij, *Calvinist Copernicans.*
2. Hooykaas, *Humanisme, science, et reforme;* Hooykaas, *Religion;* Vermij, *Secularisering en natuurwetenschap;* Jorink, "Boeck der natuere"; for a position more sympathetic to dogmatic Calvinism, see Knoeff, *Boerhaave.*
3. Hooykaas, "Rise of Modern Science."
4. Weber, *Protestant Ethic.*
5. Feuer, "Science and the Ethic," 18–19; Borkenau, "Sociology of the Mechanistic."
6. Stimson, "Puritanism and the New Philosophy."
7. Merton, *Science, Technology and Society;* for a clear statement by Merton of his intellectual debt to the British Marxists, see Merton, "Science and the Economy"; in the introduction to the reprint of his thesis in 1970, Merton expressed puzzlement at the way his thesis had become known as a "Puritanism and science" argument when he himself remained "more partial to" the sections on economic and military influences (see esp. xii–xiii).
8. Webster, *Great Instauration.*
9. For instance, Ashworth, "Habsburg Circle"; Baldwin, "Alchemy and the Society of Jesus"; Harris, "Confession-Building"; Dear, "Miracles"; Grell, "Caspar Bartholin"; Kusukawa, *Transformation of Natural Philosophy;* Shapiro, "Latitudinarianism and Science"; Jacob, "The Church"; Jacob, *Newtonians;* Jacob and Jacob, "Anglican Origins"; Hooykaas, *Religion;* Jaki, *Origin of Science;* Funkenstein, *Theology and the Scientific Imagination;* Brooke, *Science and Religion;* and Brooke, Osler, and Meer, *Science in Theistic Contexts.*
10. Stuijvenberg, "'The' Weber Thesis." On the question of periodization, see also Jacob and Kadane, "Missing."
11. For the "hedonist-libertarian" ethic, see esp. Feuer, "Science and the Ethic," and Feuer, *Scientific Intellectual.*
12. The details of Clusius's life for this chapter have been taken from Hunger, *De L'Escluse,* vol. 1. Hunger also published a short summary: Hunger, "De L'Escluse."
13. Post, *Modern Devotion;* Engen and Oberman, *Devotio moderna.*
14. For an overview of "Erasmianism," see Grafton and Jardine, *Humanism to Humanities,* 122–57.

15. Booy, *Weldaet der scholen*; Booy, *Kweekhoven der wijshied.*
16. Mak, *Rederijkers.*
17. Boheemen and Heijden, *Westlandse rederijkerskamers in de zestiende en zeventiende eeuw.*
18. Bredero, *Spanish Brabanter*, 7.
19. Duke, "Heaven in Hell's Despite," 65–66.
20. Dixhoorn and Roberts, "Edifying Youths."
21. Bouwsma, "Lawyers and Early Modern Culture."
22. For an overview, see Duke, "Heaven in Hell's Despite"; Waite, "Dutch Nobility."
23. Waite, "Anabaptist Movement in Amsterdam"; Waite, "Dutch Nobility."
24. Hunger, *De L'Escluse*, 14–15.
25. See esp. Kusukawa, "*Aspectio Divinorum Operum*," 44; and more generally Kusukawa, *Transformation of Natural Philosophy*, and Nutton, "Wittenberg Anatomy."
26. Bono, *Word of God*, 71; on Luther's view of natural history, Dannenfeldt, "Wittenberg Botanists," 223–26.
27. Helm, "Protestant and Catholic?"
28. Cañizares-Esguerra, "Iberian Science."
29. Hunger, *De L'Escluse*, 23–24.
30. Duke, "Salvation by Coercion."
31. Hunger, *De L'Escluse*, 89–93.
32. See Hunger, *De L'Escluse*, 94, for his use of the letter, written to the imperial physician Johannes Crato von Krafftheim in Vienna on 29 November.
33. Hunger, "De L'Escluse," 3; Hunger, *De L'Escluse*.
34. Mout, "Family of Love," 86.
35. Hamilton, *Family of Love*, 5; Mout, "Family of Love," 81.
36. Mout, "Family of Love," 85.
37. Rekers, *Montano*, 75–76; Mout, "Family of Love," 90.
38. Rekers, *Montano*, 176.
39. Bouwsma, *Concordia Mundi.*
40. Evans, *Rudolph.*
41. On the Brethren, see esp. Evans, *Rudolph*, 31–33.
42. Orta, *Colóquios.*
43. Orta, *Colloquies*, vii–ix; Boxer, *Two Pioneers*, 6–13; Nogueira, "Orta"; both Boxer and Nogueira rely on the biographical information developed by Augusto da Silva Carvalho, *Garcia d'Orta: comemoração do quarto centenário da sua partida para a India em 12 de Março de 1534* (Coimbra, 1934). See also Grove, "Transfer of Botanical Knowledge," 164–67.
44. I owe this to Jon Arrizabalaga and his colleague Pepe Pardo. But Carvalho (see previous note) proposed that "Ruano" was the "licenciado" Dimas Bosque, the apparent author of the "lettre au lecteur" to Orta's book: Clusius, *Aromatum et simplicium*, 16.
45. Orta, *Colloquies*, 99–112.
46. Attewell, "India and Arabic Learning," 7–13, 17–18.
47. The title page gives a date of 10 April 1563; the copy Clusius owned says that it was purchased on 6 January 1564: Orta, *Colóquios* (University Library Cambridge, Adv.d.3.21).

48. Clusius, *Aromatum et simplicium.*
49. Hunger, *De L'Escluse,* 79, 88, 93–98.
50. Orta, *Colloquies,* xiv–xix.
51. Hunger, *De L'Escluse,* 81, 118.
52. Attewell, "India and Arabic Learning," 14.
53. Nave and Imhof, *Botany in the Low Countries,* 14.
54. Parker, "New Light," 222; Feliú, "Restoration of Aranjuez," 99.
55. Murray, *Flanders and England,* 111–44.
56. Meerbeeck, *Recherches historiques de Dodoens.*
57. Pavord, *Tulip,* 61–62, 69.
58. On the origins of the "Revolt" in light of Spanish-directed government, see esp. Parker, "Dutch Revolt and Polarization"; Parker, *Dutch Revolt;* Koeningsberger, "Why Did the State General Become Revolutionary?"; Tracy, *Financial Revolution;* Tracy, *Holland under Habsburg Rule;* and Rodríguez-Salgado, *Changing Face of Empire.*
59. Hunger, *De L'Escluse,* 114–17, 119–20.
60. Israel, *Dutch Republic,* 170–78.
61. Hunger, *De L'Escluse,* 126.
62. Freedberg, "Science, Commerce, and Art," 388.
63. Pavord, *Tulip,* 63.
64. On reading "signatures" in nature, see Bono, *Word of God,* 123–66.
65. Manning, *Emblem.*
66. See esp. Evans, *Rudolph,* 167, 123–25.
67. Hendrix and Vignau-Wilberg, *Nature Illuminated,* 5; Kaufmann, *Mastery of Nature;* Neri, "From Insect to Icon," 36–46.
68. Ashworth, "Natural History and the Emblematic"; Ashworth, "Emblematic Natural History."
69. Goldthwaite, *Wealth,* 255.
70. See esp. Yates, *Astraea;* Yates, *Occult Philosophy;* Harkness, *Dee's Conversations.*
71. For a dramatic account, see Fruin, *Siege and Relief of Leyden.*
72. As a response, plans by Philip to found another Catholic university in the region were renewed, although they never brought fruit: Frijhoff, "Deventer en zijn gemiste universiteit."
73. Jurriaanse, *Founding,* 7.
74. See esp. Gelderen, *Political Thought.*
75. Israel, *Dutch Republic,* 164.
76. Translation by Otterspeer, "University of Leiden," 324–25; the printed transcription I have used is printed as appendix 1 in Kroon, *Bijdragen,* 111–13.
77. Dorsten, *Poets, Patrons, and Professors,* 13–18, 23–29, 34–38, 46–47; Yates, *French Academies.*
78. Quotations from Gelderen, "Machiavellian Moment and the Dutch Revolt," 216, 218, 220–22; and more generally, see Gelderen, *Political Thought,* and Skinner, *Foundations of Modern Political Thought.*
79. Dorsten, *Poets, Patrons, and Professors,* 6.
80. Duke, "Ambivalent Face of Calvinism."
81. Otterspeer, *Groepsportret,* 139.

82. Kaplan, "'Remnants of the Papal Yoke.'"
83. Jurriaanse, *Founding*, 13.
84. In 1618 this rule was changed and they were appointed for a set number of years; Siegenbeek van Heuklelom-Lamme and Idenburg-Siegenbeek van Heukelom, *Album scholasticum*, ix.
85. For an excellent discussion of the young university in Leiden, see Grafton, "Civic Humanism."
86. The ambitions of the university to provide a complete fourteen-year curriculum are outlined in Otterspeer, "University of Leiden."
87. Dorsten, *Poets, Patrons, and Professors*, 5; Otterspeer, *Groepsportret*, 137–38.
88. Frijhoff, *La société néerlandaise*; Frijhoff, "Amsterdamse Athenaeum," 48–52.
89. He clearly wanted students to read classical sources in full, rejecting the method of summary according to categories associated with Ramism. On the goal of Ramist pedagogy, see Grafton and Jardine, *Humanism to Humanities*, 161–200; on Lipsius famously declaring that "He will never be great, to whom Ramus is great," see Hoorn, "On Course for Quality," 78.
90. Lunsingh Scheurleer and Posthumus Meyjes, *Leiden University*, 3.
91. Woltjer, "Introduction," 3.
92. Grafton, *Scaliger*, vol. 1.
93. But see Kooi, "Popish Impudence."
94. Dorsten, *Poets, Patrons, and Professors*, 78, 106–30, esp. 125–30; Dorsten and Strong, *Leicester's Triumph*; Israel, *Dutch Republic*, 220–30; Otterspeer, *Groepsportret*, 145–48.
95. Dorsten, *Poets, Patrons, and Professors*, 168.
96. Saravia remained a committed Calvinist, and probably because of his animosity toward the nondoctrinaire intellectuals who opposed his plots later wrote to Richard Bancroft, archbishop of Canterbury, that Lipsius, for instance, was a Familist, although it is not now clear how deeply Lipsius was involved with them; Mout, "Heilige Lipsius," 203–4.
97. Mout, "Heilige Lipsius," 205; Mout, "'Which Tyrant?'"
98. Dorsten, *Poets, Patrons, and Professors*, 148–51.
99. See esp. Waszink, "Inventio in the Politic," which shows how Lipsius's (commonly used) method of collecting *sententiae* in commonplace books helped to give Tacitus a monarchical flavor in the sixteenth and seventeenth centuries, although modern analysts find in it a longing for the old Roman Republic. On Lipsius's importance in the revival of neo-Stoicism more generally, see Saunders, *Lipsius*, and Oestreich, *Neostoicism*.
100. Woltjer, "Introduction," 4.
101. Morford, *Stoics*.
102. Beukers, "Terug naar de wortels," 7–8.
103. Quoted from Latin version given in Kroon, *Bijdragen*, 11–12.
104. See Nutton, "Dr. James's Legacy," 207.
105. Kroon, *Bijdragen*, 20.
106. Kroon, *Bijdragen*, 23.

107. Nutton, "Hippocrates in the Renaissance."
108. Lindeboom, "Medical Education in the Netherlands," 203.
109. Morford, "Stoic Garden."
110. Stewart, "Early Modern Closet Discovered"; Shapin, "'The Mind Is Its Own Place.'"
111. Lipsius, *Of Constancie*, bk. 2, chap. 2, 131–32.
112. Meerbeeck, *Recherches historiques de Dodoens;* Morford, "Stoic Garden," 167.
113. Morford, "Stoic Garden," 167.
114. Veendorp and Baas Becking, *Hortus Academicus*, 25; Lindeboom, *Dutch Medical Biography*, 1491–93. On his intellectual breadth, see Lunsingh Scheurleer, "Amphithêatre," 217.
115. Andel, *Chirurgijns*, 46.
116. Heel et al., *Tulp*, 196–97.
117. Houtzager, *Medicyns, vroedwyfs en chirurgyns;* Houtzager and Jonker, *Snijkunst verbeeld.* The guild also possessed various items of naturalia: see Delft Municipal Archive, Afd. 1, no. 1981, "Chirurgÿnsgilde-boek met de ampliatien, alteratien en 'gevolge van dien zeedert den Jaeren 1584. tot den Jaeren 1749 Inclusive,'" fols. 21–23v: "Inventaris vande goedren sÿnde op de Anatomie Camer," 1619.
118. Ferrari, "Public Anatomy Lessons."
119. Pol, "Library."
120. Cole, *History of Comparative Anatomy*, 100.
121. Kroon, *Bijdragen*, 53, 56–58.
122. Lunsingh Scheurleer, "Amphithêatre," 217–20.
123. Swan, *Art, Science, and Witchcraft*, 56–58.
124. Kroon, *Bijdragen*, 55.
125. Lindeboom, *Dutch Medical Biography*, 1491–92; *NNBW*, 4:1051–53.
126. Hunger, "Paludanus," 354–56.
127. Berendts, "Clusius and Paludanus."
128. For this kind of position, see Russell, *Town and State Physician;* for an excellent Dutch example, see Lieburg, "Pieter van Foreest."
129. Schepelern, "Naturalienkabinett"; Berkel, "Citaten uit het boek der natuur," 171.
130. See also Hunger, "Paludanus," 361, and Schepelern, "Natural Philosophers and Princely Collectors," 125–26.
131. Gelder, "Leifhebbers," 264.
132. Terwen-Dionisius, "Date and Design."
133. Hunger, "Paludanus," 358.
134. Veendorp, "Cluyt"; Bosman-Jelgersma, "Dirck Outgaertsz Cluyt"; Bosman-Jelgersma, "Clusius en Clutius."
135. Hunger, *De L'Escluse*, 174; Hogelande is described as a "famous plant collector" in Hopper, "Clusius's World," 21.
136. Hunger, *De L'Escluse*, 114.
137. For information on Marie de Brimeu, see Berendts, "Clusius and Paludanus," 53–54.
138. Hunger, *De L'Escluse*, 114, 187–89.
139. Hunger, *De L'Escluse*, 189–93, 210.
140. Hunger, *De L'Escluse*, 215–16.

141. Hopper, "Clusius's World"; Swan, *Art, Science, and Witchcraft*, 51–65.
142. Veendorp, "Cluyt"; Bosman-Jelgersma, "Dirck Outgaertsz Cluyt"; Bosman-Jelgersma, "Clusius en Clutius."
143. Hunger, *De L'Escluse*, 214–15.
144. Hunger, *De L'Escluse*, 217. For an account of the botanical inventory of 1594, including details of where the specimens were set in the beds, see 219–235, and a slightly revised view in Tjon Sie Fat, "Clusius's Garden."
145. Hopper, "Clusius's World," 15.
146. For more on the collection of illustrations as probably Cluyt's, used for instruction, but possibly Clusius's, meant for the hand coloring of woodcuts, see Swan, *Clutius Watercolors;* Swan, "Lectura-Imago-Ostensio"; Ramón-Laca, "L'Ëcluse"; and Whitehead, Vliet, and Stearn, "Clusius in the Jagiellon." The most thorough study, by Jacques de Groote, attributes the collection to Jonker Karel van Sint Omars: http://www.tzwin.be/libri%20picturati.htm.
147. Bontius, *Tropische geneeskunde*, 95.
148. The magisterial account of the history of Dutch navigation, emphasizing the skills of the practitioners, is Davids, *Zeewezen en wetenschap*, on Waghenaer's *Speigel der zeevaerdt* (1584–85) and its like, see esp. 56–63.
149. The best modern accounts of Linschoten are Moer, *Linschoten*, 7–20, and Boogaart, *Civil and Corrupt Asia*, 1–7.
150. Gelder, "Paradijsvogels."
151. Corn, *Scents of Eden*, 115. As for the king's ransom: such things are mostly incomparable, but in 1194, King Richard Lionheart was ransomed for 150,000 marks, paid in silver (my thanks to Jennifer Holmes and Anne Hardy for this). A mark later equaled 13 shillings, 4 pence (or 160 pence), or two-thirds of a pound sterling. In others words, King Richard was ransomed for the equivalent of 100,000 pounds sterling. Between about 1200 and 1580, the wages of building craftsmen rose about fourfold (according to Brown and Hopkins, as presented in a table at http://privatewww.essex.ac.uk/~alan/family/N-Money.html#Brown), while the "Madre de Dios" was worth five times Richard's ransom.
152. Beekman, *Troubled Pleasures*, 43.
153. Kish, "Medicina, Mensura, Mathematica."
154. Van Berkel, in Berkel, Helden, and Palm, *History of Science in the Netherlands*, esp. 22.
155. For an account in English of these activities, see Zandvliet, *Mapping for Money*, 33–49, esp. 38–41 on Plancius and 42–46 on Claesz.
156. Burnell and Tiele, *Voyage of Linschoten*, 1:xxxii–xxxvi.
157. Linschoten, *Itinerario*. For modern editions of the text, see Burnell and Tiele, *Voyage of Linschoten;* Kern, *Itinerario;* and Terpstra, *Itinerario*.
158. For a modern edition of and commentary on the illustrations, see Boogaart, *Civil and Corrupt Asia*.
159. Berkel, "Citaten uit het boek der natuur," 174.
160. Kern, *Itinerario*, 1:89.
161. Kern, *Itinerario*, 2:24–25; Burnell and Tiele, *Voyage of Linschoten*, 2:86–87.

162. Schoute, *Geneeskunde in den dienst,* 107.
163. Beekman, *Troubled Pleasures,* 61; on the influence of these first stories on Dutch literature thereafter, see 39–79.
164. Beekman, *Troubled Pleasures,* 53–55.
165. The first embargo lasted from 1585 to 1590, the second and more effective was reimposed in 1598 and lasted until the truce of 1609; on the effectiveness of the second for hindering Dutch-Portuguese trade, see Ebert, "Dutch Trade with Brazil," 61–63.
166. His story was turned into a best-selling work of historical fiction by James Clavell, *Shōgun.* For a recent historical account, see Milton, *Samurai William.*
167. Bruijn et al., *Dutch-Asiatic Shipping,* 1–6; Israel, *Dutch Primacy,* 67–73.
168. Heniger, "Eerste Reis."
169. The text is also transcribed in Hunger, *De L'Escluse,* 1:267; the Dutch order that repeats Clusius's instructions ("Dat sy meebrenghen tusschen pampier geleyt taexkens met heur bladeren en vruchten en bloemen") is given in Schoute, *Geneeskunde in den dienst,* 49–50.
170. Berendts, "Clusius and Paludanus," 50, quoting a letter of Pinelli to Clusius, 24 October 1600.
171. Beekman, *Troubled Pleasures,* 56.
172. Gelder, "Wereld binnen handbereik," 25.
173. Fock, "In het interieur," 77, 79.
174. For Peiresc's cabinet, see Schnapper, *Géant,* 237–40; on Worm's, see Schepelern, "Museum Wormianum."
175. Gelder, "Leifhebbers," 264–66; Schepelern, "Naturalienkabinett."
176. Jong, "Nature and Art," 44, 54–56.

4. COMMERCE AND MEDICINE IN AMSTERDAM

1. See also the still penetrating Heckscher, *Rembrandt's Anatomy.*
2. In this interpretation, I follow Schupbach, *Paradox of Rembrandt.*
3. Liddell and Scott, *Greek-English Lexicon,* 885.
4. Heel et al., *Tulp,* 42.
5. For some hints about such practices, see Budge, *Amulets and Talismans;* Blécourt, "Witch Doctors, Soothsayers and Priests"; Wilson, *Magical Universe,* esp. 311–71; on midwives, Marland, "*Mother and Child Were Saved*"; Marland, *Art of Midwifery.*
6. Harley, "Historians as Demonologists"; Gijswijt-Hofstra, "European Witchcraft Debate"; Blécourt, *Termen van Toverij;* Gijswijt-Hofstra and Frijhoff, *Witchcraft in the Netherlands;* Frijhoff, "Emancipation of the Dutch Elites"; Swan, *Art, Science, and Witchcraft.*
7. Steendijk-Kuypers, *Volksgezondheidzorg te Hoorn,* 267–73; Huisman, "Itinerants"; Huisman, "Civic Roles and Academic Definitions"; Huisman, *Stadsbelang en standsbesef;* Huisman, "Shaping the Medical Market."
8. Schneider, *Nederlandse krant.*
9. Wittop Koning, "Wondermiddelen," 3, 11.
10. Lindeboom, *Dutch Medical Biography,* 83–84.

11. Mortier, "Wereldbeeld van de Gentse almanakken."
12. For a discussion in English, see esp. Eamon, *Science and Secrets.*
13. For an overview, see Copenhaver, "Natural Magic, Hermetism, and Occultism."
14. Della Porta, *Magia naturalis* (1589).
15. Baumann, *Drie eeuwen.*
16. Porto, "Liquor alkahest."
17. Among the many works on Paracelsus, see esp. Pagel, *Paracelsus,* and Webster, *Paracelsus to Newton.*
18. For a fine short summary of Paracelsus's basic views, see Newman, *Gehennical Fire,* 106–10, and Moran, *Distilling Knowledge,* 67–98.
19. For a few details on these medical translators, see Lindeboom, *Dutch Medical Biography.*
20. See esp. Trevor-Roper, "Paracelsian Movement"; Evans, *Rudolph;* Goodman, *Power and Penury;* Moran, *Alchemical World;* Halleux and Bernès, "Cour savante d'Ernest de Bavière"; and Debra Harkness, *The Jewel House of Art and Nature: Elizabethan London and the Social Foundations of the Scientific Revolution* (forthcoming).
21. Wittop Koning, *Handel in geneesmiddelen,* 60–61; Bosman-Jelgersma, *Delftse apothekers.*
22. Wittop Koning, "Voorgeschiedenis," 8.
23. Wittop Koning, *Handel in geneesmiddelen,* 38–56.
24. For a recent study of the retailing of apothecaries in London, see Patrick Wallis's forthcoming article in *Economic History Review.*
25. Gelder, "Leifhebbers," 273.
26. *Catalogus van een seer wel gestoffeerde Konstkamer / Catalogus musei instructissimi,* printed in Dutch and Latin.
27. Veen, "Met grote moeite," 65.
28. For work on local Dutch surgeons' guilds, see Hoeven, "Chirurgijn-gilde te Deventer"; Houtzager, *Medicyns, vroedwyfs en chirurgyns;* Lieburg, "Genees- en heelkunde"; Heel et al., *Tulp,* 189–215; and Houtzager and Jonker, *Snijkunst Verbeeld.*
29. Lieburg, "Genees- en heelkunde," 175–76.
30. "Lÿste van Taxatie waer naer haer de Chirurgÿns alhier voortaen sullen hebben te reguleren ontrent het schrÿven van Specificatien van haer verdient salaris," Utrecht municipal archive 41/suppl. 144: vol. 1: "Acta et decreta Collegii medici, 1706–1783," fols. 23–34.
31. Municipal Archive, The Hague, 488/6, "Bijlagen bij de 'acta collegii,' 1658–1774," Lit. B: Ordinance of 25 August 1622.
32. Bylebyl, "Medical Meaning of 'Physica'"; Cook, "Good Advice"; Santing, *Geneeskunde en humanisme;* Santing, "Doctor philosophus."
33. Fockema Andreae and Meijer, *Album studiosorum Franekerensis;* Napjus, *Hoogleeraren in de geneeskunde aan Franeker;* Jensma, Smit, and Westra, *Universiteit te Franeker;* Evers, "Illustre School at Harderwyk." A fine study of medicine in Groningen, including information on the university, is Huisman, *Stadsbelang en standsbesef;* Ketner, *Album promotorum Rheno-Trajectinae; Album studiosorum Rheno-Traiectinae;*

Vredenburch, *Schets van eene geschiedenis;* Kernkamp, *Acta et decreta;* Ten Doesschate, *Utrechtse Universiteit.*

34. Kroon, *Bijdragen,* 27.
35. Rather, "Non-Natural"; Niebyl, "Non-Naturals"; Bylebyl, "Galen on the Pulse"; Mikkeli, *Hygiene.*
36. White, "Medical Astrologers"; Westman, "Astronomer's Role"; Burnett, "Astrology and Medicine"; Grafton, *Cardano's Cosmos;* Moyer, "Astronomers' Game."
37. Broecke, "Astrology at Louvain"; Kish, "Medicina, Mensura, Mathematica"; Waterbolk, "'Reception' of Copernicus's Teachings."
38. Mout, "'Which Tyrant?'" 138–39.
39. Garin, *Astrology in the Renaissance,* 63–64; Wright, "Study in the Legitimisation of Knowledge."
40. Sherrington, *Fernel,* 22, 33–38, 40–41, 43–45.
41. Lloyd et al., *Hippocratic Writings,* 149; see also Jouanna, *Hippocrates,* 70, 215–16, 354.
42. Broecke, "Astrology at Louvain," 442–44.
43. Newman, *Pseudo-Geber;* Newman, *Gehennical Fire,* 92–106.
44. Nauert, *Agrippa,* 237; Yates, *Bruno.*
45. Shackelford, *Philosophical Path.*
46. Moran, "Privilege, Communication, and Chemiatry"; Moran, "Court Authority and Chemical Medicine"; Moran, *Chemical Pharmacy Enters the University.*
47. Ørum-Larsen, "Uraniborg"; Shackelford, "Tycho Brahe"; Evers, "Illustre School at Harderwyk," 98–100.
48. Russell, *Town and State Physician;* Lieburg, "Pieter van Foreest."
49. Moore, "'Not by Nature but by Custom.'"
50. Bontekoe, *Thee,* 155.
51. Baumann, *Beverwijck.*
52. Moulin, "Barbette"; Baumann, "Job van Meekren."
53. For some of his work, see Porto, "Liquor Alkahest," 22–24.
54. On Sylvius in Amsterdam, see Baumann, *François Dele Boe Sylvius,* 10–19; for an overview of his ideas, in English, see King, *Road to Medical Enlightenment,* 93–112. For a reconstruction of his collection of paintings at his Leiden residence and what can be inferred of his outlook from it, see Smith, "Science and Taste."
55. Bylebyl, "School of Padua"; Bylebyl, "Commentary"; Bylebyl, "Manifest and Hidden."
56. Beukers, "Terug naar de wortels."
57. Ten Doesschate, *Utrechtse Universiteit,* 5–10.
58. Lindeboom, "Medical Education in the Netherlands"; Beukers, "Clinical Teaching"; Otterspeer, *Groepsportret,* 203–4. For a careful study of the hospital itself, see Ramakers, "Caecilia Gasthuis."
59. Hellinga, "Geschiedenis Armenverzorging," 3045–46.
60. Cook, *Trials,* 101.
61. Geyl, *Geschiedenis van het Roonhuysiaansch geheim;* Aveling, *Chamberlens and the Midwifery Forceps,* 179.

62. Municipal Archive, The Hague, 488/1, "Statuta Autographa Collegii medicorum Hagiensium. anno 1658," a printed version of which is at 488/6, "Bijlagen bij de 'acta collegii,' 1658–1774"; Municipal Archive, Haarlem, Collegium Medico-Pharmaceuticum, K. en O. II, 106; K. en O. II, 112.
63. Temkin, "Role of Surgery."
64. Wittop Koning, "Voorgeschiedenis," 1–5.
65. Heel et al., *Tulp*, 58–59.
66. For this and what follows on the early Amsterdam civic garden, see Stomps, "Geschiedenis van de Amsterdamse Hortus"; the same author's piece in Brugmans, *Gedenkboek*, 393–94; Wittop Koning, "Voorgeschiedenis," 5; Seters, "Voorgeschiedenis." The hospital had a live-in apothecary appointed to it from 1610: Wittop Koning, *Handel in geneesmiddelen*, 83, 129–30.
67. Hopper, "Dutch Classical Garden," 32; Hopper, "Marot"; Bezemer-Sellers, "Clingendael"; Bezemer-Sellers, "Bentinck Garden"; Bezemer-Sellers, "Gardens of Frederik Hendrik."
68. Vries, "Country Estate Immortalized"; Jong, *Nederlandse tuin- en landschapsarchitechtuur*, 17–56.
69. Bosman-Jelgersma, "Augerius Clutius," 58–59.
70. Seters, "Voorgeschiedenis," 38–42.
71. Heel et al., *Tulp*, 118.
72. Wittop Koning, "Voorgeschiedenis," 1–3; Seters, "Voorgeschiedenis," 42; for a biographical listing by location of about 163 Amsterdam apothecaries between the late sixteenth century and 1637, see Wittop Koning, *Handel in geneesmiddelen*, 91–130.
73. Heel et al., *Tulp*, 118, 218.
74. Any such comparison is rife with difficulties, but one estimate puts the number of physicians and dentists per 1,000 population in Europe between 1960 and 2000 at 1.1, while another has the number of doctors per 1,000 in 1998 at 3.68: *Health at a Glance*, 107, and *EU Encyclopedia*, 500, respectively.
75. Pelling and Webster, "Medical Practitioners."
76. For instance, see the example of Hoeven, "Chirurgijn-gilde te Deventer."
77. Heel et al., *Tulp*, 45–47.
78. Heel et al., *Tulp*, 44, 57–58.
79. Lint, "Comment Jan de Doot," and for additional examples, Heel et al., *Tulp*, 165–87.
80. Heel et al., *Tulp*, 149–63.
81. Baumann, *François Dele Boe Sylvius*, 8.
82. Frijhoff, "Amsterdamse Athenaeum," 54–55.
83. Ottenheym, "Amsterdam Ring Canals."
84. Heel et al., *Tulp*, 49–50.
85. Heel et al., *Tulp*, 54–55.
86. For an overview, see Israel, *Dutch Republic*, 474–77; for more detail, see Brienen, *Nadere reformatie*, and Sluijs, *Puritanism en nadere reformatie*.
87. Heel et al., *Tulp*, 49, 51.
88. Heel et al., *Tulp*, 51–60.
89. Heel et al., *Tulp*, 67–72; Balbian Verster, *Burgemeesters van Amsterdam*, 51–56.

90. Cook, "Physical Methods."
91. See the introduction to Wittop Koning, *Facsimile*, 7–11.
92. Haver Droeze, *Collegium Medicum*, 3–4. But also see the criticism of this account of Tulp's dinner leading to the following moves, in Wittop Koning, "Oorsprong," 802.
93. For example, Mooij, *Plosslag*, 37–38; on this epidemic in Amsterdam, which was present in the province of Holland from the end of 1634, see Andel, "Plague Regulations in the Netherlands," and Dijkstra, *Een epidemiologische beschowing*, 27.
94. Wittop Koning, "Oorsprong"; Wittop Koning, *Facsimile*, 12–16; Heel et al., *Tulp*, 113, 58–59.
95. Lindeboom, *Dutch Medical Biography*, 1544–46.
96. The regulation is printed in Haver Droeze, *Collegium Medicum*, 4; a list of inspectors from 1637 is at Amsterdam Municipal Archive, P.A. 27/17: "Nomina et Series Inspectorum Collegii Medici, 1637–1797."
97. Heel et al., *Tulp*, 217–37; Haver Droeze, *Collegium Medicum*, 5; Wittop Koning, "Voorgeschiedenis." The pertinent statutes are kept as items 12 to 15 in Amsterdam archives, P.A. 27/14: "Stukken betreffende de oprichting 1638; met retroacta, 1550–1637," and P.A. 27/1: "Ordonantie Boeck, 1638–1718," fols. 1–8.
98. Municipal Archive, The Hague, 488/1, "Statuta Autographa Collegii medicorum Hagiensium. anno 1658," a printed version of which is at 488/6, "Bijlagen bij de 'acta collegii,' 1658–1774." These ordinances are the only medical ones from the period I know to be in Latin rather than Dutch, indicating both the more aristocratic atmosphere of The Hague and the fact that it was not technically a chartered city. Haarlem Municipal Archive, K. en O. II, 106. Municipal Archive, Utrecht, 41/suppl. 144: vol. 1: "Acta et decreta Collegii medici, 1706–1783," fols. 1–7. Also Weyde, "Collegium Medicum Ultrajectinum."
99. *Horti Amstelodamensis alphabetico ordine exhibens eas.*
100. Seters, "Voorgeschiedenis," 40–44.
101. Heel et al., *Tulp*, 203, 208, 206.
102. Beekman, *Troubled Pleasures*, 43.
103. Quoted in Schupbach, *Paradox of Rembrandt*, 33. See also Lunsingh Scheurleer, "Amphithéâtre," 220–22.
104. For anatomy as a commentary on the lesson of mortality embedded in the Delphic *nosce teipsum*, see Schupbach, *Paradox of Rembrandt*, 33–35, 90–102.
105. On the vanitas theme in Leiden art, see Rupp, "Matters of Life and Death," 272–73, and more generally Vroom, *Modest Message*.
106. Schupbach, *Paradox of Rembrandt*, esp. 27–40, and his review of the early modern literature on 'cognitio sui, cognitio Dei,' 66–84, 90–102.
107. Translation from the preface of his *Primitiae anatomicae* (1615) in Schupbach, *Paradox of Rembrandt*, 21.
108. On the struggles of the curators to replace Pauw, see Kroon, *Bijdragen*, 58–62.
109. The names are included in the list of 'Hoogleeraren en Lectoren' in Siegenbeek van Heuklelom-Lamme and Idenburg-Siegenbeek van Heukelom, *Album scholasticum*, 194–95; on Valckenburg and Florentius, see Lindeboom, *Dutch Medical Biography*, 2019, 596, respectively.

110. Kroon, *Bijdragen*, 66–68; Lunsingh Scheurleer, "Amphithéâtre," 222–23, 239–41.
111. Barge, *De oudste inventaris*, 34–55; also see Otterspeer, *Groepsportret*, 191–95.
112. Jong, "Nature and Art."
113. Lunsingh Scheurleer, "Amphithéâtre," 223–69.
114. For an appreciative view, see Weevers, *Poetry of the Netherlands*, 74–78.
115. See the translation of his encomium in Heckscher, *Rembrandt's Anatomy*, 80.
116. Colie, *"Some Thankfulnesse to Constantine"*; Jongh, *Questions of Meaning*, 47–48; Schenkeveld, *Dutch Literature*, 15–16; Regin, *Traders, Artists, Burghers*, 50–51.
117. Vondel, *Gijsbrecht van Amstel*.
118. My translation of "Auditor te disce, & dum per singula vadis, / Crede vel in minima parte latere Deum." Quoted and given a slightly different translation in Heckscher, *Rembrandt's Anatomy*, 112–13. Also see Schupbach, *Paradox of Rembrandt*, esp. 22, 31, 33.
119. Heckscher, *Rembrandt's Anatomy*, 113–15.
120. Heckscher, *Rembrandt's Anatomy*, 113–15.
121. Israel, *Dutch Republic*, 475.
122. Heel et al., *Tulp*, 85.
123. Cook, *Decline*; Cook, *Trials*; Brockliss and Jones, *Medical World*.

5. TRUTHS AND UNTRUTHS FROM THE INDIES

1. Bontius, *Tropische geneeskunde*, 128–31.
2. Bontius, *Tropische geneeskunde*, 198–99.
3. For instance, Shapin, "House of Experiment"; Biagioli, "Knowledge, Freedom, and Brotherly Love"; Golinski, "Noble Spectacle."
4. Latour, *Science in Action*; Latour, *Pandora's Hope*; Cook and Lux, "Closed Circles?"; see also the deeply flawed but provocative Law, "Methods of Long-Distance Control."
5. Shapin, *Social History of Truth*, relies heavily on the manuals of etiquette being produced by humanists to advise courtiers and rural gentlemen on how to be urban men of credit. For more general reflections, see O'Neill, *Question of Trust*.
6. Lucassen, *Migrant Labour*, 156–57.
7. Schoute, *Geneeskunde in den dienst*, 51.
8. Bruijn and Helsinga, *Muiterij*.
9. One list of fifteen ships arriving at Batavia after setting out from The Netherlands shows losses of anywhere from 30 to 2.5 percent, with an average death rate of 14.4 percent on the outward voyage alone: Schoute, *Geneeskunde in den dienst*, 42.
10. Bruijn et al., *Dutch-Asiatic Shipping*, 143–72; De Vries, "Connecting Europe and Asia," 74, 82.
11. Schoute, *Geneeskunde in den dienst*, 14–20.
12. Bruijn, "Ship's Surgeons."
13. Subrahmanyam, "Dutch Tribulations." Although the provinces of the Netherlands themselves had a law according to which any slave who set foot on Dutch soil became free whether the master desired it or not (Wessels, *Roman-Dutch Law*, 406), in Asia

the VOC adopted the practice of slavery, a practice also often found among the indigenous people.

14. Warnsinck, *Reisen van de Graaff*; Barend-van Haeften, *Oost-Indië Gespiegeld.*
15. Wills, *Pepper, Guns and Parleys*, 55, 34.
16. Bontius, *Tropische geneeskunde*, 20–21.
17. Meilink-Roelofsz, *Asian Trade and European Influence*, 207, 211.
18. Andrade, "Political Spectacle."
19. Quotation from Stapel, *Geschiedenis van Nederlandsch-Indië*, 61; see also Stapel, *Gouverneurs-generaal*, 15; Both served as governor general from 1609 to 1614.
20. Meilink-Roelofsz, "Aspects of Dutch Colonial Development," 61–62.
21. Meilink-Roelofsz, *Asian Trade and European Influence*, 202, 207–18.
22. Knaap, "Headhunting in Amboina"; Lieberman, "Comparative Thoughts," esp. 220.
23. For an account of the English attempts to hold on to Run, see Milton, *Nathaniel's Nutmeg.*
24. Meilink-Roelofsz, *Asian Trade and European Influence*, 289–92.
25. Harrison, "Europe and Asia," 4:654.
26. Kroeskamp, "De Chinezen te Batavia"; Blussé, *Strange Company.*
27. Zanden, *Rise and Decline of Holland's Economy*, 75–79.
28. Keuning, "Ambonese, Portuguese, Etc.," 368–80.
29. Coolhaas, "Notes and Comments"; Keuning, "Ambonese, Portuguese, Etc."; Chancey, "Amboyna Massacre."
30. Keuning, "Ambonese, Portuguese, Etc.," 380.
31. Keuning, "Ambonese, Portuguese, Etc.," 380–94; quotation on 394.
32. Prakash, "Dutch East India Company in the Trade of the Indian Ocean," 189.
33. The VOC's settlement on Taiwan, for instance, operated at a considerable financial deficit for more than twenty-five years before turning into a profit in the 1650s, when Ming merchants fled the Qing and set up on the island, bringing revenues that could be taxed: Veen, "How the Dutch Ran a Colony."
34. Bulbeck et al., *Southeast Asian Exports*, 19–20, quotation at 11.
35. Tuchschere, "Coffee in the Red Sea Area," esp. 51–53.
36. Glamann, *Dutch-Asiatic Trade*, 73–90, quotation from p. 73.
37. De Vries, "Connecting Europe and Asia," 64–65.
38. Mintz, *Sweetness and Power*, and esp. Stols, "Expansion of the Sugar Market." Also Smith, "Complications of the Commonplace," and Higman, "Sugar Revolution." On Javanese sugar, see Bulbeck et al., *Southeast Asian Exports*, 113.
39. Pomeranz, *Great Divergence*, 201.
40. Steensgaard, *Asian Trade Revolution*, 143.
41. The best biographical accounts of Bontius remain M. A. van Andel's introduction to vol. 10 of *Opuscula*, and Römer, *Bontius.*
42. Römer, *Bontius*, 2.
43. A remark from Bontius's *Animadversiones in Garciam ab Orta*, bk. 1: Bontius, *Tropische geneeskunde*, 42–43. This accessible edition of Bontius's work prints an English translation of the original Dutch and Latin on a facing page, but I have sometimes silently corrected the translations to make them closer to the original meaning.

44. Bontius, *Tropische geneeskunde,* ix–xi.
45. Nationaalarchief, VOC-archives, 1.04.02, inventory no. 147.
46. On the development and redevelopment of the hospital at Batavia, see Schoute, *Geneeskunde in den dienst,* 110–13.
47. Bontius, *Tropische geneeskunde,* ix–xi.
48. Lindeboom, *Dutch Medical Biography,* 858–59.
49. Römer, *Epistolae,* 2. The information on Bontius's appointment and the loss of his wife between The Netherlands and the Cape was first discovered by Pop, *Geneeskunde,* 333–34.
50. Bontius, *Tropische geneeskunde,* xv; Schoute, *Geneeskunde in den dienst,* 55–57.
51. Römer, *Bontius,* 2.
52. Bontius, *Tropische geneeskunde,* xlv.
53. Römer, *Bontius,* 4–5.
54. For accounts of the sieges, see Graaf, *Regering,* 144–63.
55. For a narrative of the early military struggles, see Chijs, *De Nederlanders te Jakarta.* For hints at indigenous views of the Dutch, Reid, "Early Categorizations," 287–88.
56. This is recounted in his *Observationes: Aliquot explurimis selectae* ("Some select observations") and reiterated in his *Dialogi: de conservanda valetudine* ("Dialogues on the preservation of health"): Bontius, *Tropische geneeskunde,* 204–7, 75.
57. Bontius, *Tropische geneeskunde,* 202–5, 118–19, 206–7, 109, 188–89.
58. Bontius reports that the disease had confined him for the four months previous to his writing the chapter on "tenesmus" in his *Methodus medendi* (Bontius, *Tropische geneeskunde,* 126–29), which is prefaced by the dedication to the Heren XVII dated 19 November 1629, placing his illness during the period of the second siege; see also 196–197.
59. Bontius, *Tropische geneeskunde,* 188–89.
60. Bontius, *Tropische geneeskunde,* 102–3.
61. Bontius, *Tropische geneeskunde,* 128–29. In his natural history (see below), he went so far as to argue that it was possible to predict where diseases would occur by noting where the plants grew that had the "manifest or occult" qualities capable of fighting them: Bontius, *Tropische geneeskunde,* 338–39.
62. Bontius, *Tropische geneeskunde,* 106–11.
63. Bontius, *Tropische geneeskunde,* 32–33.
64. Bontius, *Tropische geneeskunde,* 140–43, 180–83, 46–49.
65. After attending a forbidden gathering of ministers in 1605, the elder Durie had fled to Leiden, where he resided until his death in 1616 as the first minister of the Scots church. Andrew's elder brother, John, followed in their father's footsteps by becoming a minister and became well known for his plans to reunite all the evangelical churches during the Thirty Years' War.
66. On Durie, see Schoute, *Geneeskunde in den dienst,* 132–36.
67. It is probable that the 'Mr. Andries Derews' who, together with a minister was present at the making of Bontius's last will and testament, is a mistranscription for Mr. Andries 'Dereus,' or Andreas Dureaus: Bontius, *Tropische geneeskunde,* xlvi–xlvii.
68. Bontius, *Tropische geneeskunde,* 56–63.

69. Bontius, *Tropische geneeskunde*, 80–83.
70. This work is undated, but in a letter to his brother of 18 February 1631, Bontius wrote about the "animadversions" which were being sent with the letter, along with an explanation of why he felt compelled to correct Orta, clearly indicating that it was this treatise he had just finished: Bontius, *Tropische geneeskunde*, xlviii–xlix.
71. Bontius notes that Clusius added a comment to Orta's passages on the coconut palm, noting that its leaves were used as a medium on which to write; the first edition that contains this information is Clusius, *Exoticorum libri decem*, chap. 26.
72. Bontius, *Tropische geneeskunde*, 22–23.
73. Bontius, *Tropische geneeskunde*, 4–5, 6–7, 22–23, 28–29, 2–3, 10–11.
74. Although the letter is undated as printed in Bontius, *Tropische geneeskunde*, xlviii–li, it is dated by the editor, Andel, in his introduction, xvi–xvii.
75. Bontius, *Tropische geneeskunde*, xlviii–li.
76. Quotations from Bontius, *Tropische geneeskunde*, 102–5, 128–29, 94–95. At the time, another attack from the Matamarese was feared.
77. There is a considerable amount of repetition among Bontius's works; see, e.g., the story told of the Rhinoceros in "Notes on Da Orta," chap. 14 (on ivory) and in his natural history, bk. 5, chap. 1; or that of the excellent ability in pickling fish possessed by the Malayans, in his "Animadversions," chap. 32 (on calamus aromaticus, or sweet flag) and in his natural history, bk. 5, chap. 20.
78. Bontius, *Tropische geneeskunde*, 224–25, 226–27, 232–33, 234–35, 246–47.
79. Bontius, *Tropische geneeskunde*, 298–99, 326–27.
80. In the second letter printed in Römer, *Epistolae Jacobi Bontii*, he writes, "Ick heb een magnifycke bibliotheek wel by de twee duysent boecken meest uytgelesen autheuren, so dat ick de beste occasye heb om iet fræijs te doen als ick oyt in Europa soude connen becomen hebben," but although the letter implies that he had all his books with him, the context does not make it certain. Léonard Blussé informed me that at least one of Bontius's contemporaries in Batavia bought books from him.
81. Bontius, *Tropische geneeskunde*, xlviii–xlix.
82. Bontius, *Tropische geneeskunde*, 230–31, xxii–xxiii, 220–21, 8–9.
83. Grove, "Indigenous Knowledge," esp. 129–33.
84. Schoute, *Geneeskunde in den dienst*, 118.
85. Bontius, *Tropische geneeskunde*, 11.
86. For recent information on traditional medicines in Indonesia, see Sutarjadi, Dyatmiko, and et al., *International Congress on Traditional Medicine*.
87. Bontius, *Tropische geneeskunde*, 412–13.
88. Bontius, *Tropische geneeskunde*, 28–29, 94–95, 396–97.
89. Bontius, *Tropische geneeskunde*, 40–43, 34–35.
90. Raben, "Batavia and Columbo," 86.
91. On society in early Batavia, see Taylor, *Social World of Batavia*; Taylor, "Europese en Euraziatische vrouwen"; Blussé, *Strange Company*; and Raben, "Batavia and Columbo."
92. Bontius, *Tropische geneeskunde*, 358–59; this entry is not in the Sherard manuscript.
93. Reid, *Southeast Asia*, 53.

94. Bontius, *Tropische geneeskunde,* 220–21, xliv–xlvii.
95. Bontius, *Tropische geneeskunde,* 248–49.
96. Raben, "Batavia and Columbo," 86.
97. Bruijn, "Ship's Surgeons," 132.
98. Bontius, *Tropische geneeskunde,* 78–79.
99. Although there is a growing literature on the subject, see esp. Adas, *Machines as the Measure of Men,* and Harrison, "Medicine and Orientalism," 42–50.
100. Bontius, *Tropische geneeskunde,* 24–25, 366–67.
101. Mauss, *The Gift.* Some modern historians of science have seen the culture of gift exchanges behind the relationships of princes and courtly natural philosophers: see esp. Biagioli, *Galileo Courtier,* and Findlen, *Possessing Nature.*
102. Godelier, *Enigma,* 102.
103. Bontius, *Tropische geneeskunde,* 102–3.
104. Reid, *Southeast Asia,* 54–57.
105. Bontius, *Tropische geneeskunde,* 396–97.
106. Bontius, *Medicina Indorum.*
107. Bontius, *Account of the Diseases . . . of the East Indies.*
108. See, e.g., Jeanselme, "L'oeuvre."
109. For good overviews, see Hoboken, "Dutch West India Company"; Emmer, "WIC"; Heijer, *WIC;* Postma and Enthoven, *Riches from Atlantic Commerce;* Klooter, *Dutch in the Americas;* Prins et al., *Low Countries and New World(s).*
110. Rooden, "L'empereur," 53; Ebert, "Dutch Trade with Brazil."
111. Schmidt, *Innocence Abroad.*
112. For an excellent short overview, see De Vries, "Netherlands and the New World."
113. Boxer, *Dutch in Brazil;* Schwartz, "Commonwealth within Itself."
114. Postma, *Dutch in the Atlantic;* Thornton, *Africa and Africans.*
115. Frijhoff, "Médecin selon Jacques Cahaignes."
116. Andel, "Introduction"; Pies, *Piso,* 33–65.
117. Liebstad is southeast of Dresden, near the modern border with the Czech Republic, not the Leibstad in Switzerland sometimes said to be his birthplace, or the city of Lipstadt, between Dortmund and Bielefeld, as given in Lindeboom, *Dutch Medical Biography.*
118. Whitehead, "Biography of Marcgraf," 311.
119. Pies, *Piso,* 74.
120. Whitehead, "Biography of Marcgraf," 303, 309–10.
121. Rink, *Holland on the Hudson.*
122. Schnurmann, *Atlantische Welten;* Schnurmann, "Representative Entrepreneur."
123. Snapper, *Meditation,* 14–45.
124. Barleaus, *Rerum per Octennium;* Schmidt, *Innocence Abroad,* 252–57.
125. Piso and Marcgraff, *Historia naturalis Brasiliae.*
126. Whitehead and Boeseman, *Portrait of Dutch Seventeenth Century Brasil,* 28.
127. Whitehead and Boeseman, *Portrait of Dutch Seventeenth Century Brasil.* Also Brienen, "Art and Natural History," which I have not seen.
128. For instance, see Holthuis, *Marcgraf's (1648) Brazilian Crustacea.*

129. Schmidt, *Innocence Abroad*, 264.
130. Smith, *Business of Alchemy*, 167.
131. *Opuscula selecta Neerlandicorum de arte medica*, 14:4–17, 26–31, 18–25.
132. Personal communication from Michael Pye, who is working on a book on Dutch Brazil.
133. See part of a letter of De Laet to Lucas Holstenius, 10 October 1636, quoted in Freedberg, *Eye of the Lynx*, 286–87, and more generally Freedberg on the work of the Linceans in publishing an illustrated edition of Hernández in 1649–51 (too late for use in the 1648 Piso and Marcgraf edition), 245–304. Also Varey and Chabrán, "Medical Natural History," esp. 140, and more generally Varey, Chabrán, and Weiner, *Searching for the Secrets*.
134. Piso, *De Indiae utriusque*.
135. Whitehead, "Original Drawings," 30, summarizing the work most recently edited by E. de C. Falcão with bibliography and notes by O. Pinto, "Estudo crítico dos tralbalhos de Marcgrave e Piso . . . ," Brasiliensa Documenta, 2, with citations in Whitehead, "Original Drawings," 216.
136. Whitehead, "Biography of Marcgraf," 302, 310.
137. Pies, *Piso*, 41, 84–87; Andel, "Introduction," xviii–xxi.
138. Whitehead, "Biography of Marcgraf," 310.
139. Sherard 186, now in the Plant Sciences Library of the University of Oxford. The opening folio page, aside from some later notes at the top, is written out as the printed title page was apparently intended to be set: "Jacobi Bontii medici arcis ac civitatis Bataviae Novae in Indiis ordinarii Exoticorum Indicorum Centuria prima" ("The first century [hundred instances] of Indian exotics, by Jacob Bontius, physician-in-ordinary of the city of New Batavia in the Indies"); underneath the title is a dedication to "the most noble and generous Jacob Specx," governor general, and dated 1630. I came across it in the mid-1980s but only in recent years recognized its significance. My thanks to Harm Beukers for suggesting that I have yet another look at it.
140. Bontius, *Tropische geneeskunde*, 216–17.
141. Sherard 186, 79v–80.
142. Heel et al., *Tulp*, 160.
143. Bontius, *Tropische geneeskunde*, 238–39.
144. Bontius, *Tropische geneeskunde*, 286–87, 380–83; the latter chapter is not in the Sherard manuscript.
145. For instance, Edward Tyson later remarked: "I confess I do mistrust the whole representation" (Tyson, *Orang-Outang*, 19); and T. H. Huxley: "Bontius (1658) gives an altogether fabulous and ridiculous account and figure of an animal which he calls 'Orang-outang'; and though he says 'vidi Ego cujus effigiem hic exhibeo,' the said effigies . . . is nothing but a very hairy woman of rather comely aspect, and with proportions and feet wholly human" (Huxley, "Man-Like Apes," 11).
146. I owe these considerations—and my thanks—to an e-mail from Richard Frazier of Central Missouri State University, who was responding to the illustrations published in Cook, "Global Economies," 111–12.
147. Pomeranz, *Great Divergence*, 207.

6. MEDICINE AND MATERIALISM

1. For a history of his reputation, see Stéphane Van Damme, *Descartes*.
2. For Descartes's biography, I follow what I consider the most reliable guide: Rodis-Lewis, *Descartes: Biographie*, translated as Rodis-Lewis, *Descartes*.
3. Berkel, *Beeckman*, 43; for a condensed life in English, see Van Berkel's entries in Berkel, Helden, and Palm, *History of Science in the Netherlands*, 410–13, and Bunge et al., *Dictionary*, 1:68–74.
4. On the importance of music in early modern natural philosophy, see esp. Cohen, *Quantifying Music*; Szamosi, "Polyphonic Music"; and Gouk, *Music*; and on Beeckman, Buzon, "Science de la nature."
5. Letter of 23 April 1619, Descartes, *CSM*, 3:4.
6. Bunge et al., *Dictionary*, 73; Rodis-Lewis, *Descartes*, 25–29.
7. Jacob, *Cultural Meaning*, 52–54.
8. Letter of 24 January 1619, Descartes, *CSM*, 3:1. Rodis-Lewis, *Descartes*, 29–30.
9. Rodis-Lewis, *Descartes*, 56–58.
10. Sarasohn, "Nicolas-Claude Fabri de Peiresc"; Miller, *Peiresc's Europe*.
11. Sassen, "Reis van Gassendi."
12. Sassen, "Reis van Mersenne."
13. For this explanation, see Rodis-Lewis, *Descartes*, 77–78.
14. Schmitt, *Studies in Renaissance Philosophy and Science*.
15. Verbeek, *Descartes and the Dutch*, 7, 9, and more generally on the revival of scholasticism within Reformed theology, see Ruler, "Franco Petri Burgersdijk." For short treatments of Ames and Maccovius, Bunge et al., *Dictionary*, 1:23–24, 2:661–65.
16. Popkin, *History of Scepticism*; Sanches, *That Nothing Is Known*.
17. Descartes, *CSM*, 1:7.
18. Berkel, *Beeckman*, 130–33; Rodis-Lewis, *Descartes*, 85–88.
19. Descartes, *CSM*, 3:7.
20. Lindeboom, *Dutch Medical Biography*, 890–91; Lindeboom, *Descartes and Medicine*, 27–31.
21. Bunge et al., *Dictionary*, 2:824–26.
22. Garber, *Descartes Embodied*, 51, also 85–110; also see Garber, "Science and Certainty."
23. Colie, "*Some Thankfulnesse to Constantine*." Bunge et al., *Dictionary*, 338–39.
24. Lennon and Easton, *Cartesian Empiricism of Bayle*, 1–2; also see Hatfield, "Descartes' Physiology."
25. Quotations from letter of 15 April 1630, Descartes, *CSM*, 3:21; and part 5 of the *Discours*, Descartes, *CSM*, 1:134, 149.
26. On the importance of unnamed assistants, see Shapin, *Social History of Truth*, 379–92.
27. Napjus, *Hoogleeraren in de geneeskunde aan Franeker*, 48.
28. Rodis-Lewis, "Descartes' Life and the Development," 37.
29. Quotations from Descartes, *CSM*, 1:99, and letter to Mersenne, early June 1637, Descartes, *CSM*, 3:59.

30. Rodis-Lewis, *Descartes*, 99, 127, 158; Lindeboom, *Descartes and Medicine*, 37–41.
31. Descartes, *CSM*, 1:99.
32. French, "Harvey in Holland," 47; Lieburg, "Beeckman and Harvey"; Lieburg, "Zacharias Sylvius," 243.
33. For a good recent analysis, see Fuchs and Grene, *Mechanisation of the Heart*, 115–41.
34. Lindeboom, *Descartes and Medicine*, 42–43.
35. French, "Harvey in Holland," 72–81; Lindeboom, *Dutch Medical Biography*, 2117–18.
36. Descartes, *CSM*, 1:100.
37. Lindeboom, *Descartes and Medicine*, 37. Also see Dankmeijer, *Biologische Studies van Descartes*, partially translated as Dankmeijer, "Travaux biologiques de Descartes"; Hall, "Descartes' Physiological Method."
38. Letter of 27 February 1637, Descartes, *CSM*, 3:53.
39. From his "Descartes and Method in 1637," 33–51, in Garber, *Descartes Embodied*.
40. Descartes, *CSM*, 1:143.
41. Descartes, *CSM*, 1:143, 144.
42. Descartes, *CSM*, 1:134, 139, 151; also see Shapin, "Descartes the Doctor."
43. Rodis-Lewis, *Descartes*, 113–32.
44. Descartes, *CSM*, 2:11.
45. Descartes, *CSM*, 2:54, 55–56.
46. Descartes, *CSM*, 2:61.
47. Gassendi's are the fifth set of objections; for his praise of the sixth meditation, see Descartes, *CSM*, 2:239.
48. Sherman, *Fabric of Character*, 3.
49. Quotations from Aristotle, *Nichomachean Ethics*, 10.5; the definition of perfect happiness is at 10.7–8. Aristotle's *Rhetoric* stressed the importance of using the passions only to achieve the good: esp. 1:iii–xi (1358b–1372a).
50. Both quotations from Verbeke, "Ethics and Logic in Stoicism," 24. See also MacIntyre, *After Virtue*.
51. McAdoo, *Structure of Moral Theology*; Wood, *English Casuistical Divinity*; Hoopes, *Right Reason in the English Renaissance*; Mulligan, "Robert Boyle"; Wojcik, *Robert Boyle*.
52. Febvre, *Problem of Unbelief*. See also Wootton, "Lucien Febvre," and Davis, "Rabelais among the Censors (1940s, 1540s)." For a possible example of a real atheist, see Wootton, *Paolo Sarpi*.
53. Parel, *Machiavellian Cosmos*, esp. 87–88.
54. Descartes, *CSM*, 3:352–3.
55. Rodis-Lewis, *Descartes*, 44–46.
56. Mosse, "Religious Thought," 199.
57. See also Lennon, *Gods and Giants*.
58. See, e.g., Berman, *Reenchantment of the World*; Force, "Origins of Modern Atheism"; Febvre, *Problem of Unbelief*.
59. Descartes, *CSM*, 1:151.
60. Descartes, *CSM*, 2:43.

61. Descartes, *CSM*, 2:59–61.
62. Letter of 25 January 1638, quoted in Rodis-Lewis, *Descartes*, 127.
63. For an excellent overview, see Park and Kessler, "Psychology"; Rodis-Lewis, *Descartes*, 45.
64. Plato, *Dialogues*, 3:509–513.
65. Burton, *Anatomy of Melancholy*, 1.1.1.1, 10.10.10.
66. Verbeek, *Descartes and the Dutch*, 13.
67. Bos, *Correspondence of Regius*, 3–11.
68. For a sympathetic treatment of Voetius, see McGahagan, "Cartesianism," 31–216.
69. These have been recently reprinted in Bos, *Correspondence of Regius*, 197–248.
70. I am here following the exposition given by Verbeek, *Descartes and the Dutch*, 13–15. Also see Berkel, "Descartes in debat met Voetius."
71. Verbeek, "Descartes and the Problem of Atheism"; Ruler, *Crisis of Causality*.
72. Verbeek, *Descartes and the Dutch*, 15–19.
73. Descartes, *CSM*, 2:396.
74. Rodis-Lewis, *Descartes*, 167.
75. For a thorough account of the dispute in Utrecht, see Verbeek, *Querelle d'Utrecht*; also Berkel, "Descartes in debat met Voetius."
76. Article 188, pt. 4, and part of the title to art. 30, pt. 1: Descartes, *CSM*, 1:249, 203.
77. For the importance of Elizabeth in pushing Descartes to new conclusions, see Harth, *Cartesian Women*, 67–78.
78. Descartes to Pollot, 6 October 1642; Descartes, *CSM*, 3:214–15. Néel, *Descartes et la Princesse Élisabeth*.
79. Adam and Tannery, *Oeuvres*, 3:660.
80. Descartes, *CSM*, 3:218, 217–20, 226–29, 228.
81. Adam and Tannery, *Oeuvres*, 5:64–66.
82. Descartes, *CSM*, 3:237.
83. Letter of May or June 1645, Descartes, *CSM*, 3:249; I believe it a good guess to date it at about the time of the Battle of Naseby, which took place on 14 June; Descartes, *CSM*, 3:251.
84. Letter of June 1645, Descartes, *CSM*, 3:253–54.
85. Seneca, *Four Dialogues*, 26–27. See also Fothergill-Payne, "Seneca's Role in Popularizing Epicurus"; Joy, "Epicureanism."
86. Kraye, "Moral Philosophy," 381.
87. Sassen, "Reis van Gassendi," 271.
88. Charles-Dawbert, "Libertine"; Pintard, *Libertinage érudit*.
89. On Descartes's debt to neo-Stoicism, see Levi, *French Moralists*, esp. 257–98. Also see Dear, "Mechanical Microcosm."
90. Descartes, *CSM*, 3:257, 258, 262, 264–65.
91. Quoted in "Translator's preface" to *The Passions of the Soul*, Descartes, *CSM*, 1:325.
92. Descartes, *CSM*, 1:271, 270, 3:275, 289.
93. Letter of May 1646; Descartes, *CSM*, 3:285–88.
94. Letter 1 November 1646; Descartes, *CSM*, 3:290.
95. Åkerman, *Queen Christina*, 45; Descartes, *CSM*, 3:327.

96. Descartes, *CSM*, 1:403.
97. Bredero, *Spanish Brabanter*, 47, 78.
98. Weevers, *Poetry of the Netherlands*, 77.
99. Translation by Adriaan J. Barnouw in Stuiveling, *Sampling of Dutch Literature*, 20.
100. Rodis-Lewis, *Descartes*, 13–15, 53–54, 29.
101. Cole, *Olympian Dreams*, 27.
102. Levi, *French Moralists*; Keohane, *Philosophy and the State*; Lovejoy, *Reflections*, 129–245.
103. Descartes, *CSM*, 1:328, 333, 349, 335.
104. Descartes, *CSM*, 1:345.
105. Verbeek, "Passions et la fièvre."
106. Letter of 20 November 1647, Descartes, *CSM*, 3:324–26.
107. Descartes, *CSM*, 1:347, 348, 382.
108. Seneca, *Four Dialogues*, 12–15.
109. Descartes, *CSM*, 1:186.
110. Descartes, *CSM*, 1:341; also see the important remarks on the heart as the source of divine possession in Caciola, "Mystics, Demoniacs, and Physiology."
111. See esp. Santing, "*De Affectibus*."
112. Descartes, *CSM*, 1:345, 403.
113. Descartes, *CSM*, 1:399, 388, 392, 350, 353.
114. Fuchs and Grene, *Mechanisation of the Heart*, 138.
115. Descartes, *CSM*, 3:287–88, 1:403.
116. James, *Passion and Action*.
117. On Gassendi, see Sarasohn, "Motion and Morality"; Sarasohn, "Gassendi"; Joy, *Gassendi the Atomist*.
118. Blom, "*Felix*," 119–50, esp. 126–38. Blom points out that Burgersdijk was a Thomist rather than Augustinian "in that he presents the passions not as the dark side of man . . . but as the material upon which man can develop his potentialities" (127).
119. Bontius, *Tropische geneeskunde*, 98. Bontius gives the quotation from Horace as "qui nisi servit / Imperat; hunc fraenis, hunc tu compesce catenis," a variation on the full "Ira furor brevis est: animum rege, qui nisi paret, / imperat; hunc frenis, hunc tu compesce catena." Horace, *Second Epistles*, 1.62–63.
120. Åkerman, *Queen Christina*, 49–51.
121. Kambouchner, *L'homme des passions*, quotations on 352, 354.
122. Des Chene, *Physiologia*.
123. Bos and Krop, *Franco Burgersdijk*; Bunge et al., *Dictionary*, 181–90.
124. Blom, "*Felix*"; see also Kossmann, "Development," 92–98.
125. The best brief account is Verbeek, *Descartes and the Dutch*, 34–70; see also Ruestow, *Physics*, 36–72.
126. Beukers, "Mechanistiche Principes."
127. *Universiteit te Leuven*, 1425–1975, 133; see also Vanpaemel, *Echo's van een wetenschappelijke revolutie*.
128. See J. A. Schuster's entry on Rohault in Gillispie, *Dictionary of Scientific Biography*, 11:506–9; Vanpaemel, "Rohault."

129. Lindeboom, *Dutch Medical Biography*, 1791–94; Lindeboom, *Florentius Schuyl*, esp. 67–83.
130. Kossmann, "Development," 101.
131. See esp. Verbeek, *Descartes and the Dutch;* Colie, *Light and Enlightenment.*
132. Quoted in Boogman, "*Raison d' État* Politician Johan de Witt," 71.
133. The unusual nature of the De la Courts' views is stressed in Blom and Wildenberg, *Pieter de la Court;* also see Haitsma Mulier, "Language of Seventeenth-Century Republicanism," esp. 188–191.
134. Wildenberg, "Appreciaties van de gebroeders De la Court."
135. Smit, "Netherlands and Europe," 23–24.
136. Court, *Interest van Holland*, 21.
137. Klashorst, "'Metten schijn van monarchie getempert,'" esp. 119–120, 136.
138. Wildenberg, "Appreciaties van de gebroeders De la Court."
139. Kossmann, *Politieke theorie in Nederland*, quotation on 37; the argument is elaborated on 38–48. Also see Kossmann, *Politieke theorie en geschiedenis*, 59–92; translated as Kossmann, "Popular Sovereignty." See also Haitsma Mulier, *Myth of Venice*, 120–169, who argues for the influence of Machiavelli as well as Descartes and Hobbes, since Machiavelli also grounded politics on the passions. Also Haitsma Mulier, "Controversial Republican," and Gelderen, "Machiavellian Moment and the Dutch Revolt."
140. Smit, "Netherlands and Europe," 24.
141. Court, *Interest van Holland*, preface.
142. Kerkhoven and Blom, "De la Court en Spinoza," 160.
143. For the most careful modern biography, see Nadler, *Spinoza*, and Israel, *Radical Enlightenment*, 159–74; Klever, "Companion to Spinoza."
144. Among them, see Israel, *Radical Enlightenment*, esp. 230–41; Nadler, *Spinoza;* Grene, *Spinoza;* and Jaspers, *Spinoza.*
145. James, *Passion and Action*, 136–56.
146. Spinoza, *Ethics*, pt. 4, props. 40, 41, quoted from Spinoza, *Ethics*, 167–68.
147. Wartofsky, "Action and Passion," quotations at 352, 349.
148. Kossmann, *Politieke theorie en geschiedenis*, 50–58; Haitsma Mulier, "Language of Seventeenth-Century Republicanism," 191–93; Haitsma Mulier, "Controversial Republican," 256; Prokhovnik, *Spinoza.*
149. For other radicals around Spinoza, see Israel, *Radical Enlightenment*, 175–229.
150. The evidence for this comes from Spinoza's report to Leibniz in 1676; on the butchery, see Rowen, *John de Witt*, 861–884; on Spinoza's reaction, 885–886.
151. On the continuing influence of Hobbes and Descartes in The Netherlands, see Kossmann, *Politieke theorie*, 59–103; but even he admits that "as far as I know, there are no traces of any direct influence exercised by De la Court and Spinoza on Dutch political theory" ("Development of Dutch Political Theory," 105).
152. Bontekoe, *Thee*, 391.

7. INDUSTRY AND ANALYSIS

1. Farber, "Development of Taxidermy"; Schultze-Hagen et al., "Avian Taxidermy."
2. The famous work of Hermes Trismegistus was thought to have been written at the

beginning of Egyptian civilization; on the fascination with Egypt, see Singer, "Hieroglyphs," and Grafton, *Defenders of the Text*, 145–77.

3. Lucas, *Egyptian Materials*, 270–326; Schrader, *Observationes et historiae*, 236.
4. Partington, *History of Chemistry*, 2:98, quoting from Brasavola's *Examen omnium simplicium medicamentorum.*
5. Dannenfeldt, "Egyptian Mumia," esp. 169–71.
6. Thorndike, *History of Magic*, 8:106.
7. Dannenfeldt, "Egyptian Mumia," 173–74; Partington, *History of Chemistry*, 2:444.
8. Thorndike, *History of Magic*, 8:414.
9. For example, see Maets, *Chemia Rationalis*, 162–64, and a manuscript of his chemistry course from 1675 and 1676, British Library, Sloane MSS 1235, fols. 5–5b.
10. See esp. Park, "Criminal and the Saintly Body"; Park, "Life of the Corpse"; Gentilcore, "Contesting Illness in Early Modern Naples"; and Caciola, "Mystics, Demoniacs, and Physiology."
11. Schrader, *Observationes et historiae*, 236.
12. Browne, *Chirurgorum comes*, 710–14; he also gives the variation on Paré's recipe used by Balthasar Timeus à Guldenklee, "a man famous for Embalming."
13. Bosman-Jelgersma and Houtzager, "Balseming."
14. For what follows, I am heavily indebted to Jansma, *De Bils*.
15. "Sed fidem superat omnem, exsiccatum hominis Cadaver Recenter Mortuum Diceres tanto Theatro Dignissimum opus." The wooden plaque containing Van Horne's testimony is reproduced in Jansma, *De Bils*, 47. De Bils later claimed that he had spent forty thousand gilders on the preparations, not including accounting for his time (66).
16. Jansma, *De Bils*, 96–99.
17. Lindeboom, *Dutch Medical Biography*.
18. Jansma, *De Bils*, 48–53.
19. Jansma, *De Bils*, 48–54.
20. Lindeboom, *Dutch Medical Biography*, 909.
21. There is no mention of this or De Bils in the thorough study of medicine in Rotterdam by Lieburg, *Medisch onderwijs te Rotterdam*, but the public anatomy lessons begun there by Nicolaas Zas in 1642 had stopped for some reason in 1654 (18–19, 27).
22. Jansma, *De Bils*, 54–58.
23. Browne, *Chirurgorum comes*, 713.
24. De Bils, *Coppy of a Certain Large Act (Obligatory) of Yonker Louis de Bils*. As described by Michael Hunter, this is the first publication for which Boyle was personally responsible, undertaken with help and encouragement by Hartlib, translated from the Dutch by John Pell, with an accompanying letter by a French virtuoso, perhaps Pierre Borell: Boyle, "Large Act of Anatomy by de Bils," cxix–cxxii.
25. Barbette, *Pest-Beschryving*; see the four-page appendix "Aen de Heer J. Rykenwaert, Hoogvermaerde Geneesheer tot Rotterdam, tegens de verdediging van Jr. Louys de Bils, Heer van Koppensdamme, etc."
26. Jansma, *De Bils*, 58–67.
27. Jansma, *De Bils*, 65, 67, 68–69.
28. Jansma, *De Bils*, 77, 78–79, 83–88, 90.

29. Browne, *Chirurgorum comes,* 713.
30. Jansma, *De Bils,* 67.
31. Partington, *History of Chemistry,* 2:208; Lindeboom, *Dutch Medical Biography,* 1031.
32. Eloy, *Dictionnaire historique de la médecine,* 655–56; Gannal and Harlan, *History of Embalming,* 91–92, 96; see Clauder, *Methodus balsamandi corpora humana,* chap. 5, sec. 3 (128–40) on his view of the method of De Bils, and chap. 6 (140–81) for his own method; and for a contemporary English version, Browne, *Chirurgorum comes,* 713–14.
33. For a biography, see Schierbeek, *Jan Swammerdam.*
34. Lindeboom, *Cabinet van Jan Swammerdam;* Lindeboom, *Letters of Swammerdam,* 23.
35. The investigations into muscles by his Danish fellow student Steno became more famous: Vugs, *Leven en werk van Stensen.*
36. Swammerdam, *Book of Nature,* ii; for the original, see Swammerdam, *Biblia naturae,* sig. B.
37. Reported by Lindeboom, *Cabinet van Jan Swammerdam,* xii, from a letter of Olaus Borch to Bartholin (which I have not yet seen), which places the event around 1661–62, when Borch was in The Netherlands.
38. Schrader, *Observationes et historiae,* 237–38.
39. *Oxford English Dictionary.*
40. Dioscorides, *Greek Herbal of Dioscorides,* 49.
41. Partington, *History of Chemistry,* 2:97.
42. Yonge, *Currus triumphalis,* 50, 48–50, preface. Also see Davisson, *Philosophia pyrotechnica,* 325–26, and Davisson, *Cours de chymie,* 308.
43. Sloane 1235: "Collegium Chymicum Secretum / A / D. Carolo de Maes apud Lugdunenses," 1675 and 1676: fol. 5, "Modus Condiendi Cadavera."
44. Maets, *Chemia rationalis,* 162–63.
45. Stephan Blankaart, *Neue und besondere Manier alle verstorbene Cörper mit wenig Ukosten der Gestalt zu Balsamiren* (Hanover: Gottlieb Heinrich Grentz, 1690). I owe this reference to Tomomi Kinukawa. I should point out, however, that Professor Anthony Nuck of Leiden continued to search for the real secret of De Bils, according to a report in a letter he sent to Dr. Slare that was read to the Royal Society on 7 June 1682: Birch, *History of the RS,* 151.
46. Partington, *History of Chemistry,* 2:267, citing Libavius, *Alchemia,* 1597, 2.2.36.
47. Birch, *History of the RS,* 1:110, 840.
48. Cole, *History of Comparative Anatomy,* 445.
49. Tompsett, Wakeley, and Dobson, *Anatomical Techniques,* x; Cole, *History of Comparative Anatomy,* 445–50. See Boyle's work diaries as edited by Michael Hunter, at http://www.bbk.ac.uk/boyle/workdiaries/WDClean.html.
50. Birch, *History of the RS,* 1:327, 374, 378, 393.
51. Partington, *A History of Chemistry,* 2:494.
52. Adelmann, *Malpighi,* 2238.
53. Schrader, *Observationes et historiae,* 238–40.
54. Swammerdam, *Biblia naturae,* sigs. C, C2.

55. These "eggs" were later identified by Leeuwenhoek as follicles, not the eggs themselves, which were first seen under a microscope by Van Baer in the early nineteenth century.
56. Swammerdam, *Miraculum naturae.*
57. Lindeboom, *Cabinet van Jan Swammerdam*, xvii.
58. Swammerdam, *Bybel*, sig. C. His demonstration of the spinal marrow was appropriated by Blasius (on whom see below), in his *Anatome medulla spinalis* (1666): Fournier, "Fabric of Life," 85.
59. Luyendijk-Elshout, "Death Enlightened"; Hansen, "Resurrecting Death"; Kooijmans, *De doodskunstenaar.*
60. Lindeboom, "Jan Swammerdam als microscopist," 96–97.
61. Vermij, "Bijdrage tot Hudde."
62. Berkel, Helden, and Palm, *History of Science in the Netherlands*, 54.
63. Hacking, *Emergence of Probability*, 92–118.
64. Rowen, *John de Witt*, 411–18; Pater, *Petrus van Musschenbroek*, 5.
65. Blasius, *Anatome animalium.*
66. Slee, *Rijnsburger collegianten*, 77.
67. Lindeboom, "Collegium Privatum Amstelodamense"; Cole, *History of Comparative Anatomy*, 330–41. It would appear that the group was still meeting when John Locke was in exile in The Netherlands, a point I owe to Harm Beukers.
68. Seters, "Voorgeschiedenis," 44; Brugmans, *Gedenkboek*, 181, 394.
69. Lindeboom, *Observationes anatomicae*; Cole, *History of Comparative Anatomy*, 332.
70. Lindeboom, *Ontmoeting met Jan Swammerdam*, xv.
71. Swammerdam, *Biblia naturae*, sig. l.
72. Lindeboom, *Ontmoeting met Jan Swammerdam*, 12.
73. Swammerdam, *Historia insectorum.*
74. Ruestow, "Piety and Natural Order."
75. Ruestow, "Rise of the Doctrine of Vascular Secretion," 268.
76. On judging cloth, and the use of lenses, see Ford, "The Van Leeuwenhoek Specimens," 51; Reddy, *Rise of Market Culture*; Harte, *New Draperies.*
77. Helden, *Invention of Telescope.*
78. Fournier, "Fabric of Life," 47–49, 9–16; Ruestow, *Microscope in the Dutch Republic*, 6–10; Wilson, "Visual Surface and Visual Symbol"; Harwood, "Rhetoric and Graphics in *Micrographia*"; Dennis, "Graphic Understanding."
79. Ruestow, *Microscope in the Dutch Republic*, 9–10.
80. The second method of Hooke is given in his *Lectures and Collections* of 1678; Hooke's text describing both methods for making a single-lens microscope are reproduced in Ford, *Leeuwenhoek Legacy*, 29–30.
81. Barth, "Huygens at Work"; Fournier, "Fabric of Life," 13.
82. Zuylen, "Microscopes of Leeuwenhoek," 310.
83. Lindeboom, "Jan Swammerdam als microscopist," 95, quoting from Swammerdam's *Biblia naturae*, 89–91. Although Lindeboom argues that Swammerdam did not learn the method from Hudde until 1673 (98), the consensus among those who have studied the matter is that he did so in the 1660s: e.g., Fournier, "Fabric of Life," 89.

84. See the fine account of his painstaking labors in Ruestow, *Microscope in the Dutch Republic*, 112–17.
85. Lindeboom, "Jan Swammerdam als microscopist," 98.
86. Ford, "Van Leeuwenhoek Specimens"; Ford, *Single Lens*, 33–40; Ford, *Leeuwenhoek Legacy*.
87. Zuylen, "Microscopes of Leeuwenhoek," 320–22.
88. Wheelock, *Perspective;* Steadman, *Vermeer's Camera;* Huerta, *Giants of Delft*, 15–29.
89. Zuylen, "Microscopes of Leeuwenhoek," 326; also see Zuylen, "On the Microscopes of Leeuwenhoek"; Zuylen, "Microscopen van Antoni Van Leeuwenhoek"; Ford, *Leeuwenhoek Legacy*, 141–81.
90. Ford, "Van Leeuwenhoek Specimens"; Ford, *Leeuwenhoek Legacy*, 51–69, 83–125.
91. For a still generally sound biography, see Dobell, *Leeuwenhoek*. Also see Schierbeek, *Leeuwenhoek;* Schierbeek, *Measuring the Invisible World*.
92. He appears just above and to the right of the central figure in *The Anatomy Lesson of Dr. Cornelis Issacz. 's-Gravesande*, 1681, now in the Stedelijk Museum het Prinsenhof, Delft.
93. Hall and Hall, *CHO*, no. 2209, 18 April 1673. The virtuosi discussed Leeuwenhoek's letter at their meeting on 7 May, and Oldenburg printed it in his *Philosophical Transactions* of 19 May.
94. Wrop, *Briefwisseling van Constantijn Huygens (1608–1687)*, 6:330–31.
95. Palm, "Leeuwenhoek and Other Dutch Correspondents"; Vermij and Palm, "John Chamberlayne als vertaler."
96. Meyer, "Leeuwenhoek as Experimental Biologist"; Cole, "Leeuwenhoek's Zoological Researches"; Lindeboom, "Leeuwenhoek and Sexual Reproduction"; Ruestow, "Images and Ideas"; Ruestow, "Leeuwenhoek and Spontaneous Generation"; Fournier, "Fabric of Life," 79–91; Ruestow, *Microscope in the Dutch Republic*, 146–200.
97. Velde, "Bijdrage tot Bontekoe."
98. Molhuysen, *Bronnen tot de geschiedenis der Leidsche Universiteit*, 3:283, 300–301, 302, 314; on Bontekoe's activities as a proponent of Cartesianism more generally, see Thijssen-Schoute, *Nederlands Cartesianisme*, 276–317.
99. For a reprint and facing-page translation, see *Opuscula selecta Neerlandicorum de arte medica*, vol. 14, which includes an introduction to his life and work in English by F. M. G. de Feyfer, lii–xcix. Also see Velde, "Bijdrage tot Bontekoe." On the "tea doctor," see Schama, *Embarrassment of Riches*, 171–72.
100. Glamann, *Dutch-Asiatic Trade*, 18; it was not turning up enough, however, to be noticed before the eighteenth century in Posthumus, *Inquiry into the History of Prices in Holland*.
101. Foss, "European Sojourn," 133.
102. Bontekoe, *Thee*, 331.
103. Bontekoe, *Thee*, 453. For remarks on the beginning of "Chinoiserie" in the seventeenth century, see Miller, "Fraisse at Chantilly," and Lach, *Asia in the Making of Europe*, vol. 2.
104. Bontekoe, *Thee*, xciii.
105. Bontekoe, *Thee*, 373–75. He says this was after a period of great domestic trouble,

indicating his marriage of 1669, which ended in divorce a few (unspecified) years later.

106. Bontekoe, *Thee*, 123.
107. Bontekoe, *Thee*, 191, 129, 211, 327, 359, 377.
108. Bontekoe, *Thee*, 123.
109. Pagel, *Van Helmont.*
110. For a brief summary of Helmont's life and views, see Pagel, "Helmont, Johannes (Joan) Baptista Van"; Pagel, *Van Helmont*, esp. 19–34; Pagel, *Paracelsus to Van Helmont*; Heinecke, "Mysticism and Science of Van Helmont"; Debus, *Chemistry and Medical Debate.*
111. Webster, "Water as the Ultimate Principle."
112. Niebyl, "Sennert, Van Helmont, and Medical Ontology."
113. Bontekoe, *Thee*, 351.
114. Bontekoe, *Thee*, 225, 273.
115. Bontekoe, *Thee*, 377, 333, 245, 359.
116. Bontekoe, *Thee*, 449.
117. From his "Fondamenten, van het Acidum en Alcali," which is part 4 of the "Opbouw der medicyne," in Bontekoe, *Alle de . . . werken van de Heer Corn. Bontekoe*, 187.
118. On Sylvius, see Baumann, *François Dele Boe Sylvius*, with a shorter summary at Baumann, *Drie eeuwen*, 114–22; also King, *Road to Medical Enlightenment*, 93–112; Beukers, "Laboratorium van Sylvius"; Beukers, "Acid Spirits and Alkaline Salts." The documentation does not allow the pinpointing of when Sylvius pronounced his acid-alkali theory, but it appears in a medical disputation of 1660: Baumann, *François Dele Boe Sylvius*, 65.
119. Eamon, "New Light on Boyle." See also Debus, "Sir Thomas Browne."
120. Defilipps, "Historical Connections." But see Pelliot, *Notes on Polo*, 103–4, who notes that the Asiatic "Brazilwood" is commonly known as "sappan-wood," so that the country was first named after not the wood but the fabled Isle of Brazil, which was supposed to be in the Atlantic.
121. For the general importance of dyeing to "science," see Travis, *Rainbow Makers*; Déré, "Profil d'un artiste chimiste"; Nieto-Galan, "Calico Printing and Chemical Knowledge"; Schaffer, "Experimenters' Techniques."
122. Zanden, "Op zoek naar de 'Missing Link'"; Ogilvie and Cerman, *European Proto-Industrialization.*
123. See Posthumus, *Geschiedenis van de Leidsche lakenindustrie.*
124. For a description of work in the bleaching fields and their depictions, see Stone-Ferrier, *Images of Textiles*, 119–61.
125. English translation by Marian Fournier, in Fournier, "Fabric of Life," 86, from his *Bybel der Natuure / Biblia Naturae* (1737–38), 870–71 and 868–69, which reprints passages from the *Historia.*
126. Fournier, "Fabric of Life," 94, 88; also see Ruestow, "Piety and Natural Order."
127. Baumann, *Drie eeuwen*, 120.
128. Lindeboom, "Leeuwenhoek and Sexual Reproduction"; Ruestow, "Leeuwenhoek and Spontaneous Generation," 244–46.

129. Lindeboom, *Letters of Swammerdam*, 108.
130. Ruestow, "Piety and Natural Order," 231.
131. Lindeboom, *Ontmoeting met Jan Swammerdam*.
132. Swammerdam, *Ephemeri vita*.
133. Lindeboom, *Letters of Swammerdam*, 104–5, April 1678.
134. Lindeboom, "Leeuwenhoek and Sexual Reproduction."
135. This and other reflections are prompted by remarks in Ruestow, *Microscope in the Dutch Republic*.
136. Wilson, *Invisible World*, 254–55.
137. Fournier, "Fabric of Life," 188.
138. A sentiment ascribed to Otto Frederik Müller, Ruestow, *Microscope in the Dutch Republic*, 261.

8. GARDENS OF THE INDIES TRANSPORTED

1. Pabbruwe, *Padtbrugge*, 11; Popkin, *La Peyrère*.
2. Pabbruwe, *Padtbrugge*, 19–20.
3. Goonewardena, *Foundation of Dutch Power; Machtige eyland*.
4. Pabbruwe, *Padtbrugge*, 23, 25, 29.
5. Uragoda, *History of Medicine in Sri Lanka*, 77. Carstens arrived in Ceylon in 1636 and stayed at least through 1643.
6. Uragoda, *History of Medicine in Sri Lanka*, 69. "Stupes" are cloths dipped in a medicinal liquid and applied to an affected part.
7. Heniger, *Van Reede*, 29; Pabbruwe, *Padtbrugge*, 29–30.
8. Haan, "Cleyer," 428.
9. Wijnands, "Hortus Auriaci," 68.
10. Heniger, *Van Reede*, 29; Pabbruwe, *Padtbrugge*, 29–30. The letter was sent on 24 April 1669; I am assuming that a letter sent from Amsterdam in May 1668 would have reached Batavia no more than six months afterward.
11. For what follows on Cleyer's early life, see Haan, "Cleyer," 426–31, supplemented by Kraft, *Andreas Cleyer Tagebuch*, 34–40.
12. Kraft, "Mentzel, Couplet, Cleyer," 176. An inventory at his death in the VOC archives shows that among other things he possessed books in Latin, some of them medical (works by Calvin, Comenius, Kircher, a chemical pharmacopoeia by J. H. Cardilucius, T. Bartholin's *Actorum medicorum*, parts 3–7, Sylvius's *Opera omnia*, parts 8 and 9 of the *Miscellenea curiosa*, and several others): Bruijn, "Ship's Surgeons," 271–73.
13. For information on Johannes and Anna Ammanus—the latter being his aunt—see Haan, "Cleyer," 465–68.
14. Heniger, *Van Reede*, 31.
15. Bassani and Tedeschi, "Image of the Hottentot"; Merians, "What They Are, Who We Are."
16. Karsten, *Old Company's Garden*, 1–66.
17. Heniger, *Van Reede*, 8.
18. Heniger, *Van Reede*, 70.

19. Quoted from Heniger, *Van Reede,* 29.
20. Pabbruwe, *Padtbrugge,* 30–31.
21. They began work for Van Goens in 1668.
22. Heniger, *Van Reede,* 70, 30, quotation at 30; Heniger, "Hermann."
23. The quotation is from Christopher Sweitzer, another German employee of the VOC who served in Sri Lanka from 1676 to 1682, in Raven-Hart, *Germans in Dutch Ceylon,* 78.
24. Hermann deserves further study, but for the moment, see Heniger, "Hermann."
25. Quoted from Heniger, *Van Reede,* 29, 31.
26. Heniger, *Van Reede,* 5–25.
27. Heniger, *Van Reede,* 29; but Heniger also notes that Van Reede never referred to such letters (30).
28. Uragoda, *History of Medicine in Sri Lanka,* 68.
29. Heniger, *Van Reede,* 37–38, 41.
30. On Matthew, see Heniger, *Van Reede,* 38–39.
31. For an account of Matthew's still not fully published manuscripts, see Heniger, *Van Reede,* 105–24. Heniger concludes that Matthew's manuscripts were his "medico-botanical notebook for the consultation of relevant information on medicinal plants" (123).
32. Heniger, *Van Reede,* 30.
33. Figueiredo, "Ayurvedic Medicine in Goa," 233–35. Figueiredo notes that the testimony they wrote for Van Reede was written in Konkani, the language of Goa, and that they mentioned that their work was in keeping with the medical book titled *Nighantu.*
34. Heniger, *Van Reede,* 41–43.
35. These quotations are found both in Grove, "Transfer of Botanical Knowledge," 161–62, 168, 169, 171, and, word-for-word the same in Grove, "Indigenous Knowledge," 126, 134, 136–37, 139; the first, third, and fourth are also in Grove, *Green Imperialism,* 78, 87, 89–90, and the second implied at greater length.
36. Heniger, *Van Reede,* 30.
37. Heniger, *Van Reede,* 62.
38. Heniger, *Van Reede,* 53–54, 41–42.
39. One of the directors who took a keen interest in the botany of the East Indies, Huydecoper, was also a strong supporter of Van Reede against Van Goens: Gaastra, *Bewind en beleid,* 125, 156, 162.
40. Lindeboom, *Dutch Medical Biography,* 1041–42.
41. On Grimm, see Dorssen, "Ten Rhijne," 176; Uragoda, *History of Medicine in Sri Lanka,* 65; Heniger, *Van Reede,* 53–54; Lindeboom, *Dutch Medical Biography,* 727–28. I have not seen a copy of either version of Grimm's work, as they do not exist in the public domain in the UK nor in the Koninklijke Bibliotheek in The Hague.
42. Heniger, *Van Reede,* 53–54.
43. Ten Rhijne to Sibelius, 7 November 1680, Sloane 2729, fol. 68.
44. Heniger, *Van Reede,* 61; Van Reede, *Hortus.* Ten Rhijne had heard of Seyn's death by 1680 (Sloane 2729, letter of Ten Rhijne to Sibelius, 1680, fol. 68), which he thought a terrible blow to the plans for fully publishing "our" *Hortus.*

45. Coolhaas, *Generale missiven*, 4:294–95.
46. Heniger, *Van Reede*, 70–71, 44.
47. Raven-Hart, *Germans in Dutch Ceylon*, 10.
48. Heniger, *Van Reede*, 83.
49. Diedenhofen, "Maurits and His Gardens"; Diedenhofen, "'Belvedere.'"
50. Bezemer-Sellers, "Gardens of Frederik Hendrik," 132.
51. Jong, "Netherlandish Hesperidies," 21.
52. Prest, *Garden of Eden*; Vries, "Country Estate Immortalized," 91.
53. Quotation from Jong, "Netherlandish Hesperidies"; see also Jong, "Profit and Ornament," and Jong, *Nederlandse tuin- en landschapsarchitechtuur*, esp. 15–56.
54. Visits of Guido and Giulio de Bovio (17 December 1677) and François Maximillien Mission (October 1687). The hippopotamus specimen, unique in Europe, was made from a well-prepared skin (probably with the skull intact) supported on an iron and hardwood frame, and was still on display in good condition in 1767. See the full quotations in Gogelein, *Hortus*, 19–21.
55. Jong, *Nederlandse tuin- en landschapsarchitechtuur*, 117. Also see Veendorp and Baas Becking, *Hortus Academicus*.
56. Wijnands, *Botany of the Commelins*, 6–13.
57. Wijnands, "Hortus Auriaci," 64; Heniger, *Van Reede*, 72.
58. Tachard, *Voyage to Siam*, 51–52.
59. Heniger, *Van Reede*, 8, 70–71.
60. Haan, "Cleyer," 457.
61. Breyne, *Excoticarum*, appendix.
62. Barnett, "Academia."
63. Haan, "Cleyer," 453; Kraft, "Mentzel, Couplet, Cleyer," 163–64.
64. Ehring, "Leprosy Illustration"; it is said to show lepromatous leprosy, in the advanced stage of facies leonia. My thanks to Thomas Rütten for this reference.
65. Pas, "Earliest European Descriptions of Japan's Flora"; Muntschick, "Een Manuscript."
66. Haan, "Cleyer," 457. The work was titled *Der Orientalisch-Indianische Kunst-und Lust-Gärtner* (Dresden).
67. Wijnands, "Hortus Auriaci," 66.
68. Cleyer did so with the permission of the VOC, secured by Speelman: Haan, "Cleyer," 459–60.
69. Rookmaaker, *Zoological Exploration of Southern Africa*, 22; Wijnands, "Hortus Auriaci," 70.
70. Heniger, *Van Reede*, 69.
71. From the English translation of Guy Tachard, *Voyage de Siam des péres jésuites envoyés par le Roi aux Indes et à la Chine* (Paris, 1686): Tachard, *Voyage to Siam*, 63.
72. For the account of the journey, see Waterhouse, *Simon van der Stel's Journal*. The Dublin manuscript printed by Waterhouse appears to be a corrected copy in a fair hand of the manuscript now in the Plant Sciences Library of Oxford, Sherard 179, which would appear to be the original because of errors of orthography and spelling; it

was miscataloged and bound in the early twentieth century as "[Paul] Hermann's journal in Dutch of botanical expedition from Cape of Good Hope, Aug. 1685 to Jan.'86."

73. Gunn and Codd, *Botanical Exploration of Southern Africa;* for a printed and translated version of one of the sets of copies made from Claudius's originals (which are now missing), see Wilson, *Hove-Exalto,* and Rijssen, *Codex Witsenii.*
74. Vongsuravatana, *Jésuite à la cour de Siam,* 234.
75. Foss, "European Sojourn," 125, 130–31.
76. Kraft, "Mentzel, Couplet, Cleyer"; Demaerel, "Couplet and the Dutch."
77. Tachard, *Voyage to Siam,* 63; for an account, see Vongsuravatana, *Jésuite à la cour de Siam,* and the forthcoming study by Kapil Raj.
78. Heniger, *Van Reede,* 75.
79. Lindeboom, *Dutch Medical Biography,* 396–97; Schoute, *Geneeskunde in den dienst,* 170–73; quotation from Uragoda, *History of Medicine in Sri Lanka,* 68. Daelmans's work went through five editions in Dutch and two in German, by the last Dutch printing (1720) having been considerably enlarged to include, for example, remarks on the various sicknesses he had encountered on Ceylon, at Batavia, and throughout Coromandel. I have not been able to see a copy of any of the editions.
80. Heniger, *Van Reede,* 75–77, quotation at 77; Wijnands, "Hortus Auriaci," 66.
81. Sherard, *Paradisus Batavus.*
82. Heniger, "Hermann," 529.
83. British Library, Sloane MS 4036, fol. 21–22, printed in Wijnands, "Hortus Auriaci," 85.
84. Raven-Hart, *Germans in Dutch Ceylon,* 81.
85. Blussé, *Bitter Bonds,* 157.
86. For example, note the large semicircular orangerie constructed by Bentinck at "Sorgvliet" (which he took over from Cats): Bezemer-Sellers, "Bentinck Garden," 117.
87. Oldenburger-Ebbers, "Notes on Plants," 165.
88. Wijnands, "Hortus Auriaci," 76; Oldenburger-Ebbers, "Notes on Plants," 165–66.
89. Jong, "Netherlandish Hesperidies," 15.
90. Wijnands, "Hortus Auriaci," 67.
91. Birch, *History of the RS,* 169.
92. Oldenburger-Ebbers, "Notes on Plants," 173; also see Wijnands, "Hortus Auriaci," 78–79.
93. Wijnands, "Hortus Auriaci," 68–72. For more details, see Kuijlen, Oldenburger-Ebbers, and Wijnands, *Paradisus Batavus.*
94. Wijnands, "Hortus Auriaci," 79; Oldenburger-Ebbers, "Notes on Plants," 173.
95. Wijnands, *Botany of the Commelins,* 14–22.
96. Wijnands, "Hortus Auriaci," 70.
97. Leupe, *Rumphius;* Harting, "George Everard Rumphius"; *Rumphius gedenkboek, 1702–1902;* Sarton, "Rumphius, Plinius Indicus (1628–1702)"; Sirks, "Rumphius, the Blind Seer of Amboina"; Wit, *Rumphius Memorial Volume;* Rumphius, *Ambonese Curiosity Cabinet,* xxv–cxii; Rumphius, *Ambonsche lant-beschrijvinge,* xiii–xxxvii.
98. Wit, "Georgius Everhardus Rumphius," 7.
99. Rumphius, *Ambonese Curiosity Cabinet,* cii–cv, lxxxvi–lxxxvii, cii.

100. Rumphius, *Ambonese Curiosity Cabinet,* lxxv.
101. Harting, "George Everard Rumphius," 8; for a recent edition of the former work, see Rumphius, *Ambonsche lant-beschrijvinge.*
102. Rumphius, *Ambonese Curiosity Cabinet,* lxxv.
103. Haan, "Cleyer," 451–52.
104. Rumphius, *Amboinsche kruid-boek.*
105. Valentyn, *Oud en Nieuw Oost-Indiën.*
106. Rumphius, *D'Amboinsche rariteitkamer;* Rumphius, *Ambonese Curiosity Cabinet,* 3–5.
107. Waals, "Met boek en plaat," 224.
108. Merian, *Metamorphosis.*
109. On Merian, see Schiebinger, *The Mind Has No Sex?* 68–79; Davis, *Women on the Margins,* 140–202; and Kinukawa, "Art Competes with Nature"; on the evidence from her diary, Valiant, "Merian," 468–69.
110. Valiant, "Merian," 468–69.
111. On Schurman, see Baar et al., *Schurman,* and Baar et al., *Choosing.*
112. See esp. Saxby, *Quest for the New Jerusalem;* on the financial arrangements at Wieuwerd, 236, 245, on Merian, 264–65, on the butterflies, 384. On Van Deventer's activities at Wieuwerd, see Lieburg, *Van Deventer,* 25–44.
113. Saxby, *Quest for the New Jerusalem,* 381.
114. For details of their collections, see Bergvelt and Kistemaker, *Wereld binnen handbereik.*
115. Saxby, *Quest for the New Jerusalem,* 287; Landwehr, *Studies in Dutch Books with Coloured Plates,* 27.
116. On Surinam at the time of her visit, see Davis, *Women on the Margins,* 172–77; Saxby, *Quest for the New Jerusalem,* 273–88; and Postma, "Suriname"; Postma is the source of the figures.
117. Goedaert, *Metamorphosis Naturalis;* Landwehr, *Studies in Dutch Books with Coloured Plates,* 26.
118. This is undoubtedly the right explanation, which is offered in Waals, "Met boek en plaat," 224. It is known that "Mrs Merian" colored the plates of Rumphius's book for customers: Landwehr, *Studies in Dutch Books with Coloured Plates,* 28.
119. Valiant, "Merian," 472; Davis, *Women on the Margins,* 178–79; but see the doubts of Rumphius, *Ambonese Curiosity Cabinet,* lxxxix.
120. For one person's description, see Davis, *Women on the Margins,* 178–91.
121. Landwehr, *Studies in Dutch Books with Coloured Plates,* 28.

9. TRANSLATING WHAT WORKS

1. Dorssen, "Ten Rhijne," 149, noted that the Japanese shōgun had ordered the VOC to send a physician, but he was not able to trace the orders in the archives in Batavia; the main details are in Iwao, "Willem ten Rhijne," 16–17; more details have been discovered by Heniger, *Van Reede,* 54–56. I thank Shigehisa Kuriyama for telling me of Iwao's article and for sending me a copy. For a translation of it, I thank Tomomi

Kinukawa, who talked me through it, and Sumiyo Umekawa and Penny Barrett, who composed a written translation.

2. Indeed, Asian pilots, including Japanese ones, guided many early Dutch voyages, although by 1630 the situation had been reversed, with various employees of the VOC and free Dutch acting as guides on Japanese ships to Southeast Asia: Davids, "Navigeren in Azië," 19–22.
3. Kato, "Unification," 213–14.
4. The Treaty of Zaragoza forced the Spanish to give up their claims to the Moluccas; in 1565, however, they occupied Cebu in the Philippines and claimed that it was east of the line drawn by that treaty, and in 1571 they made Manila their main base of operations in the Far East.
5. Prakash, "Precious Metal Flows," 84–85. For a recent authoritative analysis of European-Asian exchange, see De Vries, "Connecting Europe and Asia," which concludes that "when Europeans rounded the Cape of Good Hope and sailed into Asian waters, they entered a very large market that was connected to Europe by a slender thread. The Asian market was so large, and its monetized sectors were expanding (if not growing, in the modern sense) so rapidly, that the establishment of an equilibrium with Europe was ever out of reach" (96–97).
6. For a narrative of events, see Boxer, *Christian Century*. For Japanese views of the Europeans, see Toby, "'Indianness' of Iberia."
7. Charles, "L'introduction," 93–98; Nogueira, "Luís de Almeida," 227–34.
8. Bowers, *Western Medical Pioneers*, 15–16.
9. Charles, "L'introduction," 98.
10. Nogueira, "Luís de Almeida," 235.
11. Vos, "From God to Apostate."
12. For an evocation of these events, see Endo, *Silence*.
13. Kato, "Unification," 213–21. For a recent narrative, see Milton, *Samurai William*.
14. For a fine overview of Dutch trade in Japan from 1641 to 1672, see Boxer, "Jan Compagnie in Japan," 148–55.
15. For a more detailed description of the island in English, see Goodman, *Japan*, 18–24.
16. Nogueira, "Luís de Almeida," 235.
17. Nakayama, *Japanese Astronomy*, 88–98, quotations at 91, 92.
18. Hanley, *Everyday Things*.
19. Totman, *Early Modern Japan*, 126.
20. Ooms, *Tokugawa Ideology*.
21. Otsuka, "Chinese Medicine in Japan," 328–34.
22. See, e.g., Goble, "Medicine and New Knowledge."
23. I thank Vivienne Lo for help in identifying these two figures.
24. Macé, "Évolution de la médecine japonaise." For accounts in English of Japanese medicine before 1600, see Fujikawa, *Japanese Medicine*, 1–33; Bowers, *Western Medical Pioneers*, 3–10; and Vos, "From God to Apostate."
25. Vos, "From God to Apostate," 25.
26. Nogueira, "Luís de Almeida," 235.
27. Yoshida, "Rangaku," 62–67.

28. Unless otherwise specified, the information on Schamberger is taken from Michel, *Schamberger.*
29. Bowers, *Western Medical Pioneers,* 30.
30. Beukers, "Dodonaeus in Japanese," 286–87; Bowers, *Western Medical Pioneers,* 30–31.
31. For this information I thank a former doctoral student, Tomomi Kinukawa, who relied on *Meijizen Nihon Igakushi,* 3:48, 1:24. See also Bowers, *Western Medical Pioneers,* 30.
32. See Wolfgang Michel, "Zu den im 17. Jahrhundert durch die Dejima-Chirurgen ausgestellten Ausbildungszeugnissen," *Studies in Languages and Cultures* (Faculty of Languages and Cultures, Kyushu University) 19 (2003): 137–155, at http://www.flc.kyushu-u.ac.jp/~michel/publ/aufs/69/69.htm.
33. Bowers, *Western Medical Pioneers,* 31.
34. Gulik, "Dutch Surgery in Japan," 43, and information supplied by Tomomi Kinukawa.
35. Fujikawa, *Japanese Medicine,* 42; Bowers, *Western Medical Pioneers,* 29; Beukers, "Dodonaeus in Japanese," 285; Beukers, "Invloed," 19.
36. Gulik, "Dutch Surgery in Japan," 44.
37. Beukers, "Dodonaeus in Japanese," 286–87.
38. Heniger, *Van Reede,* 55–56; Iwao, "Willem ten Rhijne," 17, citing "Daghregister van 't Comptoir Nagasackij," held in the Algemeen Rijksarchiven, The Hague, January 1676 [1667], 1 April 1668. The daily records of the Deishima factory have been published for the eighteenth century, but not yet for the seventeenth. I have not been able to check the original "Daghregister" in these instances, although I have read it for the period of Willem ten Rhijne's visit, below.
39. Iwao, "Willem ten Rhijne," 16.
40. Dam, *Beschryvinge van de Oostindische Companie,* bk. 2, pt. 1:453.
41. Boxer, "Jan Compagnie in Japan," 164–66, 180–81; Beukers, "Dodonaeus in Japanese," 287.
42. Heniger, *Van Reede,* 54–56.
43. Pieter Boddens, Daniel Godtke, and Samuel Manteau.
44. Van der Poel matriculated on 21 April 1667 and held his disputation on 2 May 1667: *Album studiosorum Lugduno Batavae;* Molhuysen, *Bronnen,* vol. 3.
45. Lindeboom gives a definite birth date of 2 January 1649 (Lindeboom, *Dutch Medical Biography,* 1622), another biographical entry gives a baptism of 19 January 1649 (*NNBW,* 6:1213), and the biography of him by D. Schoute in 1937 (*Opuscula selecta Neerlandicorum de arte medica,* 14:xl–li) says that the municipal archivist in Deventer placed the baptism on 14 January 1649; other biographical entries apparently incorrectly place his birth date about two years earlier: Dorssen, "Ten Rhijne," and L. S. A. M. von Römer's entry in *NNBW,* 9:861–63. I was unable to locate any entry on his baptism in the "Klapper hervormde dopen Deventer," Rijksarchief Deventer, although I was able to find a record of his mother's and brother's baptisms (the latter, "Berentien," on 6 May 1655).
46. Slee, *Illustre School te Deventer;* Fockema Andreae and Meijer, *Album studiosorum*

Franekerensis, no. 6918; *Album studiosorum Lugduno Batavae*, 3 March 1668; he presented a confession to the Reformed Church as a student in the Assenstraat at midwinter 1665 and took out a certificate of good standing on leaving for Leiden on 10 February 1668 ("Lidmaten-en attestatieboek," Rijksarchief Deventer).

47. Ten Rhijne, *Exercitatio physiologica in celebrem Hippocratis textum de vet. med.*; it was reprinted as Ten Rhijne, *Meditationes in magni Hippocratis textum xxiv. de veteri medicina.*
48. Amsterdam archive, P.A. 27/20: "Nomina Medicorum 1641–1753"; Carrubba, "Latin Document"; Dorssen, "Ten Rhijne," 142.
49. Ten Rhijne later dedicated his work on leprosy to Van Dam.
50. Heniger, *Van Reede*, 56. The details of his appointment are given in an article by G. F. Pop of 1869 and quoted from him by Dorssen, "Ten Rhijne," 142–43.
51. For accounts of Ten Rhijne, see Dorssen, "Ten Rhijne and Leprosy," 5–12; Dorssen, "Ten Rhijne"; *Opuscula selecta Neerlandicorum de arte medica*, 14:xl–li; Stiefvater, *Die akupunkter des Ten Rhijne*; Snelders, "Ten Rhyne."
52. "Discursus Navigatorius de Chymiae et Botaniae Antiqutate & Dignitate, Quem Anno MDCLXXIV. ix. Kalendarum Martii in Auditorio Anatomico, quod est apud Jacatrenses Batavos," Ten Rhijne, *Dissertatio de arthritide*, 193–269; on the anatomical theater, Dorssen, "Ten Rhijne," 150–52.
53. Buschof, *Moxibustion*, 3–9.
54. Busschof and Roonhuis, *Two Treatises*, 8–9, 73, 74, 75–76.
55. Algemeen Rijksarchief (hereafter: "ARA"), The Hague, Het Archief van de Nederlandse factorij in Japan (hereafter "NFJ"), "Deshima Dagregisters," vol. 87 (29 October 1673–19 October 1674).
56. Thursday, 4 October: ARA, NFJ, vol. 88 (20 October 1674–7 November 1675).
57. From the Japanese summary of the record, Iwao, "Willem ten Rhijne," 5–6.
58. Iwao, "Willem ten Rhijne," 7.
59. ARA, NFJ, vol. 88.
60. Iwao, "Willem ten Rhijne," 23–90; also see Michel, "Ten Rhijne," 81–83.
61. Iwao, "Willem ten Rhijne," 10.
62. They were: Nakayama Sakuzaemon, Nakajima Seizaemon, Namura Hachizaemon, Narabayashi Shin'uemon, Yokoyama Yosaburō, Tominanga Ichirobei, Motoki Shōdayū, and Kahuku Kichizaemon.
63. For thoughtful remarks about the use of pulse diagnosis in "Western" and East Asian medicine, see Kuriyama, *Expressiveness of the Body*, and Kuriyama, "Pulse Diagnosis"; see also Hsu, "Science of Touch, Part I," and Hsu, "Science of Touch, Part II."
64. Iwao, "Willem ten Rhijne," 23. Also see the English abstract of Otsuka, "Ten Rhyne"; Carrubba and Bowers, "First Treatise," 372.
65. My thanks to Keiko Daidoji, who drew my attention to this entry in Eishun Murakami, *Sango Binran* ("A Guide to Three Languages"). The character in question is translated into German as "geschwulsten," a swelling or tumor: see its use in Michel, "Ten Rhijne," 81.
66. Nogueira, "Luís de Almeida," 232.

67. According to Shigehisa Kuriyama, Japanese medicine in this period was coming to be very concerned with stagnation, "congealed obstructions," using terms such as *chō*, *shaku*, *ketsu*, and *kori:* Kuriyama, "Katakori."
68. Gulik, "Dutch Surgery in Japan," 42.
69. Carrubba and Bowers, "First Treatise," 374–75.
70. Iwao, "Willem ten Rhijne."
71. For an English translation of "Mantissa Schematica" and "De Acupunctura" of Ten Rhijne, *Dissertatio de arthritide*, 145–91, see Carrubba and Bowers, "First Treatise," and in German, Stiefvater, *Die akupunkter des Ten Rhijne;* Michel, "Ten Rhijne," 85–107 (for the "De acupunctura"); and Michel, "Ten Rhijne" (for "Mantissa"). For commentaries on them, see Michel, "Ten Rhijne"; Michel, "Frühe Westliche Beobachtungen"; Bivins, *Acupuncture*, 48–65; and Lu and Needham, *Celestial Lancets*, 270–76.
72. Carrubba and Bowers, "First Treatise," 376–78.
73. Carrubba and Bowers, "First Treatise," 378–86.
74. Carrubba and Bowers, "First Treatise," 376–77; the original is the "Mantissa Schematica" in Ten Rhijne, *Dissertatio de arthritide*, 145–68.
75. Carrubba and Bowers, "First Treatise," 374–75.
76. For modern thoughts on this subject, see Kuriyama, "Bloodletting."
77. Carrubba and Bowers, "First Treatise," 377, 378, citing Hippocrates, "On Winds," 4, and "Aphorisms," sec. 8, aphorism 6.
78. Carrubba and Bowers, "First Treatise," 375, 376; he believed that the text from which they never deviated was the "Daykio" (379).
79. Carrubba and Bowers, "First Treatise," 377, 386–96.
80. ARA, NFJ, vol. 88.
81. For an account of these journies, see Goodman, *Japan*, 25–31.
82. Clearly this was a subject where the Japanese thought that European surgical methods could help: one of the Japanese works on the "Caspar Method," *Komo geka*, contains a chapter on "rock in the breast"; another, by Katsuragawa Hoan—a two-volume work on surgery—also describes "rock in the breast"; and a third, by Chinzan Narabayashi (who knew Ten Rhijne), based on Paré, also contains a section on "breast cancer." My thanks to Tomomi Kinukawa for this information.
83. ARA, NFJ, vol. 88, 15 March–14 May; Iwao, "Willem ten Rhijne," 9–11.
84. "Soo wanneer daar een goede kuire wort aengedaan, het grootelÿx tot reputatie van onse genes kun streeken sal": ARA, NFJ, vol. 88, 29 May, 23 June, and quotation from 30 June; vol. 89 (7 November 1675–27 October 1676), 15 November.
85. ARA, NFJ, vol. 89, 5 December 1675, 10, 17 January 1676.
86. Spiegel, *Opera*.
87. Otsuka, "Ten Rhyne," 258; Iwao, "Willem ten Rhijne," 12. On the *Tahel Anatomia*, see Kuriyama, "Between Mind and Eye"; Kuriyama, "Visual Knowledge."
88. Camphuis later recorded that the trip to Edo that year cost much more than usual partly because it had two extra people along (six in all), one of them being Ten Rhijne: Dam, *Beschryvinge van de Oostindische Companie*, bk. 2, pt. 1:643–44.
89. ARA, NFJ, vol. 89, 27 February, 14 April.

90. ARA, NFJ, vol. 89, 20 April. The letter is written with a difficult syntax, so I am grateful for the advice of Dr. Iris Bruyn on how best to interpret it.
91. ARA, NFJ, vol. 89; the long letter was copied into the Daghregister on 20 April, and Camphuis's record of his request to "Brasman" is on 24 April.
92. Iwao, "Willem ten Rhijne," 13.
93. ARA, NFJ, vol. 89, 22, 25, 27, 28, 29, 30 April, 2, 4, 5, 6 May.
94. ARA, NFJ, vol. 89, 7, 28 June.
95. Coolhaas, *Generale missiven*, 4:159–60.
96. Dam, *Beschryvinge van de Oostindische Companie*, bk. 2, pt. 1:453.
97. Dorssen, "Ten Rhijne," 156–59.
98. Dorssen, "Ten Rhijne," 173.
99. List of works sent to Europe, in a letter of Ten Rhijne to Caspar Sibelius, from Batavia, 20 February 1680 [according to a letter of 25 March 1681, fol. 73, *sic* for 20 February 1681], fols. 52–53, and to the same, 7 November 1680, fol. 64, [November?] 1680, fol. 68, BL Sloane 2729. The last mentioned work on botany never appeared in print.
100. Haan, "Cleyer," 171.
101. Coolhaas, *Generale missiven*, 4:294–95.
102. Prakash, "Dutch East India Company," 192.
103. Veen, "How the Dutch Ran a Colony"; Andrade, "Political Spectacle."
104. For an account of his influence on the decision of the Emperor in this instance, see Attwater, *Schall*, 117–19. Attwater's book is "adapted" from Duhr, *Jésuite en Chine*, which is in turn adapted from Väth, *Johann Adam Schall von Bell.*
105. Shepherd, *Statecraft and Political Economy*, 47–90.
106. Wills, *Pepper, Guns and Parleys*, 28–104. The VOC did help the Qing to expel Cheng's son from Amoy and Quemoy.
107. Demaerel, "Couplet and the Dutch," 98–101.
108. Kraft, "Mentzel, Couplet, Cleyer," 196.
109. Demaerel, "Couplet and the Dutch," 102.
110. Golvers, *Rougement*, 530–31.
111. Kajdański, "Boym," 166–67.
112. Demaerel, "Couplet and the Dutch," 110–12; Begheyn, "Letter from Cleyer."
113. Pelliot, "Boym"; Kajdański, "Boym."
114. Boym, *Flora Sinensis*; I have consulted the British Library copy, which was once Joseph Banks's; for a description of this and other works by Boym, see Szczesniak, "Writings of Boym," 492–94. In Boym, *Briefve Relation*, 73, he announced that he was preparing a *Medicus Sinicus seu singularis Ars explorandi pulsum & prædicendi & futura Symptomata, & affectiones aegrotantium à multis ante Christum saeculis tradita, & apud Sinas conservata; quae quidem ars omnino est admirabilis & ab Europaeâ diversa.* This may be the work of Boym mentioned by Athanasius Kircher in his work on China published in 1667, although he had doubts about whether it had yet been printed: Kajdański, "Boym," 162–63. On the use Kircher made of Boym, see Szczesniak, "Writings of Boym."
115. It is possible that an anonymous work was this piece by Boym: *Les sécrets de la médecine des Chinois consistent en la parfaite connaissance du pouls* (Grenoble, 1671; trans-

lated into Italian in an edition at Milan, 1676, and into English for a London edition of 1707). For the best account of this work, see Grmek, *Reflects de la sphygmologie,* 59–64. Demaerel, "Couplet and the Dutch," 96–97; Kajdański, "Boym," 174–80; Kraft, "Mentzel, Couplet, Cleyer," 185–87. It is also just possible that at least some of the Chinese texts and charts possessed by Ten Rhijne, which he had translated in Japan, had first been collected by Boym and sent to Batavia before coming into Ten Rhijne's hands.

116. On the printing of the *Specimen,* a listing of the parts of the original manuscript, and a reproduction of the original dedication to the Heren XVII, see Kraft, "Mentzel, Couplet, Cleyer," 164–69, 189–94.
117. The charge was made by Gottlieb Siegfried Bayer in his *Museum Sinicum* (Petersburg, 1730) (see Kraft, "Mentzel, Couplet, Cleyer," 159–61), with a more accepting view taken by Kraft and Pelliot, "Boym," 145; but doubts remain about Cleyer's character, as in Kajdański, "Boym," 165–67.
118. In the British Library copy, each of the four surviving internal sections is independently paginated, and different copies of the *Specimen* apparently bound them together in different order. I follow the discussion of Szczesniak, "Writings of Boym," 508–13, and Kajdański, "Boym," both of whom also used the British Library copy. For an account of the contents based on a different copy, which apparently did not contain the study of medical simples, see Grmek, *Reflects de la sphygmologie,* 70–72. On the work's publication, which Mentzel directed through another member of the Academiae and a physician of Frankfurt, Sebastian Scheffer, see Kraft, "Mentzel, Couplet, Cleyer," 162–69.
119. Lu and Needham, *Celestial Lancets,* 277.
120. This is an independently paginated section of fifty-four pages in Cleyer, *Specimen medicinae Sinicae.* It is not mentioned on the *Specimen's* title page. For the best recent study of the book, which emphasizes the importance of this botanical treatise, see Kajdański, "Boym."
121. Kajdański, "Boym," 169–70, 180–85.
122. Grmek, *Reflects de la sphygmologie,* 72.
123. At the same time, it should be noted that Ten Rhijne has no good opinion of what was published as the *Specimen,* which he heard of by a letter of 12 December 1681: Ten Rhijne to Sibelius, 25 February 1683, fols. 130–131, Sloane 2729.
124. Szczesniak, "Floyer"; Grmek, *Reflects de la sphygmologie*; Hsu, "Science of Touch, Part I"; Hsu, "Science of Touch, Part II"; Lu and Needham, *Celestial Lancets,* 282–84.
125. Sloane 2729, letter of Ten Rhijne to Sibelius, 25 March 1681.
126. Sloane 2729, letter of Ten Rhijne to Sibelius, 25 February 1683, fols. 130–131.
127. Translated and copied into the letter books of the Royal Society: LBC.8, fols. 447–448.
128. Royal Society of London, LBC.9, 374–75, copy of letter of Ten Rhijne to Secretary Francis Aston, 25 August 1683.
129. Sloane 2729, letters of Ten Rhijne to Sibelius, 25 March 1681 and 25 February 1683, fols. 73, 130–131. This seems to confirm the account of De Haan that a German asso-

ciate of Cleyer's named Vogel left Batavia in May 1681 with a manuscript yet suggests that it was not what would be the *Specimen* of 1682 but the *Clavis* of 1686: Haan, "Cleyer," 454. Although it is often assumed that Couplet brought the manuscript with him to Europe when he returned from China in 1682, the manuscript seems to have arrived via Cleyer's postal network: Foss, "European Sojourn," 116–19; for the refutation, see Kraft, "Mentzel, Couplet, Cleyer," 170–72.

130. Kraft, "Mentzel, Couplet, Cleyer," 169–75.
131. Boym and Cleyer, *Clavis medica ad Chinarum doctrinam de pulsibus.*
132. Pelliot doubted whether the *Specimen* and the *Clavis* were both by Boym (Pelliot, "Boym"), a view supported by Grmek (Grmek, *Reflects de la sphygmologie*) and taken up by Lu and Needham, *Celestial Lancets*, 277–86. That view has definitively been laid to rest by Kajdański, "Boym." On the other hand, Szezesniak thought of the *Specimen* and the *Clavis* as two editions of Boym's works (Szczesniak, "Writings of Boym," esp. 513), which is equally false, despite some overlapping text and illustrations shared between the two.
133. Sloane 2729, Ten Rhijne to Sibelius, 25 February 1683, fol. 130–131.
134. The excerpt (apparently from early 1684) is printed in Michael Bernhard Valentini in *India literata; seu dissertationes epistolicae de plantis, arboribus, . . . etc.* (Frankfurt, 1716), 432a–b, and quoted in Szczesniak, "Writings of Boym," 516–17. Also see Kraft, "Mentzel, Couplet, Cleyer," 166–67. For an older version of Cleyer's attack on Ten Rhijne's reputation, see Dorssen, "Ten Rhijne," 215–28. Since Cleyer did not yet have a copy of the *Clavis* in hand, he presumably was referring to the *Specimen* and Scheffer's help with its publication.
135. Bruijn, "Ship's Surgeons," 125.
136. On Cleyer, see Haan, "Cleyer."
137. Dorssen, "Ten Rhijne," 174–78; Coolhaas, *Generale missiven*, 4:385, 418.
138. Letter of Ten Rhijne to Sibelius, Sloane 2729, 25 March 1681, fol. 73. The request was no doubt for Thomas Erpenius, *Arabicae linguae tyrocinium*, edited by J. Golius (Leiden, 1656), or one of the other editions; exactly which lexicon he had in mind is unknown. The other work in question is: *Les six voyages de Jean Baptiste Tavernier: qu'il a fait en Turquie, en Perse, et aux Indes, pendant l'espace de quarante ans*, first published in 1676, which Ten Rhijne later sent back since he already had a copy.
139. For instance, letter of Ten Rhijne to Sibelius, 17 December 1681, fol. 92, Sloane 2729, responding to a letter of November 1680. In a letter of 21 March 1682/3, to his friend Johannes Groenevelt, and translated and copied into the letter books of the Royal Society, Ten Rhijne explains that the printing had been stopped because of the death of the two members of the Heren XVII "who had the management of them," Backer and Kelies, and because of "the Jelous Persecutions of Some Envious Persons (who do nothing themselves, and hinder others)" (Royal Society, LBC.8, fols. 446–48).
140. Royal Society, Letter Book, LBC.8.240–242; summarized but not printed in Hall and Hall, *CHO*, 13:368. The letter was written before he knew for certain about the publication of the *Specimen* and again underlines that he considered much of what went into that book to be his own.
141. Birch, *History of the RS*, 4:119–20, 122.

142. Wrop, *Briefwisseling van Constantijn Huygens,* letters no. 7001 (6 January 1676) and no. 7011 (24 February 1676).
143. Wrop, *Briefwisseling van Constantijn Huygens,* vol. 6, *1663–1687,* 368–69, 371. His story is published in Temple, *Miscellanea,* 189–238: "An Essay upon the Cure of the Gout by Moxa. Written to Monsieur de Zulichem. Nimmeguen June 18, 1677"; Temple's story is summarized in Rosen, "Sir William Temple and the Therapeutic Use of Moxa for Gout in England" (although Rosen was unaware that "Zulichem" was Huygens).
144. A. Leeuwenhoek, letter of 14 May 1677, *Philosophical Transactions* 12, no. 136 (June 25, 1677), 899–895 [sic for 905]. Leeuwenhoek possessed a copy of Busschof's book: Palm, "Italian Influences on Antoni van Leeuwenhoek," 161–62.
145. Busschof and Roonhuis, *Two Treatises;* Buschof, *Moxibustion.*
146. Sydenham, *Tractatus de podagra,* trans. as Sydenham, *Treatise Concerning the Gout.* The passage occurs at the third paragraph from the end, probably as a late addition.
147. Bontius, *Tropische geneeskunde,* 286, my translation. For a slightly different translation, see 287, and Carrubba and Bowers, "First Treatise," 394, which is also quoted in Lu and Needham, *Celestial Lancets,* 270.
148. It is not in Sherard 186. Ten Rhijne later comments on this passage and corrects it, noting especially that the "stylus" is rather a needle, and gold rather than bronze, and that it is not driven through the organ and out the other side. See Carrubba and Bowers, "First Treatise," 394.
149. *Philosophical Transactions* 4, no. 49 (19 July 1669): 983–986, quotation at 984. It is not clear where the original account, "communicated in French by M.I.," originated. Although the account was in French, the author might have been Dutch or German.
150. Aston also wrote promptly to Robert Plot, in Oxford, informing him of Ten Rhijne's letter: Gunther, *Early Science in Oxford,* 12:26.
151. Sloane 2729, Groenevelt to Sibelius, 24 January, 31 March 1682, fols. 109, 116–117.
152. Birch, *History of the RS,* 4:140, 143.
153. Sloane 2729, 27 June 1682, fols. 122–123.
154. Sloane 2729, fol. 140 (30 May 1683). The Royal Society ordered on 9 May 1683 that a copy of the book be placed in their library: Birch, *History of the RS,* 4:204. So it was, and remains, although there are no annotations in it.
155. Ten Rhijne, *Dissertatio de arthritide.*
156. *Philosophical Transactions* 13, no. 148 (10 June 1683), 221 [sic for 222]–235, quotation from 221. The attribution of the summary to "Mr. Gold" is from a letter of Aston to Plot: Gunther, *Early Science in Oxford,* 12:35.
157. Ten Rhijne, *Verhandelinge van het Podagra en Vliegende Jicht . . . Chineese Japanse wijse om door het branden van Moxa en het steken met een Goude Naald, alle ziekten en voornamelijk het Podagra te genesen. Door den Heer W. Ten Rhyne, Med. Doct. etc.* (Amsterdam: Jan ten Hoorn, 1684).
158. De Beer, *Correspondence of John Locke,* letters 785 (4 October 1684), 2:635, and 858 (4 August 1686), 3:23.
159. Guyonnet, "Saint Sébastien, patron des acupuncteurs"; Feucht, "Akupunktur"; Rosenberg, "Ten Rhyne's De acupuntura"; Michel, "Frühe Westliche Beobachtungen."

160. Rosen, "Lorenz Heister on Acupuncture," 387.
161. For his life, see Haberland, *Kaempfer.*
162. Bowers and Carrubba, "Doctoral Thesis of Kaempfer," 303–10. This is apparently another reference to *cander.*
163. For his comments on acupuncture and moxibustion, see Kaempfer, *Exotic Pleasures,* 108–38. For his botanical studies, see Werger-Klein, "Kaempfer, Botanist."
164. "Medecijnen doctor ten dienste der E. Comp.": notarized certificate of his marriage to Juffrow Elizabeth Wassenborgh on 13 December 1681, Sloane MS 2729, fol. 96. According to a letter of Sibelius to Locke of 9 September 1687, Ten Rhijne apologized for not writing because he had been in mourning for his deceased wife (who therefore probably died in early 1687): De Beer, *Correspondence of John Locke,* letter 961, 3:266. It is known that he married again, but nothing more.
165. Ten Rhijne, *Verhandelinge van de Asiatise melaatsheid.* For an English translation, *Opuscula selecta Neerlandicorum de arte medica,* vol. 14.
166. Dorssen, "Ten Rhijne and Leprosy," 257–59; Dorssen, "Ten Rhijne," 186.
167. Ten Rhijne, *Schediasma;* Ten Rhijne, "Account of the Cape," 4:768–82.
168. Blussé, *Bitter Bonds,* 154.

10. THE REFUSAL TO SPECULATE

1. On Heidanus, see the entry by Hans van Ruler, Bunge et al., *Dictionary,* 1:397–402.
2. Sassen, "Intellectual Climate," 8; Ruestow, *Physics,* 75–88; *Willem III en de Leidse Universiteit,* 12–13; Israel, *Radical Enlightenment,* 29.
3. For an excellent account, if a bit too intent on connecting the "radical Enlightenment" to Spinoza, see Israel, *Radical Enlightenment.*
4. Quoted in Israel, *Radical Enlightenment,* 381; for the Bekker affair, see 378–92, and the entry of Andrew Fix in Bunge et al., *Dictionary,* 1:74–77.
5. Vermij, *Calvinist Copernicans,* 357.
6. Israel, *Radical Enlightenment,* esp. 286–94, 307–27.
7. See Vermij, *Bernard Nieuwentijt;* Berkel, Helden, and Palm, *History of Science in the Netherlands,* 77–79, 543–45; Vermij, *Calvinist Copernicans,* 349–58; and Vermij's entry in Bunge et al., *Dictionary,* 733–36.
8. The English translation of 1718 was heavily cut with, for instance, most of the passages demonstrating the supernatural origins of the Bible omitted; the French translation of 1725 was based on the English version; the German translation of 1722 (with another translation into German later) is said to be more faithful to Nieuwentijt's original.
9. Bunge et al., *Dictionary,* 734.
10. Sassen, "Intellectual Climate," 8.
11. Some of the notes Boerhaave wrote about his life shortly before his death were printed (in large italic letters) as part of the funeral oration printed by Schultens (Schultens, *Oratio Academica*); they were later gathered as the "Commentariolus," which is reprinted, along with an English translation, in Lindeboom, *Herman Boerhaave,* 377–86.

12. Clercq, *Sign of the Lamp.*
13. See the entry on Senguerd by Gerhard Berthold Wiesenfeldt in Bunge et al., *Dictionary,* 2:911–14.
14. Lindeboom, *Herman Boerhaave,* 379. Three of the disputations, "De mente humana," delivered from November 1686 to July 1687, are not considered philosophically important, but they do indicate ways in which he had absorbed powerful anti-Cartesian elements from Senguerd: see Thijssen-Schoute, *Nederlands Cartesianisme,* 255.
15. Sassen, "Intellectual Climate," 4–6.
16. Kegel-Brinkgreve and Luyendijk-Elshout, *Boerhaave's Orations,* 32, 40, 51, 52, 53. Also see the excellent introduction to the oration, 18–31.
17. A point noticed in both Sassen, "Intellectual Climate," 5, and Kegel-Brinkgreve and Luyendijk-Elshout, *Boerhaave's Orations,* 20.
18. Lindeboom, *Herman Boerhaave,* 379.
19. Thijssen-Schoute, *Nederlands Cartesianisme,* 54. The title of the thesis was "On nature" (*De natura*).
20. Thijssen-Schoute, *Nederlands Cartesianisme,* 255; Sassen, *Geschiedenis van de Wijsbegeerte,* 224; Lindeboom, *Herman Boerhaave,* 24.
21. On De Volder, see the entry by Gerhard Berthold Wiesenfeldt in Bunge et al., *Dictionary,* 2:1041–44.
22. Israel, *Radical Enlightenment,* 310–11.
23. Sassen, "Intellectual Climate," 10.
24. Remarks from the "Commentariolus," in Lindeboom, *Herman Boerhaave,* 379–80, 382.
25. Israel, *Radical Enlightenment,* 127–28; Kegel-Brinkgreve and Luyendijk-Elshout, *Boerhaave's Orations,* 54–55.
26. Lindeboom, *Herman Boerhaave,* 27.
27. A compilation of Hermann's botanical lessons, published by his pupil, Lothar Zumbach as *Florae Lugduno Batavae flores* in 1690: Heniger, "Some Botanical Activities," 2–3.
28. Lindeboom, *Herman Boerhaave,* 380–82.
29. Thijssen-Schoute, *Nederlands Cartesianisme,* 254.
30. "E Gelrica academia Leydam reverso accidit insonti, nec opinanti, aliquid, unde praevidebat," in Lindeboom, *Herman Boerhaave,* 382.
31. *Willem III en de Leidse Universiteit,* 13–14.
32. On Pitcairne, see esp. Guerrini, "Tory Newtonians"; Guerrini, "Pitcairne"; Brown, "Medicine in the Shadow of the *Principia.*" On Newtonianism as a philosophy that made the new science more comfortable for theologians, see esp. Jacob, "Christianity and the Newtonian Worldview."
33. "O calumnia! Quis Te pestium teterrimam peperit." For the episode on the boat, see Schultens, *Oratio academica,* 22–24.
34. Lindeboom, *Herman Boerhaave,* 382–83.
35. Ultee, "Politics of Appointment," 171–72.
36. Indeed, the oration has even been described as "Baconian": Kegel-Brinkgreve and Luyendijk-Elshout, *Boerhaave's Orations,* 55.

37. Kegel-Brinkgreve and Luyendijk-Elshout, *Boerhaave's Orations*, quotations at 68, 80, 69, 73, 78, 70, 80.
38. Kegel-Brinkgreve and Luyendijk-Elshout, *Boerhaave's Orations*, 95, 96, 99, 102.
39. Kegel-Brinkgreve and Luyendijk-Elshout, *Boerhaave's Orations*, 114.
40. Translation of the *Institutes* as Boerhaave, *Academical Lectures*, 65–66.
41. Boerhaave, *Academical Lectures*, 63, 71, 74.
42. The strongest argument for his religious concepts guiding his scientific ideas is Knoeff, *Boerhaave*, although I remain doubtful.
43. Luyendijk-Elshout, "Mechanicisme contra vitalisme," 20–22.
44. Burton, *Life and Writings of Boerhaave*, 53; Lindeboom, *Herman Boerhaave*, 386.
45. See Kegel-Brinkgreve and Luyendijk-Elshout, *Boerhaave's Orations*, 121–44.
46. Ultee, "Politics of Appointment."
47. Heniger, "Some Botanical Activities," 4–5. See also Hunger, "Boerhaave als natuurhistoricus," 36–37.
48. Quotation from Wijnands, "Hortus Auriaci," 62; for details on his activities, see esp. Heniger, "Some Botanical Activities," and Stearn, "Influence of Leyden," 147–50.
49. Sassen, *Geschiedenis van de wijsbegeerte*, 224.
50. Turner, *Essay on Gleets*.
51. "Vie de M. Herman Boerhaave," in Boerhaave, *Institutions de médecine de Mr. Herman Boerhaave, traduites du Latin en françois par M. de la Mettrie* (Paris: Huart et Briasson, 1740), 42–45.
52. Thomson, "La Mettrie."
53. Wellman, "La Mettrie's *Institutions*."
54. On Mandeville's family tree, see Kaye, *Fable*, 1:xvii, 2:380–85.
55. Roldanus, "Adraen Paets."
56. Paets held a position in the Maas admiralty from 1660, and he sat on the board from 1669 (Roldanus, "Adraen Paets," 137); Captain Bernard Verhaar served in the Rotterdam admiralty from 1628 and was still active in 1668 (Kaye, *Fable*, 2:383). See also Dekker, "Private Vices, Public Virtues," 481–83.
57. On Paets's opposition to the local Quakers, see Hull, *Benjamin Furly and Quakerism in Rotterdam*, 41–42, 146, 182–83.
58. See esp. Fix, *Prophecy and Reason*, 215–46; Slee, *Rijnsburger Collegianten*, 238–66; Israel, *Radical Enlightenment*, 342–58.
59. Kerkhoven and Blom, "De la Court en Spinoza," 160.
60. Thijssen-Schoute, *Uit de republiek der letteren*, 112–14; Hazewinkel, "Pierre Bayle à Rotterdam," 21–25.
61. For a recent account, see Israel, *Radical Enlightenment*, 331–41.
62. Horne, *Social Thought of Mandeville*, 22 (who unfortunately finds the roots of this in Jansenism rather than in Calvinism). See also Dekker, "Private Vices, Public Virtues," 493; Kaye, *Fable*, 1:xlv–lii; and Burtt, *Virtue Transformed*, 132–33, 143.
63. For an interpretation of Mandeville's views as due not so much to Republicanism, Cartesianism, and the medical and natural historical emphasis on matters of fact as to his intimate acquaintance with the works of Spinoza (although this is only inferred), see Israel, *Radical Enlightenment*, 623–27.

64. Kaye, *Fable*, 1:xviii.
65. Thijssen-Schoute, "Diffusion européene des idées de Bayle," 192.
66. Kaye, *Fable*, 1:xviii–xix; Molhuysen, *Bronnen*, 4:*206. The quotations are from Mandeville's short account of the thesis in his *Treatise on the Hypochondriack and Hysterick Passions* (1711), 120–21.
67. Dekker, "Private Vices, Public Virtues," esp. 483–84; a full account of the events is in Hazewinckel, *Geschiedenis van Rotterdam*, 1:247–63.
68. Dekker, "Private Vices, Public Virtues," 485, 494; also see the account of the affair in Israel, *Dutch Republic*, 857–58, who sees it as an example of the decline of the party of True Freedom.
69. He was named at a meeting of the College of Physicians on 17 November 1693 as one of seven practicing in London without the college's permission: "Annals of the College of Physicians of London," vol. 6, fols. 88–89; also in Clark, *History of the Royal College*, 2:450.
70. Groenevelt, *Tutus cantharidum in medicinâ usus internus*; Ward, "Unnoted Poem"; Cook, *Trials*, 199–200.
71. For the quotations in this and the next paragraph, see Kaye, *Fable*, 1:17–37.
72. Goldsmith, *Private Vices*, 44.
73. Mandeville, *Treatise the Hypochondriack and Hysterick*, xi.
74. Mandeville, *Treatise the Hypochondriack and Hysterick*, 35, 38, 60, 55, 56–58, 59, 172–205, 44, ix.
75. Mandeville, *Treatise the Hypochondriack and Hysterick*, 61–62, 71, 68.
76. Mandeville, *Treatise the Hypochondriack and Hysterick*, 332.
77. For further thoughts, see McKee, "Honeyed Words."
78. For example, *True Meaning of the Fable of the Bees*, 106.
79. Mandeville, *Treatise the Hypochondriack and Hysterick*, xiii.
80. Beukers, "Clinical Teaching."
81. Klein, "Experimental History"; Knoeff, *Boerhaave*.
82. Perhaps this was Lord Cartaret, secretary of state and Walpole's chief Whig rival.
83. Kaye, *Fable*, 1:401, 253, 260–68, 269, 272.
84. Kaye, *Fable*, 1:405, 408, 409, 411–12.
85. Kaye, *Fable*, 1:384–85.
86. See esp. Geyer-Kordesch, "Passions and the Ghost."
87. Baglivi, *Practice of Physick*, 9, 15. This is a translation of his *De praxi medica ad priscam observandi rationem revocanda* . . . (Rome: Typis Dominici Antonii Herculis, 1696).

11. CONCLUSIONS AND COMPARISONS

1. Nelson, *On the Roads*; Huff, *Rise of Early Modern Science*; Kuriyama, *Expressiveness of the Body*; G. E. R. Lloyd, *The Ambitions of Curiosity*.
2. Vries, "Country Estate Immortalized," quotations at 97.
3. Among the wealth of important studies on the regions mentioned in this and the next paragraph, I mention only a few of the best known. For Italy, see esp. Eamon, *Science and Secrets*; Meli, "Authorship and Teamwork"; and Freedberg, *Eye of the Lynx*.

4. Cañizares-Esguerra, "Iberian Science"; Pérez, *Asclepio renovado;* Bueno, *Los Señores;* Goodman, *Power and Penury.*
5. Moran, "German Prince-Practitioners"; Smith, *Business of Alchemy;* Munt, "Impact of Dutch Cartesian Medical Reformers."
6. Koerner, *Linnaeus.*
7. Wellman, *Making Science Social;* Solomon, *Public Welfare, Science and Propaganda;* Howard, "Guy de la Brosse and the Jardin."
8. Lux, *Patronage and Royal Science;* Stroup, *Company of Scientists;* Brown, *Scientific Organizations.*
9. Webster, *Great Instauration;* Frank, *Harvey;* Hunter, *Establishing the New Science;* Hoppen, *Common Scientist;* Cunningham, "Sir Robert Sibbald."
10. As an example of one historian of science who came to acknowledge this point, see Westfall, "Science and Technology."
11. Iliffe, "Foreign Bodes"; Cook and Lux, "Closed Circles?"; Secord, "Knowledge in Transit."
12. Boyle, *Usefulnesse of Experimental Naturall Philosophy,* 2:220–21.
13. For an example of studies of patronage, see Biagioli, "Galileo's System of Patronage"; Moran, *Patronage and Institutions;* and Findlen, *Possessing Nature.*
14. Nakayama, *Japanese Astronomy,* 91.

Bibliography

MANUSCRIPT SOURCES

BRITISH LIBRARY

Sloane 1235: "Collegium Chymicum Secretum / A / D. Carolo de Maes apud Lugdunenses." 1675 and 1676

Sloane 2729, containing letters of Willem ten Rhijne

PLANT SCIENCES LIBRARY OF OXFORD

Sherard 179: listed as Hermann's journal in Dutch of botanical expedition from Cape of Good Hope, August 1685 to January 1686; it is the journal of the first expedition to Namaqualand led by Simon van der Stel

Sherard 186: "Jacobi Bontii medici arcis ac civitatis Bataviae Novae in Indiis ordinarii exoticorum Indicorum centuria prima"

ROYAL SOCIETY OF LONDON

Letter Books, LBC.8 and 9

ROYAL COLLEGE OF PHYSICIANS, LONDON

"Annals of the College of Physicians of London"

ALGEMEEN RIJKSARCHIEF, THE HAGUE

Het Archief van de Nederlandse factorij in Japan (hereafter "NFJ"), "Deshima Dagregisters," vols. 87 (29 October 1673–19 October 1674), 88 (20 October 1674–7 November 1675), and 89 (7 November 1675–27 October 1676)

VOC-archives, 1.04.02, inventory no. 147

MUNICIPAL ARCHIVES

Amsterdam, "Ordonantie boeck, 1638–1718," P.A. 27/1; "Stukken betreffende de oprichting 1638; met retroacta, 1550–1637," P.A. 27/14; "Nomina et series inspectorum Collegii medici, 1637–1797," P.A. 27/17; "Nomina medicorum, 1641–1753," P.A. 27/20

Delft, Afd. 1, no. 1981, "Chirurgÿnsgilde-boek met de ampliatien, alteratien en 'gevolge van dien zeedert den Jaeren 1584. tot den Jaeren 1749 Inclusive'"
Deventer, "Klapper hervormde dopen Deventer"; "Lidmaten- en attestatieboek"
Haarlem, Collegium Medico-Pharmaceuticum, K. en O. II, 106; K. en O. II, 112
The Hague, "Statuta autographa collegii medicorum Hagiensium. anno 1658," 488/1 (a printed version of which is at 488/6, "Bijlagen bij de 'acta collegii,' 1658–1774")
Utrecht, "Vol. I: "Acta et decreta Collegii medici, 1706–1783," 41/suppl. 144

WEB SITES

Materials by Wolfgang Michel, http://www.flc.kyushu-u.ac.jp/~michel/publ/articles.html
Jacques de Groote, on the "Libri picturati," http://www.tzwin.be/libri%20picturati.html

REFERENCE SOURCES

Aa, A. J. van der, ed. *Biographisch woordenboek der Nederlanden.* 12 vols. Haarlem: J. J. van Brederode, 1858–78.

Album studiosorum academiae Lugduno Batavae, 1575–1875. The Hague: Martinus Nijhoff, 1875.

Album studiosorum academiae Rheno-Traiectinae, 1636–1886. Utrecht: J. L. Beijers, 1886.

Bunge, Wiep van, Henri Krop, Bart Leeuwenburgh, Han van Ruler, Paul Schuurman, and Michiel Wielema, gen. eds. *The Dictionary of Seventeenth- and Eighteenth-Century Dutch Philosophers.* 2 vols. Bristol: Thoemmes, 2003.

Eloy, N. F. J. *Dictionnaire historique de la médecine ancienne et moderne: ou mémoires disposés en ordre alphabétique pour servir a l'histoire de cette science.* 4 vols. Mons: H. Hoyois, 1778.

The European Union Encyclopedia and Directory, 2004. London: Europa, 2003.

Fockema Andreae, S. J., and Th. J. Meijer. *Album studiosorum Academiae Franekerensis (1585–1811, 1816–1844).* Franeker: T. Wever, 1969.

Gillispie, Charles C., gen. ed. *Dictionary of Scientific Biography.* 18 vols. New York: Scribners, 1970–90.

Health at a Glance: OECD Indicators, 2003. Paris: Organisation for Economic Co-operation and Development, 2003.

An Intermediate Greek-English Lexicon, Founded upon the Seventh Edition of Liddell and Scott's Greek-English Lexicon. 1889. Oxford: Clarendon Press, 1968.

Jensma, G. Th., F. R. H. Smit, and F. Westra, eds. *Universiteit te Franeker, 1585–1811: bijdragen tot de geschiedenis van de Friese hogeschool.* Leeuwarden: Fryske Akademy, 1985.

Kernkamp, G. W. *Acta et dectreta senatus: vroedschapsresolutien en andere bescheiden betreffende de Utrechtsche academie.* Utrecht: Broekhoff, 1936.

Ketner, F. *Album promotorum academiae Rheno-Trajectinae, 1636–1815.* Utrecht: Broekhoff, 1816.

Lindeboom, G. A. *Dutch Medical Biography: A Biographical Dictionary of Dutch Physicians and Surgeons, 1475–1975.* Amsterdam: Rodopi, 1984.

Molhuysen, P. C. *Bronnen tot de geschiedenis der Leidsche universiteit.* 4 vols. The Hague: Martinus Nijhoff, 1913–20.

Nieuw Nederlandsch Biographisch Woordenboek. 10 vols. Leiden: A. W. Sijthoff, 1911–37. [Cited as NNBW.]

Siegenbeek van Heuklelom-Lamme, C. A., and O. C. D. Idenburg-Siegenbeek van Heukelom. *Album scholasticum academiae Lugduno-Batavae, MDLXXV–MCMXL.* Leiden: E. J. Brill, 1941.

Slee, Jacob Cornelius van. *De illustre school te Deventer, 1630–1878.* The Hague: Martinus Nijhoff, 1916.

De Universiteit te Leuven, 1425–1975. Leuven: Universitaire Pers, 1976.

Vries, Jan de. *Nederlands etymologisch woordenboek.* Leiden: E. J. Brill, 1992.

PRIMARY SOURCES AND EDITIONS CITED

Adelmann, Howard B. *Marcello Malpighi and the Evolution of Embryology.* 5 vols. Ithaca, NY: Cornell University Press, 1966.

Albertus Magnus. *On Animals: A Medieval "Summa Zoologica."* Trans. and ed. K. F. Kitcell, Jr., and I. M. Resnick. 2 vols. Baltimore: Johns Hopkins University Press, 1999.

Aristotle. *Nichomachean Ethics.* Trans. J. A. K. Thomson. Baltimore: Johns Hopkins University Press, 1955.

———. *The Works of Aristotle Translated into English.* Rev. ed. Trans. Benjamin Jowett. Oxford: Oxford University Press, 1961.

Bacon, Francis. *Francis Bacon: A Selection of His Works.* Ed. Sidney Warhaft. New York: Odyssey, 1965.

Baglivi, Geo. *The Practice of Physick.* 1696. London: Andr. Bell, 1704.

Barbette, Paulus. *Pest-Beschryving.* 1655. Amsterdam: Jacob Lescailje, 1680.

Barlaeus, Caspar. *Le marchand philosophe de Caspar Barlaeus: un éloge du commerce dans Hollande du siècle d'or: étude, texte et traduction "Du Mercator sapiens."* Trans. and ed. Catherine Secretan. Paris: Honoré Champion, 2002.

———. *Rerum per octennium in Brasilia.* Amsterdam: Joannis Blaeu, 1647.

Birch, Thomas. *The History of the Royal Society of London for Improving of Natural Knowledge from Its First Rise.* Intro. A. Rupert Hall. 1756–57. London: Johnson Reprint, 1968.

Blasius, Gerard. *Anatome animalium.* Amsterdam: Joannis à Someren, 1681.

Boerhaave, Herman. *Academical Lectures.* 6 vols. London: W. Innys, 1743–51.

Bontekoe, Cornelis. *Alle de . . . werken van de heer Corn. Bontekoe.* Amsterdam: Jan ten Hoorn, 1689.

———. *Tractaat van het excellenste kruyd thee / Treatise about the Most Excellent Herb Tea.* In *Opuscula selecta neerlandicorum de arte medica,* 14:118–465. Amsterdam: Sumptibus Societatis, 1937.

Bontius, Jacobus. *An Account of the Diseases, Natural History, and Medicines of the East Indies.* London: T. Noteman, 1769.

———. *De medicina indorum.* Leiden: Franciscus Hackius, 1642.

———. *Tropische geneeskunde / On Tropical Medicine.* In *Opuscula selecta Neerlandicorum de arte medica.* Vol. 10. Amsterdam: Sumptibus Societatis, 1931.

Bos, Erik-Jan. *The Correspondence between Descartes and Henricus Regius.* Utrecht: Zeno, 2002.

Boyle, Robert. "Large Act of Anatomy by de Bils." In *The Works of Robert Boyle*, vol. 1, ed. Michael Hunter and Edward B. Davis, 41–50. London: Pickering and Chatto, 1999.

———. *Usefulnesse of Experimental Naturall Philosophy.* Oxford: Hen. Hall, 1663.

Boym, Michael. *Briefve Relation.* Paris: Sebastian Cramoisy and Gabriel Cramoisy, 1654.

———. *Flora sinensis.* Vienna: Matthaeus Rictius, 1656.

Boym, Michael, Andreas Cleyer, and Philippe Couplet. *Clavis medica ad Chinarum doctrinam de pulsibus.* [Nuremberg], 1686.

Breyne, Jakob. *Exoticarum aliarumque minus cognitarum plantarum centuria prima.* Danzig: D. F. Rhetius, 1678.

[Browne, Richard, prob. author]. *Chirurgorum comes.* London: Edw. Jones, 1687.

Burton, Robert. *Anatomy of Melancholy.* Oxford: H. Crippes, 1621.

Burton, William. *Life and Writings of Herman Boerhaave.* 2nd ed. London: H. Lintot, 1746.

Buschof, Hermann. *Erste Abhandlung über die Moxibustion in Europa: Das Genau Untersuchte und Auserfundene Podagra, Vermittelst Selbst Sicher-Eigenen Genäsung und Erlösenden Hülff-Mittels.* Ed. Michel Wolfgang. Heidelberg: Karl F. Haug, 1993.

Busschof, Herman, and Hermann Roonhuis. *Two Treatises, the One Medical, of the Gout . . . the Other Partly Chirurgical, Partly Medical.* London: H. C., 1676.

Clauder, Gabriel. *Methodus balsamandi corpora humana.* Jena: Oan Bielckium, 1679.

Cleyer, Andreas. *Specimen medicinae Sinicae.* Frankfurt: Joannis Petri Zubrodt, 1682.

Clusius, Carolus. *Aromatum, et simplicium aliquot medicamentorum apud Indos nascentium historia.* Antwerp: Christopher Plantin, 1567.

———. *Aromatum, et simplicium liquot medicamentorum apud Indos nascentium historia.* Intro. M. de Jong, D. A., and Wittop Koning, D.A. 1567. Nieuwkoop: B. de Graaf, 1963.

———. *Exoticorum libri decem: quibus animalium, plantarum, aromatum, aliorumque peregrinorum fructuum historiae describuntur.* Leiden: Plantiana Raphelengius, 1605.

Coolhaas, W. Ph. *Generale missiven van gouverneurs generaal en raden aan Heren XVII der Vereenigde Oostindische Compagnie.* Rijks Gescheidkundige Publicatiën: Groote ser., 9 vols. The Hague: M. Nijhoff, 1960–88.

[Court, Pieter Cornelis de la.] *Interest van Holland, ofte gronden van Hollands-welvaren; aangewezen door V. D. H.* Amsterdam: n.p., 1662.

Dam, Pieter van. *Beschryvinge van de Oostindische Companie.* Rijks Gescheidkundige Publicatiën: Groote ser., vols. 63, 68, 74, 76, 83, 87, 96. Ed. F. W. Stapel and C. W. Th. van Boetzelaer. The Hague: M. Nijhoff, 1927–54.

Davisson, William. *Philosophia pyrotechnica.* Paris: Joan Bessin, 1640.

———. *Le cours de chymie.* Amiens: Michel du Neuf-Germain, 1675.

De Beer, E. S., ed. *The Correspondence of John Locke.* 8 vols. Oxford: Clarendon Press, 1976–89.

De Bils, Louis. *Coppy of a Certain Large Act (Obligatory) of Yonker Louis de Bils.* Trans. John Pell. London, 1659.

Descartes, René. *Oeuvres de Descartes.* Ed. Charles Adam and Paul Tannery. Rev. ed., 1897–1913. Paris: Vrin/C.N.R.S., 1964–76.

———. *The Philosophical Writings of Descartes*. 3 vols. Ed. and trans. John Cottingham, Robert Stoothoff, and Dugald Murdoch. Cambridge: Cambridge University Press, 1985–91. [Cited as CSM.]

Dioscorides. *The Greek Herbal of Dioscorides*. Ed. Robert R. Gunther. 1934. New York: Hafner, 1959.

Endo, Shusaku. *Silence*. Trans. William Johnson. 1969. London: Peter Owen, 1996.

Engen, John van, trans., and Heiko A. Oberman, intro. *Devotio Moderna: Basic Writings*. New York: Paulist Press, 1988.

Erasmus, Desiderius. *The Praise of Folly*. Trans. Hoyt Hopewell Hudson. 1941. Princeton, NJ: Princeton University Press, 1974.

Goedaert, Johannem. *Metamorphosis naturalis*. 3 vols. Middelburgh: Jacques Fierens, 1660–69.

Groenevelt, Joannes. *Tutus cantharidum in medicinâ usus internus*. London: R. E. Prost, 1703.

Hall, A. Rupert, and Marie Boas Hall, eds. *The Correspondence of Henry Oldenburg*. 13 vols. Madison: University of Wisconsin Press; London: Mansell; Taylor and Francis, 1965–86.

Hermann, Paulus. *Paradisus Batavus, continens plus centum plantas affabré aere incisas et descriptionibus illustratas*. [Ed. William Sherard.] Leiden: Abrahamum Elzevier, 1698.

Huxley, T. H. "On the Natural History of the Man-Like Apes." In *Collected Essays*, vol. 7: *Man's Place in Nature and Other Anthropological Essays*. 1868. London: Macmillan, 1894.

Kaempfer, Engelbert. *Exotic Pleasures: Fascicle III: Curious Scientific and Medical Observations*. Trans. and intro. Robert W. Carrubba. Carbondale: Southern Illinois University Press, 1996.

Kaye, F. B. ed. *Bernard Mandeville, The Fable of the Bees: Or, Private Vices, Publick Benefits*. 1924. 2 vols. Oxford: Clarendon Press, 1957.

Kegel-Brinkgreve, E., and A. M. Luyendijk-Elshout, trans. *Boerhaave's Orations*. Sir Thomas Browne Institute. Leiden: E. J. Brill/Leiden University Press, 1983.

Le Roy, Louis. *Of the Interchangeable Course or Variety of Things in the Whole World*. Trans. R[obert] A[shley]. London: Charles Yetsweirt, 1594.

Liddell and Scott. *An Intermediate Greek-English Lexicon*. 7th ed. 1889. Oxford: Oxford University Press.

Lindeboom, G. A., ed. and comp. *Het cabinet van Jan Swammerdam (1637–1680)*. Amsterdam: Rodopi, 1980.

———, ed. *The Letters of Jan Swammerdam*. Amsterdam: Swets and Zeitlinger, 1975.

———, ed. *Observationes anatomicae Collegii privati Amstelodamensis*. Nieuwkoop: B. de Graaf, 1975.

———, ed. and comp. *Ontmoeting met Jan Swammerdam*. Ontmoetingen met mystici, 3. Kampen: J. H. Kok, 1980.

Linschoten, Jan Huygen van. *Itinerario: voyage ofte schipvaert van Jan Huygen van Linschoten naer Oost ofte Portugaels Indien, 1579–1592*. Amsterdam: Cornelis Claesz., 1596.

———. *Itinerario: voyage ofte schipvaert van Jan Huygen Van Linschoten naer Oost ofte Portugaels Indien, 1579–1592*. Ed. H. Kern. 5 vols. The Hague: Martinus Nijhoff, 1910–39.

———. *Itinerario: Voyage ofte schipvaert naer Oost ofte Portugaels Indien, 1579–1592*. Gen. ed. H. Terpstra and H. Kern, original ed. 1910–39. 2nd ed. 3 vols. The Hague: Nijhoff, 1955–57.

———. *The Voyage of John Huygen van Linschoten to the East Indies: From the Old English Translation of 1598*. Ed. Arthus Coke and P. A. Tiele. 2 vols. London: Hakluyt Society, 1885.

Lipsius, Justus. *Of Constancie*. 1594. New Brunswick, NJ: Rutgers University Press, 1939.

Lloyd, G. E. R., ed., and J. Chadwick, W. N. Mann, and et al., trans. *Hippocratic Writings*. Harmondsworth: Penguin, 1978.

[Maets, Carolus de]. *Chemia rationalis*. Leiden: Jacobum Mocquee, 1687.

Mandeville, Bernard. *A Treatise of the Hypochondriack and Hysterick Diseases. In Three Dialogues*. 2nd ed., enl. 1711. London: J. Tonson, 1730.

Merian, Maria Sibylla. *Metamorphosis insectorum Surinamensium*. Amsterdam: By the author, 1705.

Montaigne, Michele de. *The Complete Works of Montaigne*. Trans. Donald M. Frame. 1943. Stanford, CA: Stanford University Press, 1989.

Orta, Garcia da. *Colloquies on the Simples and Drugs of India*. New ed. Lisbon, 1895. Trans. with intro. Clements Markham. London: Henry Sotheran, 1913.

———. *Colóquios dos simples, e drogas he cousas mediçinais de India*. Goa: Joannes de Endem, 1563.

———. *Dell'historia de i semplici aromati, et altre cose*. Ed. Carolus Clusius. Venice: Francesco Ziletti, 1589.

Pepys, Samuel. *The Diary of Samuel Pepys*. Transcribed and ed. Latham, Robert, and Williams Matthews. 10 vols. London: G. Bell and Sons, 1972.

Piso, William. *De Indiae utriusque re naturali et medica libri quatuordecim*. Amsterdam: Lodovicum et Danielem Elzevirios, 1658.

Piso, William, and Georg Marcgraff. *Historia naturalis Brasiliae*. Leiden: Franciscum Hackium, Lud. Elzevirium, 1648.

Plato. *The Dialogues of Plato*. 3rd ed. Trans. Benjamin Jowett. Oxford: Oxford University Press, 1892.

Reede, Henricus van, tot Drakenstein. *Hortus indicus Malabaricus*. 12 vols. Amsterdam: Joannis van Someren and Joannis van Dyck, 1678–1703.

Rhijne, Willem ten. *Dissertatio de arthritide*. London: R. Chiswell, 1683.

Römer, L. S. A. M. von, ed. *Epistolae Jacobi Bontii*. Batavia: Gualtherium Kolff, 1921.

Rumphius, Georgius Everhardus. *Amboinsche kruid-boek*. Ed. Joannes Burmann. 6 parts in 4 vols. Amsterdam, The Hague, and Utrecht: François Changuion et al., 1741–55.

———. *The Ambonese Curiosity Cabinet*. Trans. and ed. E. M. Beekman. New Haven and London: Yale University Press, 1999.

———. *D'Amboinsche rariteitkamer*. Amsterdam: François Halma, 1705.

———. *De generale lant-beschrijvinge van het Ambonse gouvernement: ofwel de ambonsche lant-beschrijvinge*. Transcription, notes, glossary, and new biography by W. Buijze. The Hague: W. Buijze, 2001.

Sanches, Francisco. *That Nothing Is Known*. Ed. and trans. Douglas F. S. Thomson, intro. Elaine Limbrick. Cambridge: Cambridge University Press, 1988.

Schrader, Justus. *Observationes et historiae*. Amsterdam: Abraham Wolfgang, 1674.
Schultens, Albert. *Oratio academica*. Leiden: Johannem Luzac, 1738.
Seneca. *Four Dialogues*. Ed. C. D. N. Costa. Warminster: Aris and Phillips, 1994.
Sloane, Hans. *A Voyage to the Islands Madera, Barbados, Nieves, S. Christophers and Jamaica*. London: For the author, 1707.
Smith, Captain John. *The Complete Works*. 3 vols. Ed. Philip L. Barbour. Chapel Hill: University of North Carolina Press, 1986.
Spiegel, Adrian. *Opera*. Ed. Johannes Ant. Vander Linden. Amsterdam: Iohannem Blaeu, 1645.
Spinoza, Benedict de. *Ethics*. Trans. Andrew Boyle, rev. G. H. R. Parkinson. London: Dent, 1993.
Swammerdam, Jan. *Book of Nature; or, The History of Insects*. Trans. Thomas Flloyd, rev. and improved by notes from Reamur and others by John Hill. London: C. G. Seyffert, 1758.
———. *Bybel der natuure /Biblia naturae*. Ed. Herman Boerhaave. Latin trans. Hieronimus David Gaubius. 2 vols. Leiden: Isaak Severinus, 1737.
———. *Ephemeri vita*. Amsterdam: Abraham Wolfgang, 1675.
———. *Historia insectorum generalis*. 2 vols. Utrecht: Meinardus van Dreunen, 1669.
———. *Miraculum naturae sive uteri muliebris fabrica*. Leiden: Severinus Matthaeus, 1672.
Sydenham, Thomas. *Tractatus de podagra et hydrope*. London: Walter Kettilby, 1683.
———. *Treatise Concerning the Gout*. Trans. John Drake. London: Walter Kettilby, 1684.
[Tachard, Guy]. *A Relation of the Voyage to Siam Performed by Six Jesuits*. London: J. Robinson and A. Churchil, 1688.
[Temple, Sir William]. *Miscellanea*. London: Edw. Gellibrand, 1680.
Ten Rhijne, Willem. "An Account of the Cape of Good Hope and the Hottentotes." In *A Collection of Voyages and Travels*, 4:768–82. London: A. and J. Churchill, 1732.
———. *Dissertatio de arthritide*. London: R. Chiswell, 1683.
———. *Exercitatio physiologica in celebrem Hippocratis textum de vet. med.* Leiden: n.p., 1669.
———. *Meditationes in magni Hippocratis textum xxiv. de veteri medicina*. Leiden: J. à Schuylenburgh, 1672.
———. *Schediasma de promontorio bonae spei, ejusve tractus incolis Hottentottis*. Schaffhausen: n.p., 1686.
———. *Verhandelinge van de Asiatise melaatsheid*. Amsterdam: A. van Someren, 1687.
The Travels of Marco Polo. Trans. Ronald Latham. Harmondsworth: Penguin Books, 1958.
The True Meaning of the Fable of the Bees. London: William and John Innys, 1726.
Tyson, Edward. *Orang-Outang, Sive Homo Sylvestris: Or, The Anatomy of a Pygmie Compared with That of a Monkey, an Ape, and a Man*. London: Thomas Bennet and Daniel Brown, 1699.
Valentyn, François. *Oud en Nieuw Oost-Indiën*. Dordrecht and Amsterdam: Joannes van Braam and Gerard onder de Linden, 1724.
Vondel, Joost van den. *Gijsbrecht van Amstel*. Trans. and intro. Kristiaan P. G. Aercke. Ottawa: Dovehouse, 1991.

Warnsinck, J. C. M., ed. *Reisen van Nicolaus de Graaff: gedan naar alle gewesten des werelds beginnende 1639 tot 1687 incluis.* 1930. De Linschoten-Vereeniging 33. The Hague: Nijhoff, 1976.

Waterhouse, Gilbert, ed. *Simon van der Stel's Journal of His Expedition to Namaqualand, 1685–6.* Dublin University Press Series. London: Longmans, Green, 1932.

Wittop Koning, D. A., intro. *Facsimile of the First Amsterdam Pharmacopoeia, 1636.* Nieuwkoop: B. de Graaf, 1961.

Wrop, J. A., ed. *De briefwisseling van Constantijn Huygens (1608–1687).* The Hague: Martinus Nijhoff, 1917.

Yonge, James. *Currus triumphalis.* London: J. Martin, 1679.

SECONDARY SOURCES CITED

Adas, Michael. *Machines as the Measure of Men: Science, Technology, and Ideologies of Western Dominance.* Ithaca, NY: Cornell University Press, 1989.

Åkerman, Susanna. *Queen Christina of Sweden and Her Circle: The Transformation of a Seventeenth-Century Philosophical Libertine.* Leiden: E. J. Brill, 1991.

Alpers, Svetlana. *The Art of Describing: Dutch Art in the Seventeenth Century.* Chicago: University of Chicago Press, 1983.

Andel, M. A. van. *Chirurgijns, vrije meesters, beunhazen en kwalkzalvers: de chirurgijnsgilden en de praktijk der heelkunde (1400–1800).* 1941. The Hague: Martinus Nijhoff, 1981.

———. "Introduction." In *Opuscula selecta Neerlandicorum de arte medica.* Vol. 14. Amsterdam: Nederlansche Tijdschfit voor Geneeskunde, 1937.

———. "Plague Regulations in the Netherlands." *Janus* 21 (1916): 410–44.

Andrade, Tonio. "Political Spectacle and Colonial Rule: The *Landdag* on Dutch Taiwan, 1629–1648." *Itinerario* 21 (1997): 57–93.

Appadurai, Arjun, ed. *The Social Life of Things: Commodities in Cultural Perspective.* Cambridge: Cambridge University Press, 1986.

Ashworth, William B., Jr. "Emblematic Natural History of the Renaissance." In *The Cultures of Natural History,* ed. N. Jardine, J. A. Secord, and E. Spary, 17–37. Cambridge: Cambridge University Press, 1995.

———. "The Habsburg Circle." In *Patronage and Institutions: Science, Technology and Medicine at the European Court, 1500–1750,* ed. Bruce T. Moran, 137–67. Woodbridge: Boydell, 1991.

———. "Natural History and the Emblematic World View." In *Reappraisals of the Scientific Revolution,* ed. David C. Lindberg and Robert S. Westman, 303–32. Cambridge: Cambridge University Press, 1990.

Attewell, Guy. "India and the Arabic Learning of the Renaissance: The Case of Garcia D'Orta." M.A. thesis, Warburg Institute, London: University of London, 1997.

Attwater, Rachel. *Adam Schall: A Jesuit at the Court of China, 1592–1666.* London: Geoffrey Chapman, 1963.

Auerbach, Erich. *Mimesis: The Representation of Reality in Western Literature.* Trans. Willard R. Trask. 1957. Princeton, NJ: Princeton University Press, 1974.

Augerius Gislenus Busbequius, 1522–1591: Vlaams humanist en keizerlijk gezant. Koninklijke Vlaamse Academie voor Wetenschappen, Letteren en Schone Kunsten van België. Brussels: Paleis der Academiën, 1955.

Aveling, James H. *The Chamberlens and the Midwifery Forceps: Memorials of the Family and an Essay on the Invention of the Instrument.* 1882. New York: Arno, 1977.

Baar, Mirjam de, Machteld Löwensteyn, Marit Monteiro, and A. Agnes Sneller, eds. *Anna Maria Schurman (1607–1678): een uitzonderlijk geleerde vrouw.* Zutphen: Walburg, 1992.

———. *Choosing the Better Part: Anna Maria Van Schurman (1607–1678).* Trans. Lynne Richards. Dordrecht: Kluwer, 1996.

Backer, Christian M. E. de. *Farmacie te Gent in de late middeleeuwen: apothekers en receptuur.* Hilversum: Verloren, 1990.

Bakhtin, Mikhail M. *Rabelais and His World.* Trans. Helene Iswolsky. Cambridge, MA: MIT Press, 1968.

Balbian Verster, J. F. L. de. *Burgemeesters van Amsterdam in de zeventiende en achttiende eeuw.* Zutphen: W. J. Thieme en Cie, 1932.

Baldwin, Martha. "Alchemy and the Society of Jesus in the Seventeenth Century: Strange Bedfellows?" *Ambix* 40 (1993): 41–64.

Barend-van Haeften, Marijke. *Oost-Indië gespiegeld: Nicolaas de Graaf, een schrijvend chirurgijn in dienst van de VOC.* Zutphen: Walburg, 1992.

Barge, J. A. J. *De oudste inventaris der oudste academische anatomie in Nederland.* Leiden: H. E. Stenfert Kroese's, 1934.

Barnett, Frances Mason. "Medical Authority and Princely Patronage: The Academia Naturae Curiosorum, 1652–1693." Ph.D. diss., University of North Carolina at Chapel Hill, 1995.

Baron, Hans. *The Crisis of the Early Italian Renaissance: Civic Humanism and Republican Liberty in an Age of Classicism and Tyranny.* 2nd ed. Princeton, NJ: Princeton University Press, 1966.

Barrera, Antonio. "Local Herbs, Global Medicines: Commerce, Knowledge, and Commodities in Spanish America." In *Merchants and Marvels: Commerce, Science, and Art in Early Modern Europe,* ed. Pamela H. Smith and Paula Findlen, 163–81. New York: Routledge, 2002.

Barth, Michael. "Huygens at Work: Annotations in His Rediscovered Personal Copy of Hooke's *Micrographia.*" *Annals of Science* 52 (1995): 601–13.

Bassani, Ezio, and Letizia Tedeschi. "The Image of the Hottentot in the Seventeenth and Eighteenth Centuries: An Iconographic Investigation." *Journal of the History of Collections* 2 (1990): 157–86.

Baumann, E. D. *François Dele Boe Sylvius.* Leiden: E. J. Brill, 1949.

———. "Job van Meekren." *Nederlandsch Tijdschrift voor Geneeskunde* 67 (1923): 456–79.

———. *Johan van Beverwijck in leven en werken geschetst.* Dordrecht: Revers, 1910.

———. *Uit drie eeuwen Nederlandse geneeskunde.* Amsterdam: H. Meulenhoff, n.d.

Bayly, C. A. *Empire and Information: Intelligence Gathering and Social Communication in India, 1780–1870.* Cambridge: Cambridge University Press, 1996.

Beagon, Mary. *Roman Nature: The Thought of Pliny the Elder.* Oxford: Clarendon Press, 1992.

Beekman, E. M. *Troubled Pleasures: Dutch Colonial Literature from the East Indies, 1600–1950*. Oxford: Clarendon Press, 1996.

Begheyn, Paul. "A Letter from Andries Cleyer, Head Surgeon of the United East India Company at Batavia, to Father Philips Couplet, S.J., Missionary in China, 1669." *Lias* 20 (1993): 245–49.

Benedict, Barbara M. *Curiosity: A Cultural History of Early Modern Inquiry*. Chicago: University of Chicago Press, 2001.

Berendts, Ans. "Carolus Clusius (1526–1609) and Bernardus Paludanus (1550–1633): Their Contacts and Correspondence." *Lias* 5 (1978): 49–64.

Bergvelt, Ellinoor, and Renée Kistemaker, eds. *De wereld binnen handbereik: Nederlandse kunst- en ariteitenverzamelingen, 1585–1735*. Zwolle: Waanders /Amsterdams Historisch Museum, 1992.

Berkel, Klaas van. *Citaten uit het boek der natuur: opstellen over Nederlandsewetenschapsgeschiedenis*. Amsterdam: B. Bakker, 1998.

———. "Citaten uit het boek der natuur: zeventiende-eeuwse Nederlandse naturaliënkabinetten en de ontwikkeling van de natuurwetenschap." In *De wereld binnen handbereik: Nederlandse kunst-en rariteitenverzamelingen, 1585–1735*, ed. Ellinoor Bergvelt and Renée Kistemaker, 169–91. Zwolle: Waanders Uitgevers/Amsterdams Historisch Museum, 1992.

———. "Descartes in debat met Voetius: de mislukte introductie van het Cartesianisme aan de Utrechtse universiteit (1639–1645)." *Tijdschrift voor de Geschiedenis der Geneeskunde, Natuurwetenschappen, Wiskunde en Techniek* 7 (1984): 4–18.

———. *Isaac Beeckman (1588–1637) en de mechanisering van het wereldbeeld*. Amsterdam: Rodopi, 1983.

Berkel, Klaas van, Albert van Helden, and Lodewijk Palm, eds. *A History of Science in the Netherlands: Survey, Themes, and Reference*. Leiden: E. J. Brill, 1999.

Berman, Morris. *The Reenchantment of the World*. Ithaca, NY: Cornell University Press, 1981.

Beukers, Harm. "Acid Spirits and Alkaline Salts: The Iatrochemistry of Franciscus Dele Boë, Sylvius." *Sartoniana* 12 (1999): 39–58.

———. "Clinical Teaching in Leiden from Its Beginning until the End of the Eighteenth Century." In *Clinical Teaching Past and Present*, ed. H. Beukers and J. Moll, 139–52. Amsterdam: Rodopi, 1989.

———. "Dodonaeus in Japanese: Deshima Surgeons as Mediators in the Early Introduction of Western Natural History." In *Dodonaeus in Japan: Translation and the Scientific Mind in Tokugawa Period*. Ed. W. F. vande Walle, 282–97. Leuven: Leuven University Press, 2001.

———. "Invloed van de Zuidnederlanse chirurgie op de Noord-Nederlandse in de zestiende eeuw." Manuscript.

———. "Het laboratorium van Sylvius." *Tijdschrift voor de Geschiedenis der Geneeskunde, Natuurwetenschappen, Wiskunde en Techniek* 3 (1980): 28–36.

———. "Mechanistiche principes bij Franciscus Dele Boë, Sylvius." *Tijdschrift voor de Geschiedenis der Geneeskunde, Natuurwetenschappen, Wiskunde en Techniek* 5 (1982): 6–15.

———. "Terug naar de wortels." Oegstgeest: Drukkerij de Kempenaer, 1989.

Bezemer-Sellers, Vanessa. "Clingendael: An Early Example of a Le Nôtre Style Garden in Holland." *Journal of Garden History* 7 (1987): 1–42.

———. "The Bentinck Garden at Sorgvliet." In *The Dutch Garden in the Seventeenth Century*, ed. John Dixon Hunt, 99–129. Dumbarton Oaks Colloquium on the History of Landscape Architecture, 12. Washington, DC: Dumbarton Oaks, 1990.

———. "Condet Aurea Saecula: The Gardens of Frederik Hendrik." In *Princely Display: The Court of Frederick Hendrik of Orange and Amalia Van Solms*, comp. and ed. Marika Keblusek and Jori Zijlmans, 126–42, 223–24. The Hague: Historical Museum; and Zwolle: Waanders, 1997.

Bénézet, Jean-Pierre. *Pharmacie et médicament en Méditerranée occidentale (treizième–seizième siècles)*. Paris: Honoré Champion, 1999.

Biagioli, Mario. *Galileo Courtier: The Practice of Science in the Culture of Absolutism*. Chicago: University of Chicago Press, 1993.

———. "Galileo's System of Patronage." *History of Science* 28 (1990): 1–62.

———. "Knowledge, Freedom, and Brotherly Love: Homosociality and the Accademia Dei Lincei." *Configurations* 3 (1995): 139–66.

Bivins, Roberta. *Acupuncture, Expertise, and Cross-Cultural Medicine*. Houndsmill, Basingstoke: Palgrave, 2000.

Black, Jeremy. *The British Abroad: The Grand Tour in the Eighteenth Century*. New York: St. Martin's Press, 1992.

———. *The British and the Grand Tour*. London: Croom Helm, 1985.

———. *A Military Revolution? Military Change and European Society, 1550–1800*. Basingstoke: Macmillan Education, 1991.

Black, William George. *Folk-Medicine: A Chapter in the History of Culture*. London: Folklore Society, 1883.

Blécourt, Willem de. *Termen van Toverij: De veranderende betekenis Van Toverij in Noordoost-Nederland tussen de zestiende en twintigste eeuw*. Nijmegen: SUN, 1990.

———. "Witch Doctors, Soothsayers and Priests: On Cunning Folk in European Historiography and Tradition." *Social History* 19 (1994): 285–303.

Bloch, Dorete. "Whaling in the Faroe Islands, 1584–1994: An Overview." In *The North Atlantic Fisheries, 1100–1976: National Perspectives on a Common Resource*, ed. Poul Holm, David J. Starkey, and Jón Th. Thór, 49–61. Studia Atlantica, 1. Esbjerg: Fiskerimuseet, 1996.

Blom, H. W. "*Felix Qui Potuit Rerum Cognoscere Causas*: Burgersdijk's Moral and Political Thought." In *Franco Burgersdijk (1590–1635): Neo-Aristotelianism in Leiden*, ed. E. P. Bos and H. A. Krop, 119–50. Studies in the History of Ideas in the Low Countries. Amsterdam: Rodopi, 1993.

Blom, H. W., and I. W. Wildenberg, eds. *Pieter de la Court in zijn tijd: aspecten van een veelzijdig publicist (1618–1685)*. Amsterdam: APA—Holland University Press, 1986.

Blussé, Léonard. *Strange Company: Chinese Settlers, Mestizo Women and the Dutch in VOC Batavia*. Verhandelingen KITLV 122. Dordrecht: Foris, 1986.

———. *Bitter Bonds: A Colonial Divorce Drama of the Seventeenth Century*. Trans. Diane Webb. Princeton, NJ: Markus Wiener, 2002.

Boheemen, F. C. van, and Th. C. J. van der Heijden. *De westlandse rederijkerskamers in de zestiende en zeventiende eeuw.* Amsterdam: Rodopi, 1985.

Bonfield, Lloyd. "Affective Families, Open Elites and Strict Family Settlements in Early Modern England." *Economic History Review,* 2nd ser., 39 (1986): 341–54.

Bono, James J. *The Word of God and the Languages of Man: Interpreting Nature in Early Modern Science and Medicine,* vol. 1: *Ficino to Descartes.* Madison: University of Wisconsin Press, 1995.

Boogaart, Ernst van den. *Civil and Corrupt Asia: Images and Text in the "Itinerario" and the "Icones" of Jan Huygen van Linschoten.* Chicago: University of Chicago Press, 2003.

Boogman, J. C. "The *Raison d' État* Politician Johan de Witt." *Low Countries History Yearbook* 11 (1978): 55–78.

Boorstin, Daniel J. *The Discoverers.* New York: Random House, 1983.

Booy, Engelina Petronella de. *Kweekhoven der wijshied: basis- en vervolgonderwijs in de steden van de provincie Utrecht van 1580 tot het begin der negentiende eeuw.* Stichtse Historische Reeks, 5. Zutphen: Walburg, 1980.

———. *De weldaet der scholen: het plattelandsonderwijs in de provincie Utrecht van 1580 tot het begin der negentiende eeuw.* Stichtse Historische Reeks, 3. Utrecht: n.p., 1977.

Borkenau, Franz. "The Sociology of the Mechanistic World-Picture." Zur Soziologie des mechanistischen Weltbildes. 1932. Trans. Richard W. Hadden. *Science in Context* 1 (1987): 109–27.

Bos, E. P., and H. A. Krop, eds. *Franco Burgersdijk (1590–1635): Neo-Aristotelianism in Leiden.* Studies in the History of Ideas in the Low Countries. Amsterdam: Rodopi, 1993.

Bosman-Jelgersma, Henriëtte A., "Augerius Clutius (1578–1636), apotheker, botanicus en geneeskundige." *Kring voor de Geschiedenis van de Pharmacie in Benelux, Bulletin* 64 (1983): 55–62.

———. "Clusius en Clutius." *Kring voor de Geschiedenis van de Pharmacie in Benelux, Bulletin* 64 (1983): 6–10.

———. "Dirck Outgaertsz Cluyt." *Farmaceutisch Tijdschift voor België* 53 (1976): 525–48.

———. *Pieter van Foreest: de Hollandse Hippocrates, 1521–1597.* Heiloo: Vereniging Oud Heiloo, 1984.

———. *Vijf eeuwen Delftse apothekers: een bronnenstudie over de geschiedenis van de farmacie in een Hollandse stad.* Amsterdam: Ronald Meesters, 1979.

———, ed. *Petrus Forestus Medicus.* Krommenie: Knijnenberg, 1996.

Bosman-Jelgersma, Henriëtte A., and H. L. Houtzager. "De balseming van Prins Willem van Oranje." *Kring voor de Geschiedenis van de Pharmacie in Benelux, Bulletin* 66 (1984): 47–50.

Bossy, John. "Moral Arithmetic: Seven Sins into Ten Commandments." In *Conscience and Casuistry in Early Modern Europe,* ed. Edmund Leites, 214–34. Cambridge: Cambridge University Press, 1988.

Bourdieu, Pierre. *Distinction: A Social Critique of the Judgment of Taste.* Trans. Richard Nice. Cambridge, MA: Harvard University Press, 1984.

Bouwsma, William J. *Concordia Mundi: The Career and Thought of Guillaume Postel (1510–1581).* Cambridge, MA: Harvard University Press, 1957.

———. "Lawyers and Early Modern Culture." *American Historical Review* 78 (1973): 303–27.

Bowers, John Z. *Western Medical Pioneers in Feudal Japan*. Baltimore: Johns Hopkins University Press, 1970.

Bowers, John Z., and Robert W. Carrubba. "The Doctoral Thesis of Engelbert Kaempfer on Tropical Diseases, Oriental Medicine, and Exotic Natural Phenomena." *Journal of the History of Medicine* 25 (1970): 270–310.

Boxer, Charles R. *The Christian Century in Japan, 1549–1650*. Berkeley: University of California Press, 1951.

———. *The Dutch in Brazil, 1624–1654*. Oxford: Clarendon Press, 1957.

———. "Jan Compagnie in Japan, 1672–1674, or Anglo-Dutch Rivalry in Japan and Formosa." In *Papers on Portuguese, Dutch, and Jesuit Influences in Sixteenth- and Seventeenth-Century Japan: Writings of Charles Ralph Boxer*, comp. Michael Moscato, 147–211. Washington, DC: University Publications of America, 1979.

———. *Two Pioneers of Tropical Medicine: Garcia d'Orta and Nicolás Monardes*. London: Wellcome Historical Medical Library/Hispanic and Luso-Brazilian Councils, 1963.

Bredekamp, Horst. *The Lure of Antiquity and the Cult of the Machine: The Kunstkammer and the Evolution of Nature, Art and Technology*. German ed. 1993. Trans. Allison Brown. Princeton, NJ: Markus Wiener, 1995.

Bredero, G. A. *The Spanish Brabanter: A Seventeenth-Century Dutch Social Satire in Five Acts*. Trans. H. David Brumble, III. Binghamton, NY: Center for Medieval and Early Renaissance Texts and Studies, 1982.

Brienen, Rebecca P. "Art and Natural History at a Colonial Court: Albert Eckhout and Georg Marcgraf in Seventeenth-Century Dutch Brazil." Ph.D. diss., Northwestern University, 2002.

Brienen, T., ed. *De nadere reformatie en het gereformeerd piëtisme*. The Hague: Boekcentrum, 1989.

Brockbank, William. "Sovereign Remedies: A Critical Depreciation of the Seventeenth-Century London Pharmacopoeia." *Medical History* 8 (1964): 1–14.

Brockliss, Laurence, and Colin Jones. *The Medical World of Early Modern France*. Oxford: Clarendon Press, 1997.

Broecke, Steven Vanden. "The Limits of Influence: Astrology at Louvain University, 1520–1580." Ph.D. diss., Katholieke Universiteit Leuven, 2000.

Brooke, John Hedley. *Science and Religion: Some Historical Perspectives*. The Cambridge History of Science Series. Cambridge: Cambridge University Press, 1991.

Brooke, John Hedley, Margaret J. Osler, and Jitse M. van der Meer, eds. *Science in Theistic Contexts: Cognitive Dimensions*. Osiris. Chicago: University of Chicago Press, 2001.

Brothwell, Don, and Patricia Brothwell. *Food in Antiquity: A Survey of the Diet of Early Peoples*. Expanded ed. 1969. Baltimore: Johns Hopkins University Press, 1998.

Brotton, Jerry. *The Renaissance Bazaar: From the Silk Road to Michelangelo*. Oxford: Oxford University Press, 2002.

Brown, Christopher. *Dutch Paintings*. London: National Gallery Publications, 1983.

Brown, Harcourt. *Scientific Organizations in Seventeenth Century France (1620–1680)*. 1934. New York: Russell and Russell, 1967.

Brown, Theodore M. "Medicine in the Shadow of the *Principia.*" *Journal of the History of Ideas* 48 (1987): 629–48.

Brownstein, Daniel Abraham. "Cultures of Anatomy in Sixteenth-Century Italy." Ph.D. diss., University of California at Berkeley, 1996.

Brugmans, H., ed. *Gedenkboek van het Athenaeum en de Universiteit van Amsterdam, 1632–1932.* Amsterdam: Stadsdrukkerij, 1932.

Bruijn, Iris Diane Rosemary. "Ship's Surgeons of the Dutch East India Company in the Eighteenth Century: Commerce and the Progress of Medicine." Ph.D. diss., Rijksuniversiteit te Leiden, 2004.

Bruijn, J. R. "Dutch Fisheries: An Historiographical and Thematic Overview." In *The North Atlantic Fisheries, 1100–1976: National Perspectives on a Common Resource,* ed. Poul Holm, David J. Starkey, and Jón Th. Thór, 105–20. Studia Atlantica, 1. Esbjerg: Fiskerimuseet, 1996.

Bruijn, J. R., and E. S. van Eyck van Heslinga. *Muiterij: oproer en berechting op schepen van de VOC.* Haarlem: De Boer Maritiem, 1980.

Bruijn, J. R., F. S. Gaastra, I. Schöffer, and A. C. J. Vermeulen, assist. *Dutch-Asiatic Shipping in the Seventeenth and Eighteenth Centuries.* Vol. 1: *Introductory Volume.* Rijks Geschiedkundige Publicatiën. The Hague: Martinus Nijhoff, 1987.

Buchan, James. *Frozen Desire: The Meaning of Money.* New York: Farrar, Straus, and Giroux, 1997.

Budge, Ernest A. Wallis. *Amulets and Talismans.* New York: University Books, 1961.

Bueno, Mar Rey. *Los señores del fuego: destiladores y espagiricos en la corte de los Austrias.* Madrid: Ediciones Corona Borealis, 2002.

Bulbeck, David, Anthony Reid, Lay Cheng Tan, and Yiqi Wu, comp. *Southeast Asian Exports since the Fourteenth Century: Cloves, Pepper, Coffee, and Sugar.* Leiden: Koninklijk Instituut voor Taal-, Land-en Volkenkunde Press, 1998.

Burnett, Charles. "Astrology and Medicine in the Middle Ages." *Bulletin of the Society for the Social History of Medicine* 37 (1985): 16–18.

Burney, Ian Adnan. "Viewing Bodies: Medicine, Public Order, and English Inquest Practice." *Configurations* 2 (1994): 33–46.

Burtt, Shelly. *Virtue Transformed: Political Argument in England, 1688–1740.* Cambridge: Cambridge University Press, 1992.

Bury, J. B. *The Idea of Progress: An Inquiry into Its Origin and Growth.* Intro. Charles A. Beard. New York: Dover, 1955.

Busbecq, Ogier Ghiselin de. *The Life and Letters of Ogier Ghiselin de Busbecq: Seigneur of Bousbecque, Knight, Imperial Ambassador.* 2 vols. Ed. and trans. Charles Thornton Forster and F. H. Blackburne Daniell. London: Kegan Paul, 1881.

Buzon, Frédéric de. "Science de la nature et théorie musicale chez Isaac Beeckman (1588–1637)." *Revue d'Histoire des Sciences* 38 (1985): 97–120.

Bylebyl, Jerome J. "Commentary on Early Clinical Teaching at Padua." In *A Celebration of Medical History,* ed. Lloyd G. Stevenson, 200–211. Baltimore: Johns Hopkins University Press, 1982.

———. "Galen on the Non-Natural Causes of Variation in the Pulse." *Bulletin of the History of Medicine* 45 (1971): 482–85.

———. "The Manifest and the Hidden in the Renaissance Clinic." In *Medicine and the Five Senses*, ed. W. F. Bynum and Roy Porter, 40–60. Cambridge: Cambridge University Press, 1993.

———. "The Medical Meaning of *Physica*." In *Renaissance Medical Learning: Evolution of a Tradition*, ed. Michael R. McVaugh and Nancy G. Siraisi. *Osiris*, 2nd ser., 6 (1990): 16–44.

———. "Medicine, Philosophy, and Humanism in Renaissance Italy." In *Science and the Arts in the Renaissance*, ed. John W. Shirley and F. David Hoeniger, 27–49. Washington, DC: Folger Shakespeare Library, 1985.

———. "*De Motu Cordis*: Written in Two Stages? Response." *Bulletin of the History of Medicine* 51 (1977): 140–50.

———. "The School of Padua: Humanistic Medicine in the Sixteenth Century." In *Health, Medicine and Mortality in the Sixteenth Century*, ed. Charles Webster, 335–70. Cambridge: Cambridge University Press, 1979.

Bynum, Caroline Walker. *The Resurrection of the Body in Western Christianity*. New York: Columbia University Press, 1995.

———. "Wonder." *American Historical Review* 102 (1997): 1–26.

Caciola, Nancy. "Mystics, Demoniacs, and the Physiology of Spirit Possession in Medieval Europe." *Comparative Studies in Society and History* 42 (2000): 268–306.

Calabi, Donatella. *The Market and the City: Square, Street and Architecture in Early Modern Europe*. Trans. Marlene Klein. Aldershot: Ashgate, 2004.

Campbell, Mary B. *The Witness and the Other World: Exotic European Travel Writing, 400–1600*. Ithaca, NY: Cornell University Press, 1988.

Cañizares-Esguerra, Jorge. "Iberian Science in the Renaissance: Ignored How Much Longer?" *Perspectives on Science* 12 (2004): 86–124.

Carlino, Andrea. *Books of the Body: Anatomical Ritual and Renaissance Learning*. Trans. John Tedeschi and Anne C. Tedeschi. Chicago: University of Chicago Press, 1999.

Carreras y Artau, Tomás, and Joaquín Carreras y Artau. *Historia de la filosofía Española: filosofía Cristiana de los siglos trece al quince*. 2 vols. Madrid: Real Academia de Ciencias Exactas, Fisicas y Naturales, 1939–43.

Carrillo, Jesús. "From Mt. Ventoux to Mt. Masaya: The Rise and Fall of Subjectivity in Early Modern Travel Narrative." In *Voyages and Visions: Towards a Cultural History of Travel*, ed. Jaś Elsner and Joan-Pau Rubiés, 57–73. London: Reaktion, 1999.

Carrubba, Robert W. "The Latin Document Confirming the Date and Institution of Wilhelm ten Rhyne's M.D." *Gesnerus* 39 (1982): 473–76.

Carrubba, Robert W., and John Z. Bowers. "The Western World's First Detailed Treatise on Acupuncture: Willem ten Rhijne's *De acupunctura*." *Journal of the History of Medicine* 29 (1974): 371–97.

Chancey, Karen. "The Amboyna Massacre in English Politics, 1624–1632." *Albion* 30 (1998): 583–98.

Chaney, Edward. *The Evolution of the Grand Tour: Anglo-Italian Cultural Relations since the Renaissance*. London: Frank Cass, 1998.

———. *The Grand Tour and the Great Rebellion: Richard Lassels and "The Voyage to Italy" in the Seventeenth Century*. Geneva: Slatkine, 1985.

Charles, Pierre. "L'introduction de la médecine européene au Japon par les Portugais au seizième siècle." In *Primeiro Congresso da Historia da Expansão Portuguesa no Mundo,* 2:89–105. Lisbon: Ministério das Colónias, 1938.

Charles-Dawbert, Françoise. "Libertine, littérature clandestine et privilège de la raison." *Recherches sur le Dix-septième Siècle* 7 (1984): 45–57.

Chartier, Roger, ed. *Passions of the Renaissance.* Vol. 3 of *A History of Private Life.* Gen. ed. Phillipe Ariès and Georges Duby. Trans. Arthur Goldhammer. Cambridge, MA: Harvard University Press, Belknap, 1989.

Chaudhuri, K. N. *Asia before Europe: Economy and Civilisation of the Indian Ocean from the Rise of Islam to 1750.* Cambridge: Cambridge University Press, 1990.

Chijs, J. A. van der. *De Nederlanders te Jakatra.* Amsterdam: Frederik Muller, 1860.

Clark, George N. *A History of the Royal College of Physicians of London.* 2 vols. Oxford: Clarendon Press, 1964–66.

Clark, W. B., and Meradith T. McMunn, eds. *Beasts and Birds of the Middle Ages: The Bestiary and Its Legacy.* Philadelphia: University of Pennsylvania Press, 1989.

Clercq, Peter de. *At the Sign of the Oriental Lamp: The Musschenbroek Workshop in Leiden, 1660–1750.* Rotterdam: Erasmus, 1997.

Cohen, H. F. *Quantifying Music: The Science of Music at the First Stage of the Scientific Revolution, 1580–1650.* Dordrecht: Reidel, 1984.

Cole, F. J. *A History of Comparative Anatomy: From Aristotle to the Eighteenth Century.* London: Macmillan, 1944.

———. "Leeuwenhoek's Zoological Researches." *Annals of Science* 2 (1937): 1–46, 185–235.

Cole, John R. *The Olympian Dreams and Youthful Rebellion of René Descartes.* Urbana: University of Illinois Press, 1992.

Colie, Rosalie L. *Light and Enlightenment: A Study of the Cambridge Platonists and the Dutch Arminians.* Cambridge: Cambridge University Press, 1957.

———. *"Some Thankfulnesse to Constantine": A Study of English Influence upon the Early Works of Constantijn Huygens.* The Hague: Martinus Nijhoff, 1956.

Cook, Harold J. "Bernard Mandeville." In *A Companion to Early Modern Philosophy,* ed. Steven Nadler, 469–82. Oxford: Blackwell, 2002.

———. "Bernard Mandeville and the Therapy of the 'Clever Politician.'" *Journal of the History of Ideas* 60 (1999): 101–24.

———. "Body and Passions: Materialism and the Early Modern State." *Osiris* 17 (2002): 25–48.

———. "Boerhaave and the Flight from Reason in Medicine." *Bulletin of the History of Medicine,* 74 (2000): 221–40.

———. "The Cutting Edge of a Revolution? Medicine and Natural History near the Shores of the North Sea." In *Renaissance and Revolution: Humanists, Scholars, Craftsmen and Natural Philosophers in Early Modern Europe,* ed. J. V. Field and Frank A. J. L. James, 45–61. Cambridge: Cambridge University Press, 1993.

———. *The Decline of the Old Medical Regime in Stuart London.* Ithaca, NY: Cornell University Press, 1986.

———. "Global Economies and Local Knowledge in the East Indies: Jacobus Bontius

Learns the Facts of Nature." In *Colonial Botany: Science, Commerce, and Politics in the Early Modern World*, ed. Londa Schiebinger and Claudia Swan, 100–118, 299–302. Philadelphia: University of Pennsylvania Press, 2005.

———. "Good Advice and Little Medicine: The Professional Authority of Early Modern English Physicians." *Journal of British Studies* 33 (1994): 1–31.

———. "Living in Revolutionary Times: Medical Change under William and Mary." In *Patronage and Institutions: Science, Technology, and Medicine at the European Court, 1500–1750*, ed. Bruce T. Moran, 111–35. Woodbridge: Boydell, 1991.

———. "Medical Communication in the First Global Age: Willem ten Rhijne in Japan, 1674–1676," *Disquisitions on the Past and Present* 11 (2004): 16–36.

———. "The Moral Economy of Natural History and Medicine in the Dutch Golden Age." In *Contemporary Explorations in the Culture of the Low Countries*, ed. William Z. Shetter and Inge Van der Cruysse, 39–47. Publications of the American Association of Netherlandic Studies, 9. Lanham, Md.: University Press of America, 1996.

———. "Natural History and Seventeenth-Century Dutch and English Medicine." In *The Task of Healing: Medicine, Religion and Gender in England and the Netherlands, 1450–1800*, ed. Hilary Marland and Margaret Pelling, 253–70. Rotterdam: Erasmus, 1996.

———. "The New Philosophy and Medicine in Seventeenth-Century England." In *Reappraisals of the Scientific Revolution*, ed. David C. Lindberg and Robert S. Westman, 397–436. Cambridge: Cambridge University Press, 1990.

———. "The New Philosophy in the Low Countries." In *The Scientific Revolution in National Context*, ed. Roy Porter and Mikuláš Teich, 115–49. Cambridge: Cambridge University Press, 1992.

———. "Physical Methods." In *Companion Encyclopedia of the History of Medicine*, ed. W. F. Bynum and Roy Porter, 939–60. London: Routledge, 1993.

———. "Physicians and Natural History." In *Cultures of Natural History*, ed. Nicholas Jardine, James Secord, and Emma Spary, 91–105. Cambridge: Cambridge University Press, 1996.

———. "Physick and Natural History in Seventeenth-Century England." In *Revolution and Continuity: Essays in the History and Philosophy of Early Modern Science*, ed. P. Barker and R. Ariew, 63–80. Studies in Philosophy and the History of Philosophy, 24. Washington, DC: Catholic University of America Press, 1991.

———. "Time's Bodies: Crafting the Preparation and Preservation of Naturalia." In *Merchants and Marvels: Commerce, Science and Art in Early Modern Europe*, ed. Pamela H. Smith and Paula Findlen, 223–47. London: Routledge, 2002.

———. *Trials of an Ordinary Doctor: Joannes Groenevelt in Seventeenth-Century London*. Baltimore: Johns Hopkins University Press, 1994.

———. "Das Wissen von den Sachen." Trans. into German by Jan Neersö. In *Seine Welt Wissen: Enzyklopädien in der Frühen Neuzeit*, ed. Ulrich Johannes Schneider, 81–124. Darmstadt: WBG, 2006.

Cook, Harold J., and David S. Lux. "Closed Circles or Open Networks? Communicating at a Distance during the Scientific Revolution." *History of Science* 36 (1998): 179–211.

Coolhaas, W. Ph. "Notes and Comments on the So-Called Amboina Massacre." In *Dutch*

Authors on Asian History, ed. M. A. P. Meilink-Roelofsz, M. E. van Opstall, and G. J. Schutte,198–240. Dordrecht: Foris, 1988.

Copenhaver, Brian P. "Natural Magic, Hermetism, and Occultism in Early Modern Science." In *Reappraisals of the Scientific Revolution*, ed. David C. Lindberg and Robert S. Westman, 261–301. Cambridge: Cambridge University Press, 1990.

Corn, Charles. *The Scents of Eden: A History of the Spice Trade*. London: Kodansha International, 1999.

Cunningham, Andrew. *The Anatomical Renaissance: The Resurrection of the Anatomical Projects of the Ancients*. Aldershot: Scolar Press, 1997.

———. "Fabricius and the 'Aristotle Project' in Anatomical Teaching and Research at Padua." In *The Medical Renaissance of the Sixteenth Century*, ed. Andrew Wear, Roger K. French, and Ian M. Lonie, 195–222, 330–31. Cambridge: Cambridge University Press, 1985.

———. "Sir Robert Sibbald and Medical Education, Edinburgh, 1706." *Clio Medica* 13 (1978): 135–61.

Dankmeijer, J. *De biologische studies van René Descartes*. Leidse Voordrachten. Leiden: Universitaire Pers, 1951.

———. "Les travaux biologiques de René Descartes (1596–1650)." *Archives Internationales d'Histoire des Sciences* 16 (1951): 675–80.

Dannenfeldt, Karl H. "Egyptian Mumia: The Sixteenth-Century Experience and Debate." *Sixteenth Century Journal* 16 (1985): 163–80.

———. "Wittenberg Botanists during the Sixteenth Century." In *The Social History of the Reformation*, ed. Lawrence P. Buck and Jonathan W. Zophy, 223–48. Columbus: Ohio State University Press, 1972.

Dash, Mike. *Tulipomania: The Story of the World's Most Coveted Flower and the Extraordinary Passions It Aroused*. New York: Crown, 1999.

Daston, Lorraine. "Baconian Facts, Academic Civility, and the Prehistory of Objectivity." *Annals of Scholarship* 8 (1991): 337–63.

———. "Marvelous Facts and Miraculous Evidence in Early Modern Europe." In *Questions of Evidence: Proof, Practice, and Persuasion across the Disciplines*, ed. James Chandler, Arnold I. Davidson, and Harry Harootunian, 243–74. Chicago: University of Chicago Press, 1994.

———. "The Moral Economy of Science." *Osiris* 10 (1995): 3–24.

———. "Speechless." In *Things That Talk: Object Lessons from Art and Science*, ed. Lorraine Daston, 9–24, 375–76. New York: Zone Books, 2004.

Daston, Lorraine, and Peter Galison. "The Image of Objectivity." *Representations* 40 (1992): 81–128.

Daston, Lorraine, and Katharine Park. *Wonders and the Order of Nature, 1150–1750*. New York: Zone Books, 1998.

Davids, C. A. "Navigeren in Azië: de uitwisseling van kennis tussen Aziaten en navigatiepersoneel bij de voorcomagnieën en de VOC, 1596–1795." In *VOC en cultuur: wetenschappelijke en culturele relaties tussen Europa an Azië ten tijde vande Verenigde Oostindische Compagnie*, ed. J. Bethlehem and A. C. Meijer, 17–37. Amsterdam: Schiphouwer en Brinkman, 1993.

———. *Zeewezen en wetenschap: de wetenschap en de ontwikkeling van de navigatietechniek in Nederland tussen 1585 en 1815*. Amsterdam: De Bataafsche Leeuw, 1986.

Davis, Natalie Zemon. "Rabelais among the Censors (1940s, 1540s)." *Representations* 32 (1990): 1–32.

———. *Women on the Margins: Three Seventeenth-Century Lives*. Cambridge, MA: Harvard University Press, 1995.

Dear, Peter. "A Mechanical Microcosm: Bodily Passions, Good Manners, and Cartesian Mechanism." In *Science Incarnate: Historical Embodiments of Natural Knowledge*, ed. Christopher Lawrence and Steven Shapin, 51–82. Chicago: University of Chicago Press, 1998.

———. "Miracles, Experiments, and the Ordinary Course of Nature." *Isis* 81 (1990): 663–83.

Debus, Allen G. *Chemistry and Medical Debate: Van Helmont to Boerhaave*. Canton, MA: Science History Publications, 2001.

———. "Sir Thomas Browne and the Study of Colour Indicators." *Ambix* 10 (1962): 29–36.

Deerr, Noel. *The History of Sugar*. 2 vols. London: Chapman and Hall, 1949–50.

Defilipps, Robert A. "Historical Connections between the Discovery of Brazil and the Neotropical Brazilwood, *Cesalpinia Echinata* Lam." *Archives of Natural History* 25 (1998): 103–8.

Dekker, Rudolf M. "Dutch Travel Journals from the Sixteenth to the Early Nineteenth Centuries." *Lias* 22 (1995): 277–99.

———. *Humour in Dutch Culture of the Golden Age*. Basingstoke: Palgrave, 2001.

———. *Lachen in the gouden eeuw: een geschiedenis van de Nederlandse humor*. Amsterdam: Wereldbibliotheek, 1997.

———. "'Private Vices, Public Virtues' Revisited: The Dutch Background of Bernard Mandeville." Trans. Gerard T. Moran. *History of European Ideas* 14 (1992): 481–98.

DeLancey, Julia A. "Dragonsblood and Ultramarine: The Apothecary and Artists' Pigments in Renaissance Florence." In *The Art Market in Italy, Fifteenth–Seventeenth Centuries*, ed. Marcello Fantoni, Louisa C. Matthew, and Sara F. Matthews-Grieco, 141–50. Modena: Franco Cosimo Panini, 2003.

Delaunay, Paul. "Pierre Belon Naturaliste." *Bulletin de la Société d'Agriculture, Sciences et Arts de la Sarthe* 41 (1923–24): 13–39.

Demaerel, Paul. "Couplet and the Dutch." In *Philippe Couplet, S.J. (1623–1693): The Man Who Brought China to Europe*, ed. Jerome Heyndrickx, 87–120. Nettetal: Steyler-Verlag, 1990.

Demaitre, Luke. "Domesticity in Middle Dutch 'Secrets of Men and Women.'" *Social History of Medicine* 14 (2001): 1–25.

Demiriz, Yildiz. "Tulips in Ottoman Turkish Culture and Art." In *The Tulip: A Symbol of Two Nations*, ed. Michiel Roding and Hans Theunissen, 57–75. Utrecht and Istanbul: M. Th. Houtsma Stichting, Turco-Dutch Friendship Association, 1993.

Dennis, Michael Aaron. "Graphic Understanding: Instruments and Interpretation in Robert Hooke's *Micrographia*." *Science in Context* 3 (1989): 309–64.

Deschamps, Léon. "Pierre Belon: naturaliste et explorateur." *Revue de Géographie* 21 (1887): 433–40.

Des Chene, Dennis. *Physiologia: Natural Philosophy in Late Aristotelian and Cartesian Thought.* Ithaca, NY: Cornell University Press, 1996.

De Vries, Jan. "Connecting Europe and Asia: A Quantitative Analysis of the Cape-Route Trade, 1497–1795." In *Global Connections and Monetary History, 1470–1800*, ed. Dennis O. Flynn, Arturo Giráldez, and Richard von Glahn, 35–106. Aldershot: Ashgate, 2003.

———. *The Dutch Rural Economy in the Golden Age, 1500–1700.* New Haven and London: Yale University Press, 1974.

———. *European Urbanization, 1500–1800.* Harvard Studies in Urban History. Cambridge, MA: Harvard University Press, 1984.

———. "The Netherlands and the New World: The Legacy of European Fiscal, Monetary, and Trading Institutions for New World Development from the Seventeenth to the Nineteenth Centuries." In *Transferring Wealth and Power from the Old to the New World: Monetary and Fiscal Institutions in the Seventeenth through the Nineteenth Centuries*, ed. Michael D. Bordo and Roberto Cortés-Conde, 100–139. Cambridge: Cambridge University Press, 2001.

———. "On the Modernity of the Dutch Republic." *Journal of Economic History* 33 (1973): 191–202.

De Vries, Jan, and Ad van der Woude. *The First Modern Economy: Success, Failure, and Perseverance of the Dutch Economy, 1500–1815.* Cambridge: Cambridge University Press, 1997.

———. *Nederland, 1500–1815: de eerste ronde van moderne economische groei.* Amsterdam: Balans, 1995.

Déré, Anne Claire. "Profil d'in artiste chimiste du dix-huitième siècle: les manuscrits de Dom Malherbe." *Archives Internationales d'Histoire des Sciences* 45 (1995): 298–310.

Diedenhofen, Wilhelm. "'Belvedere,' or the Principle of Seeing and Looking in the Gardens of Johan Maurits Van Nassau-Siegen at Cleves." In *The Dutch Garden in the Seventeenth Century*, ed. John Dixon Hunt, 49–80. Dumbarton Oaks Colloquium on the History of Landscape Architecture, 12. Washington, DC: Dumbarton Oaks, 1990.

———. "Johan Maurits and His Gardens." In *Johan Maurits Van Nassau-Siegen, 1604–1679: A Humanist Prince in Europe and Brazil*, ed. Ernst van den Boogaart with H. R. Hoetink and P. J. P. Whitehead, 197–236. The Hague: Johan Maurits van Nassau, 1979.

Dijkstra, Jan Gerard. *Een epidemiologische beschowing van de Nederlandsche pest-epidemieën der zeventiende eeuw.* Amsterdam: Volharding, 1921.

Dixhoorn, Arjan van, and Benjamin Roberts. "Edifying Youths: The Chambers of Rhetoric in Seventeenth-Century Holland." *Paedagogica Historica* 39 (2003): 325–37.

Dobell, Clifford. *Antony Van Leeuwenhoek and His "Little Animals."* New York: Russell and Russell, 1958.

Dolan, Brian. *Exploring European Frontiers: British Travellers in the Age of Enlightenment.* Basingstoke: Macmillan, 2000.

———. *Ladies of the Grand Tour.* London: HarperCollins, 2001.

Dorssen, J. M. H. van. "Dr. Willem Ten Rhijne and Leprosy in Batavia in the Seventeenth Century." *Janus* 2 (1897–98): 252–60, 355–64.

———. "Willem Ten Rhijne." *Geneeskunde Tijdschrift voor Nederlands Indië* 51 (1911): 134–228.

Dorsten, J. A. van. *Poets, Patrons, and Professors: Sir Philip Sidney, Daniel Rogers, and the Leiden Humanists.* Leiden: Leiden University Press; and Oxford: Oxford University Press, 1962.

Dorsten, J. A. van, and Roy Strong. *Leicester's Triumph.* Leiden: Leiden University Press; and Oxford: Oxford University Press, 1964.

Duhr, Joseph. *Un Jésuite en Chine: Adam Schall, astronome et conseiller impérial (1592–1666).* Brussels: L'Édition Universelle; and Paris: Desclée de Brower, 1936.

Duke, A. C. "Building Heaven in Hell's Despite: The Early History of the Reformation in the Towns of the Low Countries." In *Britain and the Netherlands: Church and State since the Reformation,* vol. 7, ed. A. C. Duke and C. A. Tamse, 45–75. The Hague: Martinus Nijhoff, 1981.

Duke, Alastair. "The Ambivalent Face of Calvinism in the Netherlands, 1561–1618." In *Reformation and Revolt in the Low Countries,* 269–93. London: Hambledon, 1990.

———. "Salvation by Coercion: The Controversy Surrounding the 'Inquisition' in the Low Countries on the Eve of the Revolt." In *Reformation and Revolt in the Low Countries,* 152–74. London: Hambledon, 1990.

Dumon, Ir. A. "De betekenis van de Busbecq in de ontwikkeling van de plantkunde gedurende de zestiende eeuw." In *Augerius Gislenus Busbequius, 1522–1591: Vlaams humanist en keizerlijk gezant,* 24–37. Koninklijke Vlaamse Academie voor Wetenschappen, Letteren en Schone Kunsten van België. Brussels: Palais der Academiën, 1954.

Eamon, William. "New Light on Robert Boyle and the Discovery of Colour Indicators." *Ambix* 27 (1980): 204–9.

———. *Science and the Secrets of Nature: Books of Secrets in Medieval and Early Modern Culture.* Princeton, NJ: Princeton University Press, 1994.

Ebert, Christopher. "Dutch Trade with Brazil before the Dutch West India Company, 1587–1621." In *Riches from Atlantic Commerce: Dutch Transatlantic Trade and Shipping, 1585–1817,* ed. Johannes Postma and Victor Enthoven, 49–75. Leiden: Brill, 2003.

Ehring, Franz. "Leprosy Illustration in Medical Literature." *International Journal of Dermatology* 33 (1994): 872–83.

Elias, Norbert. *Time: An Essay.* Trans. Edmund Jephcott. 1987. Oxford: Blackwell, 1992.

Elsner, Jaś, and Joan-Pau Rubiés. "Introduction." In *Voyages and Visions: Towards a Cultural History of Travel,* ed. Jaś Elsner and Joan-Pau Rubiés, 1–56. London: Reaktion, 1999.

Emmer, P. C. "The West India Company, 1621–1791: Dutch or Atlantic?" In *Companies and Trade,* ed. Leonard Blussé and Femme Gaastra, 71–95. Leiden: Leiden University Press, 1981.

Engelhardt, Dietrich v. "Luca Ghini (Um 1490–1556) und die Botanik des 16. Jahrhunderts: Leben, Initiativen, Kontakte, Resonanz." *Medizinhistorisches Journal* 30 (1995): 3–49.

Ergun, Nilgün, and Özge Iskender. "Gardens of the Topkapi Palace: An Example of Turkish Garden Art." *Studies in the History of Gardens and Designed Landscapes* 23 (2003): 57–71.

Evans, R. J. W. *Rudolph II and His World: A Study in Intellectual History, 1576–1612.* Oxford: Clarendon Press, 1973.

Evers, Meindert. "The Illustre School at Harderwyk, 1600–1647." *Lias* 12 (1985): 81–113.

Farber, Paul. "The Development of Taxidermy and the History of Ornithology." *Isis* 68 (1977): 550–66.
Febvre, Lucien. *The Problem of Unbelief in the Sixteenth Century: The Religion of Rabelais.* 1st French ed., 1942. Trans. Beatrice Gottlieb. Cambridge, MA: Harvard University Press, 1982.
Feliú, Carmen Añon. "The Restoration of the King's Garden at Aranjuez." In *The Authentic Garden: A Symposium on Gardens,* ed. Leslie Tjon Sie Fat and Erik de Jong, 97–102. Leiden: Clusius Foundation, 1991.
Ferrari, Giovanna. "Public Anatomy Lessons and the Carnival: The Anatomy Theatre of Bologna." *Past and Present* 117 (1987): 50–106.
Feucht, G. "Die Akupunktur im Europa des Neunzehn Jahrhunderts." *Erfahrungs-Heilkunde: Zeitschrift für die Ärztliche Praxis* 25 (1976): 459–63.
Feuer, Lewis S. "Science and the Ethic of Protestant Ascetism: A Reply to Professor Robert K. Merton." *Research in Sociology of Knowledge, Sciences and Art: A Research Annual* 2 (1979): 1–23.
———. *The Scientific Intellectual: The Psychological and Sociological Origins of Modern Science.* New York: Basic Books, 1963.
Feyerabend, Paul K. *Science in a Free Society.* London: NLB, 1978.
Figueiredo, John M. de. "Ayurvedic Medicine in Goa according to European Sources in the Sixteenth and Seventeenth Centuries." *Bulletin of the History of Medicine* 58 (1984): 225–35.
Findlen, Paula. *Possessing Nature: Museums, Collecting and Scientific Culture in Early Modern Italy.* Berkeley: University of California Press, 1994.
Fischel, Walter J. "The Spice Trade in Mamluk Egypt: A Contribution to the Economic History of Medieval Islam." 1958. In *Spices in the Indian Ocean World: An Expanding World: The European Impact on World History, 1450–1800,* ed. M. N. Pearson, 157–74. Aldershot: Variorum, 1996.
Fix, Andrew C. *Prophecy and Reason: The Dutch Collegiants in the Early Enlightenment.* Princeton, NJ: Princeton University Press, 1991.
Fock, C. Willemijn. "Kunst en rariteiten in het Hollandse interieur." In *De wereld binnen handbereik: Nederlandse kunst- en rariteitenverzamelingen, 1585–1735,* ed. Ellinoor Bergvelt and Renée Kistemaker, 70–91. Zwolle: Waanders Uitgevers/Amsterdams Historisch Museum, 1992.
Force, James E. "The Origins of Modern Atheism." *Journal of the History of Ideas* 50 (1989): 153–62.
Ford, Brian J. *The Leeuwenhoek Legacy.* Bristol: Biopress; and London: Farrand, 1991.
———. *Single Lens: The Story of the Simple Microscope.* New York: Harper and Row, 1985.
———. "The Van Leeuwenhoek Specimens." *Notes and Records of the Royal Society* 36 (1981): 37–59.
Foss, Theodore Nicholas. "The European Sojourn of Philippe Couplet and Michael Shen Fuzong, 1683–1692." In *Philippe Couplet, S.J. (1623–1693): The Man Who Brought China to Europe,* ed. Jerome Heyndrickx, 121–40. Nettetal: Steyler-Verlag, 1990.
Fothergill-Payne, Louise. "Seneca's Role in Popularizing Epicurus in the Sixteenth Cen-

tury." In *Atoms,* Pneuma, *and Tranquillity: Epicurean and Stoic Themes in European Thought,* ed. Margaret J. Osler, 115–33. Cambridge: Cambridge University Press, 1991.

Fournier, Marian. *The Fabric of Life: Microscopy in the Seventeenth Century.* Baltimore: Johns Hopkins University Press, 1996.

———. "The Fabric of Life: The Rise and Decline of Seventeenth-Century Microscopy." Ph.D. diss., Universiteit Twente, 1991.

Foust, Clifford M. *Rhubarb: The Wondrous Drug.* Princeton, NJ: Princeton University Press, 1992.

Frank, Robert G., Jr. *Harvey and the Oxford Physiologists: Scientific Ideas and Social Interaction.* Berkeley: University of California Press, 1980.

Freedberg, David. *The Eye of the Lynx: Galileo, His Friends, and the Beginning of Modern Natural History.* Chicago: University of Chicago Press, 2002.

———. "Science, Commerce, and Art: Neglected Topics at the Juncture of History and Art History." In *Art in History, History in Art: Studies in Seventeenth-Century Dutch Culture,* ed. D. Freedberg and J. de Vries, 377–428. Santa Monica, CA: Getty Center for the History of Art and the Humanities, 1991.

French, Roger K. *Ancient Natural History: Histories of Nature.* London: Routledge, 1994.

———. "Berengario da Carpi and the Use of Commentary in Anatomical Teaching." In *The Medical Renaissance of the Sixteenth Century,* ed. Andrew Wear, Roger K. French, and Ian M. Lonie, 42–74, 296–98. Cambridge: Cambridge University Press, 1985.

———. *Dissection and Vivisection in the European Renaissance.* Aldershot: Ashgate, 1999.

———. "Harvey in Holland: Circulation and the Calvinists." In *The Medical Revolution of the Seventeenth Century,* ed. Roger French and Andrew Wear, 46–86. Cambridge: Cambridge University Press, 1989.

———. "The Languages of William Harvey's Natural Philosophy." *Journal of the History of Medicine* 49 (1994): 24–51.

———. "Pliny and Renaissance Medicine." In *Science in the Early Roman Empire: Pliny the Elder, His Sources and Influence,* ed. R. French and F. Greenaway, 252–81. Totowa, NJ: Barnes and Noble, 1986.

Frijhoff, Willem. "Het Amsterdamse Athenaeum in het academische landschap van de zeventiende eeuw." In *Athenaeum illustre: elf studies over de Amsterdamse Doorluchtige School, 1632–1877,* ed. E. O. G. Haitsma Mulier, C. L. Heesakers, P. J. Knegtmans, A. J. Kox, and T. J. Veen, 37–65. Amsterdam: Amsterdam University Press, 1997.

———. "Deventer en zijn gemiste universiteit: het Athenaeum in de sociaal-culturele geschiedenis van Overijssel." *Overijsselse Historische Bijdragen* 97 (1982): 45–79.

———. "The Emancipation of the Dutch Elites from the Magic Universe." In *The World of William and Mary: Anglo-Dutch Perspectives on the Revolution of 1688–89,* ed. Dale Hoak and Mordechai Feingold, 201–18. Stanford, CA: Stanford University Press, 1996.

———. "Le médecin selon Jacques Cahaignes (1548–1612): autour de deux soutenances en médecine à Caen au debut du dix-septième siecle." *Lias* 10 (1983): 193–215.

———. *La société Néerlandaise et ses gradués, 1575–1814: une recherche sérielle sur le statut des intellectuels à partir des registres universitaires.* Amsterdam: APA—Holland University Press, 1981.

Fruin, R. *The Siege and Relief of Leyden in 1574*. Trans. Elizabeth Trevelyn, intro. George Macaulay Trevelyan. The Hague: Martinus Nijhoff, 1927.

Fuchs, Thomas. *The Mechanisation of the Heart: Harvey and Descartes*. Trans. Marjorie Grene. Rochester, NY: University of Rochester Press, 2001.

Fujikawa, Y. *Japanese Medicine*. Trans. John Ruhräh. 1911. New York: Paul B. Hoeber, 1934.

Funkenstein, Amos. *Theology and the Scientific Imagination from the Middle Ages to the Seventeenth Century*. Princeton, NJ: Princeton University Press, 1986.

Gaastra, Femme S. *Bewind en beleid bij de VOC: de financiële en commerciële politiek van de bewindhebbers, 1672–1702*. Zutphen: Walburg Pers, 1989.

———. *De gescheidenis van de VOC*. 2nd ed. Zutphen: Walburg Pers, 1991.

Gadamer, Hans-Georg. *Truth and Method*. 2nd ed., rev. Trans. Garrett Barden and John Cumming, rev. Joel Weinsheimer and Donald G. Marshall. 1975. New York: Crossroad, 1989.

Galison, Peter. "Judgment against Objectivity." In *Picturing Science, Producing Art*, ed. Caroline A. Jones and Peter Galison, 327–59. New York: Routledge, 1998.

Galluzzi, Paolo. "Art and Artifice in the Depiction of Renaissance Machines." In *The Power of Images in Early Modern Science*, ed. Wolfgang Lefèvre, Jürgen Renn, and Urs Schoepflin, 47–68. Basel: Birkhäuser, 2003.

Gannal, Jean Nicolas, and R. Harlan, trans. and eds. *History of Embalming, and of Preparations in Anatomy, Pathology, and Natural History; Including an Account of a New Process for Embalming*. Philadelphia: Judah Dobson, 1840.

Garber, Daniel. *Descartes Embodied: Reading Cartesian Philosophy through Cartesian Science*. Cambridge: Cambridge University Press, 2001.

———. "Science and Certainty in Descartes." In *Descartes: Critical and Interpretive Essays*, ed. M. Hooker, 114–51. Baltimore: Johns Hopkins University Press, 1978.

Garcia Sánchez, Expiración, and Angel López y López. "The Botanic Gardens in Muslim Spain." In *The Authentic Garden: A Symposium on Gardens*, ed. Leslie Tjon Sie Fat and Erik de Jong, 165–76. Leiden: Clusius Foundation, 1991.

Garin, Eugenio. *Astrology in the Renaissance: The Zodiac of Life*. Italian ed., 1976. Trans. Carolyn Jackson, June Allen, and Clare Robertson. London: Routledge and Kegan Paul, 1983.

Gelderblom, Oscar, and Joost Jonker. "Completing a Financial Revolution: The Finance of the Dutch East India Trade and the Rise of the Amsterdam Capital Market, 1595–1612." Paper published on the Internet, 2004. http://www.iisg.nl/~lowcountries/2004-2.pdf.

Gelderen, Martin van. "The Machiavellian Moment and the Dutch Revolt: The Rise of Neostoicism and Dutch Republicanism." In *Machiavelli and Republicanism*, ed. Gisela Bock, Quentin Skinner, and Maurizio Viroli, 205–23. Cambridge: Cambridge University Press, 1990.

———. *The Political Thought of the Dutch Revolt, 1555–1590*. Cambridge: Cambridge University Press, 1992.

Gelder, Roelof van. "Leifhebbers en geleerde luiden: Nederlandse kabinetten en hun bezoekers." In *De wereld binnen handbereik: Nederlandse kunst- en rariteitenverzamelingen, 1585–1735*, ed. Ellinoor Bergvelt and Renée Kistemaker, 259–92. Zwolle: Waanders Uitgevers/Amsterdams Historisch Museum, 1992.

———. "Paradijsvogels in Enkhuisen." In *Souffrir pour parvenir: de wereld van Jan Huygen van Linschoten*, ed. Roelof van Gelder, Jan Parmentier, and Vibeke Roeper, 30–50. Haarlem: Arcadia, 1998.

———. "De wereld binnen handbereik: Nederlandse kunst- en rariteitenverzamelingen, 1585–1735." In *De wereld binnen handbereik: Nederlandse kunst- en rariteitenverzamelingen, 1585–1735*, ed. Ellinoor Bergvelt and Renée Kistemaker, 15–38. Zwolle: Waanders Uitgevers/Amsterdams Historisch Museum, 1992.

Gentilcore, David. "Contesting Illness in Early Modern Naples: *Miracolati*, Physicians and the Congregation of Rites." *Past and Present* 148 (1995): 117–48.

Geyer-Kordesch, Johanna. "Passions and the Ghost in the Machine: Or What Not to Ask about Science in Seventeenth- and Eighteenth-Century Germany." In *The Medical Revolution of the Seventeenth Century*, ed. Roger French and Andrew Wear, 145–63. Cambridge: Cambridge University Press, 1989.

Geyl, A. *De geschiedenis van het roonhuysiaansch geheim*. Rotterdam: Meidert Boogaerdt, 1905.

Gijswijt-Hofstra, Marijke. "The European Witchcraft Debate and the Dutch Variant." *Social History* 15 (1990): 181–94.

Gijswijt-Hofstra, Marijke, and Willem Frijhoff, eds. *Witchcraft in the Netherlands from the Fourteenth to the Twentieth Century*. Trans. Rachel M. J. van der Wilden-Fall. History of the Low Countries, 1. Rotterdam: Universtaire Pers, 1991.

Glamann, Kristof. *Dutch-Asiatic Trade, 1620–1740*. Copenhagen: Danish Science Press; and The Hague: Martinus Nijhoff, 1958.

Glete, Jan. *War and the State in Early Modern Europe: Spain, the Dutch Republic and Sweden as Fiscal-Military States, 1500–1660*. London: Routledge, 2002.

Gøbel, Erik. "Danish Companies' Shipping to Asia, 1616–1807." In *Ships, Sailors and Spices: East India Companies and Their Shipping in the Sixteenth, Seventeenth, and Eighteenth Centuries*, ed. Jaap R. Bruijn and Femme S. Gaastra, 99–120. Amsterdam: NEHA, 1993.

Goble, Andrew Edmund. "Medicine and New Knowledge in Medieval Japan: Kajiwara Shōzen (1266–1337) and the 'Man'anpō.'" *Nihon Ishigaku Zasshi: Journal of the Japan Society of Medical History* 47 (2001): 193–226, 432–52.

Godelier, Maurice. *The Enigma of the Gift*. Trans. Nora Scott. Cambridge: Polity Press, 1999.

Gogelein, A. J. F. *Hortus, horti, horto*. Pamphlet. Leiden: Rijksherbarium/Hortus Botanicus, 1990.

Goldgar, Anne. *Impolite Learning: Conduct and Community in the Republic of Letters, 1680–1750*. New Haven and London: Yale University Press, 1995.

———. "Nature as Art: The Case of the Tulip." In *Merchants and Marvels: Commerce, Science, and Art in Early Modern Europe*, ed. Pamela H. Smith and Paula Findlen, 324–46. New York: Routledge, 2002.

Goldsmith, M. M. *Private Vices, Public Benefits: Bernard Mandeville's Social and Political Thought*. Cambridge: Cambridge University Press, 1985.

Goldthwaite, Richard A. *Wealth and the Demand for Art in Italy, 1300–1600*. Baltimore: Johns Hopkins University Press, 1993.

Golinski, Jan. *Making Natural Knowledge: Constructivism and the History of Science*. Cambridge: Cambridge University Press, 1998.

———. "A Noble Spectacle: Phosphorus and the Public Cultures of Science in the Early Royal Society." *Isis* 80 (1989): 11–39.

Golvers, Noël. *François de Rougement, S.J., Missionary in Ch'ang-Shu (Chiang-Nan): A Study of the Account Book (1674–1676) and the Elogium*. Leuven: Leuven University Press, Ferdinand Verbiest Foundation, 1999.

Goodman, David C. *Power and Penury: Government, Technology, and Science in Philip II's Spain*. Cambridge: Cambridge University Press, 1988.

Goodman, Grant K. *Japan: The Dutch Experience*. London: Athlone, 1986.

Goody, Jack. *The Culture of Flowers*. Cambridge: Cambridge University Press, 1993.

Goonewardena, K. W. *The Foundation of Dutch Power in Ceylon, 1638–1658*. Amsterdam: Djambatan, 1958.

Gouk, Penelope. *Music, Science and Natural Magic in Seventeenth-Century England*. New Haven and London: Yale University Press, 1999.

Graaf, H. J. de. *De regering van Sultan Agung, vorst van Mataram, 1613–1645, en die van zijn voorganger Panembahan Séda-Ing-Krapjak, 1601–1613*. Verhandelingen van Het Koninklijk Instituut voor Taal-, Land- en Volkenkunde. The Hague: Martinus Nijhoff, 1958.

Grafton, Anthony. "The Availability of Ancient Works." In *The Cambridge History of Renaissance Philosophy*, gen. ed. Charles B. Schmitt, ed. Quentin Skinner, Eckhard Kessler, assoc. ed. Jill Kraye, 767–91. Cambridge: Cambridge University Press, 1988.

———. *Bring Out Your Dead: The Past as Revelation*. Cambridge, MA: Harvard University Press, 2001.

———. *Cardano's Cosmos: The Worlds and Works of a Renaissance Astrologer*. Cambridge, MA: Harvard University Press, 1999.

———. "Civic Humanism and Scientific Scholarship at Leiden." In *The University and the City: From Medieval Origins to the Present*, ed. Thomas Bender, 59–78. New York: Oxford University Press, 1988.

———. *Defenders of the Text: The Traditions of Scholarship in an Age of Science, 1450–1800*. Cambridge, MA: Harvard University Press, 1991.

———. *Joseph Scaliger: A Study in the History of Classical Scholarship*. Vol. 1: *Textual Criticism and Exegesis*. Oxford: Clarendon Press, 1983.

Grafton, Anthony, and Lisa Jardine. *From Humanism to the Humanities: Education and the Liberal Arts in Fifteenth- and Sixteenth-Century Europe*. London: Duckworth, 1986.

Granovetter, Mark S. "The Strength of Weak Ties." *American Journal of Sociology* 78 (1973): 1360–80.

Gregory, Tullio. "Pierre Charron's 'Scandalous Book.'" In *Atheism from the Reformation to the Enlightenment*, ed. Michael Hunter and David Wootton, 87–109. Oxford: Clarendon Press, 1992.

Grell, Ole Peter. "Caspar Bartholin and the Education of the Pious Physician." In *Medicine and the Reformation*, ed. Ole Peter Grell and Andrew Cunningham, 78–100. London: Routledge, 1993.

Grene, Marjorie, ed. *Spinoza: A Collection of Critical Essays*. Modern Studies in Philosophy. Garden City, NY: Doubleday, Anchor Press, 1973.

Grmek, Mirko Drazen. *Les reflets de la sphygmologie chinoise dans la médecine Occidentale.* Extrait de la Biologie Médicale, Numéro Hors Série. Paris: Specia, 1962.
Grove, Richard. *Green Imperialism: Colonial Expansion, Tropical Island Edens and the Origins of Environmentalism, 1600–1860.* Cambridge: Cambridge University Press, 1995.
———. "Indigenous Knowledge and the Significance of South-West India for Portuguese and Dutch Constructions of Tropical Nature." *Modern Asian Studies* 30 (1996): 121–43.
———. "The Transfer of Botanical Knowledge between Asia and Europe, 1498–1800." *Journal of the Japan-Netherlands Institute* 3 (1991): 160–76.
Guerrini, Anita. "Archibald Pitcairne and Newtonian Medicine." *Medical History* 31 (1987): 70–83.
———. "The Tory Newtonians: Gregory, Pitcairne, and Their Circle." *Journal of British Studies* 25 (1986): 288–311.
Gulik, T. M. van. "Dutch Surgery in Japan." In *Red-Hair Medicine: Dutch-Japanese Medical Relations,* ed. H Beukers, A. M. Luyendijk-Elshout, M. E. van Opstall, and F. Vos, 37–50. Amsterdam: Rodopi, 1991.
Gungwu, Wang. "Merchants without Empire: The Hokkien Sojourning Communities." In *The Rise of Merchant Empires: Long-Distance Trade in the Early Modern World, 1350–1750,* ed. James D. Tracy, 400–421. Cambridge: Cambridge University Press, 1990.
Gunn, M., and L. E. Codd. *Botanical Exploration of Southern Africa: An Illustrated History of Early Botanical Literature on the Cape Flora; Biographical Accounts of the Leading Plant Collectors and Their Activities in Southern Africa from the Days of the East India Company Until Modern Times.* Cape Town: A. A. Balkema for the Botanical Research Institute, 1981.
Gunther, R. T. *Early Science in Oxford.* Vol. 12. Oxford: Oxford University Press, 1939.
Guyonnet, Georges. "Saint Sébastien, patron des acupuncteurs." *Histoire de la Médecine* 8, no. 1 (1958): 65–68.
Haan, Frits de. "Uit oude notarispapieren, II: Andries Cleyer." *Tijdschrift voor Indische Taal-, Land- en Volkenkunde* 46 (1903): 423–68.
Haberland, Detlef. *Engelbert Kaempfer, 1651–1716: A Biography.* Trans. Peter Hogg. London: British Library, 1996.
Hacking, Ian. *The Emergence of Probability: A Philosophical Study of Early Ideas about Probability, Induction and Statistical Inference.* Cambridge: Cambridge University Press, 1975.
———. *Historical Ontology.* Cambridge, MA: Harvard University Press, 2002.
———. "The Participant Irrealist at Large in the Laboratory." *British Journal of the Philosophy of Science* 39 (1988): 277–94.
Hadden, Richard. *On the Shoulders of Merchants: Exchange and the Mathematical Conception of Nature in Early Modern Europe.* Albany, NY: State University of New York Press, 1994.
Haitsma Mulier, Eco. "A Controversial Republican: Dutch Views on Machiavelli in the Seventeenth and Eighteenth Centuries." In *Machiavelli and Republicanism,* ed. Gisela Bock, Quentin Skinner, and Maurizio Viroli, 247–63. Cambridge: Cambridge University Press, 1990.

———. "The Language of Seventeenth-Century Republicanism in the United Provinces: Dutch or European?" In *The Language of Political Theory in Early-Modern Europe*, ed. Anthony Pagden, 179–95. Cambridge: Cambridge University Press, 1987.

———. *The Myth of Venice and Dutch Republican Thought in the Seventeenth Century.* Trans. Gerard T. Moran. Assen: Van Gorcum, 1980.

Haks, Donald. "Family Structure and Relationship Patterns in Amsterdam." In *Rome, Amsterdam: Two Growing Cities in Seventeenth-Century Europe*, ed. Peter van Kessel and Elisja Schulte, 92–103. Amsterdam: Amsterdam University Press, 1997.

Hall, Thomas S. "Descartes' Physiological Method: Position, Principles, Examples." *Journal of the History of Biology* 3 (1970): 53–79.

Halleux, Robert, and Anne-Catherine Bernès. "La cour savante d'Ernest de Bavière." *Archives Internationales d'Histoire des Sciences* 45 (1995): 3–29.

Hamilton, Alastair. *The Family of Love.* Cambridge: James Clark, 1981.

Hanley, Susan B. *Everyday Things in Premodern Japan: The Hidden Legacy of Material Culture.* Berkeley: University of California Press, 1997.

Hansen, Julie V. "Resurrecting Death: Anatomical Art in the Cabinet of Dr. Frederik Ruysch." *Art Bulletin* 78 (1996): 663–79.

Harkness, Deborah E. *John Dee's Conversations with Angels: Cabala, Alchemy, and the End of Nature.* Cambridge: Cambridge University Press, 1999.

Harley, David. "Historians as Demonologists: The Myth of the Midwife-Witch." *Social History of Medicine* 3 (1990): 1–26.

Harris, Steven J. "Confession-Building, Long-Distance Networks, and the Organization of Jesuit Science." *Early Science and Medicine* 1 (1996): 287–318.

Harrison, J. B. "Europe and Asia." In *The New Cambridge Modern History*, vol. 4: *The Decline of Spain and the Thirty Years War, 1609–1648/9*, ed. J. P. Cooper, 644–71. Cambridge: Cambridge University Press, 1970.

Harrison, Mark. "Medicine and Orientalism: Perspectives on Europe's Encounter with Indian Medical Systems." In *Health, Medicine and Empire: Perspectives on Colonial India*, ed. Biswamoy Pati and Mark Harrison, 37–87. Hyderabad: Orient Longman, 2001.

Harrison, Peter. "Curiosity, Forbidden Knowledge, and the Reformation of Natural Philosophy in Early Modern England." *Isis* 92 (2001): 265–90.

Hart, Marjolein C. 't. "Freedom and Restrictions: State and Economy in the Dutch Republic, 1570–1670." In *Economic and Social History in the Netherlands*, 105–30. Het Nederlandsch Economisch-Historisch Archief, 4. Amsterdam: NEHA, 1993.

———. *The Making of a Bourgeois State: War, Politics and Finance during the Dutch Revolt.* Manchester: Manchester University Press, 1993.

Harte, N. B., ed. *The New Draperies in the Low Countries and England, 1300–1800.* New York: Oxford University Press, 1998.

Harth, Erica. *Cartesian Women: Versions and Subversions of Rational Discourse in the Old Regime.* Ithaca, NY: Cornell University Press, 1992.

Harting, P. "George Everard Rumphius." In *Album der natuur*, ed. P. Harting, D. Lubach, and W. M. Logeman, 1–15. Haarlem: H. D. Tjeenk Willink, 1885.

Harwood, John T. "Rhetoric and Graphics in *Micrographia*." In *Robert Hooke: New Studies*, ed. Michael Hunter and Simon Schaffer, 119–47. Woodbridge: Boydell, 1989.

Hatfield, Gary. "Descartes' Physiology and Its Relation to His Psychology." In *The Cambridge Companion to Descartes*, ed. John Cottingham, 335–70. Cambridge: Cambridge University Press, 1992.

———. "Metaphysics and the New Science." In *Reappraisals of the Scientific Revolution*, ed. David C. Lindberg and Robert S. Westman, 93–166. Cambridge: Cambridge University Press, 1990.

Haver Droeze, J. J. *Het Collegium medicum Amstelaedamense, 1637–1798*. Haarlem: De Erven F. Bohn, 1921.

Hazewinckel, H. C. *Geschiedenis van Rotterdam*. 4 vols. 1940–42. Zaltbommel: Europese Bibliotheek, 1974–75.

———. "Pierre Bayle à Rotterdam." In *Pierre Bayle: le philosophe de Rotterdam*, ed. Paul Dibon, 20–47. Amsterdam: Elsevier, 1959.

Heckscher, William S. *Rembrandt's "Anatomy of Dr. Nicolaas Tulp": An Iconological Study*. New York: New York University Press, 1958.

Heel, S. A. C. Dudok van, I. C. E. Wesdorp, T. Beijer, J. N. Keeman, Henriëtte A. Bosman-Jelgersma, and G. Nolthenius de Man. *Nicholaes Tulp: The Life and Work of an Amsterdam Physician and Magistrate in the Seventeenth Century*. 2nd ed. Trans. Karen Gribling. Amsterdam: Six Art Promotion, 1998.

Heijer, Henk den. *De geschiedenis van de WIC*. Zutphen: Walburg, 1994.

Heinecke, Berthold. "The Mysticism and Science of Johann Baptista Van Helmont (1579–1644)." *Ambix* 42 (1995): 65–78.

Helden, Albert van. *The Invention of the Telescope*. Transactions of the American Philosophical Society, vol. 67, pt. 4. Philadelphia: American Philosophical Society, 1977.

Hellinga, G. "Geschiedenis der [. . .] gasthuizen te Amsterdam." *Nederlandsch Tijdschrift voor Geneeskunde* 77, nos. 1–3 (1933): 528–43, 1410–24, 3042–54.

Helm, Jürgen. "Protestant and Catholic Medicine in the Sixteenth Century? The Case of Ingolstadt Anatomy." *Medical History* 45 (2001): 83–96.

Helms, Mary W. "Essay on Objects: Interpretations of Distance Made Tangible." In *Implicit Understandings: Observing, Reporting, and Reflecting on the Encounters between Europeans and Other Peoples in the Early Modern Era*, ed. Stuart B. Schwartz, 355–77. Cambridge: Cambridge University Press, 1994.

Hendrix, Lee, and Thea Vignau-Wilberg. *Nature Illuminated: Flora and Fauna from the Court of the Emperor Rudolph II*. Los Angeles: J. Paul Getty Museum, 1997.

Heniger, Johannes. "De eerste Nederlandse wetenschappelijke reis naar Oost-Indië, 1599–1601." *Leids Jaarboekje* 65 (1973): 27–49.

———. *Hendrik Adriaan van Reede tot Drakenstein (1636–1691) and Hortus Malabaricus: A Contribution to the History of Dutch Colonial Botany*. Rotterdam: A. A. Balkema, 1986.

———. "Some Botanical Activities of Herman Boerhaave, Professor of Botany and Director of the Botanic Garden at Leiden." *Janus* 58 (1971): 1–78.

———. "Der wissenschaftliche Nachlass von Paul Hermann." *Wissenschaftliche Zeitschrift der Martin-Luther-Universität Halle-Wittenberg* 18 (1969): 527–60.

Higman, B. W. "The Sugar Revolution." *Economic History Review* 53 (2000): 213–36.

Hingston Quiggin, A. *A Survey of Primitive Money: The Beginnings of Currency*. Intro. A. C. Haddon. London: Methuen, 1949.

Hirschman, Albert O. *The Passions and the Interests: Political Arguments for Capitalism before Its Triumph.* Princeton, NJ: Princeton University Press, 1977.

Hobhouse, Penelope. *Plants in Garden History.* London: Pavilion Books, 1992.

Hoboken, W. J. van. "The Dutch West India Company: The Political Background of Its Rise and Decline." In *Britain and the Netherlands,* vol. 1, ed. J. S. Bromley and E. H. Kossmann, 41–61. London: Chatto and Windus, 1960.

Hoeven, J. van der. "Het chirurgijn-gilde te Deventer." *Nederlands Tijdschrift voor Geneeskunde* 78, no. 2 (1934): 1547–59.

Hollingsworth, Mary. *The Cardinal's Hat: Money, Ambition and Housekeeping in a Renaissance Court.* London: Profile Books, 2004.

Holthuis, L. B. *Marcgraf's (1648) Brazilian Crustacea.* Zoologische Verhandelinen, 268. Leiden: National Natuurhistorisch Museum, 1991.

Honour, Hugh. *The European Vision of America.* Cleveland, OH: Cleveland Museum of Art, 1975.

Hoopes, Robert. *Right Reason in the English Renaissance.* Cambridge, MA: Harvard University Press, 1962.

Hoorn, Robert-Jan van. "On Course for Quality: Justus Lipsius and Leiden University." In *Lipsius in Leiden: Studies in the Life and Works of a Great Humanist on the Occasion of His 450th Anniversary,* ed. Karl Enenkel and Chris Heesakkers, 73–92. Bloemendaal: Florivallis, 1997.

Hooykaas, R. *Humanisme, science, et réforme: Pierre de la Ramée (1515–1572).* Leyden: E. J. Brill, 1958.

———. *Religion and the Rise of Modern Science.* Grand Rapids, MI: William B. Eerdmans, 1972.

———. "The Rise of Modern Science: When and Why?" *British Journal for the History of Science* 20 (1987): 453–73.

Hoppen, K. Theodore. *The Common Scientist in the Seventeenth Century: A Study of the Dublin Philosophical Society, 1683–1708.* Charlottesville: University Press of Virginia, 1970.

Hopper, Florence. "Clusius's World: The Meeting of Science and Art." In *The Authentic Garden: A Symposium on Gardens,* ed. Leslie Tjon Sie Fat and Erik de Jong, 13–36. Leiden: Clusius Foundation, 1991.

———. "Daniel Marot: A French Garden Designer in Holland." In *The Dutch Garden in the Seventeenth Century,* ed. John Dixon Hunt, 131–58. Dumbarton Oaks Colloquium on the History of Landscape Architecture, 12. Washington, DC: Dumbarton Oaks Research Library and Collection, 1990.

———. "The Dutch Classical Garden and André Mollet." *Journal of Garden History* 2 (1982): 25–40.

Horne, Thomas A. *The Social Thought of Bernard Mandeville: Virtue and Commerce in Early Eighteenth-Century England.* New York: Columbia University Press, 1978.

Hoskins, Janet. *Biographical Objects: How Things Tell the Stories of People's Lives.* New York: Routledge, 1998.

Houghton, W. E. "The English Virtuoso in the Seventeenth Century." *Journal of the History of Ideas* 3 (1942): 51–73, 190–219.

Houtzager, H. L. *Medicyns, vroedwyfs en chirurgyns: schets van de gezondheidszorg in Delft en beschrijving van het theatrum anatomicum aldaar in de zestiende en zeventiende eeuw.* Amsterdam: Rodopi, 1979.

Houtzager, H. L., and Michiel Jonker, eds. *De snijkunst verbeeld: Delftse anatomische lessen nader belicht.* Voorburg: Reinier de Graaf Groep; and Zwolle: Waanders, n.d.

Howard, Rio. "Guy de la Brosse and the Jardin des Plantes in Paris." In *The Analytic Spirit*, ed. Harry Woolf, 195–224. Ithaca, NY: Cornell University Press, 1981.

Hsu, Elizabeth. "Towards a Science of Touch, Part I: Chinese Pulse Diagnostics in Early Modern Europe." *Anthropology and Medicine* 7 (2000): 251–68.

———. "Towards a Science of Touch, Part II: Representations of the Tactile Experience of the Seven Chinese Pulses Indicating Danger of Death in Early Modern Europe." *Anthropology and Medicine* 7 (2000): 319–33.

Huberts, Fr. de Witt. *De Beul en z'n werk.* Amsterdam: Andries Blitz, 1937.

Huerta, Robert D. *Giants of Delft: Johannes Vermeer and the Natural Philosophers; The Parallel Search for Knowledge during the Age of Discovery.* Lewisburg, PA: Bucknell University Press, 2003.

Huff, Toby E. *The Rise of Early Modern Science: Islam, China, and the West.* 2nd ed. 1993. Cambridge: Cambridge University Press, 2003.

Huisman, Frank. "Civic Roles and Academic Definitions: The Changing Relationship between Surgeons and Urban Government in Groningen, 1550–1800." In *The Task of Healing: Medicine, Religion and Gender in England and the Netherlands, 1450–1800*, ed. Hilary Marland and Margaret Pelling, 69–100. Rotterdam: Erasmus, 1996.

———. "Itinerant Medical Practitioners in the Dutch Republic: The Case of Groningen." *Tractrix* 1 (1989): 63–83.

———. "Shaping the Medical Market: On the Construction of Quackery and Folk Medicine in Dutch Historiography." *Medical History* 43 (1999): 359–75.

———. *Stadsbelang en standsbesef: gezondheidszorg en medisch beroep in Groningen, 1500–1730.* Rotterdam: Erasmus, 1992.

Hull, William I. *Benjamin Furly and Quakerism in Rotterdam.* Lancaster, PA: Swarthmore College, 1941.

Hunger, F. W. T. "Bernardus Paludanus (Berent ten Broecke) (1550–1633)." *Janus* 32 (1928): 353–64.

———. "Boerhaave als natuurhistoricus." *Nederlands Tijdschrift voor Geneeskunde* 63, no. 1A (1919): 36–44.

———. "Charles de L'Escluse (Carolus Clusius), 1526–1609." *Janus* 31 (1927): 139–51.

———. *Charles de L'Escluse: Carolus Clusius, Nederlandsche kruidkundige, 1526–1609.* 2 vols. The Hague Martinus Nijhoff, 1927–42.

Hunt, Edwin S., and James M. Murray. *A History of Business in Medieval Europe, 1200–1550.* Cambridge: Cambridge University Press, 1999.

Hunt, John Dixon. "The Garden in the City of Venice: Epitome of State and Site." *Studies in the History of Gardens and Designed Landscapes* 19 (1999): 46–61.

Hunter, Michael. *Establishing the New Science: The Experience of the Early Royal Society.* Woodbridge: Boydell, 1989.

Hutchinson, G. Evelyn. "Attitudes toward Nature in Medieval England: The Alphonso and Bird Psalters." *Isis* 65 (1974): 5–37.

Iliffe, Robert. "Foreign Bodies: Travel, Empire and the Early Royal Society of London." *Canadian Journal of History* 33, 34 (1998, 1999): 357–85, 23–50.

———. "The Masculine Birth of Time: Temporal Frameworks of Early Modern Natural Philosophy." *British Journal for the History of Science* 33 (2000): 427–53.

Israel, Jonathan I. *Dutch Primacy in World Trade, 1585–1740*. Oxford: Oxford University Press, 1989.

———. *The Dutch Republic: Its Rise, Greatness, and Fall, 1477–1806*. Oxford: Clarendon Press, 1995.

———. *Radical Enlightenment: Philosophy and the Making of Modernity, 1650–1750*. Oxford: Oxford University Press, 2001.

Iwao, Seiichi. "Dutch Physician Willem ten Rhijne and Early Western Medicine in Japan." [In Japanese.] *Bulletin of the Japan-Netherlands Institute* 1 (1976): 1–90.

Jacob, James R., and Margaret C. Jacob. "The Anglican Origins of Modern Science: The Metaphysical Foundations of the Whig Constitution." *Isis* 71 (1980): 251–67.

Jacob, Margaret C. "The Church and the Formulation of the Newtonian World-View." *Journal of European Studies* 1 (1971): 128–48.

———. "Christianity and the Newtonian Worldview." In *God and Nature: Historical Essays on the Encounter between Christianity and Science*, ed. David C. Lindberg and Ronald L. Numbers, 238–55. Berkeley: University of California Press, 1986.

———. *The Cultural Meaning of the Scientific Revolution*. Philadelphia: Temple University Press, 1988.

———. *The Newtonians and the English Revolution, 1689–1720*. Ithaca, NY: Cornell University Press, 1976.

Jacob, Margaret C., and Matthew Kadane. "Missing, Now Found in the Eighteenth Century: Weber's Protestant Capitalist." *American Historical Review* 108 (2003): 20–49.

Jaki, Stanley L. *The Origin of Science and the Science of Its Origin*. South Bend, IN: Regency/Gateway, 1979.

James, Susan. *Passion and Action: The Emotions in Seventeenth-Century Philosophy*. Oxford: Clarendon Press, 1997.

Jansma, Jan Reinier. *Louis de Bils en de anatomie van zijn tijd*. Hoogeveen: C. Pet, 1919.

Jardine, Lisa. *Worldly Goods: A New History of the Renaissance*. New York: Nan A. Talese/Doubleday, 1996.

Jaspers, Karl. *Spinoza*. Ed. Hannah Arendt. Trans. Ralph Manheim. New York: Harcourt Brace Jovanovich, Harvest, 1974.

Jeanselme, E. "L'Oeuvre de J. Bontius." Paper presented at the Sixth International Congress of the History of Medicine (Leiden-Amsterdam, 18–23 July 1927), 209–222. Anvers: De Vlijt, 1929.

Jones, Ellis. "The Life and Works of Guilhelmus Fabricius Hildanus (1560–1634)." *Medical History* 4 (1960): 112–34, 196–209.

Jones, Richard F. *Ancients and Moderns: A Study of the Rise of the Scientific Movement in Seventeenth-Century England*. 1936. St. Louis, MO: Washington University Press, 1961.

Jones, Whitney R. D. *William Turner: Tudor Naturalist, Physician and Divine*. London: Routledge and Kegan Paul, 1988.

Jong, Erik de. "'Netherlandish Hesperidies': Garden Art in the Period of William and Mary, 1650–1702." In *The Anglo-Dutch Garden in the Age of William and Mary/De Gouden Eeuw Van de Hollandse Tuinkunst*. Special double issue of *Journal of Garden History*, vol. 8, nos. 2 and 3, ed. John Dixon Hunt and Erik de Jong, 15–40. London: Taylor and Francis, 1988.

———. "Nature and Art: The Leiden Hortus as 'Musaeum.'" In *The Authentic Garden: A Symposium on Gardens*, ed. Leslie Tjon Sie Fat and Erik de Jong, 37–60. Leiden: Clusius Foundation, 1991.

———. Natuur en Kunst: *Nederlandse Tuin- en Landschapsarchitechtuur, 1650–1740*. 1993. Amsterdam Thoth, 1995.

———. "For Profit and Ornament: The Function and Meaning of Dutch Garden Art in the Period of William and Mary, 1650–1702." In *The Dutch Garden in the Seventeenth Century*, ed. John Dixon Hunt, 13–48. Dumbarton Oaks Colloquium on the History of Landscape Architecture, 12. Washington, DC: Dumbarton Oaks Research Library and Collection, 1990.

Jongh, E. de. *Questions of Meaning: Themes and Motif in Dutch Seventeenth-Century Painting*. Trans. Michael Hoyle. 1995. Leiden: Primavera, 2000.

Jorink, Eric. "'Het boeck der natuere': Nederlandse geleerden en de wonderen van Gods schepping, 1575–1715." Ph.D. diss., University of Groningen, 2003.

Jouanna, Jacques. *Hippocrates*. Trans. M. B. DeBevoise. Medicine and Culture. Baltimore: Johns Hopkins University Press, 1999.

Joy, Lynn S. "Epicureanism in Renaissance Moral and Natural Philosophy." *Journal of the History of Ideas* 53 (1992): 573–83.

———. *Gassendi the Atomist: Advocate of History in an Age of Science*. Cambridge: Cambridge University Press, 1987.

Jurriaanse, M. W. *The Founding of Leyden University*. Trans. J. Brotherhood. Leiden: E. J. Brill, 1965.

Kajdański, Edward. "Michael Boym's *Medicus Sinicus*." *T'Oung Pao* 73 (1987): 161–89.

Kambouchner, Denis. *L'homme des passions: commentaires sur Descartes*. 2 vols. Paris: Albin Michel, 1995.

Kaplan, Benjamin J. "'Remnants of the Papal Yoke': Apathy and Opposition in the Dutch Reformation." *Sixteenth Century Journal* 25 (1994): 653–69.

Karsten, Mia C. *The Old Company's Garden at the Cape and Its Superintendents*. Cape Town: Maskew Miller, 1951.

Kato, Eiichi. "Unification and Adaptation: The Early Shogunate and Dutch Trade Policies." In *Companies and Trade*, ed. Leonard Blussé and Femme Gaastra, 207–29. Leiden: Leiden University Press, 1981.

Kaufmann, Thomas DaCosta. *The Mastery of Nature: Aspects of Art, Science, and Humanism in the Renaissance*. Princeton, NJ: Princeton University Press, 1993.

Kaye, Joel. *Economy and Nature in the Fourteenth Century: Money, Market Exchange, and the Emergence of Scientific Thought*. Cambridge: Cambridge University Press, 1998.

Keay, John. *The Spice Route: A History*. London: John Murray, 2005.

Keele, K. "Leonardo da Vinci's Influence on Renaissance Anatomy." *Medical History* 8 (1964): 360–70.

Kenny, Neil. *The Uses of Curiosity in Early Modern France and Germany*. Oxford: Oxford University Press, 2004.

Keohane, Nannerl O. *Philosophy and the State in France: The Renaissance to the Enlightenment*. Princeton, NJ: Princeton University Press, 1980.

Kerkhoven, J. M., and H. W. Blom. "De la Court en Spinoza: Van correspondenties en correspondenten." In *Pieter de la Court in zijn tijd: aspecten van een veelzijdig publicist (1618–1685)*, ed. H. W. Blom and I. W. Wildenberg, 137–60. Amsterdam: APA—Holland University Press, 1986.

Keuning, J. "Ambonese, Portuguese and Dutchmen: The History of Ambon to the End of the Seventeenth Century." Translated; Dutch version 1956. In *Dutch Authors on Asian History*, ed. M. A. P. Meilink-Roelofsz, M. E. van Opstall, and G. J. Schutte, 362–97. Dordrecht: Foris, 1988.

Kiers, Judikje, and Fieke Tissink. *The Golden Age of Dutch Art: Painting, Sculpture, Decorative Art*. London: Thames and Hudson, 2000.

King, Lester S. *The Road to Medical Enlightenment, 1650–1695*. New York: American Elsevier, 1970.

Kinukawa, Tomomi. "Art Competes with Nature: Maria Sibylla Merian (1647–1717) and the Culture of Natural History." Ph.D. diss., University of Wisconsin-Madison, 2001.

Kish, George. "*Medicina, Mensura, Mathematica*: The Life and Works of Gemma Frisius, 1508–1555." *Publication of the Associates of the James Ford Bell Collection*, 4. Minneapolis, MN, 1967.

Klashorst, G. O. van de. "'Metten schijn van monarchie getempert': de verdediging van het stadhouderschap in de partijliteratuur, 1650–1686." In *Pieter de la Court in zijn tijd: aspecten van een veelzijdig publicist (1618–1685)*, ed. H. W. Blom and I. W. Wildenberg, 93–136. Amsterdam: APA—Holland University Press, 1986.

Klein, P. W., and J. W. Veluwenkamp. "The Role of the Entrepreneur in the Economic Expansion of the Dutch Republic." In *Economic and Social History of the Netherlands*, 27–53. Het Nederlandsch Economisch-Historisch Archief, 4. Amsterdam: NEHA, 1993.

Klein, Ursula. "Experimental History and Herman Boerhaave's Chemistry of Plants." *Studies in the History and Philosophy of Biological and Biomedical Sciences* 34 (2003): 533–67.

Klever, W. N. A. "Spinoza's Life and Works." In *The Cambridge Companion to Spinoza*, ed. Don Garrett, 13–60. Cambridge: Cambridge University Press, 1996.

Klooster, Wim. *The Dutch in the Americas, 1600–1800: A Narrative History with the Catalogue of an Exhibition*. Providence, RI: John Carter Brown Library, 1997.

Knaap, Gerrit. "Headhunting, Carnage and Armed Peace in Amboina, 1500–1700." *Journal of the Economic and Social History of the Orient* 46 (2003): 165–92.

Knoeff, Rina. *Herman Boerhaave (1668–1738): Calvinist Chemist and Physician*. Amsterdam: Koninklijke Nederlandse Akademie van Wetenschappen, 2002.

Koeningsberger, H. G. "Why Did the States General of the Netherlands Become Revolutionary in the Sixteenth Century?" In Koeningsberger, *Politicians and Virtuosi: Essays in Early Modern History*, 63–76. London: Hambledon, 1986.

Koerner, Lisbet. *Linnaeus: Nature and Nation*. Cambridge, MA: Harvard University Press, 1999.

Kooi, Christine. "Popish Impudence: The Perseverance of the Roman Catholic Faithful in Calvinist Holland, 1572–1620." *Sixteenth Century Journal* 26 (1995): 75–85.

Kooijmans, Luuc. *De doodskunstenaar: de anatomische lessen van Frederik Ruysch*. Amsterdam: Bert Bakker, 2004.

Kossmann, E. H. "The Development of Dutch Political Theory in the Seventeenth Century." In *Britain and the Netherlands*, vol. 1, ed. J. S. Bromley and E. H. Kossmann, 91–110. London: Chatto and Windus, 1960.

———. *Politieke theorie en geschiedenis: verspreide opstellen en voordrachten*. Amsterdam: Bert Bakker, 1987.

———. *Politieke theorie in het zeventiende-eeuwse Nederland*. Verhandeling der Koninklijke Nederlandse Akademie van Wetenschappen, Afd. Letterkunde. Amsterdam: N.V. Noord-Hollandsche Uitgevers Maatschappij, 1960.

———. "Popular Sovereignty at the Beginning of the Dutch Ancien Regime." *Low Countries History Yearbook* 14 (1981): 1–28.

Kraft, Eva S. *Andreas Cleyer tagebuch des kontors zu Nagasaki auf der insel Deschima*. Bonner Zeitschrift für Japanologie. Bonn, 1985.

———. "Christian Mentzel, Philippe Couplet, Andreas Cleyer und die Chinesische Medizin: Notizen aus Handschriften des Siebseltn Jahrhunderts." In *Fernöstliche Kultur: Wolf Haenisch Zugeeignet von Sienem Marburger Studienkreis*, 158–96. Marburg: N. G. Elwert, 1975.

Kranenburg, H. A. H. *De zeevisscherij van Holland in den tijd der Republiek*. Amsterdam: H. J. Paris, 1946.

Kraye, Jill. "Moral Philosophy." In *The Cambridge History of Renaissance Philosophy*, gen. ed. Charles B. Schmitt, ed. Quentin Skinner, Eckhard Kessler, assoc. ed. Jill Kraye, 303–86. Cambridge: Cambridge University Press, 1988.

Kroeskamp, H. "De Chinezen te Batavia (±1700) als exempel voor de Cristenen van West-Europa?" *Indonesië: tweemaandelijks tijdscrift gewijd aan het Indonesisch cultuurgebied* 6 (1952/3): 346–71.

Kroon, J. E. *Bijdragen tot de geschiedenis van het geneeskundig onderwijs aan de Leidsche Universiteit, 1575–1625*. 1911. The Hague: J. Couvreur, [c. 1920].

Kuhn, Thomas S. *The Structure of Scientific Revolutions*. 1962. International Encyclopedia of Unified Science. Chicago: University of Chicago Press, 1967.

Kuijlen, J., C. S. Oldenburger-Ebbers, and D. O. Wijnands. *Paradisus Batavus: bibliografie van plantencatalogi van onderwijstuinen, particuliere tuinen en kwekerscollecties in de Noordelijke en Zuidelijke Nederlanden (1550–1839)*. Wageningen: Pudoc, 1983.

Kuriyama, Shigehisa. "Between Mind and Eye: Japanese Anatomy in the Eighteenth Century." In *Paths to Asian Medical Knowledge*, ed. Charles Lesley and Allan Young, 21–43. Berkeley: University of California Press, 1992.

———. *The Expressiveness of the Body and the Divergence of Greek and Chinese Medicine*. New York: Zone Books, 1999.

———. "Interpreting the History of Bloodletting." *Journal of the History of Medicine* 50 (1995): 11–46.

———. "The Japanese Complaint of Katakori and the Puzzle of Local Diseases." Unpublished paper.

———. "Pulse Diagnosis in the Greek and Chinese Traditions." In *History of Diagnostics: Proceedings of the Ninth International Symposium on the Comparative History of Medicine East and West*, ed. Yosio Kawakita, 43–67. Osaka: Taniguchi Foundation, 1987.

———. "Visual Knowledge in Classical Chinese Medicine." In *Knowledge and the Scholarly Medical Traditions*, ed. Don Bates, 205–34. Cambridge: Cambridge University Press, 1995.

Kusukawa, Sachiko. "*Aspectio divinorum operum*: Melanchthon and Astrology for Lutheran Medics." In *Medicine and the Reformation*, ed. Ole Peter Grell and Andrew Cunningham, 33–56. London: Routledge, 1993.

———. *The Transformation of Natural Philosophy: The Case of Philip Melanchthon*. Cambridge: Cambridge University Press, 1995.

Lach, Donald F. *Asia in the Making of Europe*. 2 vols. in 5. Chicago: University of Chicago Press, 1965–77.

Lachmann, Richard. *Capitalists in Spite of Themselves: Elite Conflict and Economic Transitions in Early Modern Europe*. Oxford: Oxford University Press, 2000.

Landes, David S. *Revolution in Time: Clocks and the Making of the Modern World*. Cambridge, MA: Belknap Press of Harvard University Press, 1983.

Landwehr, John. *Studies in Dutch Books with Coloured Plates Published 1662–1875: Natural History, Topography and Travel Costumes and Uniforms*. The Hague: Dr. W. Junk, 1976.

Lane, Frederic C. "The Mediterranean Spice Trade: Further Evidence of Its Revival in the Sixteenth Century." In *Spices in the Indian Ocean World*, ed. M. N. Pearson, 111–20. An Expanding World: The European Impact on World History, 1450–1800. Aldershot: Ashgate, 1996.

———. "Pepper Prices before da Gama." In *Spices in the Indian Ocean World*. ed. M. N. Pearson, 85–92. An Expanding World: The European Impact on World History, 1450–1800. Aldershot: Ashgate, 1996.

———. *Venice: A Maritime Republic*. Baltimore: Johns Hopkins University Press, 1973.

Latour, Bruno. *Pandora's Hope: Essays on the Reality of Science Studies*. Cambridge, MA: Harvard University Press, 1999.

———. *Science in Action: How to Follow Scientists and Engineers through Society*. Cambridge, MA: Harvard University Press, 1987.

Latour, Bruno, and Steve Woolgar. *Laboratory Life: The Construction of Scientific Facts*. 1979. Princeton, NJ: Princeton University Press, 1986.

Laughran, Michelle A. "Medicating with or without 'Scruples': The 'Professionalization' of the Apothecary in Sixteenth-Century Venice." *Pharmacy in History* 45 (2003): 95–107.

Law, John. "On the Methods of Long-Distance Control: Vessels, Navigation and the Portuguese Route to India." In *Power, Action and Belief: A New Sociology of Knowledge?* ed. John Law, 234–63. Sociological Review Monograph, 32. London: Routledge, 1986.

Le Goff, Jacques. "Merchant's Time and Church's Time in the Middle Ages." In *Time, Work, and Culture in the Middle Ages*, 29–42. French ed. 1977. Trans. Arthur Goldhammer. Chicago: University of Chicago Press, 1980.

Lennon, Thomas M. *The Battle of the Gods and Giants: The Legacies of Descartes and Gassendi, 1655–1715*. Princeton, NJ: Princeton University Press, 1993.

Lennon, Thomas M., and Patrica Ann Easton. *The Cartesian Empiricism of François Bayle*. New York: Garland, 1992.

Lesger, Clé. "Intraregional Trade and the Port System in Holland, 1400–1700." In *Economic and Social History in the Netherlands*. Het Nederlandsch Economisch-Historisch Archief, vol. 4, 186–218. Amsterdam: NEHA, 1993.

Leupe, P. A. *Georgius Everardus Rumphius: Ambonsch natuurkundige der zeventiende eeuw*. Amsterdam: C. G. van der Post, 1871.

———. "Letter Transport Overland to the Indies by the East India Company in the Seventeenth Century." Dutch version, 1870. In *Dutch Authors on Asian History*, ed. M. A. P. Meilink-Roelofsz, M. E. van Opstall, and G. J. Schutte, 77–90. Dordrecht: Foris, 1988.

Levi, Anthony. *French Moralists: The Theory of the Passions, 1585 to 1649*. Oxford: Clarendon Press, 1964.

Lieberman, Victor. "Some Comparative Thoughts on Premodern Southeast Asian Warfare." *Journal of the Economic and Social History of the Orient* 46 (2003): 215–25.

Lieburg, Martin van. *Nieuw licht op Hendrik van Deventer (1651–1724)*. Rotterdam: Erasmus, 2002.

Lieburg, M. J. van. "De genees- en heelkunde in de Noordelijke Nederlanden, gezien vanuit de stedelijke en chirurgijnsgilde-ordonnanties van de zestiende eeuw." *Tijdschrift voor de Geschiedenis der Geneeskunde, Natuurwetenschappen, Wiskunde en Techniek* 6 (1983): 169–84.

———. "Isaac Beeckman and His Diary-Notes on William Harvey's Theory on Bloodcirculation (1633–1634)." *Janus* 69 (1982): 161–83.

———. *Het medisch onderwijs te Rotterdam (1467–1967)*. Amsterdam: Rodopi, 1978.

———. "Pieter van Foreest en de rol van de stadsmedicus in de Noord-Nederlandse steden van de zestiende eeuw." In *Pieter van Foreest: een Hollands medicus in de zestiende eeuw*, ed. H. L. Houtzager, 41–72. Amsterdam: Rodopi, 1989.

———. "Zacharias Sylvius (1608–1664), Author of the *Praefatio* to the First Rotterdam Edition (1648) of Harvey's *De motu cordis*." *Janus* 65 (1978): 241–57.

Lindeboom, G. A. "Het Collegium privatum Amstelodamense (1664–1673)." *Nederlands Tijdschrift voor Geneeskunde* 119, no. 32 (1975): 1248–54.

———. *Descartes and Medicine*. Amsterdam: Rodopi, 1979.

———. *Florentius Schuyl (1619–1669) en zijn betekenis voor het Cartensianisme in de geneeskunde*. The Hague: Martinus Nijhoff, 1974.

———. *Herman Boerhaave: The Man and His Work*. Foreword by E. Ashworth Underwood. London: Methuen, 1968.

———. "Jan Swammerdam als microscopist." *Tijdschift voor de Geschiedenis der Geneeskunde, Natuurwetenschappen, Wiskunde en Techniek* 4 (1981): 87–110.

———. "Leeuwenhoek and the Problem of Sexual Reproduction." In *Antoni van Leeuwenhoek, 1632–1723*, ed. L. C. Palm and H. A. M. Snelders, 129–52. Amsterdam: Rodopi, 1982.

———. "Medical Education in the Netherlands, 1575–1750." In *The History of Medical Education*, ed. C. D. O'Malley, 201–16. Berkeley: University of California Press, 1970.

Lindeman, Ruud, Yvonne Scherf, and Rudolf M. Dekker, comp. *Reisverslagen van Noord-Nederlanders uit het zestiende tot begin negentiende eeuw.* Rotterdam: Universiteitsdrukkerij Erasmus Universiteit, 1994.

Linebaugh, Peter. “The Tyburn Riot against the Surgeons.” In *Albion’s Fatal Tree: Crime and Society in Eighteenth-Century England*, ed. Douglas Hay, Peter Linebaugh, John G. Rule, E. P. Thompson, and Cal Winslow, 65–117. New York: Pantheon, 1975.

Lint, J. de. “Comment Jan de Doot, forgeron, s’opéra d’un calcul de la vessie.” *Asculape* 18 (1928): 50–53.

Lloyd, G. E. R. *The Ambitions of Curiosity: Understanding the World in Ancient Greece and China.* Cambridge: Cambridge University Press, 2002.

Lombard, Denys. “Questions on the Contact between European Companies and Asian Societies.” In *Companies and Trade*, ed. Leonard Blussé and Femme Gaastra, 179–87. Leiden: Leiden University Press, 1981.

Lonie, Ian M. “The ‘Paris Hippocratics’: Teaching and Research in Paris in the Second Half of the Sixteenth Century.” In *The Medical Renaissance of the Sixteenth Century*, ed. Andrew Wear, Roger K. French, and Ian M. Lonie, 155–74, 318–26. Cambridge: Cambridge University Press, 1985.

Lorch, Maristella de P. “The Epicurean in Lorenzo Valla’s *On Pleasure.*” In *Atoms, Pneuma, and Tranquillity: Epicurean and Stoic Themes in European Thought*, ed. Margaret J. Osler, 89–114. Cambridge: Cambridge University Press, 1991.

Lovejoy, Arthur O. *Reflections on Human Nature.* Baltimore: Johns Hopkins University Press, 1961.

Lu, Gwei-Djen, and Joseph Needham. *Celestial Lancets: A History and Rationale of Acupuncture and Moxa.* 1980. New intro. Vivienne Lo. London: RoutledgeCurzon, 2002.

Lucas, Alfred. *Ancient Egyptian Materials and Industries.* 4th ed. Rev. J. R. Harris. London: Edward Arnold, 1962.

Lucassen, Jan. *Migrant Labour in Europe, 1600–1900: The Drift to the North Sea.* Trans. Donald A. Bloch. London: Croom Helm, 1987.

Lugli, Adalgisa. *Naturalia et mirabilia: Il collezionismo enciclopedico nelle wunderkammern d’Europa.* Milan: Gabriele Mazzotta, 1983.

Lunsingh Scheurleer, Th. H. “Un amphithéâtre d’anatomie moralisée.” In *Leiden University in the Seventeenth Century: An Exchange of Learning*, ed. Th. H. Lunsingh Scheuleer and G. H. M. Posthumus Meyjes, 217–77. Leiden: Universitaire Pers/E. J. Brill, 1975.

Lunsingh Scheurleer, Th. H., and G. H. M. Posthumus Meyjes, eds. *Leiden University in the Seventeenth Century: An Exchange of Learning.* Leiden: Universitaire Pers Leiden/E. J. Brill, 1975.

Lux, David S. *Patronage and Royal Science in Seventeenth-Century France: The Académie de Physique in Caen.* Ithaca, NY: Cornell University Press, 1989.

Luyendijk-Elshout, Antonie M. “Death Enlightened: A Study of Frederik Ruysch.” *Journal of the American Medical Association* 212, no. 1 (1970): 121–26.

———. “Mechanicisme contra vitalisme: de school van Herman Boerhaave en de beginselen van het leven.” *Tijdschrift voor de Geschiedenis der Geneeskunde, Natuurwetenschappen, Wiskunde en Techniek* 5 (1982): 16–26.

MacDonald, Michael. "*The Fearefull Estate of Francis Spira*: Narrative, Identity, and Emotion in Early Modern England." *Journal of British Studies* 31 (1992): 32–61.

Macé, Mieko. "Évolution de la médecine japonaise face au modèle chinois: des origines jusqu'au milieu du dix-huitième siècle—l'autonomie par la synthèse." In *Cipango: Cahiers d'Études Japonaises*, 111–60. Paris: Publications Langues'O, 1992.

MacFarlane, Alan. *The Culture of Capitalism*. Oxford: Blackwell, 1987.

MacGregor, Arthur, "Collectors and Collections of Rarities in the Sixteenth and Seventeenth Centuries." In *Tradescant's Rarities: Essays on the Foundation of the Ashmolean Museum, 1683*, ed. A. MacGregor, 70–97. Oxford: Clarendon Press, 1983.

———, ed. *Tradescant's Rarities: Essays on the Foundation of the Ashmolean Museum, 1683, with a Catalogue of the Surviving Early Collections*. Oxford: Clarendon Press, 1983.

Het machtige eyland: Ceylon en de V.O.C. The Hague: SDU, 1988.

MacIntyre, Alasdair. *After Virtue: A Study in Moral Theory*. Notre Dame, IN: University of Notre Dame Press, 1981.

Maczak, Antoni. *Travel in Early Modern Europe*. Trans. Ursula Phillips Cambridge: Polity Press, 1995.

Magalhães Godinho, Vitorino. "Le repli vénitien et égyptien et la route du Cap, 1496–1533." In *Spices in the Indian Ocean World*, ed. M. N. Pearson, 93–110. An Expanding World: The European Impact on World History, 1450–1800. Aldershot: Variorum, 1996.

Mak, J. J. *De rederijkers*. Amsterdam: P. M. van Kampen en Zoon, 1944.

Manning, John. *The Emblem*. London: Reaktion, 2002.

Marland, Hilary, ed. *The Art of Midwifery: Early Modern Midwives in Europe*. London: Routledge, 1993.

———. *"Mother and Child Were Saved": The Memoirs (1693–1740) of the Frisian Midwife Catharina Schrader*. Trans. Hilary Marland. Amsterdam: Rodopi, 1987.

Marnef, Guido. *Antwerp in the Age of Reformation: Underground Protestantism in a Commercial Metropolis, 1550–1577*. Trans. J. C. Grayson. Baltimore: Johns Hopkins University Press, 1996.

Martines, Lauro. *Power and Imagination: City-States in Renaissance Italy*. 1979. New York: Vintage Books, 1980.

Masson, Georgina. *Italian Gardens*. London: Thames and Hudson, 1966.

Mauss, Marcel. *The Gift: Forms and Functions of Exchange in Archaic Studies*. Trans. Ian Cunnison, intro. E. E. Evans-Pritchard. New York: W. W. Norton, 1967.

McAdoo, H. R. *The Structure of Caroline Moral Theology*. London: Longmans, Green, 1949.

McCusker, John J., and Cora Gravesteijn. *The Beginnings of Commercial and Financial Journalism: The Commodity Price Currents, Exchange Rate Currents, and Money Currents of Early Modern Europe*. NEHA, 3rd ser., 11. Amsterdam: NEHA, 1991.

McGahagan, T. A. "Cartesianism in the Netherlands, 1639–1675: The New Science and the Calvinist Counter-Reformation." Ph.D. diss., University of Pennsylvania, 1976.

McKee, Francis. "Honeyed Words: Bernard Mandeville and Medical Discourse." In *Medicine in the Enlightenment*, ed. Roy Porter, 223–54. Amsterdam: Rodopi, 1995.

Meadow, Mark A. "Merchants and Marvels: Hans Jacob Fugger and the Origins of the

Wunderkammer." In *Merchants and Marvels: Commerce, Science, and Art in Early Modern Europe*, ed. Pamela H. Smith and Paula Findlen, 182–200. New York: Routledge, 2002.

Meerbeeck, P. J. van. *Recherches historiques et critiques sur la vie et les ouvrages de Rembert Dodoens (Dodonaeus)*. 1841. Utrecht: HES, 1980.

Meijizen Nihon Igakushi. [In Japanese.] 1955–64. 4 vols.

Meilink-Roelofsz, M. A. P. *Asian Trade and European Influence in the Indonesian Archipelago between 1500 and about 1630*. The Hague: Martinus Nijhoff, 1962.

———. "Aspects of Dutch Colonial Development in Asia in the Seventeenth Century." In *Britain and the Netherlands in Europe and Asia*, ed. J. S. Bromley and E. H. Kossmann, 56–82. London: Macmillan; New York: St. Martin's Press, 1968.

Meli, Domenico Bertoloni. "Authorship and Teamwork around the Cimento Academy: Mathematics, Anatomy, Experimental Philosophy." *Early Science and Medicine* 6 (2001): 65–95.

Melion, Walter S. *Shaping the Netherlandish Canon: Karel van Mander's Schilder-Boeck*. Chicago: University of Chicago Press, 1991.

Merians, Linda E. "What They Are, Who We Are: Representations of the 'Hottentot' in Eighteenth-Century Britain." *Eighteenth-Century Life* 17 (1993): 14–39.

Merton, Robert K. "Science and the Economy of Seventeenth Century England." *Science and Society* 3 (1939): 3–27.

———. *Science, Technology and Society in Seventeenth Century England*. 1938. New York: Harper and Row, 1970.

Meyer, A. W. "Leeuwenhoek as Experimental Biologist." *Osiris* 3 (1937): 103–22.

Michel, Wolfgang. "Frühe Westliche Beobachtungen zur Moxibustion und Akupunktur." *Sudhoffs Archiv* 77 (1993): 193–222.

———. *Von Leipzig nach Japan: Der Chirurg und Handelsmann Caspar Schamberger (1623–1706)*. Munich: Iudicium, 1999.

———. "Willem ten Rhijne und die Japanische Medizin (I)." *Studien zur Deutschen und Französischen Literatur* 39 (1989): 75–125.

———. "Willem ten Rhijne und die Japanische Medizin (II)." *Studien zur Deutschen und Französischen Literatur* 40 (1990): 57–103.

Mikkeli, Heikki. *Hygiene in the Early Modern Medical Tradition*. Humaniora Series, 305. Helsinki: Finnish Academy of Science and Letters, 1999.

Miller, Peter N. *Peiresc's Europe: Learning and Virtue in the Seventeenth Century*. New Haven and London: Yale University Press, 2000.

Miller, Susan. "Jean-Antoine Fraisse at Chantilly: French Images of Asia." *East Asian Library Journal* 9 (2000): 1–77.

Milroy, James, and Lesley Milroy. "Linguistic Change, Social Network and Speaker Innovation." *Journal of Linguistics* 21 (1985): 339–84.

Milroy, Lesley. *Language and Social Networks*. 2nd ed. Oxford: Basil Blackwell, 1987.

Milroy, Lesley, and Sue Margrain. "Vernacular Language Loyalty and Social Network." *Language in Society* 9 (1980): 43–70.

Milroy, Lesley, and James Milroy. "Social Network and Social Class: Toward an Integrated Sociolinguistic Model." *Language in Society* 21 (1992): 1–26.

Milton, Giles. *Nathaniel's Nutmeg: Or, The True and Incredible Adventures of the Spice Trader Who Changed the Course of History*. New York: Farrar, Straus, and Giroux, 1999.

———. *Samurai William: The Adventurer Who Unlocked Japan*. London: Hodder and Stoughton, 2002.

Mintz, Sidney W. *Sweetness and Power: The Place of Sugar in Modern History*. New York: Viking, 1985.

Moch, Leslie Page. *Moving Europeans: Migration in Western Europe since 1650*. Bloomington: Indiana University Press, 1992.

Moer, A. van der. *Een zestiende-eeuwse Hollander in het verre oosten en het hoge noorden: leven, werken, reizen en avonturen van Jan Huyghen van Linschoten (1563–1611)*. The Hague: Martinus Nijhoff, 1979.

Mokyr, Joel. *The Gifts of Athena: Historical Origins of the Knowledge Economy*. Princeton, NJ: Princeton University Press, 2002.

Momigliano, Arnaldo. "Ancient History and the Antiquarian." *Journal of the Warburg and Courtauld Institutes* 13 (1950): 285–315.

Mooij, Annet. *De polsslag van de stad: 350 jaar academische geneeskunde in Amsterdam*. Amsterdam: Arbeiderspers, 1999.

Moore, Cornelia Niekus. "'Not by Nature but by Custom': Johan van Beverwijck's *Van de Wtnementheyt des Vrouwelicken Geslachts*." *Sixteenth Century Journal* 25 (1994): 633–51.

Moran, Bruce T. *The Alchemical World of the German Court: Occult Philosophy and Chemical Medicine in the Circle of Moritz of Hessen (1572–1632)*. Sudhoffs Archiv, suppl. 29. Stuttgart: Franz Steiner Verlag, 1991.

———. *Chemical Pharmacy Enters the University: Johannes Hartmann and the Didactic Care of "Chymiatria" in the Early Seventeenth Century*. Madison, WI: American Institute of the History of Pharmacy, 1991.

———. "Court Authority and Chemical Medicine: Moritz of Hessen, Johannes Hartmann, and the Origin of Academic *Chemiatria*." *Bulletin of the History of Medicine* 63 (1989): 225–46.

———. *Distilling Knowledge: Alchemy, Chemistry, and the Scientific Revolution*. Cambridge, MA: Harvard University Press, 2005.

———. "German Prince-Practitioners: Aspects in the Development of Courtly Science, Technology, and Procedures in the Renaissance." *Technology and Culture* 22 (1981): 253–74.

———. "Privilege, Communication, and Chemiatry: The Hermetic-Alchemical Circle of Moritz of Hessen-Kassel." *Ambix* 32 (1985): 110–26.

———, ed. *Patronage and Institutions: Science, Technology and Medicine at the European Court, 1500–1750*. Woodbridge: Boydell, 1991.

Morford, Mark. "The Stoic Garden." *Journal of Garden History* 7 (1987): 151–75.

———. *Stoics and Neostoics: Rubens and the Circle of Lipsius*. Princeton, NJ: Princeton University Press, 1991.

Mortier, Marijke. "Het wereldbeeld van de Gentse almanakken, zeventiende en achttiende eeuw." *Tijdschrift voor Sociale Geschiedenis* 35 (1984): 267–90.

Mosse, George L. "Changes in Religious Thought." In *The New Cambridge Modern His-*

tory, vol. 4: *The Decline of Spain and the Thirty Year's War, 1609–48/59*, ed. J. P. Cooper, 169–201. Cambridge: Cambridge University Press, 1970.

Motley, John Lothrop. *History of the United Netherlands: From the Death of William the Silent to the Synod of Dort*. 4 vols. London: J. Murray, 1860–67.

———. *The Rise of the Dutch Republic: A History*. 3 vols. London: J. Chapman, 1856.

Moulin, Daniel de. "Paul Barbette, M.D.: A Seventeenth-Century Amsterdam Author of Best-Selling Textbooks." *Bulletin of the History of Medicine* 59 (1985): 506–14.

Mout, M. E. H. N. "'Heilige Lipsius, bid voor ons.'" *Tijdschrift voor Geschiedenis* 97 (1984): 195–206.

Mout, Nicolette. "The Family of Love (Huis des Liefde) and the Dutch Revolt." In *Britain and the Netherlands*, vol. 7, ed. A. C. Duke and C. A. Tamse, 76–93. The Hague: Martinus Nijhoff, 1981.

———. "'Which Tyrant Curtails My Free Mind?': Lipsius and the Reception of De Constantia (1584)." In *Lipsius in Leiden: Studies in the Life and Works of a Great Humanist on the Occasion of His 450th Anniversary*, ed. Karl Enenkel and Chris Heesakkers, 123–40. Bloemendaal: Florivallis, 1997.

Moyer, Ann. "The Astronomers' Game: Astrology and University Culture in the Fifteenth and Sixteenth Centuries." *Early Science and Medicine* 4 (1999): 228–50.

Mueller, Reinhold C. *The Venetian Money Market: Banks, Panics, and the Public Debt, 1200–1500*. Baltimore: Johns Hopkins University Press, 1997.

Mukerji, Chandra. *From Graven Images: Patterns of Modern Materialism*. New York: Columbia University Press, 1983.

Muldrew, Craig. *The Economy of Obligation: The Culture of Credit and Social Relations in Early Modern England*. Basingstoke: Macmillan; and New York: St. Martin's Press, 1998.

Mulligan, Lotte. "Robert Boyle, 'Right Reason,' and the Meaning of Metaphor." *Journal of the History of Ideas* 55 (1994): 235–57.

Munro, John H. "Bullionism and the Bill of Exchange in England, 1272–1663: A Study in Monetary Management and Popular Prejudice." In *The Dawn of Modern Banking*, 169–215. Center for Medieval and Renaissance Studies, University of California, Los Angeles. New Haven and London: Yale University Press, 1979.

Munt, Annette Henriette. "The Impact of Dutch Cartesian Medical Reformers in Early Enlightenment German Culture 1680–1720." Ph.D. diss., University College London, 2004.

Muntschick, Wolfgang. "Ein Manuscript von Georg Meister, dem Kunst- und Lustgärtner, in der British Library." *Medizin-Historisches Journal* 19 (1984): 225–32.

Murray, John J. *Flanders and England: A Cultural Bridge; The Influence of the Low Countries on Tudor-Stuart England*. Antwerp: Fonds Mercator, 1985.

Nadler, Steven. *Spinoza: A Life*. Cambridge: Cambridge University Press, 1999.

Nakayama, Shigeru. *A History of Japanese Astronomy: Chinese Background and Western Impact*. Cambridge, MA: Harvard University Press, 1969.

Napjus, J. W. *De hoogleeraren in de geneeskunde aan de hogeschool en het Athenaeum te Franeker*. Ed. G. A. Lindeboom. Nieuwe Nederlandse Bijdragen tot de Geschiedenis der Geneeskunde en der Natuurwetenschappen, 15. Amsterdam: Rodopi, 1985.

Nauert, Charles G., Jr. *Agrippa and the Crisis of Renaissance Thought.* Illinois Studies in the Social Sciences, 55. Urbana: University of Illinois Press, 1965.

———. "Humanists, Scientists, and Pliny: Changing Approaches to a Classical Author." *American Historical Review* 84 (1979): 72–85.

Nave, F. de, and D. Imhof, eds. *Botany in the Low Countries (End of the Fifteenth Century—ca. 1650): Plantin-Moretus Museum Exhibition.* The Plantin-Moretus Museum and the Stedelijk Prentenkabinet Publication 27. Antwerp: Snoek-Ducaju and Zoon, 1993.

Néel, Marguerite. *Descartes et la Princesse Élisabeth.* Paris: Éditions Elzévir, 1946.

Nelson, Benjamin. *On the Roads to Modernity: Conscience, Science, and Civilisation; Selected Writings,* ed. Toby E. Huff. Totowa, NJ: Rowman and Littlefield, 1981.

Neri, Janice L. "From Insect to Icon: Joris Hoefnagel and the 'Screened Objects' of the Natural World." In *Ways of Knowing: Ten Interdisciplinary Essays,* ed. Mary Lindemann, 23–51. Leiden: Brill, 2004.

Newman, William R. *Gehennical Fire: The Lives of George Starkey, an American Alchemist in the Scientific Revolution.* Cambridge, MA: Harvard University Press, 1994.

———. *The "Summa Perfectionis" of Pseudo-Geber.* Leiden: E. J. Brill, 1991.

Niebyl, Peter H. "The Non-Naturals." *Bulletin of the History of Medicine* 45 (1971): 486–92.

———. "Sennert, Van Helmont, and Medical Ontology." *Bulletin of the History of Medicine* 45 (1971): 115–37.

Nieto-Galan, Agustí. "Calico Printing and Chemical Knowledge in Lancashire in the Early Nineteenth Century: The Life and 'Colours' of John Mercer." *Annals of Science* 54 (1997): 1–28.

Nogueira, Fernando A. R. "Garcia de Orta, Physician and Scientific Researcher." In *The Great Maritime Discoveries and World Health,* ed. Mário Gomes Marques and John Cule, 251–63. Lisbon: Escola Nacional de Saúde Pública, 1991.

———. "Luís de Almeida and the Introduction of European Medicine in Japan." In *The Great Maritime Discoveries and World Health,* ed. Mário Gomes Marques and John Cule, 227–36. Lisbon: Escola Nacional de Saúde Pública, 1991.

North, Douglass C., and Robert Paul Thomas. *The Rise of the Western World: A New Economic History.* Cambridge: Cambridge University Press, 1973.

Nutton, Vivian. "Dr. James's Legacy: Dutch Printing and the History of Medicine." In *The Bookshop of the World: The Role of the Low Countries in the Book-Trade, 1473–1941,* ed. Lotte Hellinga, Alistair Duke, Jacob Harskamp, and Theo Hermans, 207–17. 't Goy-Houten: HES and De Graaf, 2001.

———. "Hellenism Postponed: Some Aspects of Renaissance Medicine, 1490–1530." *Sudhoffs Archiv* 81 (1997): 158–70.

———. "Hippocrates in the Renaissance." In *Die Hippokratischen Epidemien: Theorie—Praxis—Tradition,* gen. ed. Gerhard Baader and Rolf Winau, 420–39. Sudhoffs Archiv, suppl. 27. Stuttgart: Franz Steiner Verlag, 1989.

———. "Pieter van Foreest and the Plagues of Europe: Some Observations on the *Observationes.*" In *Pieter van Foreest: een Hollands medicus in de zestiende eeuw,* ed. H. L. Houtzager, 25–39. Amsterdam: Rodopi, 1989.

———. "'Prisci Dissectionum Professores': Greek Texts and Renaissance Anatomists." In *The Uses of Greek and Latin: Historical Essays*, ed. A. C. Dionisotti, Anthony Grafton, and Jill Kraye, 111–26. London: Warburg Institute, 1988.

———. "Wittenberg Anatomy." In *Medicine and the Reformation*, ed. Ole Peter Grell and Andrew Cunningham, 11–32. London: Routledge, 1993.

Oestreich, Gerhard. *Neostoicism and the Early Modern State*, ed. Brigitta Oestreich and H. G. Koenigsberger, trans. David McLintock. Cambridge: Cambridge University Press, 1982.

Ofek, Haim. *Second Nature: Economic Origins of Human Nature*. Cambridge: Cambridge University Press, 2001.

Ogborn, Miles. "Streynsham Master's Office: Accounting for Collectivity, Order and Authority in Seventeenth-Century India." *Cultural Geographies* 13 (2006): 127–55.

Ogilvie, Brian W. *The Science of Describing: Natural History in Renaissance Europe*. Chicago: University of Chicago Press, 2006.

Ogilvie, Sheilagh C., and Markus Cerman, eds. *European Proto-Industrialization*. Cambridge: Cambridge University Press, 1996.

Oldenburger-Ebbers, Carla. "Notes on Plants Used in Dutch Gardens in the Second Half of the Seventeenth Century." In *The Dutch Garden in the Seventeenth Century*, ed. John Dixon Hunt, 159–73. Dumbarton Oaks Colloquium on the History of Landscape Architecture, 12. Washington, DC: Dumbarton Oaks Research Library and Collection, 1990.

Olmi, Giuseppe. "From the Marvellous to the Commonplace: Notes on Natural History Museums (Sixteenth-Eighteenth Centuries)." In *Non-Verbal Communication in Science prior to 1900*, ed. Renato G. Mazzolini, 235–78. Florence: Leo S. Olschki, 1993.

———. "Science-Honour-Metaphor: Italian Cabinets of the Sixteenth and Seventeenth Centuries." In *The Origins of Museums: The Cabinet of Curiosities in Sixteenth- and Seventeenth-Century Europe*, ed. Oliver Impey and Arthur MacGregor, 5–16. Oxford: Clarendon Press, 1985.

O'Malley, Charles D. *Andreas Vesalius of Brussels, 1514–1564*. Berkeley: University of California Press, 1964.

O'Neill, Onora. *A Question of Trust: The BBC Reith Lectures, 2002*. Cambridge: Cambridge University Press, 2002.

Ooms, Herman. *Tokugawa Ideology: Early Constructs, 1570–1680*. Princeton, NJ: Princeton University Press, 1985.

Opsomer, J. E. "Un botaniste trop peu connu: Willem Quackelbeen (1527–1561)." *Bulletin de la Société Royale de Botanique de Belgique* 93 (1961): 113–30.

———. "Notes complémentaires sur les plantes envoyées de Turquie en 1557 par le botaniste Quackelbeen." *Bulletin de la Société Royale de Botanique de Belgique* 103 (1970): 5–10.

Ørum-Larsen, Asger. "Uraniborg—the Most Extraordinary Castle and Garden Design in Scandinavia." *Journal of Garden History* 10 (1990): 97–105.

Otsuka, Yasuo. "Chinese Traditional Medicine in Japan." In *Asian Medical Systems: A Comparative Study*, ed. Charles Leslie, 322–40. Berkeley: University of California Press, 1976.

———. "Willem ten Rhyne in Japan." [In Japanese with English abstract.] In *Rangaku in Japanese Culture*, 251–59. Tokyo: Tokyo University Press, 1971.

Ottenheym, Koen. "The Amsterdam Ring Canals: City Planning and Architecture." In *Rome, Amsterdam: Two Growing Cities in Seventeenth-Century Europe,* ed. Peter van Kessel and Elisja Schulte, 33–49. Amsterdam: Amsterdam University Press, 1997.

Otterspeer, Willem. *Groepsportret met dame: het bolwerk van de vrijheid, de Leidse Universiteit, 1575–1672.* Amsterdam: Bert Bakker, 2000.

———. "The University of Leiden—an Eclectic Institution." *Early Science and Medicine* 6 (2001): 324–33.

Pabbruwe, H. J. *Dr Robertus Padtbrugge (Parijs 1637—Amersfoort 1703), dienaar van de Verenigde Oost-Indische Compagnie, en zijn familie.* Kloosterzande: Drukkerij Duerinck, 1996.

Pagden, Anthony. *European Encounters with the New World: From Renaissance to Romanticism.* New Haven and London: Yale University Press, 1993.

Pagel, Walter. *From Paracelsus to Van Helmont: Studies in Renaissance Medicine and Science.* Ed. Marianne Winder. Collected Studies Series, 235. London: Variorum, 1986.

———. "Helmont, Johannes (Joan) Baptista Van." In *Dictionary of Scientific Biography,* vol. 6, ed. C. C. Gillispie, 253–59. New York: Scribners, 1972.

———. *Jean Baptista Van Helmont: Reformer of Science and Medicine.* Cambridge: Cambridge University Press, 1982.

———. *Paracelsus: An Introduction to Philosophical Medicine in the Era of the Renaissance.* New York: S. Karger, 1958.

Palm, L. C. "Italian Influences on Antoni van Leeuwenhoek." In *Italian Scientists in the Low Countries in the Seventeenth and Eighteenth Centuries,* ed. C. S. Maffioli and L. C. Palm, 147–63. Amsterdam: Rodopi, 1989.

———. "Leeuwenhoek and Other Dutch Correspondents of the Royal Society." *Notes and Records of the Royal Society* 43 (1989): 191–207.

Palmer, Richard. "Medical Botany in Northern Italy in the Renaissance." *Journal of the Royal Society of Medicine* 78 (1985): 149–57.

———. "Pharmacy in the Republic of Venice in the Sixteenth Century." In *The Medical Renaissance of the Sixteenth Century,* ed. Andrew Wear, Roger K. French, and Ian M. Lonie, 100–117, 303–12. Cambridge: Cambridge University Press, 1985.

Parel, Anthony J. *The Machiavellian Cosmos.* New Haven and London: Yale University Press, 1992.

Park, Katharine. "The Criminal and the Saintly Body: Autopsy and Dissection in Renaissance Italy." *Renaissance Quarterly* 47 (1994): 1–33.

———. "The Life of the Corpse: Division and Dissection in Late Medieval Europe." *Journal of the History of Medicine* 50 (1995): 111–32.

———. "The Organic Soul." In *The Cambridge History of Renaissance Philosophy,* gen. ed. Charles B. Schmitt, ed. Quentin Skinner, Eckhard Kessler, assoc. ed. Jill Kraye, 464–84. Cambridge: Cambridge University Press, 1988.

———. "Relics of a Fertile Heart: The 'Autopsy' of Clare of Montefalco." In *The Material Culture of Sex, Procreation, and Marriage in Premodern Europe,* ed. Anne L. McClanan and Karen Rosoff Encarnación, 115–33. Basingstoke: Palgrave, 2001.

Park, Katharine, and Eckhard Kessler. "The Concept of Psychology." In *The Cambridge History of Renaissance Philosophy,* gen. ed. Charles B. Schmitt, ed. Quentin Skinner,

Eckhard Kessler, assoc. ed. Jill Kraye, 455–63. Cambridge: Cambridge University Press, 1988.

Parker, Geoffrey. *The Dutch Revolt.* Rev. ed., 1977. Harmondsworth: Penguin, 1985.

———. "The Dutch Revolt and the Polarization of International Politics." In *The General Crisis of the Seventeenth Century*, ed. Geoffrey Parker and Lesley M. Smith, 57–82. London: Routledge and Kegan Paul, 1978.

———. *The Military Revolution: Military Innovation and the Rise of the West, 1500–1800.* Cambridge: Cambridge University Press, 1988.

———. "New Light on an Old Theme: Spain and the Netherlands, 1550–1650." *European History Quarterly* 15 (1985): 219–37.

Parker, John. *The World for a Marketplace: Episodes in the History of European Expansion.* Minneapolis: Associates of the James Ford Bell Library, 1978.

Parry, J., and M. Bloch, eds. *Money and the Morality of Exchange.* Cambridge: Cambridge University Press, 1989.

Partington, J. R. *A History of Chemistry.* Vol. 2. London: Macmillan, 1961.

Pas, Peter W. van der. "The Earliest European Descriptions of Japan's Flora." *Janus* 61 (1974): 281–95.

Pater, C. de. *Petrus Van Musschenbroek (1692–1761): Een Newtoniaans natuuronderzoeker.* Utrecht: Drukkerij Elinkwijk, 1979.

Pavord, Anna. *The Tulip.* New York: Bloomsbury, 1999.

Pearson, M. N., ed. *Spices in the Indian Ocean World.* An Expanding World: The European Impact on World History, 1450–1800. Aldershot: Ashgate, 1996.

Peitz, William. "The Problem of the Fetish." *Res: Anthropology and Aesthetics* 9, 13, 16 (1985–88): 5–17, 23–45, 105–23.

Pelling, Margaret, and Charles Webster. "Medical Practitioners." In *Health, Medicine and Mortality in the Sixteenth Century*, ed. Charles Webster, 165–235. Cambridge: Cambridge University Press, 1979.

Pelliot, Paul. "Michel Boym." *T'Oung Pao* 31 (1935): 95–151.

———. *Notes on Marco Polo.* Vol 1. Paris: Imprimerie Nationale, Librairie Adrien-Maisonneuve, 1959.

Pérez, Miguel López. *Asclepio renovado: alquimia y medicina en la España moderna (1500–1700).* Madrid: Ediciones Corona Borealis, 2003.

Pies, Eike. *Willem Piso (1611–1678): Begründer der Kolonialen Medizin und Leibarzt des Grafen Johann Moritz von Nassau-Siegen in Brasilien: Eine Biographie.* Düsseldorf: Interma-orb, 1981.

Pintard, René. *Le libertinage érudit dans la première moitié du dix-septième Siècle.* Paris: Boivin, 1943.

Pocock, J. G. A. *The Machiavellian Moment: Florentine Political Thought and the Atlantic Republican Tradition.* Princeton, NJ: Princeton University Press, 1975.

Pol, Elfriede Hulshoff. "The Library." In *Leiden University in the Seventeenth Century: An Exchange of Learning*, ed. Th. H. Lunsingh Scheurleer and G. H. M. Posthumus Meyjes, 395–459. Leiden: Universitaire Pers Leiden / E. J. Brill, 1975.

Pollmann, Judith. *Religious Choice in the Dutch Republic: The Reformation of Arnoldus Buchelius, 1565–1641.* Manchester: Manchester University Press, 1999.

Pomata, Gianna. "Menstruating Men: Similarity and Difference of the Sexes in Early Modern Medicine." In *Generation and Degeneration: Tropes of Reproduction in Literature and History from Antiquity through Early Modern Europe*, ed. Valeria Finucci and Kevin Brownless, 109–52. Durham, NC: Duke University Press, 2001.

Pomata, Gianna, and Nancy G. Siraisi, eds. *Historia: Empiricism and Erudition in Early Modern Europe*. Cambridge, MA: MIT Press, 2005.

Pomeranz, Kenneth. *The Great Divergence: China, Europe, and the Making of the Modern World Economy*. Princeton, NJ: Princeton University Press, 2000.

Pomian, Krzysztof. *Collectors and Curiosities: Paris and Venice, 1500–1800*. Trans. Elizabeth Wiles-Portier. Cambridge: Polity in association with Basil Blackwell, 1990.

Pop, G. F. *De geneeskunde bij het Nederlandsche zeewezen (geschiedkundige nasporingen)*. Weltevreden-Batavia: G. Kolff, 1922.

Popkin, Richard H. *Isaac la Peyrère (1596–1676): His Life, Work, and Influence*. Leiden: E. J. Brill, 1987.

———. *The History of Scepticism from Erasmus to Descartes*. Rev. ed., 1964. New York: Harper and Row, 1968.

Porto, Paulo A. "'*Summus Atque Felicissimus Salium*': The Medical Relevance of the *Liquor Alkahest*." *Bulletin of the History of Medicine* 76 (2002): 1–29.

Post, Regnerus Richardus. *The Modern Devotion: Confrontation with Reformation and Humanism*. Leiden: E. J. Brill, 1968.

Posthumus, N. W. *De geschiedenis van de Leidsche lakenindustrie*. 3 vols. The Hague: Martinus Nijhoff, 1908–39. 3 vols.

———. *Inquiry into the History of Prices in Holland*. 2 vols. Leiden: E. J. Brill, 1946–64.

Postma, Johannes. *The Dutch in the Atlantic Slave Trade, 1600–1815*. Cambridge: Cambridge University Press, 1990.

———. "Suriname and Its Atlantic Connections, 1667–1795." In *Riches from Atlantic Commerce: Dutch Transatlantic Trade and Shipping, 1585–1817*, ed. Johannes Postma and Victor Enthoven, 287–322. Leiden: Brill, 2003.

Postma, Johannes, and Victor Enthoven, eds. *Riches from Atlantic Commerce: Dutch Transatlantic Trade and Shipping, 1585–1817*. Leiden: Brill, 2003.

Prakash, Om. "The Dutch East India Company in the Trade of the Indian Ocean." In *India and the Indian Ocean, 1500–1800*, ed. Ashin Das Gupta and M. N. Pearson, 185–200. New Delhi: Oxford University Press, 1987.

———. "Precious Metal Flows in Asia and World Economic Integration in the Seventeenth Century." In *The Emergence of a World Economy*, ed. W. Fischer, R. Marvin McInnis, and J. Schneider, Part 1: 1500–1850, 83–96. Weisbaden: Franz Steiner Verlag, 1986.

Prest, John. *The Garden of Eden: The Botanic Garden and the Re-Creation of Paradise*. New Haven and London: Yale University Press, 1981.

Prins, Johanna, Bettina Brandt, Timothy Stevens, and Thomas Shannon, eds. *The Low Countries and the New World(s): Travel, Discovery, Early Relations*. Lanham, MD: University Press of America, 2000.

Proctor, Robert. "Anti-Agate: The Great Diamond Hoax and the Semiprecious Stone Scam." *Configurations* 9 (2001): 381–412.

Prokhovnik, Raia. *Spinoza and Republicanism*. Houndmills: Palgrave, 2004.
Raben, Remco. "Batavia and Columbo: The Ethnic and Spatial Order of Two Colonial Cities, 1600–1800." Ph.D. diss., Rijksuniversiteit te Leiden, 1996.
Ramakers, Rozemarijn. "Het Caecilia Gasthuis te Leiden: onderwijs instituut en verpleeg-inrichting, 1636–1799." Ph.D. diss., Rijksuniversiteit te Leiden, 1989.
Ramón-Laca, L. "Charles de l'Écluse and *Libri Picturati* A. 16–30." *Archives of Natural History* 28 (2001): 195–243.
Ramsay, G. D. *The City of London in International Politics at the Accession of Elizabeth Tudor*. Manchester: Manchester University Press, 1975.
Rather, L. J. "The 'Six Things Non-Natural': A Note on the Origin and Fate of a Doctrine and a Phrase." *Clio Medica* 3 (1968): 337–47.
Raven-Hart, R., trans. and ed. *Germans in Dutch Ceylon*. Colombo: Ceylon Government Press, 1953.
Read, Conyers. *Mr. Secretary Walsingham and the Policy of Queen Elizabeth*. Vol. 1. Oxford: Clarendon Press, 1925.
Reddy, William M. *Money and Liberty in Modern Europe: A Critique of Historical Understanding*. Cambridge: Cambridge University Press, 1987.
———. *The Navigation of Feeling: A Framework for the History of Emotions*. Cambridge: Cambridge University Press, 2001.
———. *The Rise of Market Culture: The Textile Trade and French Society, 1750–1900*. Cambridge: Cambridge University Press, 1984.
Reeds, Karen Meier. *Botany in Medieval and Renaissance Universities*. Harvard Dissertations in the History of Science. New York: Garland, 1991.
Regin, Deric. *Traders, Artists, Burghers: A Cultural History of Amsterdam in the Seventeenth Century*. Assen: Van Gorcum, 1976.
Reid, Anthony. "Early Southeast Asian Categorizations of Europeans." In *Implicit Understandings: Observing, Reporting, and Reflecting on the Encounters between Europeans and Other Peoples in the Early Modern Era*, ed. Stuart B. Schwartz, 268–94. Cambridge: Cambridge University Press, 1994.
———. *Southeast Asia in the Age of Commerce, 1450–1680*. Vol.1: *The Lands below the Winds*. New Haven and London: Yale University Press, 1988.
Rekers, Bernard. *Benito Arias Montano, 1527–1598: Studie over een groep spiritualistische humanisten in Spanje en de Nederlanden, op grond van hun briefwisseling*. Groningen: V. R. B. Kleine, 1961.
Rink, Oliver A. *Holland on the Hudson: An Economic and Social History of Dutch New York*. Ithaca, NY: Cornell University Press, 1986.
Roberts, Michael. *The Military Revolution, 1560–1660*. Belfast: M. Boyd, 1956.
Rodis-Lewis, Geneviève. "Descartes' Life and the Development of His Philosophy." In *The Cambridge Companion to Descartes*, ed. John Cottingham, 21–57. Cambridge: Cambridge University Press, 1992.
———. *Descartes: Biographie*. Paris: Calmann-Lévy, 1995.
———. *Descartes: His Life and Thought*. Trans. Jane Marie Todd. Ithaca, NY: Cornell University Press, 1998.

Rodríguez-Salgado, M. J. *The Changing Face of Empire: Charles V, Philip II and Habsburg Authority, 1551–1559*. Cambridge: Cambridge University Press, 1988.

Roldanus, C. W. "Adriaen Paets, een republikein uit de nadagen." *Tijdschrift voor Geschiedenis* 50 (1935): 134–66.

Römer, L. S. A. M. von. *Dr. Jacobus Bontius*. Bijblad op het Geneeskundig Tijdschrift voor Nederlandsch-Indië. Utigegeven door de Vereeniging tot Bevordering der Geneeskundige Wetenschappen in Nederlandsch-Indië. Batavia: G. Kolff, 1932.

Rooden, Peter T. van. "Contantijn L'Empereur's Contacts with the Amsterdam Jews and His Confutation of Judaism." In *Jewish-Christian Relations in the Seventeenth Century: Studies and Documents*, ed. J. van den Berg and Ernestine G. E. van der Wall, 51–72. Dordrecht: Kluwer, 1988.

Rookmaaker, L. C. *The Zoological Exploration of Southern Africa, 1650–1790*. Rotterdam: A. A. Balkema, 1989.

Rosen, George. "Lorenz Heister on Acupuncture: An Eighteenth Century View." *Journal of the History of Medicine* 30 (1975): 386–88.

———. "Sir William Temple and the Therapeutic Use of Moxa for Gout in England." *Bulletin of the History of Medicine* 44 (1970): 31–39.

Rosenberg, Dorothy B. "Wilhelm ten Rhyne's 'De acupunctura': An 1826 Translation." *Journal of the History of Medicine* 34 (1979): 81–84.

Rossi-Reder, Andrea. "Wonders of the Beast: India in Classical and Medieval Literature." In *Marvels, Monsters, and Miracles: Studies in the Medieval and Early Modern Imaginations*, ed. Timothy S. Jones and David A. Sprunger, 53–66. Kalamazoo, MI: Studies in Medieval Culture, 2002.

Rowen, Herbert H. *John de Witt, Grand Pensionary of Holland, 1625–1672*. Princeton, NJ: Princeton University Press, 1978.

Ruestow, Edward G. "Images and Ideas: Leeuwenhoek's Perception of the Spermatozoa." *Journal of the History of Biology* 16 (1983): 185–224.

———. "Leeuwenhoek and the Campaign against Spontaneous Generation." *Journal of the History of Biology* 17 (1984): 225–48.

———. *The Microscope in the Dutch Republic: The Shaping of Discovery*. Cambridge: Cambridge University Press, 1996.

———. "Piety and the Defense of Natural Order: Swammerdam on Generation." In *Religion, Science, and Worldview: Essays in Honor of Richard S. Westfall*, ed. Margaret J. Osler and Paul Lawrence Farber, 217–41. Cambridge: Cambridge University Press, 1985.

———. *Physics at Seventeenth and Eighteenth Century Leiden*. The Hague: Martinus Nijhoff, 1973.

———. "The Rise of the Doctrine of Vascular Secretion in the Netherlands." *Journal of the History of Medicine* 35 (1980): 265–87.

Ruggiero, Guido. *Binding Passions: Tales of Magic, Marriage, and Power at the End of the Renaissance*. New York: Oxford University Press, 1993.

Ruler, J. A. van. *The Crisis of Causality: Voetius and Descartes on God, Nature and Change*. Leiden: E. J. Brill, 1995.

———. "Franco Petri Burgersdijk and the Case of Calvinism within the Neo-Scholastic

Tradition." In *Franco Burgersdijk (1590–1635): Neo-Aristotelianism in Leiden*, ed. E. P. Bos and H. A. Krop, 37–65. Studies in the History of Ideas in the Low Countries. Amsterdam: Rodopi, 1993.

Rumphius Gedenkboek, 1702–1902. Haarlem: Koloniaal Museum, 1902.

Rupp, Jan C. C. "Matters of Life and Death: The Social and Cultural Conditions of the Rise of Anatomical Theatres, with Special Reference to Seventeenth Century Holland." *History of Science* 28 (1990): 263–87.

Russell, Andrew W., ed. *The Town and State Physician in Europe from the Middle Ages to the Enlightenment*. Wolfenbüttel: Herzog August Bibliothek, 1981.

Sahlins, Marshall. *Culture and Practical Reason*. Chicago: University of Chicago Press, 1976.

Salisbury, Joyce E. *The Beast Within: Animals in the Middle Ages*. New York: Routledge, 1994.

———, ed. *The Medieval World of Nature: A Book of Essays*. New York: Garland, 1993.

Santing, Catrien. "*De Affectibus Cordis et Palpitatione*: Secrets of the Heart in Counter-Reformation Italy." In *Cultural Approaches to the History of Medicine: Mediating Medicine in Early Modern and Modern Europe*, ed. Willem de Blécourt and Cornellie Usborne, 11–35. Houndsmills: Palgrave, 2003.

———. "*Doctor philosophus*: humanistische geleerdheid en de professionalisering van de vroegmoderne medicus." In *Medische geschiedenis in regionaal perspectief: Groningen, 1500–1900*, ed. Frank Huisman and Catrien Santing, 23–48. Rotterdam: Erasmus, 1997.

———. *Geneeskunde en humanisme: een intellectuele biographie van Theodoricus Ulsenius (c. 1460–1508)*. Rotterdam: Erasmus, 1992.

Sarasohn, Lisa T. "Epicureanism and the Creation of a Privatist Ethic in Early Seventeenth-Century France." In *Atoms, Pneuma, and Tranquillity: Epicurean and Stoic Themes in European Thought*, ed. Margaret J. Osler, 175–95. Cambridge: Cambridge University Press, 1991.

———. "Motion and Morality: Pierre Gassendi, Thomas Hobbes and the Mechanical World-View." *Journal of the History of Ideas* 46 (1985): 363–79.

———. "Nicolas-Claude Fabri de Peiresc and the Patronage of the New Science in the Seventeenth Century." *Isis* 84 (1993): 70–90.

Sargent, Thomas J., and François R. Velde. *The Big Problem of Small Change*. Princeton, NJ: Princeton University Press, 2002.

Sarton, George. *The History of Science and the New Humanism*. Cambridge, MA: Harvard University Press, 1937.

———. "Rumphius, Plinius Indicus (1628–1702)." *Isis* 27 (1937): 242–57.

Sassen, Ferd. *Geschiedenis van de wijsbegeerte in Nederland*. Amsterdam: Elsevier, 1959.

———. "The Intellectual Climate in Leiden in Boerhaave's Time." In *Boerhaave and His Time: Papers Read at the International Symposium in Commemoration of the Tercentenary of Boerhaave's Birth*, ed. G. A. Lindeboom, 1–16. Leiden: E. J. Brill, 1970.

———. "De reis van Marin Mersenne in de Nederlanden (1630)." In *Mededelingen van de Koninklijke Vlaamse Academie voor Wetenschappen*. Letteren en Schone Kunsten Van België, Klasse der Letteren, vol. 16, no. 4. Brussels: Paleis der Academiën, 1964.

———. "De reis van Pierre Gassendi in de Nederlanden (1628–1629)." In *Mededelingen der Koninklijke Nederlandse Akademie van Wetenschappen*. Afd. Letterkunde, Nieuwe Reeks, vol. 23, no. 10, 263–307. Amsterdam: N.V. Noord-Hollandsche Uitgevers Maatschappij, 1960.

Saunders, Jason Lewis. *Justus Lipsius: The Philosophy of Renaissance Stoicism*. New York: Liberal Arts Press, 1955.

Sawday, Jonathan. *The Body Emblazoned: Dissection and the Human Body in Renaissance Culture*. London: Routledge, 1995.

Saxby, T. J. *The Quest for the New Jerusalem: Jean de Labadie and the Labadists, 1610–1744*. Dordrecht: Kluwer, 1987.

Schaffer, Simon. "Experimenters' Techniques, Dyers' Hands, and the Electric Planetarium." *Isis* 88 (1997): 456–83.

Schama, Simon. *The Embarrassment of Riches: An Interpretation of Dutch Culture in the Golden Age*. New York: Knopf, 1987.

Schenkeveld, Maria A. *Dutch Literature in the Age of Rembrandt: Themes and Ideas*. Amsterdam: John Benjamins, 1991.

Schepelern, H. D. "The Museum Wormianum Reconstructed: A Note on the Illustration of 1655." *Journal of the History of Collections* 2 (1990): 81–85.

———. "Naturalienkabinett oder Kunstkammer: Der Sammler Bernhard Paludanus und Sein Katalogmanuskript in der Königlichen Bibliothek in Kopenhagen." *Nordelbingen, Beiträge zur Kunst- und Kulturgeschichte* 50 (1981): 157–82.

———. "Natural Philosophers and Princely Collectors: Worm, Paludanus, and the Gottorp and Copenhagen Collections." In *The Origins of Museums: The Cabinet of Curiosities in Sixteenth- and Seventeenth-Century Europe*, ed. O. Impey and A. MacGregor, 121–27. Oxford: Clarendon Press, 1985.

Schiebinger, Londa. *The Mind Has No Sex? Women in the Origins of Modern Science*. Cambridge, MA: Harvard University Press, 1989.

Schierbeek, A. *Antoni van Leeuwenhoek: leven en werken*. 2 vols. Lochem: De Tijdstroom, 1950–51. 2 vols.

———. *Jan Swammerdam: zijn leven en zijn werken*. Lochem: De Tijdstroom, 1946.

———. *Measuring the Invisible World: The Life and Works of Antoni van Leeuwenhoek FRS*. Intro. M. Rooseboom. New York: Abelard-Schuman, 1959.

Schmidt, Benjamin. *Innocence Abroad: The Dutch Imagination and the New World, 1570–1670*. Cambridge: Cambridge University Press, 2001.

Schmitt, Charles B. *Studies in Renaissance Philosophy and Science*. London: Variorum Reprints, 1981.

Schnapper, Antoine. *Le géant, la licorne et la tulipe: collections françaises au dix-septième siècle*. Paris: Flammarion, 1988.

Schneider, Maarten. *De Nederlandse krant, 1618–1978: van "nieuwstydinghe" tot dagblad*. 4th ed. In collaboration with Joan Hemels. 1943. Baarn: Het Weredvenster, 1979.

Schnurmann, Claudia. *Atlantische Welten: Engländer und Niederländer im Amerikanisch-Atlantischen Raum, 1648–1713*. Köln: Böhlau, 1998.

———. "Representative Atlantic Entrepreneur: Jacob Leisler, 1640–1691." In *Riches from Atlantic Commerce: Dutch Transatlantic Trade and Shipping, 1585–1817*, ed. Johannes Postma and Victor Enthoven, 259–83. Leiden: Brill, 2003.

Schoute, Dirk. *De geneeskunde in den dienst der Oost-Indische Compagnie in Nederlandsch-Indië*. Amsterdam: J. H. de Bussy, 1929.

Schultz, Bernard. *Art and Anatomy in Renaissance Italy*. Ann Arbor, MI: UMI Research Press, 1985.

Schultze-Hagen, Karl, Frank Steinheimer, Ragnar Kinzelbach, and Christoph Gasser. "Avian Taxidermy in Europe from the Middle Ages to the Renaissance." *Journal für Ornithologie* 144 (2003): 459–78.

Schulz, Eva. "Notes on the History of Collecting and of Museums in the Light of Selected Literature of the Sixteenth to the Eighteenth Century." *Journal of the History of Collections* 2 (1990): 205–18.

Schupbach, William. *The Paradox of Rembrandt's "Anatomy of Dr. Tulp."* Medical History Supplements, 2. London: Wellcome Institute for the History of Medicine, 1982.

Schwartz, Stuart B. "A Commonwealth within Itself: The Early Brazilian Sugar Industry, 1550–1670." In *Tropical Babylons: Sugar and the Making of the Atlantic World, 1450–1680*, ed. Stuart B. Schwartz, 158–200. Chapel Hill: University of North Carolina Press, 2004.

Secord, James A. "Knowledge in Transit." *Isis* 95 (2004): 654–72.

Segal, Sam. "The Tulip Portrayed: The Tulip Trade in Holland in the Seventeenth Century." In *The Tulip: A Symbol of Two Nations*, ed. Michiel Roding and Hans Theunissen, 9–24. Utrecht: M. Th. Houtsma Stichting; and Istanbul: Turco-Dutch Friendship Association, 1993.

Seifert, Arno. *Cognitio Historica: Die Geschichte als Namengeberin der Frühnneuzeitlichen Emperie*. Berlin: Duncker and Humblot, 1976.

Seters, W. H. van. "De voorgeschiedenis der stichting van de eerste Amsterdamse hortus botanicus." In *Zes en veertigste jaarboek genootschap Amstelodamum*, 34–45. Amsterdam, 1954.

Shackelford, Jole. "Tycho Brahe, Laboratory Design, and the Aim of Science: Reading Plans in Context." *Isis* (1993): 211–30.

———. *A Philosophical Path for Paracelsian Medicine: The Ideas, Intellectual Context, and Influence of Petrus Severinus: 1540/2–1602*. Copenhagen: Museum Tusculanum Press, University of Copenhagen, 2004.

Shank, Michael H. *"Unless You Believe, You Shall Not Understand": Logic, University, and Society in Late Medieval Vienna*. Princeton, NJ: Princeton University Press, 1988.

Shapin, Steven. "Descartes the Doctor: Rationalism and Its Therapies." *British Journal for the History of Science* 33 (2000): 131–54.

———. "The House of Experiment in Seventeenth-Century England." *Isis* 79 (1988): 373–404.

———. "'The Mind Is Its Own Place': Science and Solitude in Seventeenth-Century England." *Science in Context* 4 (1991): 191–218.

———. *The Scientific Revolution*. Chicago: University of Chicago Press, 1996.

———. *A Social History of Truth: Civility and Science in Seventeenth-Century England*. Chicago: University of Chicago Press, 1994.

Shapiro, Alan E. *Fits, Passions and Paroxysms: Physics, Method, and Chemistry in Newton's*

Theory of Colored Bodies and Fits of Easy Reflection. Cambridge: Cambridge University Press, 1993.

Shapiro, Barbara J. *A Culture of Fact: England, 1550–1720*. Ithaca, NY: Cornell University Press, 2000.

———. "Latitudinarianism and Science in Seventeenth Century England." *Past and Present* 40 (1968): 16–41.

———. "Law and Science in Seventeenth-Century England." *Stanford Law Review* 21 (1968–69): 727–66.

———. *Probability and Certainty in Seventeenth-Century England: A Study of the Relationship between Natural Science, Religion, History, Law, and Literature*. Princeton, NJ: Princeton University Press, 1983.

Shepherd, John Robert. *Statecraft and Political Economy on the Taiwan Frontier, 1600–1800*. Stanford, CA: Stanford University Press, 1993.

Sherman, Nancy. *The Fabric of Character: Aristotle's Theory of Virtue*. Oxford: Clarendon Press, 1989.

Sherrington, Charles S. *The Endeavour of Jean Fernel*. Cambridge: Cambridge University Press, 1946.

Silver, Larry, and Pamela H. Smith. "Splendor in the Grass: The Powers of Nature and Art in the Age of Dürer." In *Merchants and Marvels: Commerce, Science, and Art in Early Modern Europe*, ed. Pamela H. Smith and Paula Findlen, 29–62. New York: Routledge, 2002.

Simmel, Georg. *On Individuality and Social Forms: Selected Writings*. Ed. and intro. Donald N. Levine. Chicago: University of Chicago Press, 1971.

Singer, Thomas C. "Hieroglyphs, Real Characters, and the Idea of Natural Language in English Seventeenth-Century Thought." *Journal of the History of Ideas* 50 (1989): 49–70.

Siraisi, Nancy G. *The Clock and the Mirror: Girolamo Cardano and Renaissance Medicine*. Princeton, NJ: Princeton University Press, 1997.

———. *Taddeo Alderotti and His Pupils: Two Generations of Italian Medical Learning*. Princeton, NJ: Princeton University Press, 1981.

Sirks, M. J. "Rumphius, the Blind Seer of Amboina." Trans. Lily M. Perry. In *Science and Scientists in the Netherlands Indies*, 295–308. New York: Board for the Netherlands Indies, Surinam, and Curaçao, 1945.

Skinner, Quentin. *The Foundations of Modern Political Thought*. 2 vols. Cambridge: Cambridge University Press, 1978.

Slee, J. C. van. *De Rijnsburger collegianten*. Haarlem: De Erven F. Bohn, 1895.

Sluijs, C. A. van der. *Puritanisme en nadere reformatie*. Kampen: De Groot Goudriaan, 1989.

Smit, J. W. "The Netherlands and Europe in the Seventeenth and Eighteenth Centuries." In *Britain and the Netherlands in Europe and Asia*, ed. J. S. Bromley and E. H. Kossmann, 13–36. London: Macmillan; and New York: St. Martin's Press, 1968.

Smith, Pamela. *The Body of the Artisan: Art and Experience in the Scientific Revolution*. Chicago: University of Chicago Press, 2004.

———. *The Business of Alchemy: Science and Culture in the Holy Roman Empire*. Princeton, NJ: Princeton University Press, 1994.

———. "Science and Taste: Painting, Passions, and the New Philosophy in Seventeenth-Century Leiden." *Isis* 90 (1999): 421–61.

Smith, Wesley D. *The Hippocratic Tradition.* Ithaca, NY: Cornell University Press, 1979.

Smith, Woodruff D. "Complications of the Commonplace: Tea, Sugar, and Imperialism." *Journal of Interdisciplinary History* 23 (1992): 259–78.

Snapper, I. *Meditations on Medicine and Medical Education.* New York: Grune and Stratton, 1956.

Snelders, H. A. M. "Ten Rhyne, Willem." In *Dictionary of Scientific Biography*, genl. ed. C. C. Gillispie, 13:282–83. New York: Charles Scribner's Sons, 1976.

Solomon, Howard. *Public Welfare, Science, and Propaganda in Seventeenth-Century France: The Innovations of Théophraste Renaudot.* Princeton, NJ: Princeton University Press, 1972.

Solomon, Julie Robin. *Objectivity in the Making: Francis Bacon and the Politics of Inquiry.* Baltimore: Johns Hopkins University Press, 1998.

Soltow, Lee, and Jan Luiten van Zanden. *Income and Wealth Inequality in the Netherlands, Sixteenth–Twentieth Century.* Amsterdam: Het Spinhuis, 1998.

Spierenburg, Pieter. *The Spectacle of Suffering: Executions and the Evolution of Repression: From a Preindustrial Metropolis to the European Experience.* Cambridge: Cambridge University Press, 1984.

Stagl, Justin. *A History of Curiosity: The Theory of Travel, 1550–1800.* Chur: Harwood Academic, 1995.

Stannard, Jerry. "Natural History." In *Science in the Middle Ages*, ed. David Lindberg, 429–60. Chicago: University of Chicago Press, 1978.

Stapel, F. W. *Geschiedenis van Nederlandsch-Indië.* Amsterdam: J. M. Meulenhoff, 1930.

———. *De gouveneurs-generaal van Nederlandsch-Indië in beeld en woord.* The Hague: W. P. Stockum and Zoon, 1941.

Steadman, Philip. *Vermeer's Camera Obscura: Uncovering the Truth behind the Masterpieces.* Oxford: Oxford University Press, 2001.

Stearn, William T. "The Influence of Leyden on Botany in the Seventeenth and Eighteenth Centuries." *British Journal for the History of Science* 1 (1962): 137–58.

Stearns, Carol Z., and Peter N. Stearns, eds. *Emotion and Social Change: Toward a New Psychohistory.* New York: Holmes and Meier, 1988.

Steele, Robert. *Mediaeval Lore from Bartholomaeus Anglicus.* Preface William Morris. New York: Cooper Square, 1966.

Steendijk-Kuypers, J. *Volksgezondheidszorg in de zestiende en zeventiende eeuw en Hoorn: een bijdrage tot de beeldvorming van sociaal-geneeskundige structuren in een stedelijke samenleving.* Rotterdam: Erasmus, 1994.

Steensgaard, Niels. *The Asian Trade Revolution of the Seventeenth Century: The East India Companies and the Decline of the Caravan Trade.* Chicago: University of Chicago Press, 1974.

Stegeman, Saskia. "How to Set Up a Scholarly Correspondence: Theodorus Janssonius van Almeloveen (1657–1712) Aspires to Membership of the Republic of Letters." *Lias* 20 (1993): 227–43.

Steneck, Nicholas H. *Science and Creation in the Middle Ages: Henry of Langenstein (d. 1397) on Genesis*. Notre Dame, IN: University of Notre Dame Press, 1976.
Stewart, Alan. "The Early Modern Closet Discovered." *Representations* 50 (1995): 76–100.
Stiefvater, Erich W. *Die Akupunkter des Ten Rhijne*. Ulm: Karl F. Haug, 1955.
Stimson, Dorothy. "Amateurs of Science in Seventeenth Century England." *Isis* 31 (1939–40): 32–47.
———. "Puritanism and the New Philosophy in Seventeenth Century England." *Bulletin of the History of Medicine* 3 (1935): 321–34.
Stols, Eddy. "The Expansion of the Sugar Market in Western Europe." In *Tropical Babylons: Sugar and the Making of the Atlantic World, 1450–1680*, ed. Stuart B. Schwartz, 237–88. Chapel Hill: University of North Carolina Press, 2004.
Stomps, Th. J. "De geschiedenis van de Amsterdamse Hortus." *Ons Amsterdam* 3, no. 8 (August 1951): 206–14.
Stone-Ferrier, Linda A. *Images of Textiles: The Weave of Seventeenth-Century Dutch Art and Society*. Ann Arbor, MI: UMI Research Pres, 1985.
Stoye, John. *English Travellers Abroad, 1604–1667: Their Influence in English Society and Politics*. Rev. ed. New Haven and London: Yale University Press, 1989.
Stroup, Alice. A *Company of Scientists: Botany, Patronage, and Community at the Seventeenth-Century Parisian Royal Academy of Sciences*. Berkeley: University of California Press, 1990.
Stuart, David. *The Plants That Shaped Our Gardens*. London: Francis Lincoln, 2002.
Stuiveling, Garmt. A *Sampling of Dutch Literature: Thirteen Excursions into the Works of Dutch Authors*. Trans. and adapted James Brockway. Hilversum: Radio Nederland Wereldomroep, n.d.
Stuijvenberg, J. H. van. "'The' Weber Thesis: An Attempt at Interpretation." *Acta Historiae Neerlandicae* 8 (1975): 50–66.
Subrahmanyam, Sanjay. "Dutch Tribulations in Seventeenth-Century Mrauk-U." In *From the Tagus to the Ganges: Explorations in Connected History*, 200–247. New Delhi: Oxford University Press, 2005.
Subrahmanyam, Sanjay, and Luís Filipe F. R. Thomaz. "Evolution of Empire: The Portuguese in the Indian Ocean during the Sixteenth Century." In *The Political Economy of Merchant Empires: State Power and World Trade, 1350–1750*, ed. James D. Tracy, 298–331. Cambridge: Cambridge University Press, 1991.
Sutarjadi, Prof. Dr., Wahjo Dyatmiko, and et al., eds. *Proceedings of the International Congress on Traditional Medicine and Medicinal Plants, October 15–17, 1990, Denpasar-Bali, Indonesia*. Surabaya: Yayasan Widya Husada, Fakultas Farmsi, Universitas Airlangga, 1993.
Swan, Claudia. *Art, Science, and Witchcraft in Early Modern Holland: Jacques de Gheyn II (1565–1629)*. Cambridge: Cambridge University Press, 2005.
———. *The Clutius Botanical Watercolors: Plants and Flowers of the Renaissance*. New York: Harry N. Abrams, 1998.
———. "Lectura-Imago-Ostensio: The Role of the *Libri Picturati* A.18–A.30 in Medical Instruction at the Leiden University." In *Natura-cultura: l'interpretazione del mondo fisico*

nei testi e nelle immagini, ed. Giuseppe Olmi, Lucia Tongiorgi Tomasi, and Attilio Zanca, 189–214. Florence: Leo S. Olschki, 2000.

Swetz, Frank J. *Capitalism and Arithmetic: The New Math of the Fifteenth Century*. Trans. David Eugene Smith. La Salle, IL: Open Court, 1987.

Szamosi, Geza. "Polyphonic Music and Classical Physics: The Origin of Newtonian Time." *History of Science* 28 (1990): 175–91.

Szczesniak, Boleslaw. "John Floyer and Chinese Medicine." *Osiris* 11 (1954): 127–56.

———. "The Writings of Michael Boym." *Monumenta Serica: Journal of Oriental Studies* 14 (1949–55): 481–538.

Taylor, Jean Gelman. "Europese en Euraziatische vrouwen in Nederlands-Indië in de VOC-Tijd." In *Vrouwen in de Nederlandse koloniën*, ed. J. Reijs et al. Zevende Jaarboek voor Vrouwengeschiedenis, 10–33. Nijmegen: SUN, 1986.

———. *The Social World of Batavia: European and Eurasian in Dutch Asia*. Madison: University of Wisconsin Press, 1983.

Temkin, Owsei. "The Role of Surgery in the Rise of Modern Medical Thought." *Bulletin of the History of Medicine* 25 (1951): 248–59.

Ten Brink, J. *Dirck Volckertsen Coornhert en zijne wellevenskunst*. Amsterdam: Binger, 1860.

Ten Doesschate, G. *De Utrechtse Universiteit en de geneeskunde, 1636–1900*. Nieuwkoop: B. de Graaf, 1963.

Terpstra, H. *De opkomst der westerkwartieren van de Oost-Indische Compagnie (Suratte, Arabië, Perzië)* The Hague: Martinus Nijhoff, 1918.

Terwen-Dionisius, Else M. "Date and Design of the Botanical Garden in Padua." *Journal of Garden History* 14 (1994): 213–35.

Thijssen-Schoute, C. L. "La diffusion européenne des idées de Bayle." In *Pierre Bayle: le philosophe de Rotterdam*, ed. Paul Dibon, 150–95. Amsterdam: Elsevier, 1959.

———. *Nederlands Cartesianisme*. 1954. Utrecht: Hes Uitgevers, 1989.

———. *Uit de republiek der letteren: elf studiën op het gebied der ideeëngeschiedenis van de gouden eeuw*. The Hague: Martinus Nijhoff, 1967.

Thomas, Nicholas. *Entangled Objects: Exchange, Material Culture, and Colonialism in the Pacific*. Cambridge, MA: Harvard University Press, 1991.

Thompson, E. P. "The Moral Economy of the English Crowd in the Eighteenth Century." *Past and Present* 50 (1971): 76–136.

Thomson, Ann. "La Mettrie, lecteur et traducteur de Boerhaave." *Dix-Huitième Siècle* 23 (1991): 23–29.

Thorndike, Lynn. *A History of Magic and Experimental Science*. 8 vols. New York: Columbia University Press, 1923–58.

Thornton, John. *Africa and Africans in the Making of the Atlantic World, 1400–1800*. 2nd ed. Cambridge: Cambridge University Press, 1998.

Throsby, David. *Economics and Culture*. Cambridge: Cambridge University Press, 2001.

Tjessinga, J. C. *Enkele gegevens omtrent Adriaan Pauw en het slot van Heemstede*. 2 vols. Heemstede: Vereniging Oud-Heemstede-Bennebroek, 1948–49.

Tjon Sie Fat, Leslie. "Clusius' Garden: A Reconstruction." In *The Authentic Garden: A Symposium on Gardens*, ed. Leslie Tjon Sie Fat and Erik de Jong, 3–12. Leiden: Clusius Foundation, 1991.

Toby, Ronald P. "The 'Indianness' of Iberia and Changing Japanese Iconographies of Other." In *Implicit Understandings: Observing, Reporting, and Reflecting on the Encounters between Europeans and Other Peoples in the Early Modern Era*, ed. Stuart B. Schwartz, 323–51. Cambridge: Cambridge University Press, 1994.

Tompsett, D. H., foreword by Cecil Wakeley, J. Dobson, historical intro. *Anatomical Techniques*. Edinburgh: E. and S. Livingstone, 1956.

Totman, Conrad. *Early Modern Japan*. Berkeley: University of California Press, 1993.

Tracy, James D. A *Financial Revolution in the Habsburg Netherlands: Renten and Renteniers in the County of Holland, 1515–1565*. Berkeley: University of California Press, 1985.

———. *Holland under Habsburg Rule, 1506–1566: The Formation of a Body Politic*. Berkeley: University of California Press, 1990.

Travis, Anthony S. *The Rainbow Makers: The Origins of the Synthetic Dyestuffs Industry in Western Europe*. Bethlehem, PA: Associated University Presses, 1993.

Trevor-Roper, Hugh R. "The Paracelsian Movement." In *Renaissance Essays*, 149–99. London: Fontana Press, 1986.

Tribby, Jay. "Cooking (with) Clio and Cleo: Eloquence and Experiment in Seventeenth-Century Florence." *Journal of the History of Ideas* 52 (1991): 417–39.

Tuchscherer, Michel. "Coffee in the Red Sea Area from the Sixteenth to the Nineteenth Century." In *The Global Coffee Economy in Africa, Asia, and Latin America, 1500–1989*, ed. William Gervase Clarence-Smith and Steven Topik, 50–66. Cambridge: Cambridge University Press, 2003.

Tuck, Richard. "Grotius and Selden." In *The Cambridge History of Political Thought, 1450–1700*, ed. J. H. Burns, with Mark Goldie, 499–529. Cambridge: Cambridge University Press, 1991.

———. "The 'Modern' Theory of Natural Law." In *The Languages of Political Theory in Early-Modern Europe*, ed. Anthony Pagden, 99–119. Cambridge: Cambridge University Press, 1987.

———. *Philosophy and Government, 1572–1651*. Cambridge: Cambridge University Press, 1993.

Ultee, Maarten. "The Politics of Professorial Appointment at Leiden, 1709." *History of Universities* 9 (1990): 167–94.

Uragoda, C. G. A *History of Medicine in Sri Lanka From the Earliest Times to 1948*. Colombo: Sri Lanka Medical Association, 1987.

Valiant, Sharon. "Maria Sibylla Merian: Recovering an Eighteenth-Century Legend; Essay Review." *Eighteenth-Century Studies* 26 (1993): 467–79.

Van Damme, Stéphane. *Descartes*. Paris: Presses de Sciences Politiques, 2002.

Vanpaemel, Geert. *Echo's van een wetenschappelijke revolutie: de mechanistische natuurwetenschap aan de Leuvense Artesfaculteit (1650–1797)*. Verhandelingen van de Koninklijke Academie voor Wetenschappen, Letteren, en Schone Kunsten van België, Klasse der Wetenschappen. Brussels: Paleis der Academiën, 1986.

———."Rohault's *Traité de Physique* and the Teaching of Cartesian Physics." *Janus* 71 (1984): 31–40.

Varey, Simon, and Rafael Chabrán. "Medical Natural History in the Renaissance: The Strange Case of Francisco Hernández." *Huntington Library Quarterly* 57 (1994): 124–51.

Varey, Simon, Rafael Chabrán, and Dora B. Weiner, eds. *Searching for the Secrets of Nature: The Life and Works of Dr. Francisco Hernández*. Stanford, CA: Stanford University Press, 2000.

Väth, Alfons. *Johann Adam Schall von Bell S.J.: Missionar in China, Kaiserlicher Astronom und Ratgeber am Hofe von Peking, 1592–1666*. 2nd ed. In collaboration with Louis van Hee, S.J. 1933. Monumenta Serica Monograph Series, 25. Nettetal: Steyler-Verlag, 1991.

Veen, Ernst van. "How the Dutch Ran a Seventeenth-Century Colony: The Occupation and Loss of Formosa, 1624–1662." *Itinerario* 20 (1996): 59–77.

Veen, Jaap van der. "Met grote moeite en kosten: de totstandkoming van seventiendeeeuwse verzamelingen." In *De wereld binnen handbereik: Nederlandse kunst- en rariteitenverzamelingen, 1585–1735*, chief ed. Ellinoor Bergvelt and Renée Kistemaker, 51–69. Zwolle: Waanders Uitgevers/Amsterdams Historisch Museum, 1992.

Veendorp, H. "Dirck Outgerszoon Cluyt, ±1550–1598, hortulanus, imker en apotheker te Leiden." *N.R.C. Ochtendblad*, 3 December 1939.

Veendorp, H., and L. G. M. Baas Becking. *Hortus Academicus Lugduno Batavus, 1587–1937: The Development of the Gardens of Leyden University*. Haarlem: Enschedaiana, 1938.

Velde, A. J. J. van de. "Bijdrage tot de studie der werken van den geneeskundige Cornelis Bontekoe." *Koninklijke Vlaamsche Academie voor Taal en Letterkunde, Verslagen en Mededeelingen* (1925): 3–48.

Verbeek, Theo, *Descartes and the Dutch: Early Reactions to Cartesian Philosophy, 1637–1650*. Carbondale: Southern Illinois University Press, 1992.

———. "Descartes and the Problem of Atheism: The Utrecht Crisis." *Nederlands Archief voor Kerkgeschiedenis* 71 (1991): 211–23.

———. "Les passions et la fièvre: l'idée de la maladie chez Descartes et quelques cartésiens néerlandais." *Tractrix* 1 (1989): 45–61.

———, ed. and trans. *La querelle d'Utrecht: René Descartes et Martin Schoock*. Paris: Les Impressions Nouvelles, 1988.

Verbeke, Gerard. "Ethics and Logic in Stoicism." In *Atoms, Pneuma, and Tranquillity: Epicurean and Stoic Themes in European Thought*, ed. Margaret J. Osler, 11–24. Cambridge: Cambridge University Press, 1991.

Verlinden, Charles. *The Beginnings of Modern Colonization: Eleven Essays with an Introduction*. Trans. Yvonne Freccero. Ithaca, NY: Cornell University Press, 1970.

Vermeij, Geerat J. *Nature: An Economic History*. Princeton, NJ: Princeton University Press, 2004.

Vermij, R. H. "Bijdrage tot de bio-bibliografie van Johannes Hudde." *Gewina* 18 (1995): 25–35.

———. *The Calvinist Copernicans: The Reception of the New Astronomy in the Dutch Republic, 1575–1750*. Amsterdam: Koninklijke Nederlandse Akademie van Wetenschappen, 2002.

———. *Secularisering en natuurwetenschap in de zeventiende en achttiende eeuw: Bernard Nieuwentijt*. Amsterdam: Rodopi, 1991.

———, ed. and comp. *Bernard Nieuwentijt: een zekere, zakelijke wijsbegeerte*. Geschiedenis van de Wijsbegeerte in Nederland, 12. Baarn: Ambo, 1988.

Vermij, R. H., and L. C. Palm. "John Chamberlayne als vertaler van Antoni van Leeuwenhoek." *Gewina* 15 (1992): 234–42.

Vidal, Fernando. "Brains, Bodies, Selves, and Science: Anthropologies of Identity and the Resurrection of the Body." *Critical Inquiry* 28 (2002): 930–74.

Vivenza, Gloria. "Renaissance Cicero: The 'Economic' Virtues of *De Officiis* I, 22 in Some Sixteenth Century Commentaries." *European Journal of the History of Economic Thought* 11 (2004): 507–23.

Vongsuravatana, Raphaël. *Un Jésuite à la cour de Siam.* Paris: Éditions France-Empire, 1992.

Vos, Frits. "From God to Apostate: Medicine in Japan before the Caspar School." In *Red-Hair Medicine: Dutch-Japanese Medical Relations*, ed. H. Beukers, A. M. Luyendijk-Elshout, M. E. van Opstall, and F. Vos, 9–26. Amsterdam: Rodopi, 1991.

Vredenburch, W. C. A. Baron van. *Schets van eene geschiedenis van het Utrechtsche studentenleven.* Utrecht: A. Oosthoek, 1914.

Vries, Willemien B. de. "The Country Estate Immortalized: Constantijn Huygens' *Hofwijck.*" In *The Dutch Garden in the Seventeenth Century*, ed. John Dixon Hunt, 81–97. Dumbarton Oaks Colloquium on the History of Landscape Architecture, 12. Washington, DC: Dumbarton Oaks Research Library and Collection, 1990.

Vroom, N. R. A. *A Modest Message: As Intimated by the Painters of the "Monochrome Banketje."* 2 vols. Transl. Peter Gidman. Schiedam: Interbook International, 1980.

Vugs, J. G. *Leven en werk van Niels Stensen (1638–1686): onderzoeker van het zenuwstelsel.* Leiden: Universitaire Pers, 1968.

Waals, Jan van der. "Met boek en plaat: het boeken- en atlaassenbezit van verzamelaars." In *De wereld binnen handbereik: Nederlandse kunst- en rariteitenverzamelingen, 1585–1735*, chief ed. Ellinoor Bergvelt and Renée Kistemaker, 205–31. Zwolle: Waanders Uitgevers/Amsterdams Historisch Museum, 1992.

Waite, Gary K. "The Anabaptist Movement in Amsterdam and the Netherlands, 1531–1535: An Initial Investigation into Its Genesis and Social Dynamics." *Sixteenth Century Journal* 18 (1987): 249–65.

———. "The Dutch Nobility and Anabaptism, 1535–1545." *Sixteenth Century Journal* 23 (1992): 458–85.

Wake, C. H. H. "The Changing Pattern of Europe's Pepper and Spice Imports, ca. 1400–1700." *Journal of European Economic History* 8 (1979): 361–403.

———. "The Volume of European Spice Imports at the Beginning and End of the Fifteenth Century." *Journal of European Economic History* 15 (1986): 621–35.

Ward, H. Gordon. "An Unnoted Poem by Mandeville." *Review of English Studies* 7 (1931): 73–76.

Wartofsky, Marx. "Action and Passion: Spinoza's Construction of a Scientific Psychology." In *Spinoza: A Collection of Critical Essays*, ed. Marjorie Grene, 329–53. Garden City, NY: Anchor/Doubleday, 1973.

Waszink, Jan. "Inventio in the Politica: Commonplace-Books and the Shape of Political Theory." In *Lipsius in Leiden: Studies in the Life and Works of a Great Humanist on the Occasion of His 450th Anniversary*, ed. Karl Enenkel and Chris Heesakkers, 141–62. Bloemendaal: Florivallis, 1997.

Waterbolk, E. H. "The 'Reception' of Copernicus's Teachings by Gemma Frisius (1508–1555)." *Lias* 1 (1974): 225–42.

Watson, Gilbert. *Theriac and Mithridatium: A Study in Therapeutics*. London: Wellcome Historical Medical Library, 1966.

Weber, Max. *The Protestant Ethic and the Spirit of Capitalism*. First German ed. 1904. Trans. Talcott Parsons, foreword R. H. Tawney. 1930. New York: Scribner's, 1958.

Webster, Charles. *The Great Instauration: Science, Medicine and Reform, 1626–1660*. 1975. New York: Holmes and Meier, 1976.

———. *From Paracelsus to Newton: Magic and the Making of Modern Science*. Cambridge: Cambridge University Press, 1982.

———. "Water as the Ultimate Principle in Nature: The Background to Boyle's *Sceptical Chymist*." *Ambix* 13 (1966): 96–107.

Wee, Herman van der. *The Growth of the Antwerp Market and the European Economy, Fourteenth–Sixteenth Centuries*. 3 vols. The Hague: Martinus Nijhoff, 1963.

———. "Structural Changes in European Long-Distance Trade, and Particularly in the Re-Export Trade from South to North, 1350–1750." In *The Rise of Merchant Empires: Long-Distance Trade in the Early Modern World, 1350–1750*, ed. James D. Tracy, 14–33. Cambridge: Cambridge University Press, 1990.

Weevers, Theodoor. *Poetry of the Netherlands in Its European Context, 1170–1930*. London: Athlone Press, 1960.

Welch, Evelyn. "The Art of Expenditure: The Court of Paola Malatesta Gonzaga in Fifteenth-Century Mantua." *Renaissance Studies* 16 (2002): 306–17.

Wellisch, Hans (Hanan). "Conrad Gessner: A Bio-Bibliography." *Journal of the Society for the Bibliography of Natural History* 7 (1975): 151–247.

Wellman, Kathleen. "La Mettrie's *Institutions de Medecine*: A Reinterpretation of the Boerhaavian Legacy." *Janus* 72 (1985): 283–303.

———. *Making Science Social: The Conferences of Théophraste Renaudot, 1633–1642*. Norman: University of Oklahoma Press, 2003.

Werger-Klein, K. Elke. "Engelbert Kaempfer, Botanist at the VOC." In *Engelbert Kaempfer: Werk und Wirkung*, ed. Detlef Haberland, 39–60. Stuttgart: Franz Steiner, 1993.

Weschler, Lawrence. *Boggs: A Comedy of Values*. Chicago: University of Chicago Press, 1999.

Wessels, J. W. *History of the Roman-Dutch Law*. Grahamstown, Cape Colony: African Book, 1908.

Westfall, Richard S. "Science and Technology during the Scientific Revolution: An Empirical Approach." In *Renaissance and Revolution: Humanists, Scholars, Craftsmen and Natural Philosophers in Early Modern Europe*, ed. J. V. Field and Frank A. J. L. James, 63–72. Cambridge: Cambridge University Press, 1993.

Westman, Robert S. "The Astronomer's Role in the Sixteenth Century: A Preliminary Study." *History of Science* 18 (1980): 105–47.

Weyde, A. J. van. "Collegium Medicum Ultrajectinum." *Nederlandsch Tijdschrijft voor Geneeskunde* 66, no. 2 [B] (1922): 2600–2608.

Wheelock, Arthur K. *Perspective, Optics, and Delft Artists around 1650*. New York: Garland, 1977.

Whitaker, Katie. "The Culture of Curiosity." In *The Cultures of Natural History*, ed. N. Jardine, J. A. Secord, and E. Spary, 75–90. Cambridge: Cambridge University Press, 1996.

White, Lynn, Jr. "Medical Astrologers and Late Medieval Technology." *Viator* 6 (1975): 295–308.

Whitehead, P. J. P. "The Biography of Georg Marcgraf (1610–1643/4) by His Brother Christian, Translated by James Petiver." *Journal of the Society for the Bibliography of Natural History* 9 (1979): 301–14.

———. "The Original Drawings for the *Historia Naturalis Brasiliae* of Piso and Marcgrave (1648)." *Journal of the Society for the Bibliography of Natural History* 7 (1976): 409–22.

Whitehead, P. J. P., and M. Boeseman. *A Portrait of Dutch Seventeenth Century Brasil: Animals, Plants and People by the Artists of Johan Maurits of Nassau*. Koninklijke Nederlandse Akademie van Wetenschappen, Verhandelingen, Afd. Natuurkunde, 2nd ser., vol. 87. Amsterdam: North-Holland, 1989.

Whitehead, P. J. P., G. van Vliet, and W. T. Stearn. "The Clusius and Other Natural History Pictures in the Jagiellon Library, Kraków." *Archives of Natural History* 16 (1989): 15–32.

Wijnands, D. O. *The Botany of the Commelins: A Taxonomical, Nomenclatural and Historical Account of the Plants Depicted in the Moninckx Atlas and in the Four Books by Jan and Caspar Commelin on the Plants in the Hortus Medicus Amstelodamensis, 1682–1710*. Rotterdam: A. A. Balkema, 1983.

———. "Commercium Botanicum: The Diffusion of Plants in the Sixteenth Century." In *The Authentic Garden: A Symposium on Gardens*, ed. Leslie Tjon Sie Fat and Erik de Jong, 75–84. Leiden: Clusius Foundation, 1991.

———. "Hortus Auriaci: The Gardens of Orange and Their Place in Late Seventeenth-Century Botany and Horticulture." In *The Anglo-Dutch Garden in the Age of William and Mary / De gouden eeuw van de Hollandse tuinkunst*. Special double issue of *Journal of Garden History*, vol. 8, nos. 2 and 3, ed. John Dixon Hunt and Erik de Jong, 61–86. London: Taylor and Francis, 1988.

Wildenberg, I. W. "Appreciaties van de Gebroeders De la Court ten tijde van de Republiek." *Tijdschrift voor Geschiedenis* 98 (1985): 540–56.

Willem III en de Leidse Universiteit: Catalogus van de tentoonstelling gehouden in het Academisch Historisch Museum te Leiden. Exhibition catalog. Prepared by W. Otterspeer and L. van Poelgeest. Leiden: Academisch Historisch Museum, 1988.

Williams, Wes. "'Rubbing Up against Others': Montaigne on Pilgrimage." In *Voyages and Visions: Towards a Cultural History of Travel*, ed. Jaś Elsner and Joan-Pau Rubiés, 101–23. London: Reaktion, 1999.

Wills, John E., Jr. *Pepper, Guns, and Parleys: The Dutch East India Company and China, 1622–1681*. Cambridge, MA: Harvard University Press, 1974.

Wilson, Catherine. *The Invisible World: Early Modern Philosophy and the Invention of the Microscope*. Princeton, NJ: Princeton University Press, 1995.

———. "Visual Surface and Visual Symbol: The Microscope and the Occult in Early Modern Science." *Journal of the History of Ideas* 49 (1988): 85–108.

Wilson, M. L., Th. Toussaint van Hove-Exalto, and W. J. J. van Rijssen, eds. *Codex Witsenii: Annotated Watercolours . . . in the Country of Namaqua Undertaken in 1685–6*. Cape Town: Iziko Museums; and The Netherlands: Davidii Media, 2002.

Wilson, Stephen. *The Magical Universe: Everyday Ritual and Magic in Pre-Modern Europe.* London: Hambledon, 2000.
Wit, H. C. D. de, "Georgius Everhardus Rumphius." In *Rumphius Memorial Volume,* ed. H. C. D. de Wit, 1–26. Baarn: Uitgeverij en Drukkerij Hollandia, 1959.
———, ed. *Rumphius Memorial Volume.* Baarn: Uitgeverij en Drukkerij Hollandia, 1959.
Wittop Koning, D. A. *De handel in geneesmiddelen te Amsterdam tot omstreeks, 1637.* Ph.D. diss. Purmerend: J. Muusses, 1942.
———. "De oorsprong van de Amsterdamse pharmacopee." *Pharmeceutisch Weekblad* 85 (28 October 1950): 801–3.
———. "De voorgeschiedenis van het Collegium Medicum te Amsterdam." *Jaarboek Amstelodamum* (1947): 1–16.
———. "Wondermiddelen." *Kring voor de Geschiedenis van de Pharmacie in Benelux* 67 (1985): 1–17.
Wojcik, Jan W. *Robert Boyle and the Limits of Reason.* Cambridge: Cambridge University Press, 1997.
Woltjer, J. J. "Introduction." In *Leiden University in the Seventeenth Century: An Exchange of Learning,* ed. Th. H. Lunsingh Scheurleer and G. H. M. Posthumus Meyjes, 1–19. Leiden: Universitaire Pers Leiden / E. J. Brill, 1975.
Wood, Thomas. *English Casuistical Divinity during the Seventeenth Century: With Special Reference to Jeremy Taylor.* London: SPCK, 1952.
Woodall, Joanne. "In Pursuit of Virtue." In *Virtus: Virtuositeit en kunstliefhebbers in de Nederlanden,* ed. Jan de Jong, Dulcia Meijers, Mariët Westermann, and Joanna Woodall, 7–24. Zwolle: Waanders Uitgevers, 2004.
Wootton, David. "Lucien Febvre and the Problem of Unbelief in the Early Modern Period." *Journal of Modern History* 60 (1988): 695–730.
———. *Paolo Sarpi: Between Renaissance and Enlightenment.* Cambridge: Cambridge University Press, 1983.
Wright, Peter W. G. "A Study in the Legitimisation of Knowledge: The 'Success' of Medicine and the 'Failure' of Astrology." In *On the Margins of Science: The Social Construction of Rejected Knowledge,* ed. Roy Wallis. Sociological Review Monograph, no. 27, 85–101. Keele: University of Keele, 1979.
Wrigley, E. A. "A Simple Model of London's Importance in Changing English Society and Economy, 1650–1750." *Past and Present* 37 (1967): 44–70.
Yates, Frances A. *Astraea: The Imperial Theme in the Sixteenth Century.* London: Routledge and Kegan Paul, 1975.
———. *The French Academies of the Sixteenth Century.* 1947. London: Routledge, 1988.
———. *Giordano Bruno and the Hermetic Tradition.* 1964. New York: Vintage Books, 1969.
———. *The Occult Philosophy in the Elizabethan Age.* London: Routledge and Kegan Paul, 1979.
Yoshida, Tadashi. "The Rangaku of Shizuki Tadao: The Introduction of Western Science in Tokugawa Japan." Ph.D. diss., Princeton University, 1974.
Zagorin, Perez. "Francis Bacon's Concept of Objectivity and the Idols of the Mind." *British Journal for the History of Science* 34 (2001): 379–93.
Zanden, Jan Luiten van. "Economic Growth in the Golden Age: The Development of the

Economy of Holland, 1500–1650." In *Economic and Social History in the Netherlands*. Het Nederlandsch Economisch-Historisch Archief, vol. 4, 5–26. Amsterdam: NEHA, 1993.

———. "Op zoek naar de 'Missing Link': hypothesen over de opkomst van Holland in de late middeleeuwen en de vroegmoderne tijd." *Tijdschrift voor Sociale Geschiedenis* 14 (1988): 359–86.

———. *The Rise and Decline of Holland's Economy: Merchant Capitalism and the Labour Market*. Manchester: Manchester University Press, 1993.

Zandvliet, Kees. *The Dutch Encounter with Asia, 1600–1950*. Zwolle: Rijksmuseum Amsterdam / Waanders, 2002.

———. *Mapping for Money: Maps, Plans and Topographic Paintings and Their Role in Dutch Overseas Expansion during the Sixteenth and Seventeenth Centuries*. Amsterdam: Batavian Lion International, 1998.

Zuylen, J. van. "De microscopen van Antoni van Leeuwenhoek." In *Van Leeuwenhoek herdacht*, ed. H. L. Houtzager and L. C. Palm. Serie-Uitgave van het Genootschap Delfia Batavorum, no. 8, 57–69. Amsterdam: Rodopi, 1982.

———. "The Microscopes of Antoni van Leeuwenhoek." *Journal of Microscopy* 121 (March 1981): 309–28.

———. "On the Microscopes of Antoni van Leeuwenhoek." *Janus* 68 (1981): 159–98.

INDEX